Innovative and Integrated Technologies for the Treatment of Industrial Wastewater

EU Report Series

Innovative and Integrated Technologies for the Treatment of Industrial Wastewater

Antonio Lopez, Claudio Di Iaconi, Giuseppe Mascolo and Alfieri Pollice

IWA Publishing
London • New York

Published by

IWA Publishing
Alliance House
12 Caxton Street
London SW1H 0QS, UK
Telephone: +44 (0)20 7654 5500
Fax: +44 (0)20 7654 5555
Email: publications@iwap.co.uk
Web: www.iwapublishing.com

First published 2012

British Library Cataloguing in Publication Data
A CIP catalogue record for this book is available from the British Library

Library of Congress Cataloging-in-Publication Data
A catalog record for this book is available from the Library of Congress

ISBN13: 9781843393436
ISBN: 1843393433

Contents

Foreword ix
Contributors xi

Chapter 1
Aerobic granular biomass processes *1*
1.1 Background 1
1.1.1 Formation and morphology of aerobic granular sludge 2
1.1.2 Modelling of aerobic granular sludge 3
1.1.3 Aerobic granular sludge in practice 4
1.2 Laboratory scale experiments 4
1.2.1 Introduction 4
1.2.2 Objectives 5
1.2.3 Method description & lab scale reactor setup 6
1.2.4 Analytical methods 7
1.2.5 Results 9
1.3 NEREDA treatment of food industry wastewater 33
1.3.1 Introduction 33
1.3.2 NEREDA treating brewery wastewater 33
1.3.3 NEREDA treating food industry wastewater 36
1.4 Unifed treatment of abattoir wastewater 38
1.4.1 Introduction 38
1.4.2 Laboratory-scale experiences 38
1.4.3 Pilot-scale experiences – plant design 41
1.4.4 Pilot-scale experiences – results 43
1.4.5 Results of pilot-scale experiences 57
1.4.6 Conclusions 63
1.5 SBBGR treatment of landfill leachates 64
1.5.1 Introduction 64
1.5.2 Materials and methods 65
1.5.3 Results 71
1.5.4 Conclusions 85
References 86

Chapter 2
Integrated processes ***91***
2.1 Background 91
2.2 Recalcitrant wastewater treatment by integrating solar advanced oxidation processes and immobilised biomass reactor 94
2.2.1 Introduction 94
2.2.2 Wastewater containing inhibiting/toxic compounds (pesticides) 98
2.2.3 Wastewater containing a large amount of biodegradable organic compounds in addition to small concentrations of recalcitrant compounds (pharmaceuticals) 105
2.2.4 Conclusions 112
2.3 Fundamental studies on immobilized photo fenton catalyst 112
2.3.1 Introduction 112
2.3.2 Optimisation of photocatalyst preparation and lab scale experiments 113
2.3.3 Applications: adaptation to compound parabolic collector photoreactors 122
2.3.4 Photocatalytic degradation of compound mixtures 127
2.3.5 Conclusions 130
2.4 MBR/AOP treatment of pharmaceutical wastewater 130
2.4.1 Introduction 130
2.4.2 Preliminary investigations 131
2.4.3 Laboratory scale studies 135
2.4.4 Conclusions 146
2.5 Biozo treatment of landfill leachate 147
2.5.1 Introduction 147
2.5.2 Materials and methods 149
2.5.3 Analytical methods 151
2.5.4 Results 152
2.5.5 Combining biological wastewater treatment with controlled ozonation – the BIOZO system 166
2.5.6 Conclusions 171
References 172

Chapter 3
Membrane-based processes ***179***
3.1 Background 179
3.2 Recovery of phenolic compounds with membrane contactors 180
3.2.1 Introduction 180
3.2.2 Membrane performance 181
3.2.3 Membrane contactor modules 182
3.2.4 Flow in spacer-filled channels 184
3.2.5 Mass transfer investigations inside spiral wound modules 189
3.2.6 Mass transfer investigations inside spiral wound modules 191
3.2.7 Conclusions 192
3.3 Membrane chemical reactor design 193
3.3.1 Introduction 193
3.3.2 Description of the technology 195
3.3.3 Factors influencing design: results 197
3.3.4 Conclusions 209
References 211

Chapter 4
Tailoring and assessment of wastewater treatment processes ***213***
4.1 Background 213
4.2 Methodology 214
4.2.1 Specific objectives 214
4.2.2 General approach 214
4.2.3 Data and software requirements 215

4.2.4 Environmental assessment methodology ... 216
4.2.5 Economic assessment ... 219
4.2.6 Assessment strategy ... 219
4.3 Food industry wastewater ... 220
4.3.1 The assessed treatment case ... 220
4.3.2 Results of the assessment ... 226
4.3.3 Conclusions of the assessment of the food wastewater treatment ... 231
4.3.4 Uncertainties ... 231
4.4 Landfill Leachate ... 232
4.4.1 The assessed treatment case ... 232
4.4.2 Results of the environmental assessment ... 237
4.4.3 Economic assessment ... 242
4.4.4 Uncertainties ... 246
4.4.5 Conclusions ... 246
4.5 Pharmaceutical industry wastewater – the standard treatment case ... 247
4.5.1 Properties of nalidixic acid ... 247
4.5.2 Mathematically modelled treatments – photo-fenton + IBR and membrane contactor + IBR ... 248
4.5.3 Treatment by the reference technology and by the membrane bioreactor + ozonation (MBR + AOP) ... 251
4.5.4 Result of the environmental assessment ... 252
4.5.5 Economic assessment ... 254
4.5.6 Conclusion ... 256
References ... 256
Appendix 1 ... 258
Appendix 2 ... 259
Appendix 3 ... 266

Foreword

Industrial wastewater streams have often very different characteristics from domestic sewage, such as high salinity, significant toxicity due to different compounds, large variabilities in composition, pH, temperature etc. These differences make it very difficult to successfully treat many industrial effluents in conventional (domestic) wastewater treatment plants. However, the alternative approach typically requires significantly more complex treatment trains with associated higher capital and operating costs, increased operator skill levels and often significant environmental impacts.

Optimal solutions to tackle these challenging waste streams need to look at both the specific local situation in each industrial facility and at industry-wide approaches to better manage the wastewater pollutants. At both scales, the characterisation and minimisation of the key pollutant sources needs to be prioritised to enable the identification of effective up-front solutions such as separation of high-strength streams or reduction/elimination of problematic compounds in the production process. Nevertheless, this rarely eliminates the need for selective and targeted treatment solutions in most industries.

In many cases, the provision of specific source control actions and pre-treatment processes can significantly improve the biodegradability of the remaining pollutants. These may include the use of more environmentally benign or biodegradable substances, stream segregation, often in combination with selective removal of key contaminants or partial chemical treatment to increase bio-availability of the remaining compounds.

In this context, the European Commission, through its Sixth Framework Programme (FP6), co-financed the project "Innovative and integrated technologies for the treatment of industrial wastewater", with the acronym innowatech (www.innowatech.org).

The aim of this project was to assess, investigate and enhance promising innovative technologies and processes aimed at tackling some of the regular and relevant problems encountered when treating industrial wastewater, e.g.:

- Inefficient biological treatment with limited operational flexibility and stability;
- High sludge production and treatment/disposal costs;
- Occurrence of recalcitrant and/or toxic compounds impacting on overallplant performance;
- Lack of technologies for selective removal or recovery of raw materials and/or priority organic pollutants;
- Non-optimalintegration and adaptation of treatment options for specific processes.

The technologies chosen for further investigations in this project were selected on the basis of their broad applicability across various sectors, their ability to specifically target key pollutants or characteristics of common industrial wastewaters or based on their innovative concepts or features.

The INNOVATECH project was structured into six work packages (WPs), with three focused on selected key technology areas such as aerobic granular sludge systems (WP1), integrated advanced oxidation and biological processes (WP2); and membrane-enhanced industrial wastewater treatment processes (WP3). A number of different wastewater types were tested in these processes, including from food and meat processing, pharmaceutical and chemical industries, and fresh and mature landfill leachates. The remaining three WPs were mainly targeting end-users application and implementation aspects, as well as technology transfer and project management tasks. This included ecological and economic evaluations using life cycle assessment (LCA) and costing (LCC) methodologies (WP4), technology dissemination and exploitation within the project and across the industry sectors (WP5), and overall project coordination and management (WP6).

The main achievements from the project activities can be summarised as follows:

- The aerobic granular sludge systems have been shown to be effective in treating biodegradable industrial wastewater streams (e.g. from food/beverage or meat processing industries). This novel technology was shown to require a smaller footprint and to produce less excess sludge compared to conventional alternatives.
- When treating complex wastewaters (e.g. mature municipal landfill leachates or from pharmaceutical industries), the integration of biological and chemical treatment processes results in synergistic effects that increase treatment efficiency and generate costs savings.
- The exploitation of solar light in photochemical reactors achieves savings in chemical and energy use without reducing the treatment effectiveness for the removal of toxic pollutants such as those present in pharmaceutical or pesticide containing wastewater.
- Innovative methods for immobilizing photo-catalysts on the surface of commercial polymers have created a technology that can be used at near neutral pH to enhance the performance of photo-Fenton or similar processes for a range of wastewater pollutants. Further benefits were achieved through reduced chemical sludge production and the ability to regenerate the catalyst under solar irradiation.
- Innovative silicone coated, selective membranes were developed for use in a Membrane Contactor, which was optimised for the cost-effective recovery of organic acids and bases even in small- and medium-scale applications typically used in chemical industries.
- Membrane coupled photo-catalysis reactors (i.e. Membrane Chemical Reactors) were further developed and optimised to demonstrate their performance with a number of industrial wastewaters.
- Development and implementation of conceptual models of existing treatment systems and of innovative treatment solutions developed in this project have allowed comparisons of the sustainability performance of the different technologies investigated in this project.

Overall, these novel and innovative achievements will directly benefit industrial wastewater treatment systems in improving their removal performance, effluent quality, operational stability, energy efficiency and particularly their capital and operating cost bases. At the same time, these improvements will have significant environmental benefits as well. Hence the results from this project, once implemented and optimised in full-scale installations, will have the potential to directly enhance the performance and competitiveness of a range of European industry sectors.

This book reports in detail the activities and results from the various elements of this project together with the technical and operational benefits and outcomes. This should directly assist in the transfer of the findings and technologies into the broader industrial and economic communities across Europe and the world.

Antonio Lopez
Jurg Keller

Contributors

Chapter 1

Jurg Keller
Shane Watts
Advanced Water Management Centre
Gehrmann Building, Research Rd
The University of Queensland
Brisbane QLD 4072, Australia

Merle de Kreuk
Waterboard Hollandse Delta
P.O. Box 469
Johan de Wittstraat 40
3300 AL Dordrecht
The Netherlands

Maite Pijuan
ICRA - Institut Català de Recerca de l'Aigua
Edifici H_2O- Parc Científic i Tecnològic de la Universitat de Girona
Emili Grahit 101- 17003 Girona, Spain

Claudio Di Iaconi
Guido Del Moro
Annalisa Mancini
Istituto di Ricerca sulle Acque
Consiglio Nazionale delle Ricerche
Via F. De Blasio 5
70132 Bari, Italy

Achim Ried
ITT Water & Wastewater Herford GmbH
Boschstrasse 4
32051 Herford, Germany

Simona Rossetti
Marco De Sanctis
Istituto di Ricerca sulle Acque
Consiglio Nazionale delle Ricerche
Via Salaria km 29,300
00016 Monterotondo (RM), Italy

Andreas Giesen
DHV BV
P.O. Box 1132
3800 BC Amersfoort, The Netherlands

Mark M. C. van Loosdrecht
Mario Pronk
Delft University of Technology
Julianalaan 67
2628 BC Delft, The Netherlands

Chapter 2

Isabel Oller
Sixto Malato
Ana Zapata
Carla Sirtori
Plataforma Solar de Almería (CIEMAT)
Carretera Senés, Km 4
04200, Tabernas (Almería), Spain

Isabel Pariente
Rey Juan Carlos University
Chemical and Environmental Department
C/Tulipan s/n, 28933, Mostoles, Madrid, Spain

Félicien Mazille
Cesar Pulgarin
Institute of Chemical Sciences and Engineering, SB, GGEC
Ecole Polytechnique Fédérale (EPFL) de Lausanne
1015 Lausanne, Switzerland

Daniela Cassano
Sara Diomede
Piera Ielpo
Giuseppe Laera
Antonio Lopez
Giuseppe Mascolo
Angela Pinto
Alfieri Pollice
Carlo Salerno
Natascha Michel
Istituto di Ricerca sulle Acque
Consiglio Nazionale delle Ricerche
Via F. De Blasio 5
70132 Bari, Italy

Benedek Gy. Plósz
Christian Vogelsang
Department of Environmental Technology
Norwegian Institute for Water Research (NIVA)
Gaustadalléen 21, NO-0349 Oslo, Norway

Chapter 3

B. Jefferson
M. Pidou
O. Autin
J. MacAdam
S. A. Parsons
School of Water Sciences
Cranfield University
Bedforshire MK43 0AL, UK

C. Bayer
T. Melin
M. Follmann
Department of Chemical Engineering
RWTH Aachen University
Turmstr. 46
52056 Aachen, Germany

F. P. Cuperus
I. M. Wienk
Solsep BV
St Eustatius 65,
7333 NW Apeldorm
The Netherlands

Chapter 4

Mats Almemark
Mats Ek
Uwe Fortkamp
Christian Junestedt
Klara Westling
IVL Swedish Environmental Research Institute
P.O. Box 21060
SE - 10031 Stockholm
Sweden

Maria Ekenberg
AnoxKaldnes AB
Klosterängsvägen 11A
SE-226 47 Lund
Sweden

Chapter 1

Aerobic granular biomass processes

Shane Watts, Merle de Kreuk, Maite Pijuan, Claudio Di Iaconi, Achim Ried, Simona Rossetti, Guido Del Moro, Annalisa Mancini, Marco De Sanctis, Andreas Giesen, Mario Pronk, Mark M.C. van Loosdrecht and Jurg Keller

1.1 BACKGROUND

At the end of the 1990s, research on biofilm structure and formation (van Loosdrecht *et al.*, 1995, 1997a) and on the role of storage polymers (van Loosdrecht *et al.*, 1997b; Krishna and van Loosdrecht, 1999) resulted in the idea of growing aerobic granules without carrier material on readily biodegradable substrates in a Sequencing Batch Reactor (SBR) (Morgenroth *et al.*, 1997; Beun *et al.*, 1999; Dangcong *et al.*, 1999). In these studies, the conversion of readily biodegradable COD into a substrate yielding a lower maximal growth rate facilitated granule formation and it was shown to be possible to grow stable granular sludge (Figure 1.1) with integrated COD and nitrogen removal. In 1998, an international patent was submitted and granted (Heijnen and van Loosdrecht, 1998) and later extended to include a description of anaerobic feeding process (van Loosdrecht and De Kreuk, 2004).

Aerobic granule formation has been excessively studied worldwide (Figure 1.2). Many theories about aerobic granule formation have made their way into different studies. Analysis of a cross section of literature published in the last decade shows that substrate type, COD and N-load, superficial gas velocity or shear stress and oxygen concentration are important parameters. An important secondary parameter for the formation and maintenance of dense granules is the growth rate of the organisms that is influenced by cycle configuration or loading rates (De Kreuk *et al.*, 2005a). An extensive overview of various parameters that are important for anaerobic and aerobic granule formation has been given by Liu and Tay (2004a).

The first workshop on aerobic granular sludge was organised in Munich, Germany in 2004 (Bathe *et al.*, 2005). Despite the long-term application of anaerobic granular sludge in wastewater treatment, aerobic granular sludge is a relatively new observation and so during this workshop a definition of aerobic granular sludge was developed: granules making up aerobic granular activated sludge are to be understood as aggregates of microbial origin, which do not coagulate under reduced hydrodynamic shear, and which settle significantly faster than activated sludge flocs (De Kreuk *et al.*, 2005b).The main conclusion from the workshop was: "all participants agreed that up till then a lot of basic knowledge was gained on aerobic granule formation and that the pioneering stage was finished. Researchers should continue with specific research topics (e.g. factors influencing granule strength and formation, microbial diversity, conversion processes, pathogen removal, pilot- and demonstration-scale studies) and that the technology should be put into practice. Within a couple of years, this workshop should be repeated to see how research went from there; hopefully with new insights and with new applications."

In September 2006, a second aerobic granular sludge workshop was held in Delft, The Netherlands, of which a summary of the discussion outcomes is given below.

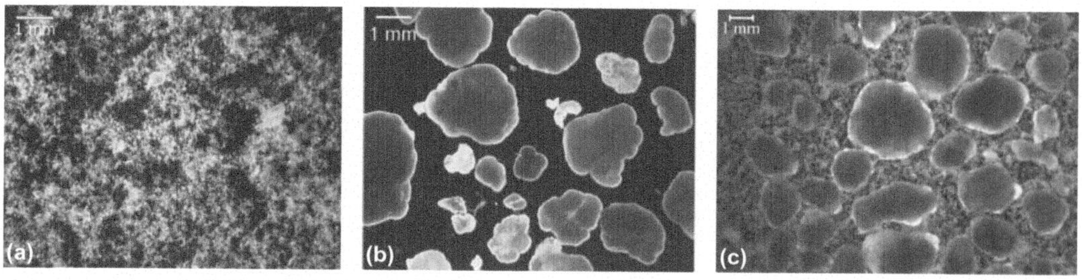

Figure 1.1 Activated sludge from a wastewater treatment plant (a) and aerobic granular sludge cultivated in a laboratory scale reactor (b) and in a pilot plant (c)

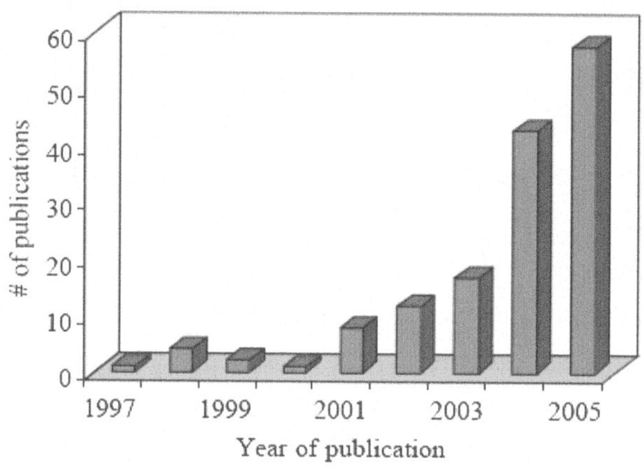

Figure 1.2 Number of publications about aerobic granular sludge per year (Source: Web of Science)

The definition of aerobic granular sludge developed in the first workshop was further expanded with explanations of the statement components:

1. Aggregates of microbial origin: speaking of granular activated sludge in the statement implies that aerobic granules need to contain active microorganisms and cannot only consist of components of microbial origin (such as proteins, EPS, etc.). The microbial population in aerobic granular sludge are to be expected more or less similar to those in activated sludge and/or biofilms, thus there is no need to describe specific groups of microorganisms in the definition. Furthermore, this part implies that no carrier material is intentionally involved or added; the aggregate is formed without the addition of such carrier material.
2. No coagulation under reduced hydrodynamic shear: this describes the difference in behaviour between activated sludge and aerobic granular sludge. Activated sludge flocs tend to coagulate when they settle (when liquid-sludge mixture is not aerated or stirred), whilst granules do not coagulate and settle as separate units.
3. Which settle significantly faster than activated sludge flocs: this means that SVI_{10} (SVI after 10 min of settling) in combination with SVI_{30} should be used as suggested by Schwarzenbeck *et al.* (2004a). The difference between the values gives an excellent indication about the granule formation and indicates the extent of thickening after settling.
4. The minimum size of the granules should be such that the biomass still fulfils point three. This minimum size was set to 0.2 mm, which was decided based on past measurements. This limit could be adjusted per case/granule type, as long as the other demands of the definition hold.
5. Sieving is considered a proper method to harvest granules from activated sludge tanks or from aerobic granule reactors, which also determines certain strength of the required biomass matrix.

When an aggregate fulfils all characteristics as described above, it can be called aerobic granular sludge. This simplifies the interpretation of experimental results and clarifies when to refer to aerobic granular sludge, activated sludge or biofilms.

1.1.1 Formation and morphology of aerobic granular sludge

Many factors have been cited as responsible for the formation and stability of aerobic granular sludge, but it is undecided amongst scientists which is the dominant factor. Workshop discussions focussed on this topic and several parameters were highlighted: a) use or appearance of specific self-aggregating cultures; b) selection by settling velocity; c) applied shear stress; d) growth rate of the organisms; e) substrate gradients inside the granules; f) formation of extra-cellular polymeric substances (EPS).

Formation of aerobic granular sludge in laboratory experiments mostly occurs with a strict selection regime for well settling sludge by applying increasingly shorter settling times. The aggregate forming organisms are maintained in the reactor, while other organisms are washed out with the effluent. To enhance the start-up of an aerobic granular sludge reactor, the use of specific self-aggregating cultures was suggested by the researchers of Nanyang University (Singapore). Non-pathogenic cultures with self-aggregating abilities were selected and added to a reactor during stat-up. This shortened the start-up time considerably (3 days instead of 9 days without specific inoculum). The selection for specific organisms to enhance granule formation and stability has also been applied by other researchers, e.g. the use of phosphate or glycogen accumulating organisms (Dulekgurgen *et al.*, 2003; De Kreuk and van Loosdrecht, 2004) or nitrifying organisms (Liu *et al.*, 2004a). However, the workshop attendees agreed that, based on published research, substrate gradients inside the granules are very important as well and that sharp decreasing substrate gradients inside the granules should be avoided. This can be achieved by using the ability of some microorganisms to convert readily degradable substrates into storage polymers and/or using organisms with low actual growth rates. EPS production by slow growing organisms enhance the granule formation (Liu *et al.*, 2004b; McSwain *et al.*, 2005, presented research results of National Taiwan University) and is seen as an important aspect by the attending researchers as well. Applied shear stress and settling time has been an important topic for studies in the past and still leads to discussions. At the aerobic granular sludge workshop 2006, several researchers showed their results about this topic (Northwestern University, USA; University of Beijing, China; Istanbul Technical University, Turkey). Shear stress is difficult to quantify and is often related to superficial gas velocity. However, factors such as stirring should not be underestimated and reported as well.

At the workshop there was a general agreement on the positive effects of hydraulic selection pressure on granule formation and stability. When aerobic granular sludge is used in combination with a membrane for effluent filtration, the selection pressure for aggregates by settling totally disappears. Even with this effluent filtration, very dense and large flocs were obtained, having a positive effect on the membrane fouling. The high density flocs only met the definition of aerobic granular sludge (above) during short periods of the total experiment, particularly during the periods when autotrophic organisms were present (results presented by INSA, France).

1.1.2 Modelling of aerobic granular sludge

Mathematical modelling has proven very useful to study complex processes, such as the aerobic granular sludge systems (Beun *et al.*, 2001; Lübken *et al.*, 2005). Biological processes in the granules are determined by concentration gradients of oxygen and diverse substrates. The concentration profiles are the result of many factors, including diffusion coefficients, conversion rates, granule size, biomass spatial distribution and density. All these factors influence each other and so the effect of individual factors cannot be studied experimentally. Moreover, experiments in granular sludge reactors take many weeks to reach steady state. A good computational model for the granular sludge process provides significant insight into the most important factors that affect the nutrient removal rates and in the distribution of different microbial populations inside the granules. Further, models could also be used for process optimisation and for scale-up and design (e.g. hydraulics) to full-scale.

Aerobic granular sludge systems can be modelled in different ways, using different modelling tools depending on the fields of interest. When the overall reactor behaviour is described (substrate removal or sludge production), traditional biofilm modelling can be used, as in AQUASIM (Reichert, 1998). Such models have already been developed by the groups in Munich and Delft. The effect of process parameters on the nutrient removal rates could be reliably evaluated with such models. Influence of oxygen concentration, temperature, granule diameter, sludge loading rate and cycle configuration have been analysed. Oxygen penetration depth in combination with the position of the autotrophic biomass played a crucial role in the conversion rates of the different components and thus on overall nutrient removal efficiencies (de Kreuk *et al.*, 2006).

When a more detailed insight into microbial or EPS distribution inside the granule is desired or to study factors influencing granule shape (presence of filamentous outgrowth, biofilm structure modelling), individually based modelling can be used (Kreft *et al.*, 2001). Such a model for aerobic granular sludge was presented by TU Delft and Universidade Nova de Lisboa, Portugal, to describe an aerobic granular sludge SBR (Xavier *et al.*, Submitted). This multi-scale model described the granular sludge reactor in detail, from the metabolism of microbial groups, through the spatial structure of granules to the dynamics of the whole reactor. However, simulations were computationally demanding and use will be restricted to a description of observed trends. With the model a preferential distribution of species along radial distances was shown, which were more heterogeneous than in strict layers. This heterogeneous structure and growth in micro colonies underlined the difficulty of representative micro-electrode measurements. Also, the accumulation of inert material in the cores of the granules was shown (Figure 1.3). Mainly the outer layers of the granule will be eroded, which contain less inert material. Therefore, aerobic granular sludge is expected to contain more inert material resulting from biomass decay than activated sludge.

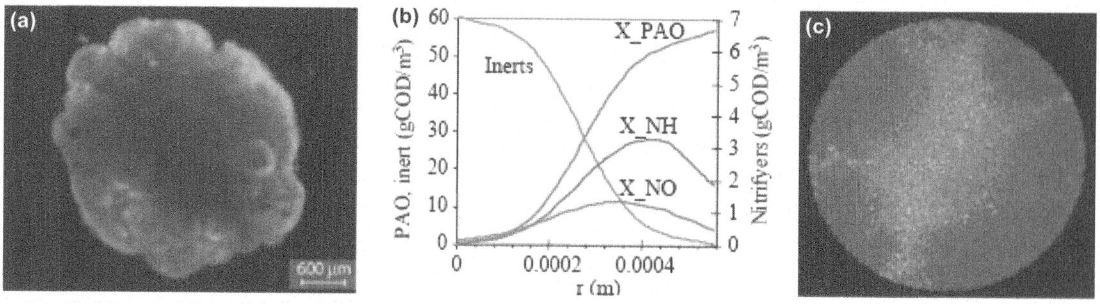

Figure 1.3 (a) Stained granule section (green ¼ live cells; red ¼ death cells); (b) biomass distribution inside a matured granule according a simulation with an ASM-like aerobic granular sludge model (de Kreuk *et al.*, 2006); (c) biomass distribution inside a mature granule according a simulation with a 2-D individually based aerobic granular sludge model (red ¼ PAO; green ¼ NH4$^+$-oxidisers; yellow ¼ NO_2^{2-} oxidisers; grey ¼ inert, Xavier *et al.*, Submitted). Subscribers to the online version of Water Science and Technology can access the colour version of this figure from http://www.iwaponline.com/wst

Waseda University (Japan) and TU Delft developed a similar multi-scale model with nitrifying granules. This model showed that within the nitrifying granules, EPS producing heterotrophic organisms grew on the products from cell lysis. These heterotrophic organisms denitrified a portion of the produced nitrate and located mainly inside the granules. At this position, they excreted EPS, which strengthened the structure of the granule and enhanced the growth of the total granule. The nitrifying organisms, growing mainly in the outer layers, detached more readily, which led to a smaller sludge residence time for these species.

Different effects of shear stress were not taken into account in the models presented at the workshop. Dr. Ivanov, Nanyang University, Singapore pointed out that mechanical compaction of granules by shear stress caused by particle-particle collisions, air-bubbles or mechanical stirring, may affect the distribution of microorganisms, detachment and porosity of the granular sludge. Changing porosity could affect the diffusion in the pores and detached portions from the granules could start new granules, both leading to a different microbial composition of the granule then described with current models.

1.1.3 Aerobic granular sludge in practice

In wastewater treatment, the activated sludge process is the dominant system. Biology of these systems has been optimised and the limits of the system have been reached. However, sludge settleability and washout from clarifiers will remain a point of attention. Major concerns of water authorities are: minimising costs of wastewater treatment; meeting effluent requirements; being prepared for future developments in effluent demands; land availability; environmental aspects such as smell and noise; and energy consumption. Aerobic granular sludge could (partly) meet these concerns. Following successful aerobic granular pilot-plant studies at two sewage treatment plants in The Netherlands by DHV, the first full-scale municipal wastewater treatment plant is planned (presented by water board "Hollandse Delta"). As end-users, water boards in The Netherlands see it as their (public) responsibility to cooperate with consultancies and universities to develop this innovation and invest in full-scale applications.

Several wastewaters from industrial and municipal origin have been used to cultivate aerobic granules and to study the treatment process and the results were presented during the workshop. The first was a pilot plant study (two reactors of 1.5 m^3 each) with the Nereda system treating sewage at a Dutch wastewater treatment plant (presented by DHV Water, The Netherlands) and the second, a demonstration scale sequencing batch biofilter granular reactor (SBBGR, 3.1 m^3) treating sewage at an Italian wastewater treatment plant (presented by IRSA).

1.2 LABORATORY SCALE EXPERIMENTS

1.2.1 Introduction

Most conventional municipal wastewater treatment plants have the disadvantage of large area requirement and high surplus biomass production. In order to overcome these disadvantages, compact systems are developed to decrease the footprint and reduce the sludge production, like the biofilm airlift reactor. The main factor for a compact system is the improvement of the settleability of the sludge (e.g. by growing the biomass as biofilms on carrier material or as granules or dense flocs without support material). Granular sludge is well known in anaerobic systems (such as UASB, IC and EGSB reactors), while in aerobic systems (continuous fed aerated fluidised beds and airlift reactors) sludge is grown on carrier particles to obtain high settling velocities (Nicolella *et al.*, 2000).

Research has shown that it is also possible to grow stable granulated sludge under aerobic conditions and the advantages of a discontinuously over a continuously fed system were shown (Morgenroth *et al.*, 1997; Beun *et al.*, 1999; Dangcong *et al.*, 1999). Besides the flexible operation of these sequenced batch airlift reactor (SBAR) systems, the stability and density of the granules were higher, while different aerobic and anaerobic processes (integrated COD, N and P removal) can occur in the same unit simultaneously and at high efficiency (De Kreuk *et al.*, 2005a). The high settling velocity of the granules makes the use of a traditional settler or 3-phase separator unnecessary, so the installation can be built very compact. All processes take place in one reactor: from influent feeding and conversion of the pollution until settling of the aerobic granular sludge and effluent discharge. The area requirement of a wastewater treatment plant with the granular SBAR will be approximately 25% of the area requirement of a conventional activated sludge system (De Bruin *et al.*, 2004). This one-reactor system leads to lower pump-capacity (no recycle flows) and more efficient aeration and thereby results in at least 30% less energy use during operation. Also the needed material to construct an installation is reduced. This will all lead to less investment and operational costs. All these advantages will lead to a large market for the application of aerobic granular sludge and this new technology (marketed by DHV, The Netherlands under the brand name Nereda^{TM}) can compete with activated sludge systems or other compact techniques (as MBR).

1.2.2 Objectives

Aerobic granular sludge is a young technology, up till now grown in well-controlled laboratory scale reactors and pilot plants (De Bruin *et al.*, 2005). Only limited information has been published on laboratory scale reactors treating industrial effluent (malting (Schwarzenbeck *et al.*, 2004b), dairy (Arrojo *et al.*, 2004; Schwarzenbeck *et al.*, 2005) and pharmaceutical effluent (Inizan *et al.*, 2004)) and domestic sewage (De Kreuk and van Loosdrecht 2006). However, the specific influences of process parameters important for industrial applications have not yet been published. Presently, two full-scale plants are realized in (food) industry and with more plants expected in the coming years. However, scaling-up the aerobic granular sludge reactor (GSBR or Nereda^{TM}) from a 3-litre laboratory scale to a full-scale application exhibits some bottlenecks and unknown aspects. In order to be able to compare an aerobic granular sludge system to other new techniques for effluent treatment, some aspects still have to be studied at laboratory scale and compared to results obtained in practice. The experimental work is organized in the next three phases.

Particulate and Polymeric COD

This phase examines conversion of particulate and polymeric COD and its influence on granule morphology. Industrial effluents, in particular food industry, can contain a high amount of slowly biodegradable or polymeric substrates. From experiments with municipal wastewater it is known that particulate COD can be converted, but the mechanism is unclear. These substances might also influence the morphology of the granules, as has been shown in biofilm research. Therefore, the influence of this particular influent on the performance of an aerobic granular sludge reactor has to be researched in a laboratory scale reactor. The results of the laboratory experiment would be compared with full-scale applications in food industry.

High Temperature

Industrial effluent can have elevated temperatures, which may influence the granule composition, morphology and substrate conversion processes. Aerobic granule formation at lower temperatures has been studied previously (De Kreuk *et al.*, 2005b; De Kreuk *et al.*, 2007), but the influence of higher temperatures is unknown. For example the effluent of the first full-scale aerobic granular sludge reactor (NEREDATM) in industry fluctuates between 35°C and 41°C degrees. Results from this plant indicate that temperatures above 40°C are critical for changes in the system (unpublished data). Therefore, the influence of high temperatures was studied in the laboratory (30°C to 50°C) and results verified with observation from the full-scale industrial applications.

Toxic Compounds

Xenobiotics can inhibit the processes in the biological plants treating industrial wastewater. Previous research showed that microbial granules could be more resistant to wastewaters containing toxic compounds than suspended microbial sludge (Fang, 2000). It has been reported that granules are more resistant to the toxic effect of phenol than flocculated sludge, because of the compact structure of the granules (Jiang *et al.*, 2003). Liu *et al.* (2005a) found out that for initial phenol concentrations up to 10 mg/L, the nitrifying activity of aerobic granules was recovered after phenol degradation. Liu *et al.* (2009) proposed that the granular structure could provide microbial cells with beneficial protection from phenol toxicity, mainly at a longer exposure time. Generally, all the studies reported that nitrifying bacteria embedded in microbial granules might have a better ability to resist to the shock of toxic compounds in wastewaters than suspended nitrifying bacteria.

Fluctuations of the influent pH and the resulting changes in reactor pH restrict the operation of a biological treatment plant. Previous work showed that the optimal pH for the nitrification process is between 7.0 and 8.2 (Antoniou *et al.*, 1990; Hall 1974; Loveless and Painter, 1968; Painter and Loveless, 1983; Shammas, 1986). Those studies were conducted with activated sludge and/or pure cultures. Smolders *et al.* (1994) highlighted that the rate of P-release under anaerobic conditions was increased as the pH was increased. Similarly, Bond *et al.* (1998) stated that a SBR operated without pH control in the anaerobic phase exhibited an improved P-removal comparing with a SBR which had pH control. So far, the effects of pulse pH shocks on nitrifier and PAO populations in granular sludge have not been investigated.

Thus, it is of practical interest to investigate the effects of pH and 2-fluorophenol (2 - Fluor phenol) on nitrification and phosphate uptake by aerobic granules.

1.2.3 Method description & lab scale reactor setup

Experiments were performed with 2 to 4 granule sequencing batch reactors (GSBR), operated in parallel. Influent or temperature was varied in the different reactors. The experimental set-up of the reactors is shown in Figure 1.4. The reactors were operated as bubble columns with a working volume of 3 litres (diameter 6.25 cm). The effluent was extracted at a height of 50 cm resulting in an exchange ratio of approximately 0.5.

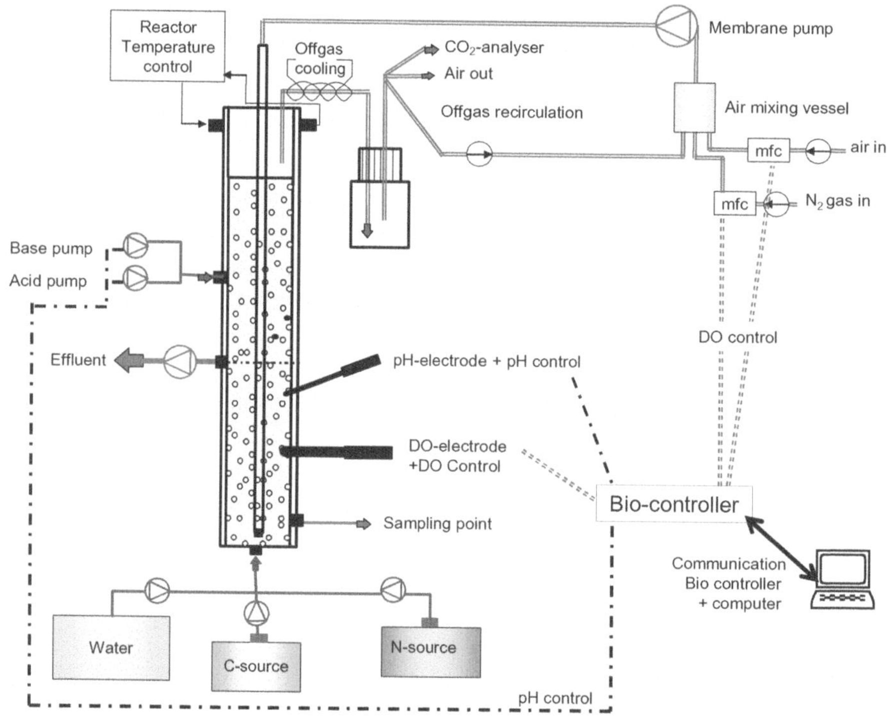

Figure 1.4 Set up of the laboratory scale GSBR's

A sparger with an airflow of 4 l/min (87.5 m^3/hr) was used for aeration. The oxygen was measured on-line continuously with an optional continuous CO_2 measurement. DO control was achieved via the off-gas recycle. The membrane pump recycled 4 l/min of off-gas and via a control based on the DO measurement, nitrogen gas or fresh air was dosed into the main air stream by mass flow controllers. In this way, the oxygen level could be kept constant during the total cycle. The total recycle flow that is dosed to the reactor always remains 4 l/min.

The pH was measured on-line continuously and was controlled with 1 M NaOH or 1 M HCl dosing. Depending on the influent, it was controlled between 6.8 and 7.2 (acetate based influent) or between 7.5 and 7.8 (glucose or starch based influent).

The reactors were operated at room temperature. During the elevated temperature experiments, the temperature was controlled via a water-jacket around the reactor and a temperature measurement in the reactor.

The synthetic influent wastewater consisted of 2 media. Medium A: C-source with a final concentration of 3.96 g COD/L; 3.6 mM $MgSO_4 \cdot 7H_2O$; 4.7 mM KCl. Medium B: 35.4 mM NH_4Cl; 4.2 mM K_2HPO_4; 2.1 mM KH_2PO_4; 10 ml/l trace element solution. From both media approximately 150 ml was dosed to the reactor together with 1.3 litre of tap water resulting in a COD-load of 1.6 kg COD/$m^3 \cdot$ day, a COD/N ratio of 7.9 and a phosphate concentration of

20 mg P/l. The different C-sources that were used were: maltose, soluble and particulate starch for Experiment 1 (particulate and soluble COD), and acetate for Experiments 2 (high temperature studies) and 3 (toxic compound studies). In order to dissolve the starch it was heated at 95°C until the liquid was clear. Both starch vessels (particulate and soluble) were continuously stirred. The final composition of both used media as presented in Table 1.1.

Table 1.1 Particle size distribution of wastewater (municipal) and starch solutions

	Soluble	Colloidal	Supra colloidal	Settleable	Reference
Size (μm) - examples:	<0.08	0.08–1	1–100	>100	(Sophonsiri and Morgenroth, 2004)
Raw influent wastewater	24	31		45	(Guellil *et al.*, 2001)
Primary effluent	46	9	19	26	(Sophonsiri and Morgenroth, 2004)
Secondary effluent	28	3	20	49	
Soluble starch	99	1	0	0	This study
Particulate starch	0	1	89	10	This study

The xenobiotics that will be used during the third set of experiments will be selected in a later stage of the project. This choice depends on the industrial plant that can be monitored in order to compare the results of the laboratory tests and a situation from practice. Presently, no industrial plant has been realised yet with a problem of special (toxic) influent. If such influent characteristic is encountered for before the start of this third phase, a choice will be made based upon avoiding overlap with research results from other groups and taking expected situations in future into consideration. The cycle arrangement is given in Figure 1.5. The total cycle time was 3 hours, consisting of a fill, aeration, settling and effluent phase. During start up, or depending on the particle shape and settleability, the settling time was varied. For example, from 15 minutes (settling velocity of 2 m/h) at start down to 6 minutes (settling velocity of 5 m/h) and 3 minutes. A short settling time of 3 minutes allowed for the selection of particles with a settling velocity larger than 10 m/hr which remained in the reactor. In all systems, the feeding was anaerobic and the influent was fed during 60 minutes from the bottom of the reactor, resulting in a plug-flow through the bed.

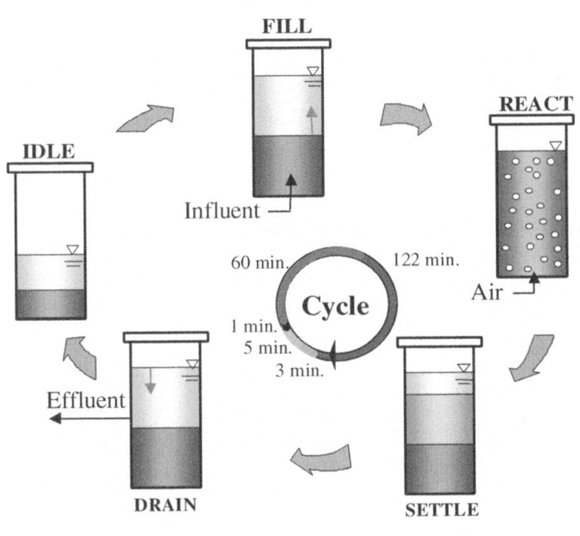

Figure 1.5 Cycle configuration of SBBC1 and SBAR2

1.2.4 Analytical methods

The measurements that were performed are given in Table 1.2. Dry weight in the reactor and ash content of the granules; TOC, biomass concentration, NH_4^+, NO_2^- and NO_3^- concentration in the bulk liquid and the PHB content of the granules were measured as described in Beun *et al.* (2001). Starch was coloured with an iodine complex and measured spectrophotometrically at a wavelength of 570 nm. N_2O in the off-gas was measured using gas-chromatography.

PO_4^- concentration in the bulk liquid was determined spectrophotometrically by use of standard test kits (Dr. Lange type LCK). Phosphate in the granules was measured according to the method described by Uhlmann *et al.* (1990). The SVI8 was determined by reading the height of the settled bed in the reactor after 8 minutes settling (after the settling and effluent

Table 1.2 Analytical methods performed in the laboratory scale reactors

Measurement	Method	Result	Remarks
On-line measurements:			
pH	pH electrode	pH profile in the reactor	
DO	DO electrode	% of dissolved O_2 profile	
CO_2	Off-gas analyser	CO_2 profile	Optional (because of the gas-recycle this is only useful during cycle measurements)
Regular measurements:			
Dry weight	24 hrs. drying at 105° C	g TSS/L	From these values the C_X in the effluent and the reactor can be calculated, together with the solid retention time (SRT), the biomass production and the yield (COD produced/ COD consumed)
Ash content	1 hr. in oven at 600°C	% ash of TSS and g VSS/L	
Effluent biomass	±250 ml filtration over 0.45 µm filter and determining dry weight and ash	g TSS/L, % ash and g VSS/L	
TOC (total organic carbon) in the effluent	TOC measurement of a settled and a filtered effluent sample	mg TOC/L	
Volume of effluent	Collect effluent of measured cycle and weigh	L of effluent	This is necessary for some calculations
Bed height	Reading from the scale on the reactor	L of settled biomass in reactor	An approximate SVI can be calculated from this, only with 5 minutes settling instead of half an hour.
NH_4 in the effluent	Dr. Lange test	mg/l NH_4-N	Check for nitrification and denitrification
NO_3/NO_2 in effluent and after feast period or filling period (check)	Merck indicators and precise measurement with Dr. Lange tests	Concentration of both components	
PO_4-P in effluent and after filling (SBAR2)	Dr. Lange test	mg/l PO_4-P	Check for phosphate uptake and release
Less frequent measurements:			
Biomass density	Dextran blue test (Jimenez *et al.*, 1988), (Garrido *et al.*, 1997)	g biomass/L granules	
Size and shape of granules	Image analysis	Average diameter, size distribution, shape factor (1), aspect ratio(2)	(1) 0 = irregular; 1 = round and smooth (2) 0 = flat (line); 1 = round
Cycle measurements:			
Acetate concentration	Chromatography	mg acetate/L	With these measurements, together with the on-line measurements, profiles of the conversions during the cycle can be created and mass balances obtained. Furthermore, some rates and yields can be determined.
PO_4-P	Dr. Lange test	mg/l PO_4-P	
NH_4, NO_3, NO_2	Spectrophotometry at 630 nm in auto-analyser	mM N/L	
PHB concentration	Freeze dry granules, grind, extract PHB, measure with GC	%PHB and %PHV	
Glycogen (not always and if, only in SBAR2)	Granules are destroyed and the supernatant is analyzed by HPLC	mg glucose/L, calculated to C-moles of glycogen/L	
Composition of the granules	Measuring C,H,N,S with HPLC	%C, %H, %N, %S of the biomass	

withdrawal phases) and calculated from the settled bed volume and the dry weight in the reactor. Calculations, including penetration depth into the granule, ammonium oxidation rate, nitrogen removal efficiency and ratio between acetate rich biomass and aerobic biomass were performed as described in Mosquera-Corral *et al.* (2005).

Morphology of the granules (particle diameter, aspect ratio and shape factor), density and ash content was measured according to Beun *et al.* (2001).

FISH was performed on thin layers of the granules in order to determine the location of the organisms within the granule structure.

Penetration depth of particulate COD could be located by colouring the starch particles with an Iodine solution. Cryosectioning the granules and microscopic observation gave information about the degradation of these starch particles.

Oxygen profiles could be measured with micro-sensors obtained from Unisense (Aarhus, Denmark), Figure 1.6. Sensors consisted of a gold plated platinum cathode and a silver/silver chloride anode. The tip diameter is 10–20 μm. The data acquisition system consisted of a picoammeter (Unisense, Denmark; model PA2000), an A/D converter (Pico Technology Ltd, UK; model ADC-101) and a computer to log the data and control the microsensor motion. Motion control consisted of a motorized two-dimensional stage (Phytron Inc., USA, model MT-65) and a motorized micromanipulator (Unisense, Denmark, model MM3M), controlling the position of the micro-electrode in three axis.

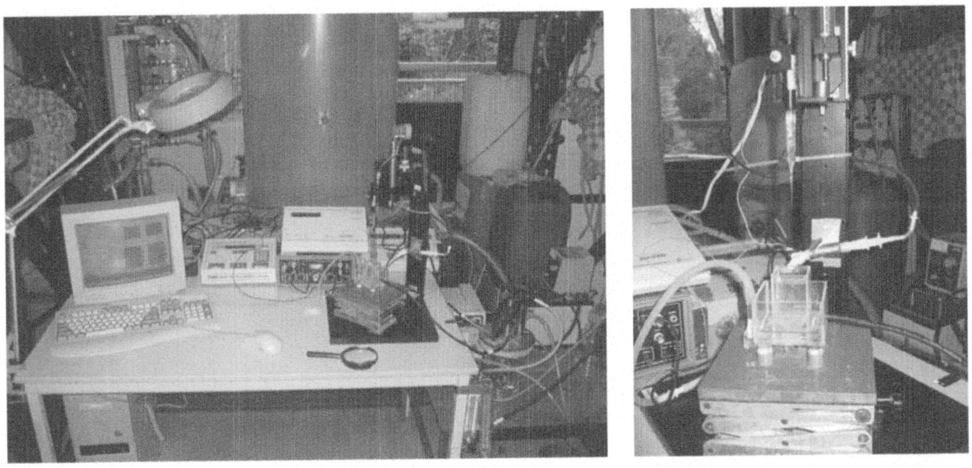

Figure 1.6 Set-up for micro-electrode measurements

Oxygen uptake rates (OUR) could be measured in a small stirred vessel, which was closed from its surrounding. The oxygen concentration in the vessel was measured with an oxygen electrode and logged in time. The OUR was determined from the slope of the oxygen profile, divided by the biomass concentration in the vessel.

1.2.5 Results

1.2.5.1 Particulate and Polymeric COD

1.2.5.1.1 Reactor Performance

For the first granule cultivation on maltose a reactor was started up with acetate for 52 days. After steady state operation, the influent was changed to maltose having an equivalent COD concentration as in the later feeding with starch (3.96 g COD/L), for another 126 days, after which the reactor had to be stopped due to technical problems and fungi growth.

These fungi (identified as *Aspergillus niger* by microscopy, Figure 1.7a and b) could be removed by natamycin addition to a final concentration of 40 ppm (Oostendorp 1981; Brothers and Wyatt 2000). Without addition of this antifungal agent, granules became totally covered with fungi and washed-out. When this agent was dosed with the influent water (80 ppm in influent) for 24 hours, fungi growth could be inhibited at once and smooth granules could be cultivated on maltose (Figure 1.7c).

During the second start-up period, maltose and particulate starch (reactor 1, PS) and maltose (reactor 2, SS) were used as carbon sources. After granule formation and finding the right method to control the fungi infections, reactor 2 was changed to soluble starch-maltose solution after 191 days of operation (reactor 2, SS). Reactor 1 start-up was with a particulate starch and maltose mix in the ratio of 1:1, with aerobic granular sludge from a pilot plant at a municipal wastewater treatment in The Netherlands. Day 0 in the report is the day that the influent COD was composed of soluble or particulate starch and maltose and the final addition of granules from the pilot plant had been taken place. Reactor 1 and reactor 2 were changed to 100% (respectively particulate and soluble) starch as influent COD component after 151 days (particulate starch) and 94 days (soluble starch) of operation. Hereafter the experiments were continued for another 171 days (PS) and 210 days (SS).

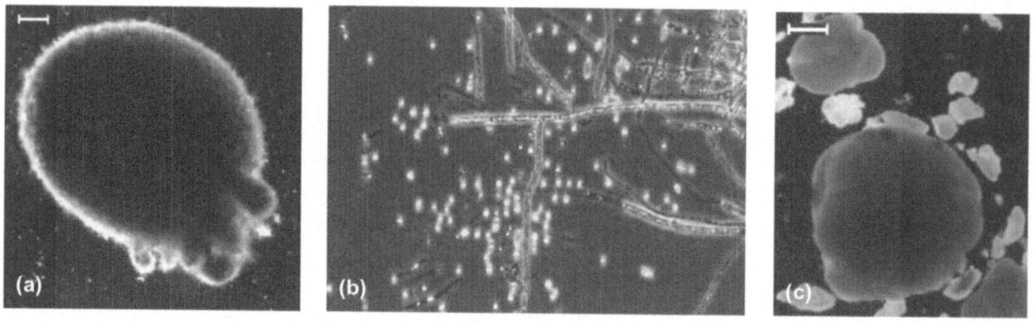

Figure 1.7 Aerobic granular sludge cultivated on maltose, covered with fungi (a), detected fungi (100x, b) and without fungi growth (c) Size bar = 1 mm

The particulate starch fed reactor performance is shown in Figure 1.8. Nitrogen removal fluctuated highly, depending on the oxygen control (fluctuating between 20% and 30% of the oxygen saturation concentration). Ammonia was totally consumed by nitrifying bacteria: after 20 days after start up, ammonia concentration in the effluent dropped to zero and stayed low throughout the total experiment. The total nitrogen removal had an average value of 31% when the reactor was only fed with particulate substrate. During the period with feeding of particulate starch and maltose the average N-removal capacity was lower at the start of the experiment (26%), but at the end before changing influent (day 89–151) the average N-removal capacity was 42%. The low de-nitrification can be due to the low penetration depth of the starch which leads to a lack of COD availability inside the granule (anoxic conditions) for simultaneous nitrification/de-nitrification (De Kreuk, Heijnen *et al.* 2005).

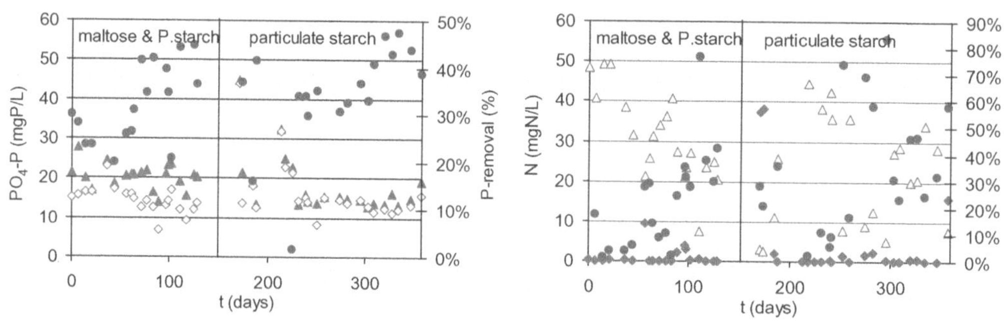

Figure 1.8 Influent, effluent and concentrations after feeding, during reactor operation of Reactor 1 (Particulate starch) (•-N and P - removal efficiency, Δ-NO_3-N effluent, ♦-NH_4-N effluent; ▲-PO_4-P after feeding, ◇-PO_4-P effluent)

This effect is stronger when only particulate starch is fed than when a mixture of maltose and starch is fed to the reactor (31% N-removal versus 42%). During the start of the experiment the oxygen concentration varied between 20% and 40% due to problems with the oxygen control and gas recycle, which could also lead to some reduction of the de-nitrification efficiency.

The removal of phosphate was on average 35% and did not change too much with changing influent. The starch removal had an average value of 99%, of which most was removed during the anaerobic period, which can be partly absorption to the granules, anaerobic hydrolysis and anaerobic cell internal storage, by glycogen accumulating organisms (GAO), G-bacteria, or phosphate accumulating organisms (PAO). From the FISH analysis performed by IRSA Rome, no GAO were detected and only 2% to 4% of the organisms were detected as PAO in both reactors (candidatus Accumulibacter), which later changed to 6–7% for reactor 1 and 1–3% for reactor 2. As this is too low to be able to cause the observed P removal and because of the low ash concentration of the granules (no P removal by precipitation) and the fact that almost no COD could be measured after the anaerobic period, we expect some other glycogen storing organism, such as G-bacteria, to dominate over the PAO. This species was not studied by FISH but would be a logical explanation since the reactor was fed with a sugar-based compound.

Similar conditions could be observed for the reactor fed with soluble starch (Figure 1.9). During the first period with a maltose and starch feed, the N-removal was 16% on average. When soluble starch was used as the sole carbon source, the

N-removal fluctuated highly. This was due to difficulties with a stable oxygen control and fluctuating granule morphology. The average total N-removal efficiency still approximated 59%. All ammonia was converted to nitrate. During the growth on maltose-soluble starch, the phosphate removal was 65%, which dropped to an average value of 46% during the period with soluble starch as influent. COD removal was near 99%, with most COD removed during the anaerobic period.

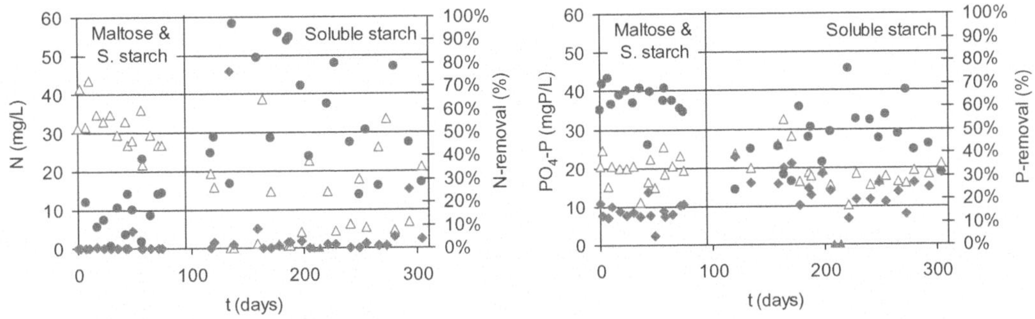

Figure 1.9 Influent, effluent and concentrations after feeding, during reactor operation of Reactor 2 (Soluble starch) (•-N and P - removal efficiency, Δ-NO_3-N effluent, ♦-NH_4-N effluent; Δ-PO_4-P after feeding, ♦-PO_4-P effluent)

The biomass growth yield measured in the reactors was extremely low. For reactor 1 during growth on maltose and particulate starch and on particulate starch solely, the biomass yield (weight of biomass produced per weight of COD consumed) was 0.25 g COD/g COD and 0.22 g COD/g COD, respectively. In reactor 2 biomass yields of 0.28 g COD/g COD (maltose and soluble starch) and 0.24 g COD/g COD (soluble starch) were measured. Standard biomass yields for activated sludge and PAO (pH 7) and biologically phosphate removing aerobic granular sludge grown on acetate are respectively 0.53, 0, 5 and 0.33 g COD/g COD (Smolders *et al.*, 1994; Metcalf & Eddy, 2003; De Kreuk and van Loosdrecht, 2004). Values in this study might be this low, because a part of the starch is removed by protozoa (Figure 1.10) and part is adsorbed to the granule without being immediately consumed. However, adsorbed starch is not measured in the effluent, so it is included as consumed COD in the yield calculation. The specific contribution of protozoa on the starch removal could not be quantified, but it has been observed that protozoa were incorporating small starch particles.

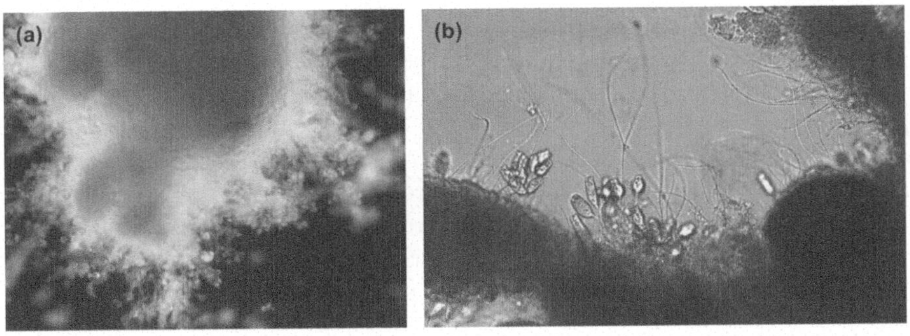

Figure 1.10 Protozoa growth on granules (magnification 7, 5x PS feed (a) 100x, (b) SS feed)

The low growth yield resulted in a very low average suspended solids concentration (dry weight) in the reactors, respectively 6.5 g TSS/L (ash content 17%) and 3.6 g TSS/L (ash content 12%) for the growth on particulate starch and maltose and solely particulate starch. For the second stage of the experiment, with growth on 100% particulate starch, this dry weight concentration decreased from 5 to 2 g TSS/L. Because of the decreasing dry weight in the reactor, the sludge age (solid retention time, SRT) was only 10 days in this second stage. In reactor 2, the average dry weight concentration was 17.2 g TSS/L (ash content 25%) when soluble and maltose were fed and only 7.3 g TSS/L (ash content 12%) when only soluble starch was used. The corresponding average SRT was 38 days and 21 days respectively. Also these values are much lower than the 70 days found in aerobic granular sludge reactors fed with acetate (De Kreuk *et al.*, 2005a), caused by decreasing biomass concentrations in the reactor and thus biomass wash out with the effluent.

1.2.5.1.2 Microscopy analysis

The presence of particulate matter in the influent can favour the growth of the filaments. With particulate starch the granular sludge gradually became fluffy, and filamentous biomass can be observed near the surface of the granular sludge. This is because particulate organic matter is attached to a local position on the surface, which could be detected by microscopy by colouring the starch particles with iodine solution. Granules taken right after the feeding period had many particles adsorbed to the granule surface, of which a large fraction were still present at the end of the aerobic period (Figure 1.11). Hydrolysis of this starch would give a constant low COD concentration at this position, which locally increases biomass growth rate. We could consider this as a locally extended feast period. We know from long feast periods in aerobic granular sludge as well as from biofilm research that this could lead to irregular structures (Picioreanu *et al.*, 2000; McSwain *et al.*, 2004; Mosquera-Corral *et al.*, 2005).

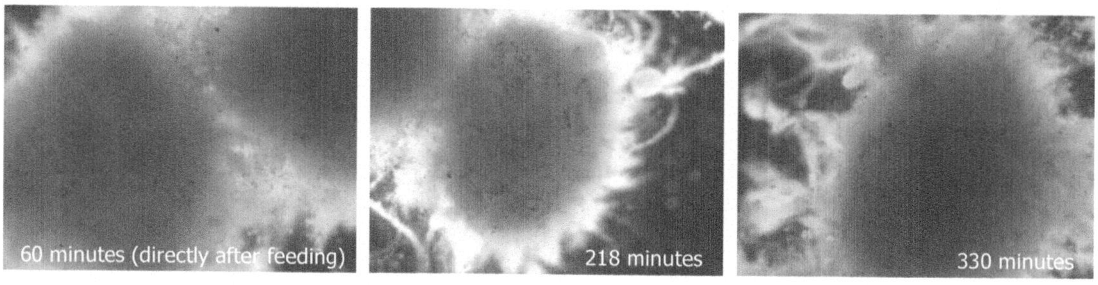

Figure 1.11 Starch adsorption and slow degradation, illustrated by iodine coloured granules (colors starch particles purple) after feeding particualte starch, after 3.5 and 5.5 hours aeration

In aerobic granules fed with particulate starch, finger shaped outgrowths could be observed, while in the granules fed with soluble starch, these fingers were detected rarely during the first operational period. When the granules matured and the reactor was operated over one year, also the granules from the reactor fed with particulate starch became more irregular. Finally, these fingers will detach more easily (by shear forces) than that smooth granules break, leading to many small granules besides the normal granules size distributions and leading to increased biomass wash-out. In the reactor fed with particulate starch, image analysing showed that approximately 30% of the total number of granules belongs to the fraction 0.2 to 0.3 mm, while 35% belongs to the fraction 2.0 to 3.0 mm, with an average diameter of 1.6 mm. To compare: for the reactor fed with soluble starch during the period with smooth and regular granules, approximately 10% of the total number of granules belongs to the fraction 0.2 to 0.3 mm, while 49% belongs to the fraction 2.5 to 3.5 mm, with an average diameter of 2.2 mm.

Presence of filaments and irregular structures can cause problems for reactor stability as they can hinder settling, leading to biomass wash-out. This is illustrated by the increasing SVI when maltose is removed from the influent (34 to 59 mL/g and 18 to 45 mL/g for reactor 1 (particulate starch) and 2 (soluble starch), respectively. SVI values are still very good compared to activated sludge, but higher than the average 20 mL/g SVI for acetate fed granules (De Kreuk *et al.*, 2005a). Also, the small particles tend to wash out quicker due to slower settling, leading to shorter sludge ages.

When the granules are compared with granules obtained from a full-scale Nereda® plant treating food industrial wastewater (Figure 1.12), we can observe that the irregular structures resemble those of the full-scale plant, but that these granules are much smaller. Also the influent to the Nereda® plant contains suspended solids, however, the majority of its incoming COD is soluble (on average 10 to 20% suspended solids in influent).

1.2.5.1.3 Microbial diversity

In order to determine the differences in diversity of the aerobic granular sludge grown on starch and grown in food industry, we performed a PCR and DGGE analysis. While the patterns of the DGGE bands of the different samples were compared, this was only a preliminary analysis.

While all the patterns were different for the DGGE analysis (Figure 1.13), there was some correlation between the full-scale food industry plant (labelled 3) and the reactor fed with particulate starch (labelled R1) on specific groups. As the population dynamics depend on many process parameters (including feeding pattern, aeration time, loading rate and temperature), comparisons can only underline the comparable results observed between the two systems rather than explain the differences that can be caused by feed.

1.2.5.1.4 Starch hydrolysis

Both soluble and particulate substrates appear to be rapidly removed from bulk liquid during the anaerobic period, as shown by the low COD analysis at the end of the anaerobic period (typically 0 mg/L). This is mainly due to adsorption of the starch to the granule surface, which needs to be hydrolysed before consumption by the bacteria. Goel *et al.* (1998a, 1998b) and San Pedro *et al.* (1994) report on the hydrolysis of starch under anaerobic and aerobic conditions. Under anaerobic conditions, fewer enzymes are produced to hydrolyse the starch, but when enzymes are present, the activity is not affected by changing the electron acceptor conditions. To verify this, starch hydrolysis batch tests were performed with soluble COD under aerobic and anaerobic conditions. All starch available was consumed during the first hours, independent of aerobic or anaerobic conditions (Figure 1.14). The average soluble starch removal rate (averaged over 3 measurements) was 5.8 g starch/g VSS.day for anaerobic conditions and 6.9 g starch/g VSS.day for aerobic conditions. This value is measured by dosing a starch solution to a certain amount of biomass (20 gram of wet granules in 200 mL liquid). Therefore we cannot speak purely of hydrolysis rate, but rather about adsorption rate of the starch to the granule. This underlines that the substrate is adsorbed and could be hydrolysed and consumed during the anaerobic period as well, favouring the growth of PAO, GAO and G-bacteria that can convert the hydrolysis products into cell-internal storage polymers under anaerobic conditions. However, the hydrolysis rate is too low to hydrolyse all starch during the anaerobic period, favouring the growth of fast growing heterotrophic organisms and filaments during the aerobic period, and so reducing settleability.

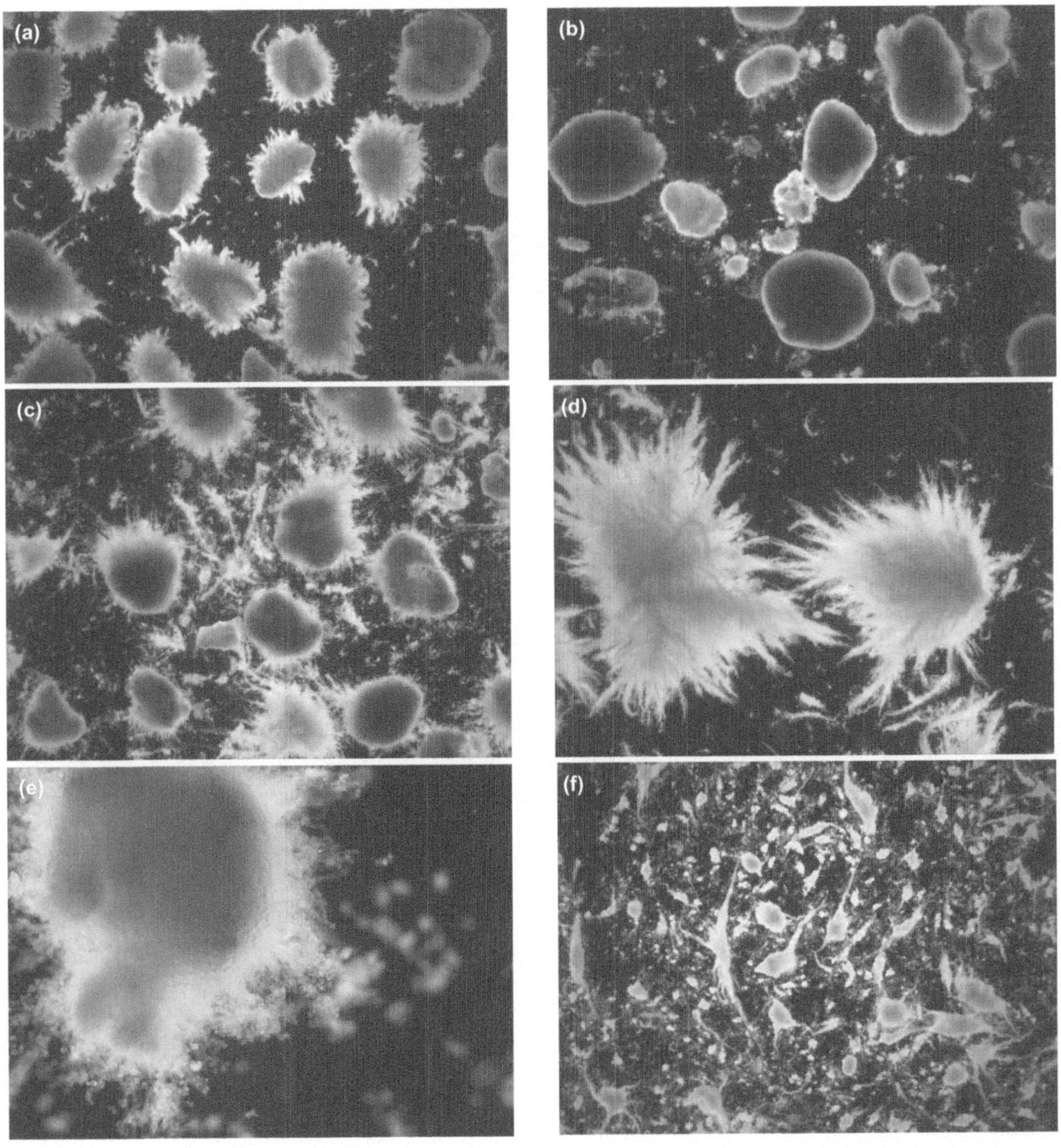

Figure 1.12 Granules present in the reactor fed with particulate starch (a, c) and granules present in the reactor fed with soluble starch during the first operation period (b) and changed granules during a later stage (d), Protozoa blooming on the granule surface of granules fed with particulate starch (e); Granules present in Nereda® systems fed with influent with suspended solids (f) (food industry, The Netherlands)

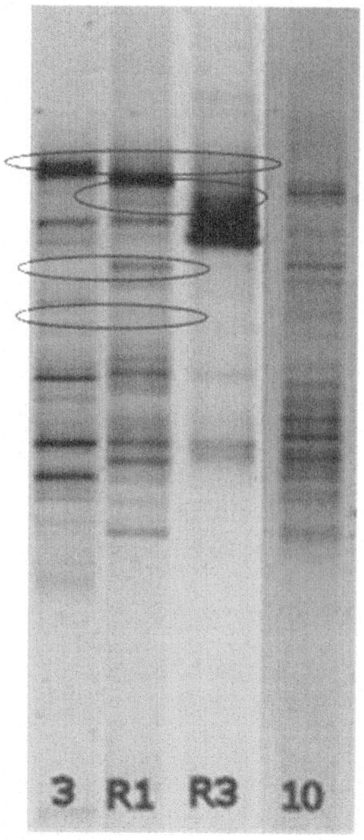

Figure 1.13 DGGE of granules from Nereda in food industry (3), from the laboratory fed with particulate starch (R1), from the reactor fed with soluble COD (R3) and from a pilot plant Nereda fed with raw municipal sewage (10)

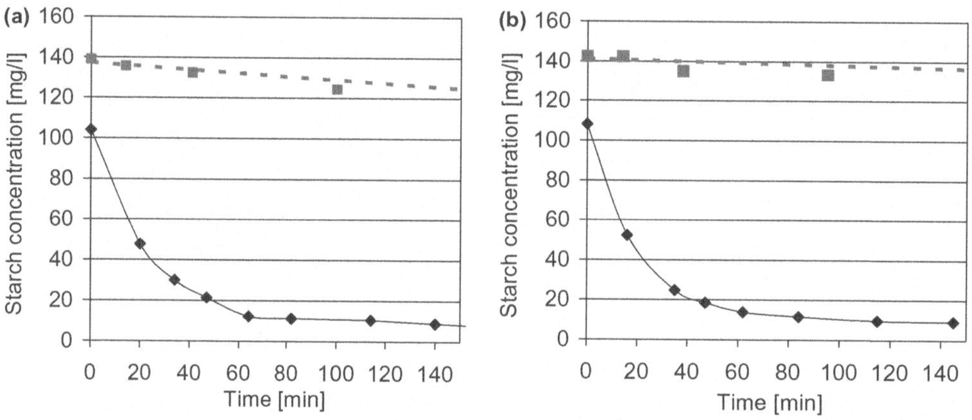

Figure 1.14 Soluble Starch removal under aerobic (a) and anaerobic conditions (b), with biomass (blue diamonds) or with effluent only (red squares)

In order to determine the mechanism of the hydrolytic enzyme activity (in the bulk liquid or within the granule itself), tests with effluent of both reactors were performed in order to determine the portion of hydrolysis taking place in the bulk liquid due to excreted enzymes, as well as tests to determine the hydrolysis rate in presence of the granular sludge. The starch degradation rate in the bulk liquid was only a fraction of the rate observed with the presence of biomass (Figure 1.14). The hydrolysis rates are given in Table 1.3. These results indicate that adsorption of the starch onto the granule, followed by hydrolysis and immediate consumption of the hydrolysed products is more likely to occur than the hydrolysis of starch in the surroundings of the granules. Enzyme activity seems to be granule bound, which is in line with the data found in activated sludge by Karahan (2006). However, the starch hydrolysis rate of this study was much lower than that of previous studies where activated sludge was used as the biomass source (Goel *et al.*, 1998c; Karahan *et al.*, 2006), implying the hydrolytic enzyme quantity secreted from granular sludge is lower than that secreted from activated sludge.

Microscopic examination of the starch adsorption to the granules shows a clear difference between particulate and soluble starch adsorption (Figure 1.15). The particulate starch is incorporated in between the filaments on the outside of the granules and adsorbs on the surface of the granules, without covering the total granule. In contrast, soluble starch forms a film on the tips of the finger-type structures. When starch is coloured with iodine, starch particles were observed throughout the total aerobic period of a cycle, while the purple stain from the film on the fingers disappears quickly (within a few minutes) after iodine and soluble starch addition to the granules under the microscope.

Table 1.3 Starch hydrolysis rates with enzymes in the bulk liquid (no biomass)

	Average hydrolysis rate (g starch/day)	Standard deviation
Anaerobic rate with granules grown on particulate starch	0.17	0.02
Anaerobic rate with granules grown on soluble starch	0.15	0.03
Aerobic rate with granules grown on particulate starch	0.07	0.004
Aerobic rate with granules grown on soluble starch	0.12	0.005

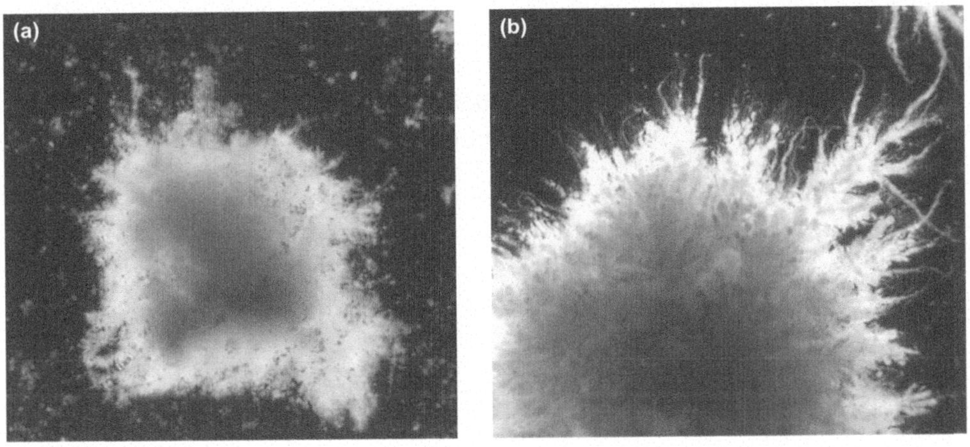

Figure 1.15 Different adsorption of suspended starch (granule from reactor 1, a) and soluble starch (granule from reactor 2, b). Magnification was 7.5x

1.2.5.1.5 Oxygen uptake rate during the cycle

Oxygen uptake rates (OUR) were measured on samples taken from the reactor throughout the aeration phase in order to determine if there was a decreased OUR at the moment that cell-external starch is depleted, indicating the duration of cell-external substrate availability and thus indicating the extended feast phase. This was studied in reactor 1, fed with particulate starch and reactor 2, fed with soluble starch. The results of the measurements are shown in Figure 1.16. Reactor 1 showed a gradual decrease in OUR, indicating that all reduced components in the effluent are oxidized in a first order kinetics. During this period, the effluent ammonium was around 1 mg/L, so oxygen uptake was needed for nitrification as well as for the catabolism by heterotrophic organisms. The measurements are in line with the microscopic observations, in which it was shown that during the total aeration phase, starch was present as particles attached to the granule. Also, the availability of this starch for heterotrophic growth was indicated by the presence of the filamentous organisms.

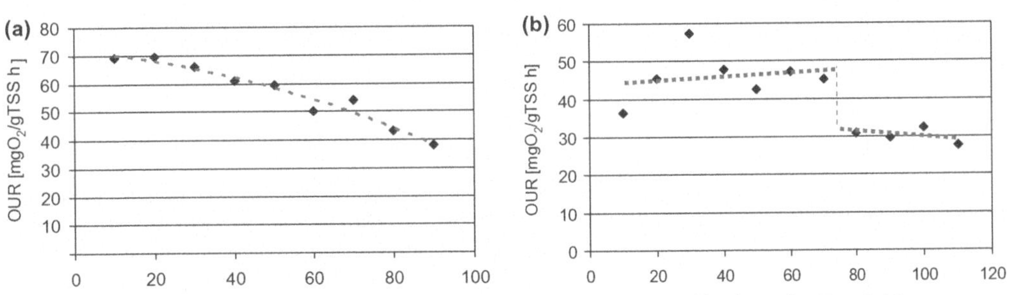

Figure 1.16 Oxygen uptake rate of the granules throughout the aeration phase from reactor 1 (particulate starch, a) and reactor 2 (soluble starch, b)

The OUR during the aeration phase for reactor 2 showed two different rates. During the first 70 minutes, a higher rate was measured than during the last 50 minutes of the cycle. This indicates that the starch had been consumed by half-way into the aeration phase. Also in this reactor, the effluent ammonium levels were approximately 1 mg/L, so this sudden decrease of the OUR was not caused by complete nitrification of the ammonium but rather the complete oxidation of the (adsorbed) soluble starch.

1.2.5.1.6 Oxygen penetration measured with micro-electrodes

In order to study the oxygen penetration depth into the granules during different stages of the aerobic granular sludge, granules were taken from the reactor at the end of the aeration phase and aerated further 2 hours before measurement commenced. A granule was fixed in a flow cell and the oxygen penetration depth measured with an oxygen electrode attached to a micromanipulator. First the oxygen penetration depth was measured with the addition of a trace element solution but without the addition of ammonium or starch (blanc-day 1, Figure 1.17b). After this, ammonium was added to a final concentration of 53 mg N/L and an oxygen profile was measured again (Figure 1.17b). Thereafter, the whole system was washed and filled with trace element solution and a blanc was measured (blanc-day1, Figure 1.17a) followed by soluble starch addition (to a final concentration 200 mg COD/L) after which another profile was measured (Figure 1.17a). The granule was left overnight with a soluble starch solution and aeration and the following day the whole procedure was repeated (day 2) but with the addition of particulate starch rather than soluble.

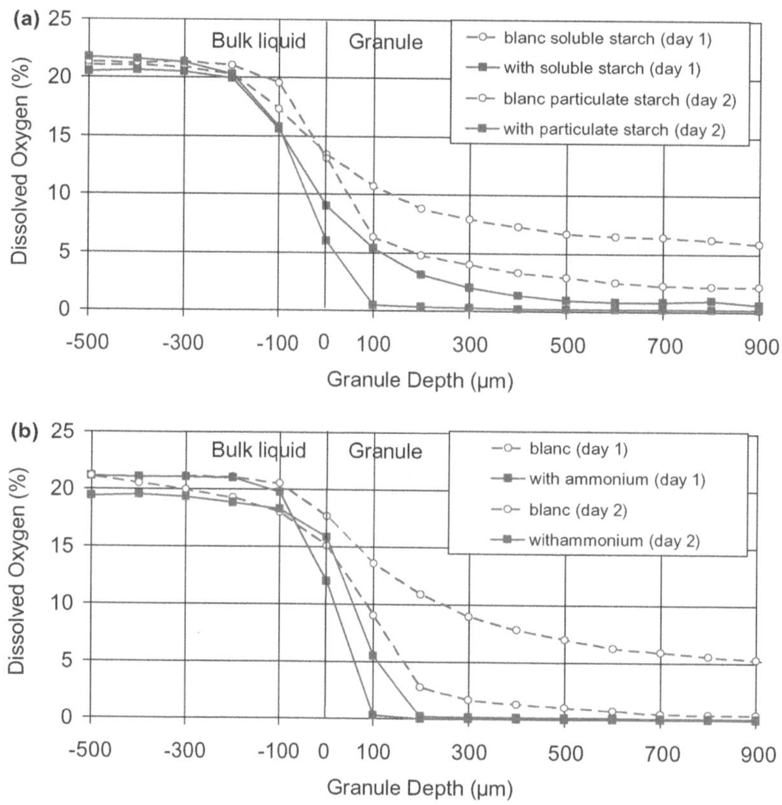

Figure 1.17 Oxygen penetration depth in an aerobic granule fed with soluble and particulate starch (a) and fed with ammonium (b)

The oxygen consumption due to ammonium oxidation decreased after one night aerated with soluble starch. On the first day, the difference in the oxygen profile caused by ammonium addition was 7.5 mg/100 µm, but decreased to 4.25 mg/100 µm for the second day. The oxygen consumption in the granule without addition of ammonium or starch (blancs) was also higher on the second day, implying part of the soluble starch was stored and still available during the experiment, consuming oxygen for the heterotrophic organisms. From the difference between blanc-day two in the ammonium curve and blanc-day 2 in the starch curve, it can be observed that the starch is being consumed and that the granule becomes more exhausted from its substrate over time. The ammonium oxidizing organisms also suffered from a night without ammonium and with oxygen, since their activity on day two was lower.

From the addition with starch, it can be seen that dosing both soluble and particulate starch led to an activity increase, which was slightly higher for the particulate starch. However, the activity increase from dosing ammonium was considerably higher, indicating that the ammonium oxidation was a faster process than starch degradation and consumption (which is limited by the hydrolysis process).

Starch addition also increased the boundary layer compared to ammonium addition. Since this increase was not caused by a larger flux (the oxygen consumption was lower), it may have been caused by protozoan activity surrounding the granules. After the last addition of particulate starch (on day 2), the granule was coloured with iodine and is shown in Figure 1.18. Several parts of the granule were covered in protozoa and it was difficult to determine where the surface of the granule started. It is likely the protozoa play a role in part of the suspended substrate removal process and differentiating between the oxygen consumption due to protozoan activity and that of bacteria on the granule is difficult.

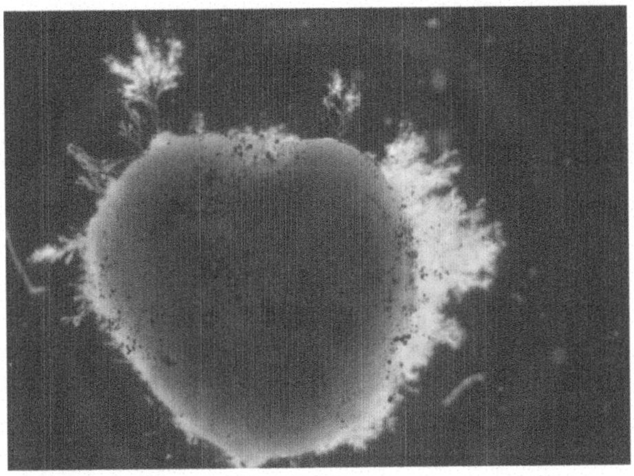

Figure 1.18 Aerobic granule after microelectrode measurements

The shift in oxygen concentration between day 1 and day 2 in Figure 1.17a also shows that there is less oxygen availability for the ammonium oxidisers when particulate starch is available. Given the fact that particulate starch is available throughout the cycle and all ammonium is oxidised during the aeration phase, the activity of the ammonium oxidisers has to be located in the most outside layer.

1.2.5.1.7 Particulate removal in industrial practice

In industrial applications, suspended solids typically represent a fraction of the total COD concentration in the influent. Therefore, the previous experiments with growth solely on particulate starch need to be seen as extreme situations. However, results in morphology and suspended solid removal of industrial plants can be compared to a large extent.

Two industrial examples of granules from Nereda® reactors were examined, both treating food industry wastewaters. The influent of Industry 1 (a cheese processing industry), largely consisted of soluble and colloidal/suspended COD, mainly as fatty acids and fats. Average influent concentrations were: 2.8 g total COD/L and 1.6 g soluble COD/L; 110 mg total N/L, with 60 mg NH_4-N/L; and, 24 mg PO_4-P/L. The reactor typically operated with temperature fluctuations between 30°C and 40°C. The SVI30 of the biomass was typically 50 mL/g during the steady state in the reported period (114 days) and the dry weight fluctuated between 4 and 9 g TSS/L (dry weight fluctuations were mainly caused by influent variability). COD-removal efficiencies were 93% for the total COD and 94% for the soluble COD. Despite the fraction of suspended COD, the granules that developed in the reactor were small and round (Figure 1.19b). However, also flocculated biomass coexisted with the granules (Figure 1.19a). From time to time a protozoa bloom was observed, attached to small granules (Figure 1.20).

Industry 2 also contained considerable influent suspended solids (35% of suspended COD). A flotation unit (DAF) was placed upstream of the reactor for partial removal of these solids. The granular sludge morphology in this reactor was similar to the granules grown in the laboratory (Figure 1.21) and to those observed by Schwarzenbeck *et al.* (2004a, 2004b, 2005), grown at laboratory scale with an influent containing barley dust (mimicking malting wastewater) and dairy wastewater collected from industry. They observed a fluffy outgrowth and sludge settleability decreased because of this morphology.

Comparing the results of Schwarzenbeck (laboratory scale granules grown with industrial wastewater) to the laboratory scale starch experiments reveals many similarities between the granular sludges produced and particle removal processes. The influent characteristics of the experiment with the dairy wastewater were quite similar to those of Industry 1, containing partly soluble and partly suspended COD. The granules formed by Schwarzenbeck in the laboratory scale reactor were larger than those observed in the industry. The main differences between the two reactors were exchange ratio and temperature as well as possible higher granule breakage at full scale due to reactor configuration. Moreover, sampling is more difficult on large scale, with possible breakage occurring during the sampling process or a specific fraction is taken from the reactor due to stratification. On the other hand, COD removal efficiencies were similar.

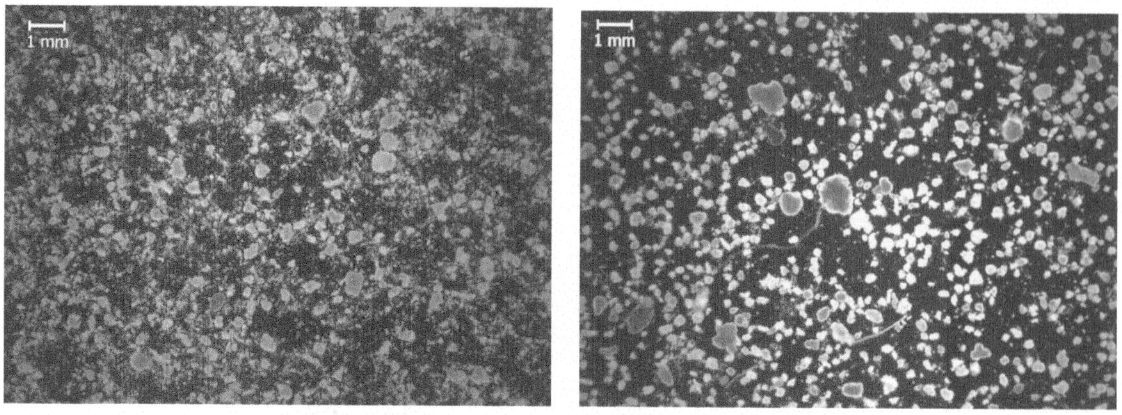

Figure 1.19 Aerobic granular sludge from Industry 1, total biomass (a) and all granules >0.2 mm (b). Size bar = 1 mm

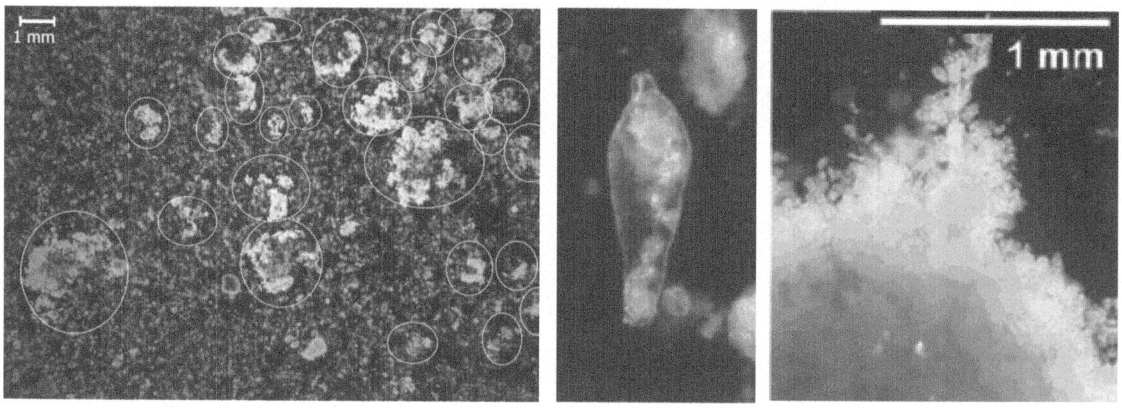

Figure 1.20 Protozoa bloom in reactor Industry 1 (e.g. Vorticella (a), Rotifiers (b)) and granule covered in protozoa (c, Schwarzenbeck 2004a)

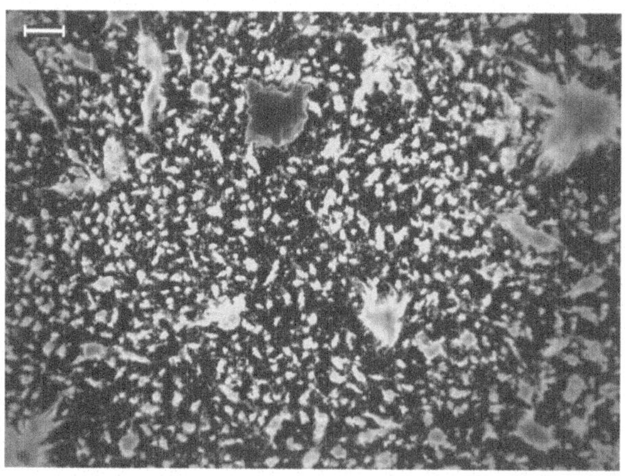

Figure 1.21 Granules from industry 2 (size bar = 1 mm)

Protozoa were observed during the experiments with barely dust, as was the case with the particulate starch laboratory reactor and Industry 1. For Industry 2, the levels of protozoa were considerably lower, although these granules resembled more the fluffy outgrowth due to slow hydrolysis of the particulate starch or slowly biodegradable substrate, probably present during aeration. Schwarzenbeck (2004b) proved that the ciliates ingested fluorescent beads, contributing to suspended solid removal and thus contributing to better effluent quality.

1.2.5.1.8 Conclusions of the particulate and polymeric COD investigations

From the laboratory scale experiments with particulate starch we concluded that aerobic granular sludge can be maintained on solely particulate substrate, although the settling characteristics decrease. When slowly biodegradable substrates with large molecular structures are fed to an aerobic granular sludge reactor, they will be adsorbed to the granules and hydrolysed on the granule surface. This leads to an extended feast phase during the aeration period, leading to finger-type structures and the growth of filaments. In parallel to other research in biofilms and aerobic granular sludge, this has likely to do with strong substrate gradients (oxygen and substrate) inside the granule.

Particulate starch in this research was removed by protozoa covering the surface of the granule as well bacteria on the granule. Quantification of this particulate substrate removal could not be done in this reactor, since the particles were also adsorbed to the granule surface and further hydrolyzed. The very low growth yield in the particulate starch reactor indicated, however, that at least a significant fraction of the starch was removed via catabolism.

When granular sludge is used in a system treating mainly particulate substrate or at least slowly biodegradable substrates, it is most likely that good nutrient removal (N and P) will be more difficult to achieve. Reasons are that the substrate will not penetrate to and be stored in the regions inside the granules were it could be used for denitrification during the aeration phase. Most of the substrate stays on the granule surface and will be oxidised directly after hydrolysis when oxygen is present. Only the hydrolysed and stored part during the anaerobic phase can contribute to denitrification. Ammonium oxidation and the availability of nitrifying organisms during the aeration phase is not limiting in such a system. For phosphate removal we could conclude the same, since phosphate accumulating organisms also need a substrate rich anaerobic period in which they can consume sufficient substrate to outcompete other heterotrophs (as in normal biological P-removal). Also this will not be reached when the suspended COD/dissolved COD or slowly biodegradable COD/readily biodegradable substrate ratios is too high. Therefore in practice, it would be advantageous to remove suspended solids from the influent in order to decrease this ratio when nutrient removal is required.

1.2.5.2 Results of the high temperature experiments

1.2.5.2.1 Granule Formation at 30°, 40° and 50°C

For the 30°C temperature study a temperature controlled bubble column was inoculated with 2 L of activated sludge from the De Harnaschpolder municipal wastewater treatment plant (The Netherlands). Settling time was decreased stepwise from 15 minutes to the final 3 minute length over approximately two weeks. This resulted in a selection for sludge with settling velocities greater than 2 m/h to 12 m/h, respectively. During the first three months of operation, temperature was monitored on-line in order to control the temperature range within the reactor bulk liquid. Subsequently, a new control system was installed and the temperature was measured offline once a week to ensure correct operation. During this period, the temperature varied from 28.5°C at the start of aeration to 31°C at the end of the cycle. This range was achieved by pre-heating the influent prior to feeding and with water jacket heating on the reactor.

Gabus *et al.* (2008) studied granule formation at 30°C, reporting that granules could not be formed at these elevated temperatures in a reactor set-up similar to the system used in this study, although granules grown at 20°C could be maintained at 30°C. Contrary to this report, in the system described here, granules first appeared after 10 days and persisted until the end of the experiment on day 187 (Figure 1.22). Granules were light coloured and even transparent during the first weeks of development (Figure 1.23).

The density of these granules slowly increased, reaching a biomass density of 198 kg/m^3 by the end of the experiment. This value higher than values reported by De Kreuk *et al.* (2005a) at 20°C.

The dry weight concentration in the reactor increased during the experiment from 4 g TSS/l to 14 g TSS/l (Figure 1.24) and fluctuated throughout the experiment. These fluctuations were mainly due to the mixing characteristics of the bubble column in combination with the extremely good settling properties of the aerobic granules, resulting in an inadequately mixed reactor. In order to obtain a representative sample for dry-weight measurements, a sample was collected from both the top and bottom of the reactor. However, having a few more or less granules in a sample already results in a relatively high deviation from the actual biomass concentration.

Parallel to the 30°C experiment, granule formation was also studied at 40°C. This reactor was started with the flocculated and small (<0.1 mm) suspended solids from an industrial biofilm reactor treating molasses containing wastewater (DSM, Delft, The Netherlands). A small amount of activated sludge from the De Harnaschpolder municipal wastewater treatment plant (The Netherlands) was added to increase biodiversity. The temperature in the bulk liquid was controlled at 40°C (varying from 39.5°C during feeding to 40.5°C at the end of aeration) with the help of pre-heating the influent (similar to the 30°C experiment).

Settling time was reduced from 15 to 3 minutes over three weeks, corresponding to a required settling velocity of 2 m/h to 12 m/h, respectively. Due to the lower oxygen solubility at this higher temperature, the dissolved oxygen concentration setpoint in the reactor was altered from 20% (1.3 mg/L) to 40% (2.6 mg/L) on day 148. After 144 days the dissolved oxygen concentration was raised to 100% (6.4 mg/L), in order to stimulate aerobic processes (such as nitrification).

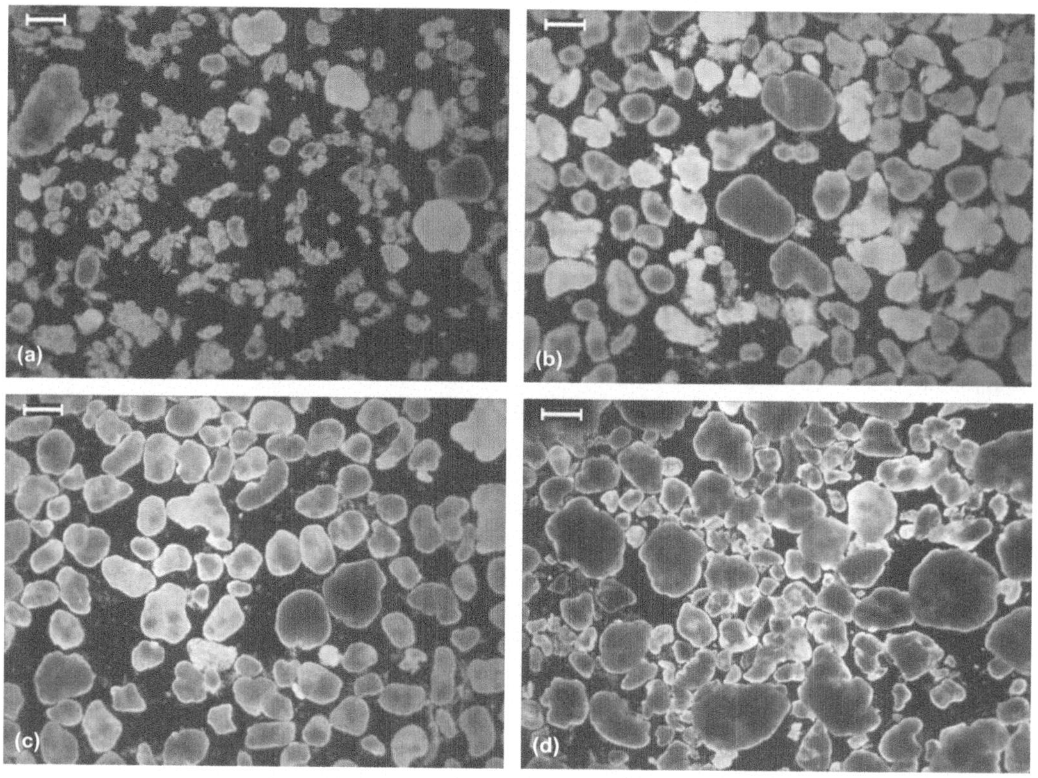

Figure 1.22 Granular sludge formed at 30°C after 25 days (a), 49 days (b), 70 days (c) and 187 days of operation (d) (size bar is 1 mm)

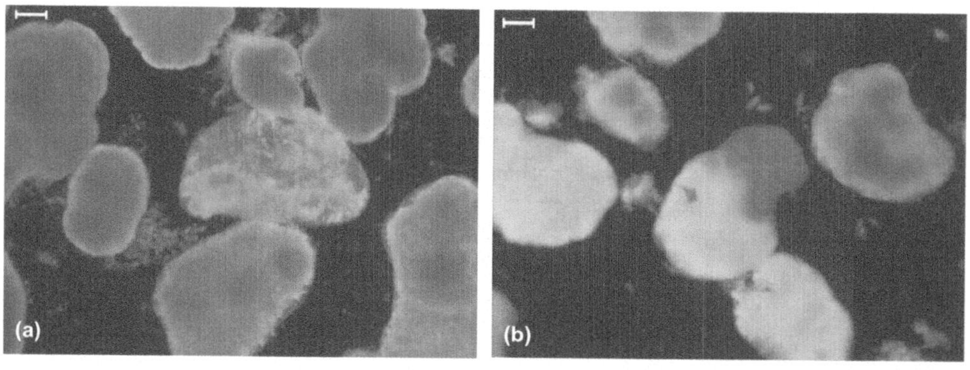

Figure 1.23 Granular sludge formed at 30°C after 70 days (a) (size bar is 250µm) and 25 days (b) (size bar is 1 mm)

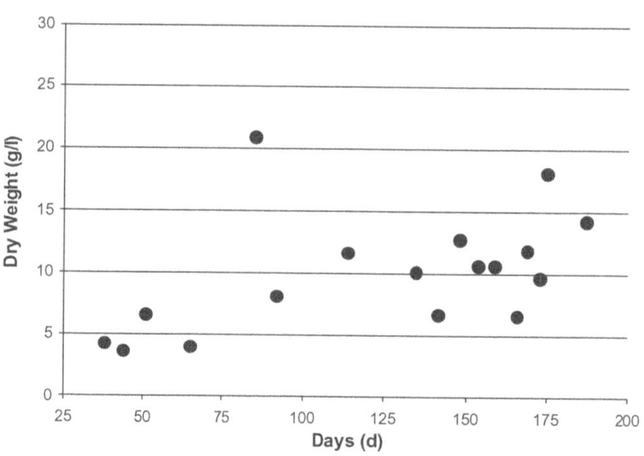

Figure 1.24 Development of the dry weight concentration at 30°C

Small granules first appeared after 4 days (Figure 1.25a). On day 20 (Figure 1.25b), the selection pressure was increased by decreasing the settling time to 3 minutes and so particles with settling velocities lower than 12 m/h were washed out of the system.

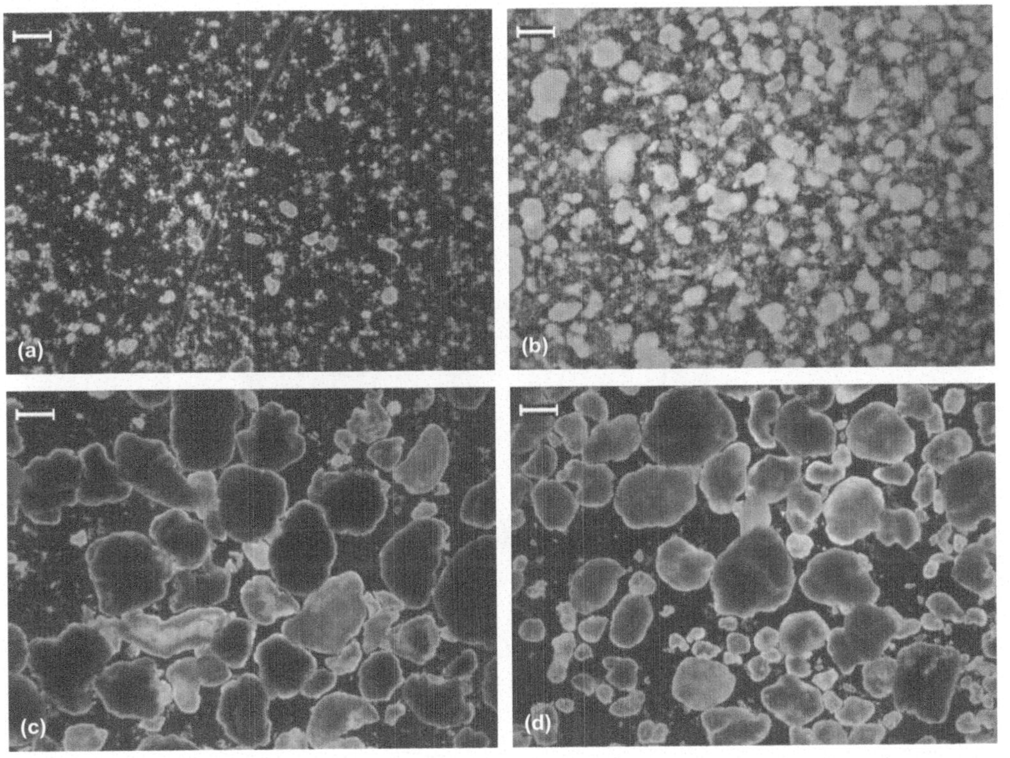

Figure 1.25 Granular sludge formed at 40°C. after 4 days (a), 20 days (b), 100 days (c) and 150 days of operation (d) (size bar is 1 mm)

During the start-up phase, many Zooglea sp. were observed on the granule surfaces, forming their specific finger-type structures (Figure 1.26). As these organisms have been observed in many laboratory- and pilot-scale reactors during start-up, their appearance may be a trigger for granule formation. As soon as the minimum required settling velocity for aggregates was increased to 12 m/h (via a 3 minute settle period), these Zooglea structures disappeared. The reason for this could be two-fold. Zooglea grow under high COD concentrations in combination with long SRTs and low biomass concentrations (Eikelboom and van Buijsen 1983). Once the granules mature and the granular biomass increases, there is an increase in the overall acetate uptake rate (and therefore a decrease in the concentration) which may negatively impact on the growth of the Zooglea and thus they disappear from the system. The second cause for their loss may be that the open structure of the Zooglea hinders the settleability of the aggregates and so they are washed out of the system at low settling times.

The dry weight concentration increased during the experiment from 0.39 g TSS/l to 24 g TSS/l (Figure 1.27). Dry weight information during the first weeks is absent due to the biomass concentration being too low to sample without disturbing the reactor performance. On day 150, 0.5 L of a mixed sample (including biomass) was collected for inoculating the 50°C reactor, resulting in a sudden decrease in the biomass concentration.

The observed density of the granules grown at 40°C was 108 kg/m^3, which was lower than at 30°C. This was probably due to the higher overall growth rates of the biomass at this high temperature and the absence of the slow growing nitrifying population, which usually form dense clusters inside the granules.

The inoculum used for the 50°C reactor consisted of 150 ml aerobic granular sludge from a Nereda® pilot reactor (Epe, The Netherlands) and 500 ml of granules from the 40°C reactor. Two sources were used in order to ensure the presence of sufficient microbial biodiversity. An aerobic granular inoculum was used (as compared to the 2 previous temperature studies) in order to determine if the granules could be maintained at this high temperature or if they would disintegrate and wash out. The dissolved oxygen concentration was controlled at 100% (5.5 mg/L) to avoid anaerobic fermentation processes within the granules, which could proliferate under this higher temperature with too low a dissolved oxygen concentrations. The temperature was raised stepwise from 40°C to 50°C over 8 days.

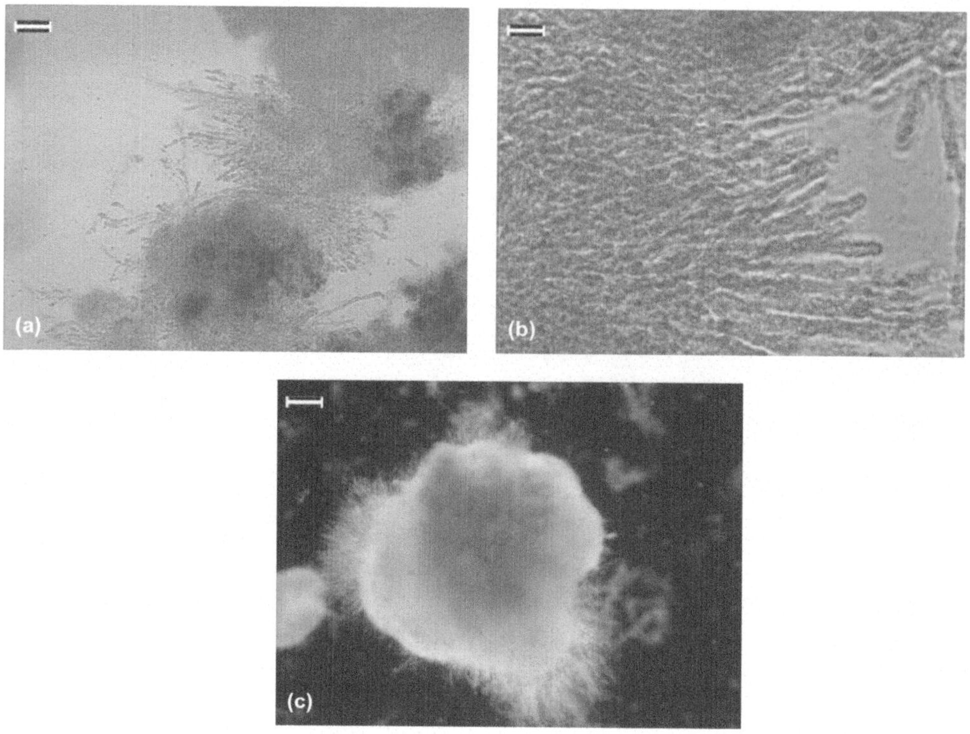

Figure 1.26 Zooglea sp. & finger type structures at 40°C after 37 days (a) (size bar is 75 μm), 37 days (b) (size bar is 18 μm), 100 days (c) (size bar is 250 μm)

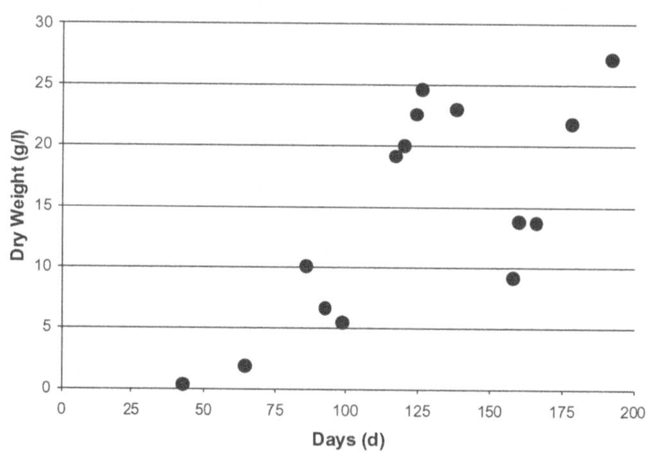

Figure 1.27 Development of the dry weight concentration at 40°C

The dry weight decreased over a period of 25 days (Figure 1.28), during which the granule appearance changed dramatically in shape and size (Figures 1.29 and 1.30). The newly formed granules were 'fluffy' and irregular with finger-type structures (Figure 1.30). These finger-type structures could indicate low COD uptake during feeding, although measurements showed that all of the COD was removed from the bulk liquid anaerobically. Filamentous growth with COD storage ability during the feeding phase may also be responsible for the 'fluffy' appearance

The density of the granules after 40 days was 63 kg/m^3. The granules had higher ash content (55%) than previously observed and a black precipitate formed in the bottom of the reactor. Analysis of the precipitate revealed it consisted mainly of carbonates and phosphates, most likely bound to magnesium and/or calcium. Analysis of offline measurements for phosphate showed that chemical precipitation was the main method of phosphate removal at 50°C. No phosphate removal could be measured during the aeration phase, indicating the precipitation process mainly occurred during feeding phase.

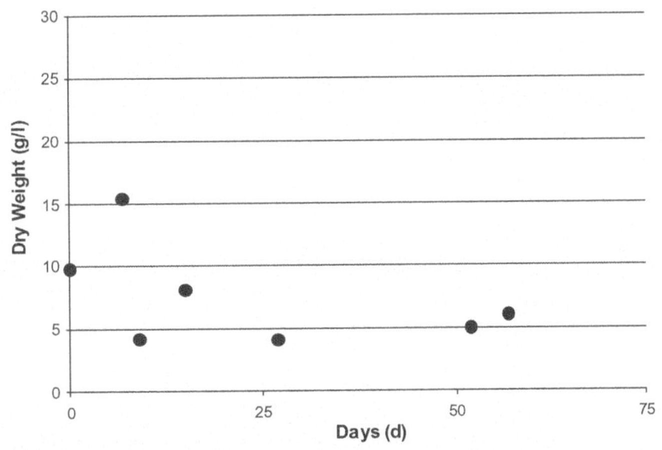

Figure 1.28 Development of the dry weight concentration at 50°C

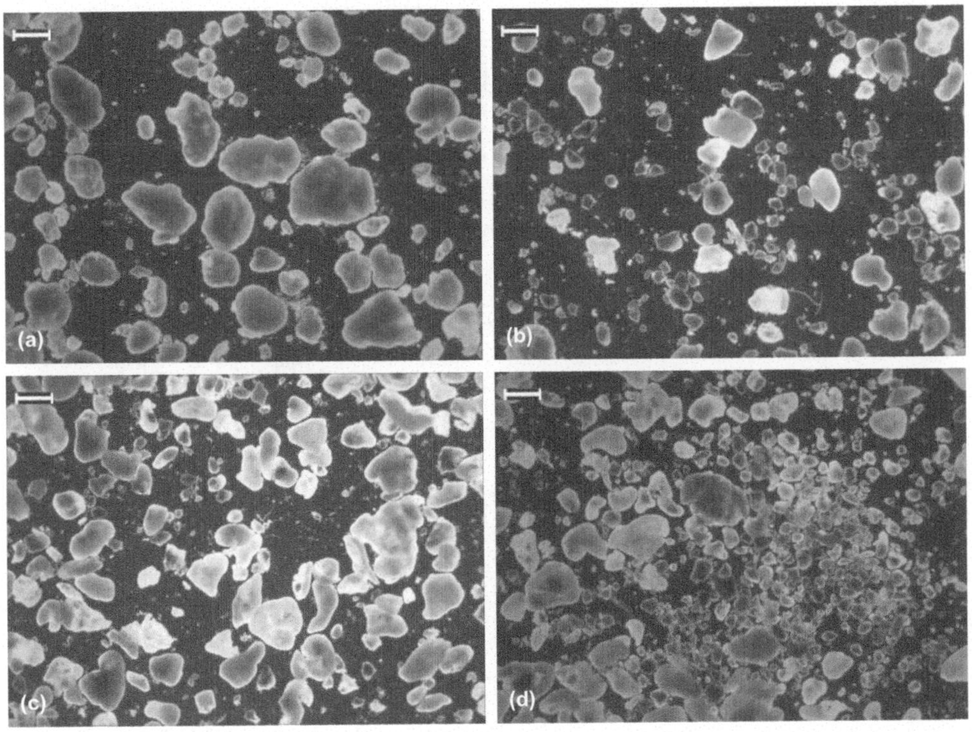

Figure 1.29 Granular sludge formed at 50°C after 0 days (a), 3 days (b), 7 days (c) and 14 days of operation (d) (size bar is 1 mm)

1.2.5.2.2 Comparison of the granule characteristics grown at different temperatures

When granule formation at low, moderate and high temperatures are compared, it has been observed that the optimum temperature for granule formation (determined according to the duration of the start up period) was at 20°C, with the appearance of the first granules at this temperature occurring after 3 days of operation. These granules were fully grown after 21 days, showing a steady state density of 109 g VSS/L biomass (De Kreuk and van Loosdrecht, 2004).

The final granule characteristics depend upon the operation temperature (Table 1.4). Granule formation at low temperatures (8°C) was unsuccessful, due to the low activity of the organisms involved. The low activity resulted in a slow uptake of COD during the anaerobic period, leading to a prolonged feast period during the aerated period, which caused granule instability and biomass wash out. However, when the reactor was started at 20°C, the temperature could be decreased to 8°C for an extended period, without affecting the granule stability.

Granule morphology changes with increasing temperatures of 30°C and above and the granules become more unstable with increasing temperature. Sludge volume indices after 8 minutes of settling (SVI8) were lower at the higher temperatures. The low SVI8 at 50°C was mainly due to the presence of precipitates, leading to compact and dense

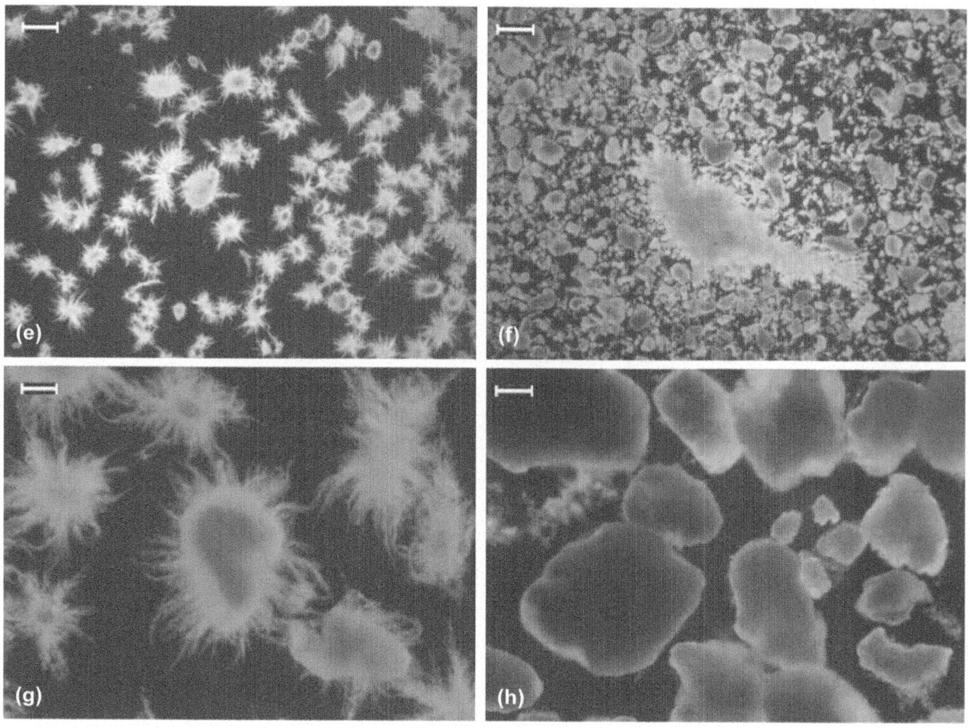

Figure 1.30 Granular sludge formed at 50°C after 60 days (e), 44 days (f) of operation (size bar is 1 mm). Finger-type growth (size bar is 150 μm) and Chemical Precipitation (h) (size bar is 250 μm)

Table 1.4 Average granule characteristics during long term temperature experiments

	8°C*, **	20°C	30°C	40°C	50°C
Dry Weight (g VSS/l)	18	18	9.5	15	2.5
Ash Content (%)	39	40	20	40	55***
Density (g TSS/L granule)	74	109	198	108	63
SVI8 (mL/g)	14	15	6	6	9
SRT (days)	95	70	63	30	15

* *Data from (De Kreuk et al., 2005b)*
** *Longer cycle (4 hrs) and thus lower loading rates*
*** *High ash content due to external precipitates*

structures next to the biomass granules. A high biomass density corresponds to a low SVI and so a well packed bed (such as at 30 and 40°C) will lead to low SVIs.

The ash content of the biomass at 30°C was the lowest. This was due to the PAO population being outcompeted to a large extent resulting in less poly-phosphate in the cells. At the higher temperatures, more precipitates were formed due to decreased solubility of various compounds such as $Ca_3(PO_4)_2$ and $CaCO_3$.

The solid retention time of the biomass in a reactor is often used as a direct measurement for the overall growth rate of the biomass. However, with granular biomass, it is unclear which part of the biomass is washed out: total granules, eroded biomass from the surface of the granules, parts of broken granules or bacteria that grew in suspension or in flocs. However, it could be observed that at higher temperatures, there was an increase in growth rates and a decrease in the SRT. The decreased SRT, in combination with the low biomass concentration and the high effluent turbidity, was partly due to suspended growth (biomass which cannot be retained in the system) and also the absence of protozoa. Protozoa grazing on the biomass will result in a higher assumed SRT, since this SRT is calculated from the washed out solids and ignores the free suspended solids consumed by protozoa.

In conclusion, it can be stated that aerobic granules can be formed under mesophilic conditions. Under thermophilic conditions, aerobic granules can be maintained, however, their appearance is fluffy and the effluent suspended solids concentration will be high. In this respect, aerobic granular sludge at thermophilic conditions behaves in a similar way to activated sludge flocs, although, in a granular sludge reactor the biomass morphology is more stable and biomass retention is improved.

1.2.5.2.3 COD and Nutrient Removal Performance at 30°, 40° and 50°C

For the 30°C temperature experiment, once the granules were formed and high COD removal efficiencies were achieved; the influent COD was typically all consumed during the anaerobic period and converted to storage polymers, mainly poly-hydroxyalkanoates (PHA), as can be seen from the cycle measurements performed. Phosphate and nitrogen removal efficiencies (ammonium and nitrite oxidation and denitrification) increased during the first month to an average of 99% (Figure 1.31) with a dissolved oxygen concentration of 25% of the saturation (1.9 mg/L).

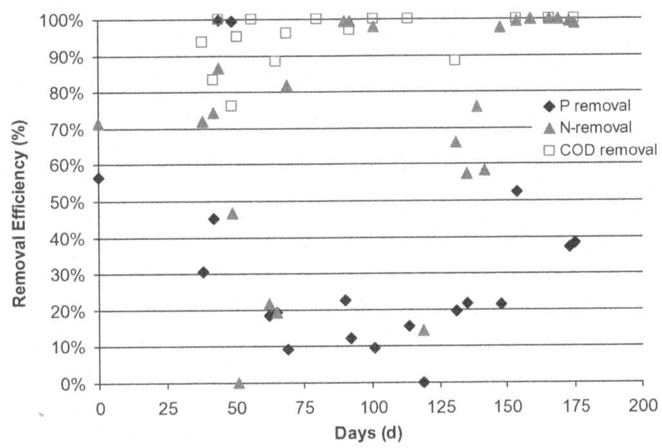

Figure 1.31 Nitrogen, Phosphate and COD conversions at 30°C

On day 56, the pH in the reactor accidently increased to 13 for 1.5 hours and although most of the granules survived and the granule characteristics did not change macroscopically, both the nitrogen and phosphate removal efficiencies decreased immediately. The addition of suspended nitrifying sludge from a laboratory scale SHARON reactor resulted in the complete recovery of the nitrogen removal performance. However, the phosphate removal efficiency remained low (20%). On day 121, the reactor pH decreased to a value of 2 as a result of a broken pH electrode. This again resulted in a strong decrease of the nitrogen removal efficiency and also negatively impacted the partially recovered phosphate uptake. Following this pH shock, the nitrifying activity was again recovered using an inoculum from the laboratory scale SHARON reactor.

The phosphate removal efficiency never completely recovered after the second pH shock. It appears the pH accidents assisted the glycogen accumulating organisms (GAO) present in out-competing the PAO in the system. Lopez-Vazquez *et al.* (2006) reported that GAO have a higher substrate uptake rate at 30°C compared to PAO. However, PAO have a lower maintenance requirement with temperatures under 30°C. Since the reactor was operated at 30°C, there was no clear advantage for GAO or PAO in this granular sludge system and the dominance of the two organisms was highly dependent upon the inoculum in this case. The initial phosphate removal efficiencies indicated that it was possible for the PAO to compete with GAO at this temperature. This was also supported by the slow increase of phosphate removal efficiency following the pH shock on day 156 until the end of the reactor operation at day 187 and it appears the PAO were growing back into the system.

Several cycle measurements were performed with the reactor operated at 30°C and one cycle measurement was performed at a reactor temperature that was operated for two days at 23°C (short term temperature decrease). Typical concentration profiles of the measured components are shown in Figure 1.32. Phosphate was released during the anaerobic feeding period (first 60 minutes) and taken up during the aerobic period. However, the aerobic period wasn't sufficiently long enough for complete uptake of the phosphate and so the phosphate slowly accumulated in the reactor. The cellular PHA concentration increased during the feeding period and decreased during aeration. From mass balancing, it was concluded that most of the incoming COD was converted into PHA (of which 63% was PHB and 27% PHV).

During the aerobic phase, nitrite accumulation occurred. However, over the whole aerobic period, all nitrogen was removed via simultaneous nitrification and denitrification (SND) as has been observed at 20°C in earlier studies (De Kreuk *et al.*, 2005a). Accumulation of nitrite could have been due to the higher temperature dependency coefficient (θ) for ammonium oxidizing bacteria (AOB) and nitrite oxidizing bacteria (NOB), which have typical values of 0.094 and 0.061, respectively (Hao *et al.*, 2002).

If the nitrification rates of this study of granules grown at 30°C are compared to those of 8°C to 20°C, some differences in behaviour at high temperatures can be seen (Figure 1.33). Despite the ASM model comment that temperature dependencies are considered until 25°C, the changed ammonium conversion rate follows the typical nitrifier pattern better than the

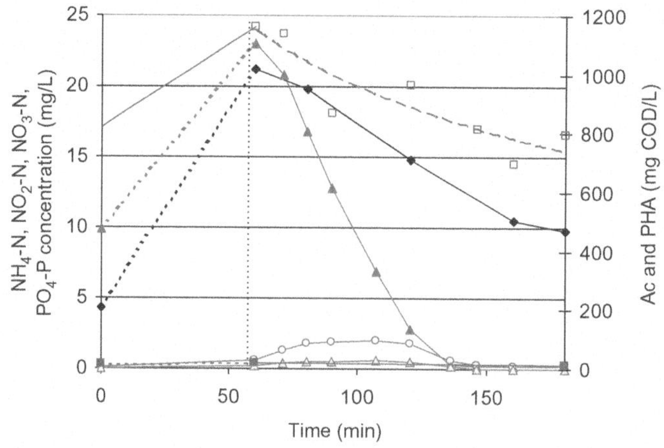

Figure 1.32 Typical concentration profiles during a cycle at 30°C (■ Acetate (mg COD/L); □ PHA (mg COD/L); ◆ PO_4-P (mg P/L); ▲ NH_4-N (mg N/L); ○ NO_2-N (mg N/L); Δ NO_3-N (mg N/L))

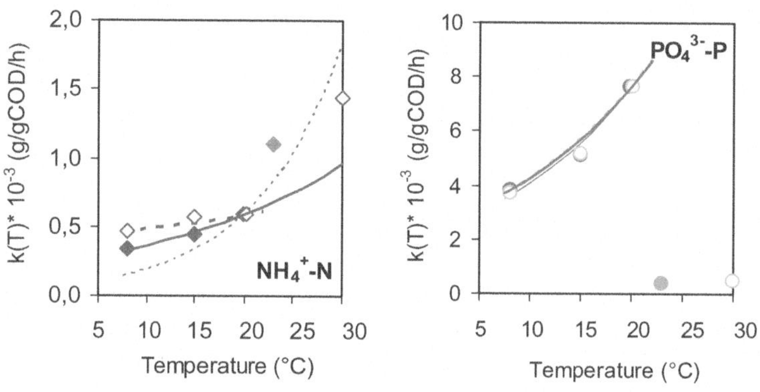

Figure 1.33 NH_4-N and PO_4-P consumption rates at different temperatures (◇, ○ and dashed line measured and modeled rates in adapted systems; ◆, ● and solid line measured and modeled rates in short term changes from 20°C to 15°C and 8°C and ◆, ● measured rates at short term changes from 30°C to 23°C; dotted line are the modeled rates according to the ASM2 model). Data are obtained from this study and from de Kreuk *et al.* (2005a)

adapted temperature dependency for granules at low temperatures. This makes sense, since conversion rates in aerobic granular sludge is highly dependent on the oxygen diffusion gradient into the granules. The lower temperature dependency for granules in adapted systems was caused by an increased oxygen penetration depth at lower temperatures, due to lower aerobic microbial activity in the outer layers. An increased aerobic zone causes an increased volume in which nitrification can take place, resulting in higher nitrification rates or a nitrifier enriched biomass. At higher temperatures (30°C), mechanisms are the opposite where higher activity due to higher temperatures at the surface causes a lower oxygen penetration depth into the granules, so a smaller aerobic volume is available for aerobic processes to occur (and at the same time a larger volume is available for anoxic de-nitrification processes). Therefore, at higher temperatures, the rate increase is slightly lower than could be expected from flocculated systems as activated sludge. If the temperature is decreased for a short period, the higher penetration depth of oxygen would lead to a larger volume of biomass under aerobic conditions and thus the overall activity decrease will be less than in an activated sludge system.

For the phosphate consumption rates, these effects are not present, since the phosphate accumulating organisms are outcompeted at higher temperatures. Therefore, a large decrease in P-uptake rate was observed at 30°C and also the short term temperature decrease was not long enough for the PAO to grow back into the system, resulting in an unchanged low phosphate uptake rate.

For the 40°C experiment, during start-up, the biomass concentration in the reactor remained low, resulting in low COD removal efficiencies. Only after 1.5 month of operation were high COD removal efficiencies achieved, which remained until the end of the experiment (Figure 1.34).

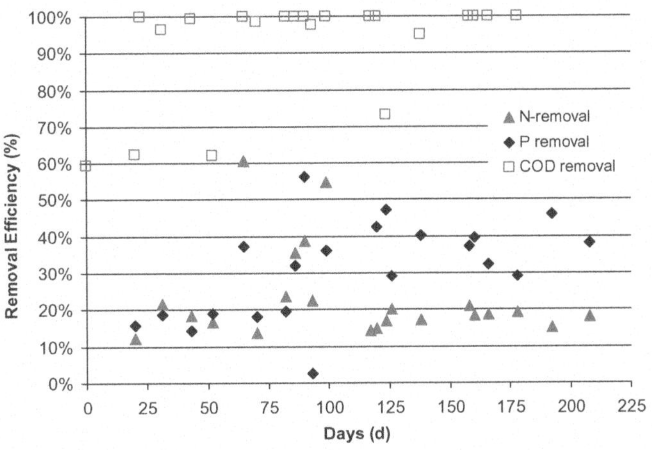

Figure 1.34 Nitrogen, Phosphate and COD conversions at 40°C

Nitrogen removal efficiencies at 40°C remained low (20%) with a dissolved oxygen concentration of 40% of saturation (2.6 mg/L). From day 144 the oxygen concentration was increased to 100% of its saturation concentration (6.4 mg/L) in order to increase nitrification by increasing the total aerobic volume of the biomass. Unfortunately this did not give any positive results and nitrogen removal remained low (20%) for the remainder of the experiment. The phosphate removal efficiencies also showed similar results, with only 30–40% P-removal achieved.

These low removal efficiencies resulted in relatively high effluent nutrient concentrations (typically 42 mg N/L and 18 mg P/L. These results were expected, since at 45°C, a transition from mesophilic to thermophilic organisms occurs. At 40°C many conversion processes are known to be inhibited to some extent, such as nitrification which generally does not occur at temperatures above 42°C. Where ammonium oxidising reactors (such as the SHARON system, in which the nitrite oxidation step is avoided) are operated at temperatures from 30 to 40°C, for process stability, often a temperature of 35°C is chosen (Hellinga *et al.*, 1998; van Kempen *et al.*, 2001).

The experiment at 50°C mainly focussed on granule formation, since ammonium oxidation, and biological phosphate removal at these high temperatures has not been reported in literature. However, all nutrient removal efficiencies were measured throughout the duration of the experiment (Figure 1.35).

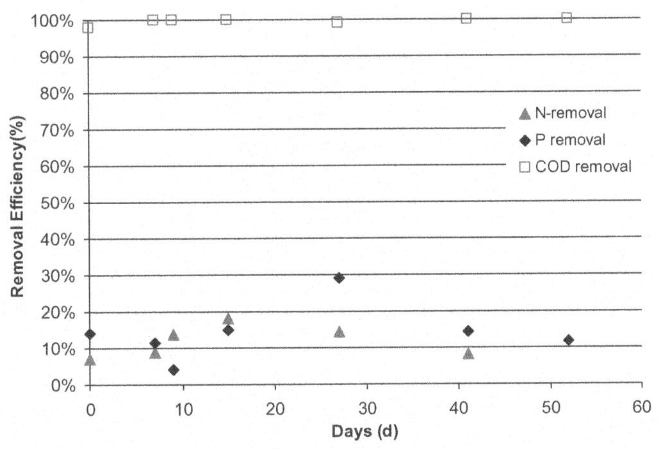

Figure 1.35 Nitrogen, phosphate and COD conversions at 50°C

Even though the biomass concentration in the bulk liquid remained low (on average, 5 g/L), COD was typically removed during the anaerobic feeding. Nitrogen and phosphate removal efficiencies remained low at 10%. Any phosphate removal mainly occurred via precipitation.

1.2.5.2.4 Comparison of Nutrient Removal at Different Temperatures

A comparison of the nitrogen removal efficiencies at temperatures ranging from 8°C till 50°C reveals an increasing efficiency up to 30°C (Figure 1.36) with higher temperatures resulting in a steep decrease in the removal efficiency. Although ammonium uptake has been reported at 40°C, it is likely that this would require acclimatised biomass. In these experiments, the temperature was increased rapidly. A slower build-up of the temperature may have yielded a higher removal efficiency.

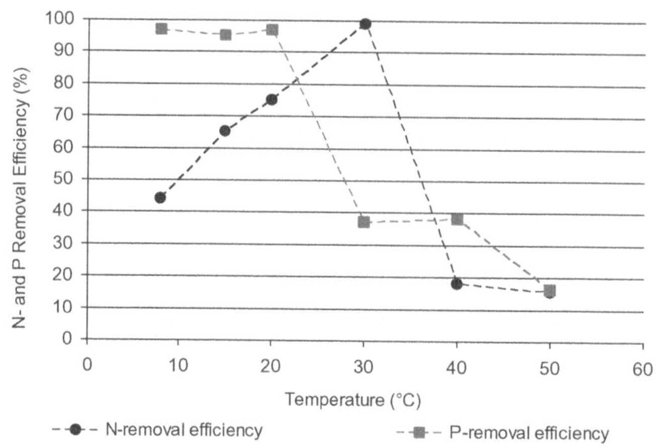

Figure 1.36 Nitrogen and Phosphate removal efficiencies at different temperatures

Phosphate removal efficiencies were not affected by temperature until 30°C was reached. However, the rapid drop at 30°C was the result of pH control problems (on days 56 and 121 of the experiment) as described above with the critical competition between PAO and GAO populations being one of the reasons why phosphate accumulation did not recover as quickly as at other temperatures.

At temperatures lower than 30°C, phosphate accumulating bacteria have a competitive advantage compared to the glycogen accumulating organisms (Figure 1.37). During other experiments at lower temperatures, biomass typically recovered quickly after pH shocks and the phosphate removal was restored within two weeks. A way of restoring the phosphate removal at 30°C would be the addition of biologically phosphate removing sludge from a wastewater treatment plant or from a plant with aerobic granular sludge exhibiting biological phosphorous removal. However, it was chosen not to do so, in order to be able to compare this study to other experiments where the reactors were also not re-inoculated with PAO.

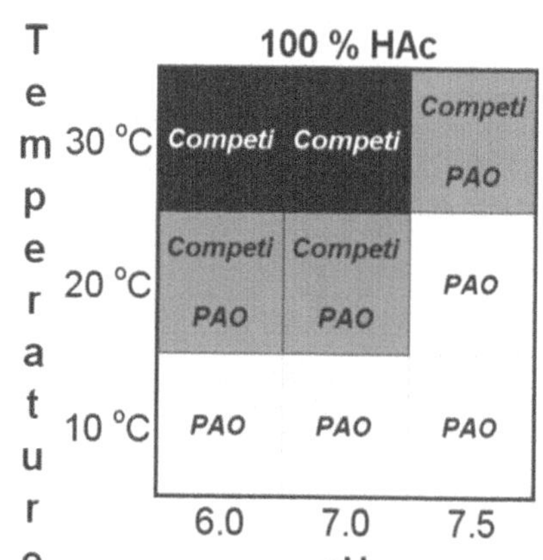

Figure 1.37 Competition between GAO (Competibacter) and PAO. Grey areas: condition in which dominant organism depends on inoculum, black area's dominance of GAO, white area's: dominance of PAO (Lopez *et al.*, 2008)

1.2.5.2.5 Elevated Temperatures in Industrial Practice

The industrial plants that are currently operated according to the Nereda® technology are operated at ambient temperatures, ranging from 15°C to 28°C degrees. Therefore, no full-scale installations are available for comparison with this study. Moreover, the influents with a higher temperature are typically characterised with high particulate COD concentrations, which influences the granule morphology severely. However, the conclusions of this study showed that for future treatment plants at high temperature, aerobic granular sludge could be a promising alternative to other technologies, mainly to avoid sludge bulking and decrease suspended growth.

Another issue learned from practice is that in industry, cleaning campaigns often occur once a week on Fridays, just before the weekend operation. Cleaning with very hot water can increase the temperature in the reactors suddenly to over 40°C, if the water is not managed adequately. While the granule structure may not be affected by these short temperature fluctuations, they should be avoided, since they will influence the bacterial population within the granules, as can be seen from this study where conventional nitrogen removal quickly deteriorated at temperatures above 40°C. Recently, autotrophic organisms have been found which are capable of ammonium removal at temperatures above 42°C. However, these organisms will not populate a reactor where these temperatures are achieved only as shocks.

1.2.5.2.6 Conclusions of Temperature Studies

From the laboratory experiments performed in this research, we can conclude that aerobic granules can be formed at all tested temperatures, under mesophilic conditions as well as being maintained at thermophilic temperatures. At higher temperatures, more biomass grows in suspension and less or no protozoa are found in the system, resulting in lower measured SRTs and higher effluent suspended solids concentrations. In these kinds of systems, post-treatment would be required for further solids polishing. Granular biomass is more stable at elevated temperatures than reported for activated sludge in literature.

Part of the reactor operation at 30°C showed a different competition between phosphate accumulating organisms and glycogen accumulating organisms than reported in literature for biologically P removing systems. It seems that PAO can proliferate in a reactor operated at 30°C fed with acetate as carbon source, contrary to activated sludge systems. However, disturbance of the reactor results in an immediate deterioration of the P-removal. At higher temperatures, phosphate is only partly removed by temperature induced precipitation, rather than by biological processes.

Nitrogen removal inhibition at higher temperatures is comparable to those observed in activated sludge, with ammonium oxidising bacteria showing lower inhibitory temperature dependency than the nitrite oxidising bacteria, resulting in nitrite accumulation during the cycle.

At all temperatures, the influent COD is removed (stored) anaerobically and aerobic granules could therefore be useful for the removal of low concentrations of COD under thermophilic conditions, when anaerobic treatment is not possible, or as a polishing step after anaerobic treatment.

1.2.5.3 Results of Batch Tests

1.2.5.3.1 Fluor Phenol Batch Tests

The laboratory scale experiments on the short term effect of several concentrations of 2-Fluor phenol on aerobic granule activity were carried out in batch tests. In order to perform a comparison, the same experiments were performed with activated sludge from the De Harnasch polder municipal wastewater treatment plant (The Netherlands) and crushed granules (Figure 1.38). The granular sludge was crushed in order to mimic the activated sludge, but having the same microbial population as the granular sludge.

Results of the effect of 2-Fluor phenol on the Ammonium Oxidizer Bacteria (AOB) activity are shown in Figure 1.39. The ammonium uptake rates obtained for activated sludge, granular sludge and crushed granules, when the 2-Fluor phenol was absent, were 1.654, 1.379 and 1.473 mg N/g VSS.h, respectively. AOB in the activated sludge and crushed granular sludge were completely inhibited by the 2-Fluor phenol. Previous studies on the effects of phenol on the nitrification process with activated sludge indicated that the increase in the initial phenol concentration led to a decrease in the nitrification percentage, with accumulation of nitrite and ammonium (Amor *et al.*, 2005). However, AOB in the granular sludge still retained 15% of their activity in the presence of 10 mg/L of 2-Fluor phenol. Liu *et al.* (2005b) have studied the influence of phenol on nitrification by aerobic granules and reported that nitrifying bacteria in microbial granules may have a better capacity to resist shock loadings of toxic compounds in wastewater than suspended nitrifying bacteria.

Figure 1.40 shows Phosphate Accumulator Organisms (PAO) were less affected by the 2-Fluor phenol than AOB. Nevertheless, granular sludge was more resistant to the toxic compound than activated sludge and crushed granular sludge. At 5, 10, 15 and 25 mg/L concentrations, activities greater that 100% were observed for granular sludge when compared to the blanks (i.e. no addition). A possible explanation for this could be that when the 2-Fluor phenol kills or inhibits the outer layer of the granule where the nitrifiers are located, allowing for deeper penetration of the oxygen to the PAO in the core. Since the 2-Fluor phenol concentration is then very low, it doesn't affect the PAO. However, due to the higher oxygen availability, PAO rates increase. This illustrates the difference between granules (with high diffusion limitations of oxygen and probably some binding of 2-Fluor phenol to the gel matrix) and activated sludge (with lower diffusion limitations of oxygen).

1.2.5.3.2 Salinity Batch Tests

The short term effect of salt on aerobic granule activity experiments were performed as described for 2-Fluor phenol. Overall, the AOB activity decreased with salt concentration for all sludges examined. In this case, the AOB activity was

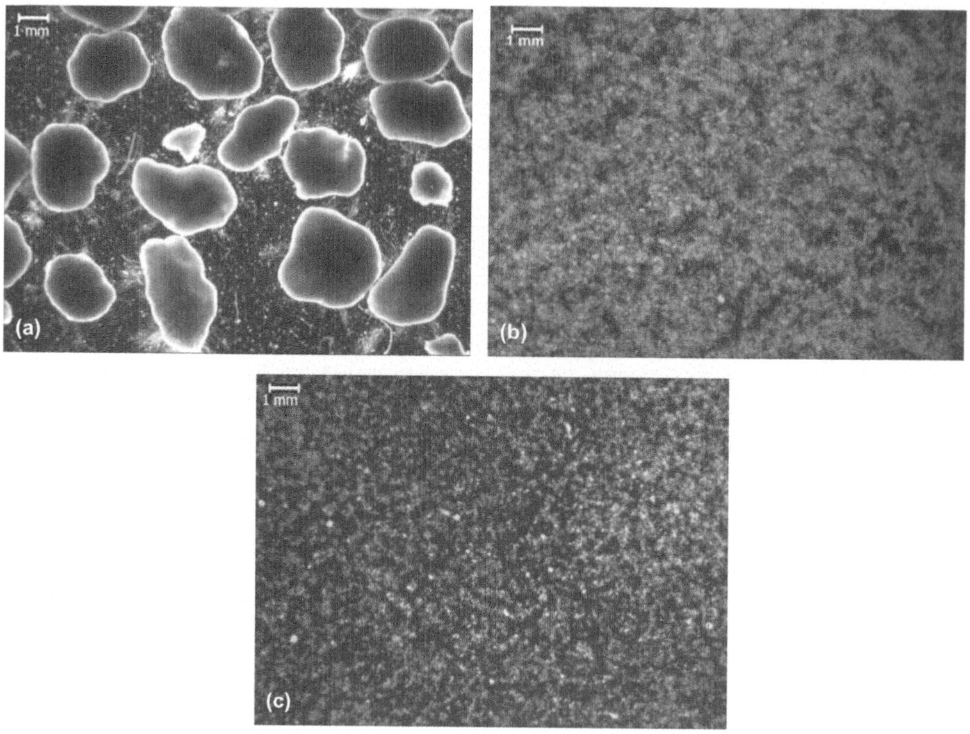

Figure 1.38 Inoculums for the batch tests: Aerobic granular sludge (a), Activated sludge (b) and Crushed granular sludge (c) Size bar = 1 mm

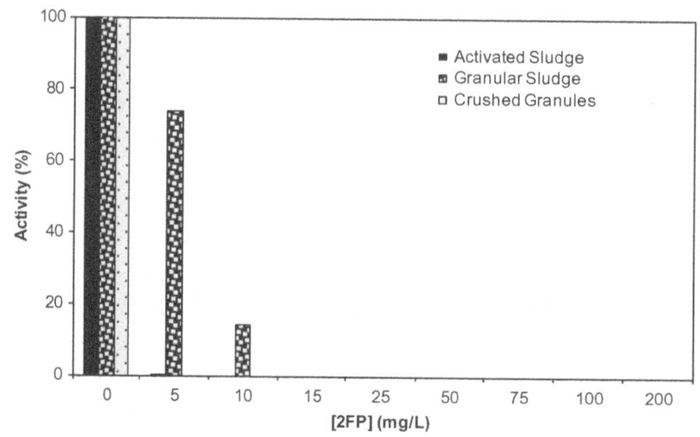

Figure 1.39 Effects of 2-Fluor phenol on the AOB activity

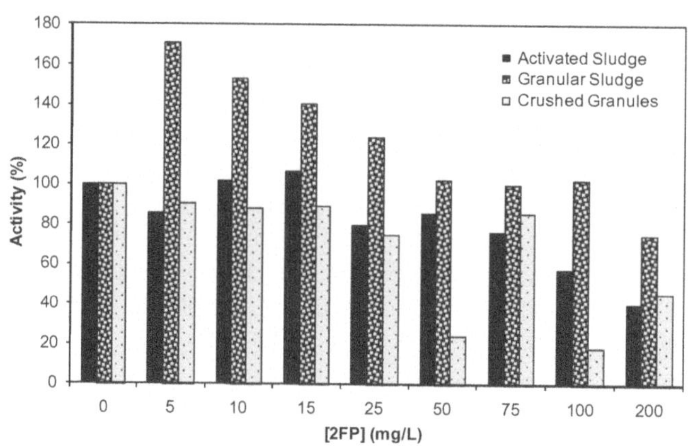

Figure 1.40 Effects of 2-Fluor phenol on the PAO activity

more affected by the salt in granular sludge (whole and crushed) than in activated sludge (Figure 1.41). The activities at 0 mg/L of NaCl were 1.235, 1.102 and 1.737 mg N/g VSS.h for activated sludge, granular sludge and crushed granular sludge, respectively. In the presence of 5 and 10 g/L of NaCl, activated sludge had higher activities, 78% and 43% respectively, than granular sludge (62% and 37%) and crushed granular sludge (53% and 11%). Ludzack and Noran (1965) found that high percentages of salt compromise the correct operation of conventional aerobic wastewater treatment at chloride concentrations above 5–8 g Cl^-/L. The results of the crushed granular sludge can be directly compared with the granular sludge as they have the same AOB population. From Figure 1.41, it can be observed that granular sludge is more protected against salt shocks than crushed granules at NaCl concentrations below 20 g/L. This suggests that the granule matrix has a shielding effect against salt toxicity

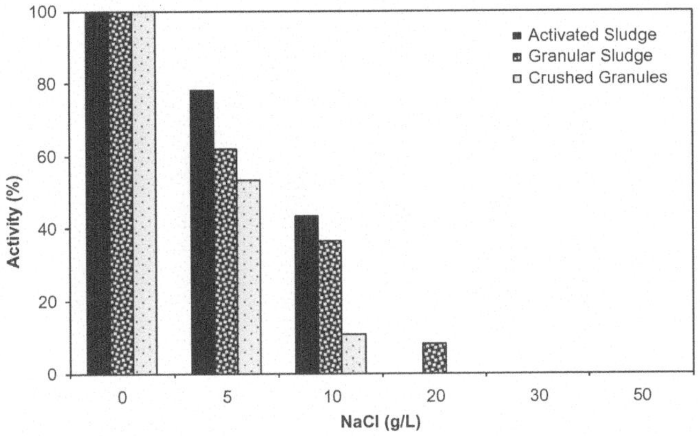

Figure 1.41 Effects of salt on the AOB activity

The effect of salt on the PAO activity was lower for the granular sludge (whole and crushed) than for activated sludge (Figure 1.42). For concentrations higher that 10 g NaCl/L, the activated sludge PAO activity became inactive whereas the granular sludge PAO still retained some activity (ca. 7% at 20 g NaCl/L). The PAO activity was consistently higher for the crushed granules compared to the whole granules. When PAO are present as free cells, they better access to oxygen in the bulk liquid as compared to when they are located in a granular structure. In the layered structure of granules, PAO populations are located inside the granule with the nitrifiers located towards the surface resulting in limited oxygen availability for the PAO. Van't Riet and Tramper (1991) stated that the presence of salt affects the maximum solubility of oxygen transfer rate directly, which can result in limited oxygen availability. This may also explain the results obtained, i.e. as the oxygen solubility decreases with the increasing salt concentration, PAO activity will also decrease.

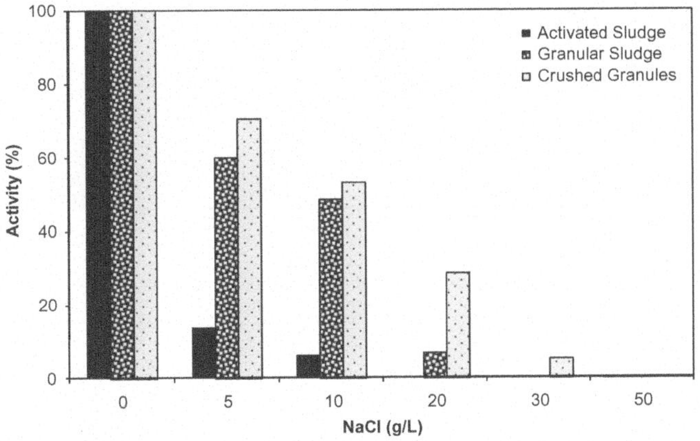

Figure 1.42 Effects of salt on the PAO activity

1.2.5.3.3 pH Batch Tests

The pH 7 was considered as the optimal pH for AOB activity and so all the activities were calculated based on that assumption. It is known that for lower pH values, acetate is toxic to the microorganisms and for higher pH, ammonium becomes toxic. The results of this study (Figure 1.43) indicated that at low pH values, AOB are more active when integrated into a granular structure (67% at pH 5 and 85% at pH 6) than in a looser structure (activated sludge: 9% at pH 5 and 47% at pH 6; crushed granular sludge: 14% at pH 5 and 51% at pH 6). As for the 2-Fluor phenol and salt batch tests, the shielding effect of the granule matrix can be seen. However, for higher pH values (>7), AOB of the crushed granules had higher activities (95%–99.9%) than the granular sludge (0%–81%) and activated sludge (35%–90%) AOB populations. It seems that AOB in the looser structures (as flocs or crushed granules) can better cope with the toxicity of the ammonium than in the granular structure.

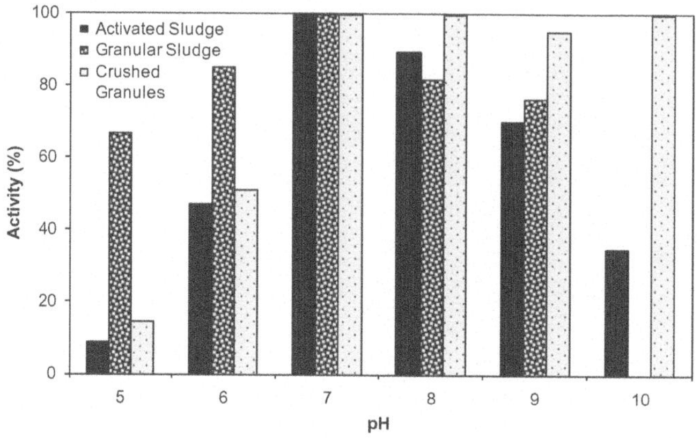

Figure 1.43 Effects of pH on the AOB activity

The granular sludge PAO activity was generally more resistant to pH shocks than both the activated sludge and crushed granular sludge (Figure 1.44). The protection of the granular structure on PAO activity was most noticeable at pH values <7 (granular sludge: 92–98%; crushed granules: 41–53%; activated sludge: 7–29%). Higher pH values (>7) had less of an effect on the PAO activity.

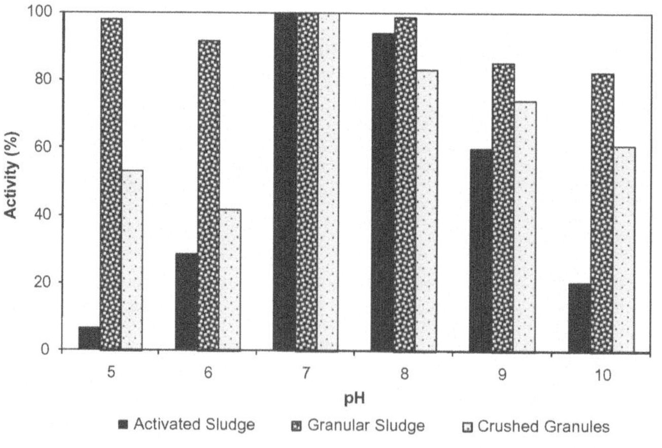

Figure 1.44 Effects of pH on the PAO activity

1.2.5.3.4 Conclusions of Batch Tests

This study showed that 2-Fluor phenol, salt and pH have a higher effect on AOB and PAO for activated sludge than for granular sludge (whole and crushed). Overall, granular sludge was the less affected by the toxic compounds tested, suggesting that the compact structure of the granule provides a protection mechanism for the microorganism from shock loadings. This indicates that aerobic granular sludge may be a promising technology for N and P removal from wastewaters containing similar toxic compounds.

1.3 NEREDA TREATMENT OF FOOD INDUSTRY WASTEWATER

1.3.1 Introduction

Nereda® is a patented innovative and advanced biological wastewater treatment technology that purifies water using the unique features of "aerobic granular biomass". Contrary to conventional processes, the purifying bacteria concentrate naturally in compact granules, with superb settling properties.

As a result of the large variety of biological processes that simultaneously take place in the granular biomass, Nereda® is capable of producing excellent effluent quality. Even when not particularly targeted, extensive biological phosphorus and nitrogen reduction is an intrinsic attribute of this technology, resulting generally in chemical-free operation.

These unique process features translate into compact, energy saving and easy to operate Nereda® installations for both industrial and municipal wastewater treatment. Nereda® presents attractive new solutions for green field installations and retrofitting or extending conventional activated sludge plants. The technology is also highly recommended for performance and capacity upgrades of existing SBR-facilities.

In the following, the results gained treating wastewater produced at a brewery and at a food industry are reported.

1.3.2 NEREDA Treating Brewery Wastewater

The wastewater used as feed for the Nereda™ pilot-plant originated from a brewery. At the time of the study, the brewery was treating the wastewater with anaerobic and aerobic treatment processes (Figure 1.45). The anaerobic process consisted UASB reactors and the aerobic system was a carrousel with simultaneous chemical precipitation of phosphate. For breweries, anaerobic pre-treatment is a very common mode of operation.

The Nereda™ pilot-plant received the anaerobically treated wastewater and was tested as a possible replacement of the carousel system and settling tanks shown in Figure 1.45. The average influent composition to the pilot-plant is shown in Table 1.5.

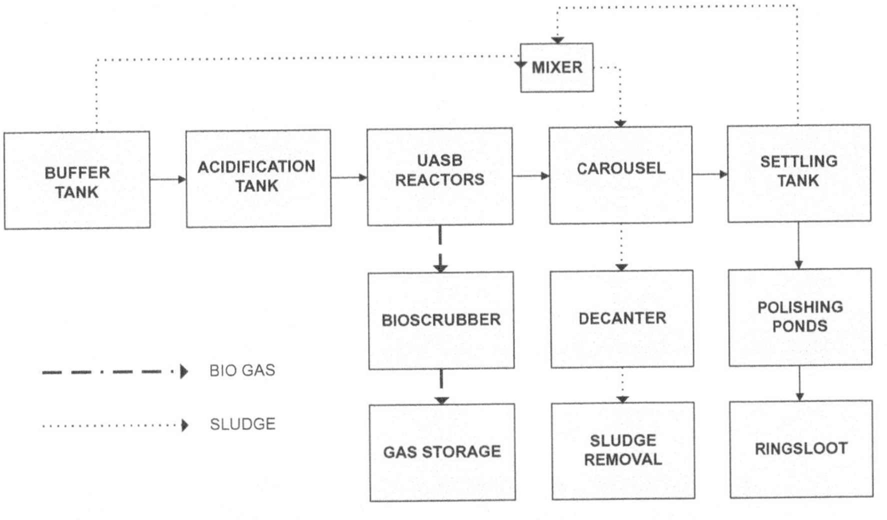

Figure 1.45 Process scheme current wastewater treatment

1.3.2.1 Setup of Nereda at brewery

The reactor of the Nereda was a 75 L, 4 m high, circular tank, operated in fed batch mode and fully automated. Air was supplied through fine bubble aerators on the tank base. The oxygen concentration was maintained at a constant level by a control system. Typical Nereda cycles consisting of simultaneous feed/discharge in upflow mode, anaerobic reaction time, aerobic reaction time, settling and sludge discharge were used in operation.

Pilot trials at the brewery were conducted over an 18 month period. Optimisation of the plant set-up (Figure 1.46) occurred continuously during this operational period.

1.3.2.2 Results

The UASB effluent had a lot of entrained suspended solids which consisted mainly of anaerobic granules and inorganic material. The solids initially had a negative impact on the efficiency and stability of the Nereda process. This issue was resolved by pre-treating the process water in a settle tank, where most of the suspended solids were removed.

Table 1.5 Average parameters of the anaerobically pre-treated wastewater used as influent for the pilot-plant test with a Nereda^{TM}

Parameter	Value	Unit
Flow	4776	m^3/day
COD tot	650	mg/L
COD dissolved (GF/A)	440	mg/L
P-tot	16	mg/L
PO_4-P	10	mg/L
N-tot	45	mg/L
NH_4-N	20	mg/L
TSS (GF/A)	240	mg/L
pH	6.3	
Temperature	25	°C

Figure 1.46 Setup of Nereda aerobic granular sludge system at the brewery

The reactor was operated from the start in a constant mode and was inoculated with the floccular sludge from the on-site aerobic wastewater treatment plant. No special start-up procedure was used and after a relatively short period (4 months), stable operation was achieved with excellent sludge settling characteristics (SVI30 < 30 ml/g, SVI5 < 50 ml/g). Figure 1.47 shows an image of the granules which developed in the system.

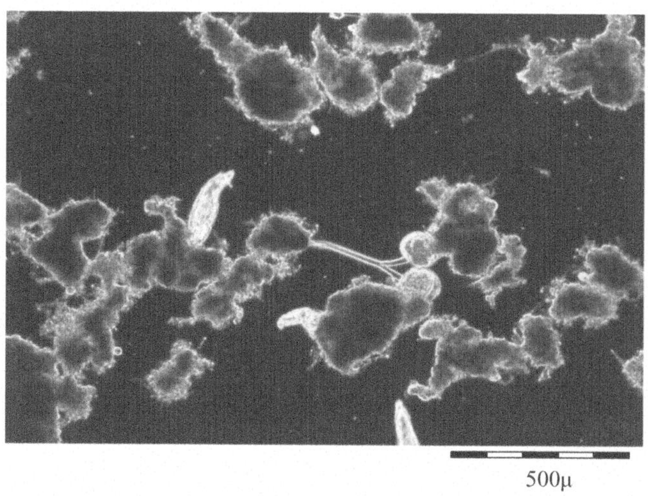

Figure 1.47 Photo of aerobic granular sludge formed on brewery wastewater

A COD load of 0.15 kg COD/kg sludge.day was accomplished, while maintaining good nutrient removal. This COD load was 3 times higher than the load applied to the on-site aerobic treatment system at the time of the study.

Excellent nitrogen removal was achieved during the study (Table 1.6). However, due to the set-up and fluctuating efficiency of the UASB system, stable phosphate removal could not be achieved. The competition between phosphate and glycogen accumulating bacteria may have been negatively influenced by higher temperatures and needs further investigation.

Table 1.6 Average effluent characteristics of the Nereda brewery reactor

Parameter	Nereda^{TM} effluent values	Unit
COD total	46	mg/L
COD dissolved	41	mg/L
P-tot	2	mg/L
PO_4-P	1	mg/L
N-tot	3,7	mg/L
NH_4-N	0,33	mg/L
TSS	12	mg/L
pH	7,7	
Temperature	25	°C

Phosphate present in the sludge granules existed in the forms of organic phosphate, polyphosphate (formed by PAOs) and metal precipitates. In order to achieve a low total phosphorous concentration in the effluent, it was necessary to minimise effluent suspended solids as much as possible.

To enable biological P-removal, adequate amounts of volatile fatty acids (VFA) should be available in the influent. According literature[1], 20 mg total COD is required to remove 1 mg PO_4-P via the Bio-P process. The efficiency of P removal is thus dependent on the COD/P ratio. Figure 1.48 shows that in the current set-up, the observed minimal ratio required to achieve a low P effluent was 40 mg COD_{VFA}/mg P which was considerably higher than previously reported. It is likely this was caused by competition between PAO and GAO populations for the same VFA substrate.

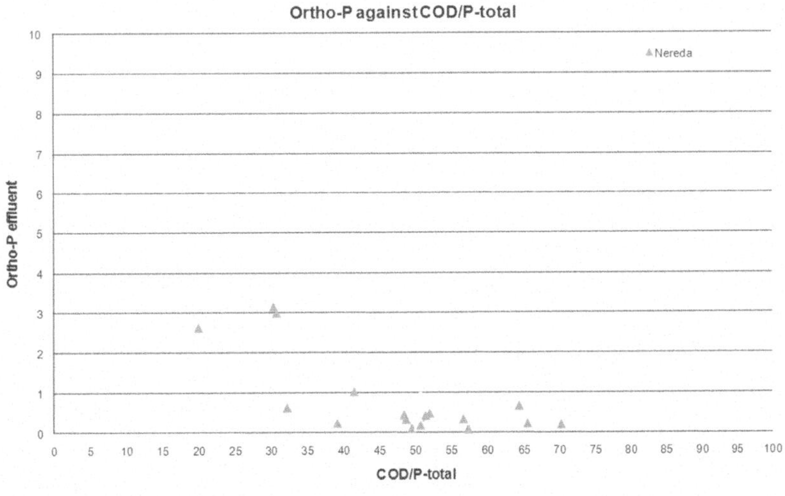

Figure 1.48 Nereda ortho-phosphate effluent quality vs COD/P ratio

1.3.2.3 Conclusions of brewery wastewater

Aerobic granular sludge with the Nereda technology has been shown to be suitable for COD and nutrient removal from anaerobically pre-treated brewery wastewater. The system operated well with a three times higher loading rate compared to the existing on-site activated sludge process. At scale-up, this would result in a significant reduction in the land required for the process since the treatment reactor could be smaller and there is no need for secondary clarifiers.

[1]STOWA. (2001). Handboek biologische fosfaatverwijdering. Hageman fulifilment. ISBN: 9057731274.

For the start-up of the process, no special procedures where required to obtain the granular sludge. The in-growth of nitrifying biomass occurred at a rate typically observed for activated sludge plants.

During the tests, it became clear that a high influx of solids can limit the process performance. A simple tilted plate settler would be sufficient to reduce this solids load and allow the formation of the granular sludge to occur.

The frequently observed issue of competition between phosphate and glycogen accumulating organisms at higher temperatures also occurred in this study. Only at high COD/P ratios could good phosphorous be achieved.

Good nitrogen removal was obtained without the need of sludge recycling pumps as in many conventional activated sludge systems.

Currently, the end user is considering upscaling the pilot-plant to a volume of 1.5 m^3 and incorporating an improved control system (with online N and P measurements) to verify process stability of P-removal and better mimic full-scale operation. A demonstration-scale plant is also being considered.

1.3.3 NEREDA Treating Food Industry Wastewater

The second industrial Nereda evaluated was a full-scale plant which treated the process water of a convenience food factory. This factory produced a diversity of chilled pre-cooked convenience foods. During this production and the cleaning of the machines, a wastewater stream (150 m^3/d) with a high dissolved COD was generated. T process water also contained fat and solid waste. Details of the process water are provided in Table 1.7.

Table 1.7 Reported parameters (average) of the convenience food industry's process water.

Parameter	Value	Unit
Flow	120–180	m^3/day
COD tot	5000	mg/L
COD dissolved (GF/A)	3500	mg/L
N-tot	7	mg/L
TSS (GF/A)	500–1000	mg/L
pH	6	
Temperature	15–35	°C

1.3.3.1 Setup of Nereda Reactor

Pre-treatment of the wastewater was undertaken, prior to treatment within the reactor. This involved fine screening (for solids removal) and dissolved aeration flotation or DAF (for removal of fats and fine solids).

Following pre-treatment the wastewater still contained high COD concentrations (on average 4000 mg/L). A large fraction of the COD was easily degradable sugars (starch) and some lipids. The wastewater was deficient in nitrogen and phosphate, which were consequently added in the form of a ready-made nutrient solution, before the biological treatment.

The biological treatment consisted of an influent buffer (150 m^3) and one reactor (250 m^3) (Figure 1.49). During production the buffer tank received a continuous flow from the DAF unit. Prior to entering the buffer tank, the nutrients were dosed to the pre-treated stream. The Nereda reactor was fed batch wise from the buffer tank with varying volumes dependent upon the volume discharged from the industrial process

A typical cycle in the reactor was six hours consisting of: simultaneous feed and discharge, mixed anaerobic reaction phase, aerobic reaction phase, sludge discharge and a settling phase. The simultaneous feed/discharge involved feeding the reactor while allow treated wastewater (from the previous cycle) to overflow at the reactor surface into an effluent tank prior to municipal sewer discharge. The sludge discharged following aeration passed into a sludge holding tank with the volume discharged dependent upon the sludge growth.

1.3.3.2 Results

During start up and operation, optimisation of the plant was required to achieve good, stable operation. These included optimisation of the DAF pre-treatment process (to prevent periodic breakthroughs of oil and grease which had negative impacts on the sludge condition) and nutrient dosage and control. Once these processes were optimised a stable sludge could be produced.

Figure 1.49 Nereda wastewater treatment at convenience food industry

Effluent of the Nereda system is shown in Table 1.8. Excellent total COD removal was achieved (97% removal). This was the combined result of excellent soluble COD removal while maintaining minimal suspended solids concentrations.

Table 1.8 Average effluent quality of the Nereda convenience food industry plant.

Parameter	Nereda^{TM} effluent values	Unit
COD total	150	mg/l
COD dissolved	75	mg/l
P-tot		mg/l
PO_4-P	1	mg/l
N-tot	4	mg/l
NH_4-N		mg/l
TSS	40	mg/l
pH	7.5*	
Temperature	15–35	°C

* Reactor pH 6.5–7.5

The temperature varied over the year of the study, but no effects of temperature variation could be observed on the formation or stability of the aerobic granules.

1.3.3.3 Conclusions of food industry wastewater

Aerobic granular sludge is a new technology that can significantly reduce space and energy requirements for industrial wastewater treatment, as well as operating costs. This study showed that an excessive loading of the Nereda reactor with suspended solids and grease/oil was detrimental to the process and should be avoided. For this wastewater, a simple pre-treatment step involving dissolved air filtration and a tilted plate settler was adequate to avoid these issues.

Granular sludge formation in the Nereda system proved to be rapid without the need of special start-up procedures. The polymeric organic substrates (such as starch) which have been shown to lead to sub-optimal granulation in the laboratory did not present problems with this food industry application.

The food industry company, where the Nereda system was located, have chosen to move the system to a new site for further testing, indicating their trust in the system. The direct reuse of effluent by the food processing factory further indicates the satisfactory effluent characteristics of the plant.

1.4 UNIFED TREATMENT OF ABATTOIR WASTEWATER

1.4.1 Introduction

The meat processing industry requires large quantities of water, much of which is discharged as wastewater containing high levels of COD and nutrients such as nitrogen (N) and phosphorus (P). Over the past two decades, biological COD and N removal from abattoir wastewater has received much greater attention than has the biological P removal. Reliable biological COD and nitrogen removal systems have been successfully designed for abattoir wastewater treatment. However, P removal continues to be achieved primarily through chemical precipitation, which incurs high costs for chemicals and for the handling and disposal of the chemical sludge produced. Biological P removal is potentially a cheaper and more environmentally sustainable option.

One of the main obstacles for achieving Bio-P removal from abattoir wastewater is the high concentration of nitrogen in the wastewater (TN around 200 $g \cdot m^{-3}$ or higher). When complete nitrification is achieved, which is a pre-requisite for a high-level removal of N, the equivalent level of nitrate formed is highly detrimental for Bio-P removal. Establishing anaerobic conditions in the treatment systems, which is required for Bio-P removal, is a significant challenge as nitrate needs to be removed completely before anaerobic conditions can be created. Denitrification will also compete with the Bio-P process for the limited amount of volatile fatty acids (VFA) present in the abattoir wastewater.

The Unifed system is a patented feed delivery process for Sequencing Batch Reactors (SBRs). Feed wastewater is evenly distributed through the sludge blanket on the horizontal plane near the base of the tank. No mixing or aeration occurs at this time. This method of feed delivery locally maximises the organic fraction of the wastewater (particularly volatile fatty acids which are typically limited in concentrations) within the sludge blanket, thereby creating a localised anaerobic zone necessary for Bio-P removal. This ensures maximal exposure of the limited organics to the carbon storing bacteria, such as poly-phosphate accumulating organisms (PAOs) and glycogen accumulating organisms (GAOs).

Laboratory and pilot scale trials have shown the Unifed feed delivery system is a very appropriate technology for treating abattoir wastewater with >90% TN and TP removal achieved via simultaneous nitrification, denitrification and phosphorus removal with floccular sludge. Aerobic granules represent the next evolutionary step in the treatment of this complex wastewater, allowing for intensification of the nutrient removal processes.

1.4.2 Laboratory-Scale Experiences

1.4.2.1 Development of aerobic granules from floccular sludge with abattoir wastewater

Two different start ups were initially performed by seeding a sequencing batch reactor (SBR) with floccular sludge. In order to obtain aerobic granules and select for fast settling microorganisms, a strategy of progressively reducing the settling time was applied to the reactor. This approach is commonly used and has been demonstrated to enhance granulation. However, during the application of this strategy, a reduction in biomass within the reactor occurred on both occasions (Figure 1.50). An increase in particle size was only observed when the biomass concentration decreased from 3g MLSS/L to levels lower than 1 g MLSS/L. Although the 90th and the 50th percentiles were increasing, suggesting an increase of the size of the granules present, the biomass concentration never recovered.

Biodegradable COD removal was achieved in both runs with an efficiency of 99% even when low levels of biomass were present. During the first run (Figure 1.51a), the volumetric exchange ratio (VER) was set and maintained at a constant 33%. However, N removal deteriorated over time due to a decrease of biomass in the SBR, causing an accumulation of NH_4^+ in the reactor thus inhibiting the bacteria present. The system could not be recovered and was stopped after 80 days of operation. During run 2 (Figure 1.51b) the initial VER applied to the SBR was 17% in order to avoid NH_4^+ accumulation in the reactor. During the first 25 days of operation, 90% N removal was achieved. The wastewater loading was slightly increased, increasing the VER to 25%. However, the system could not cope with this increase, partially because the biomass concentration was also decreasing, and N removal decreased. Although VER was reduced again, the performance did not substantially improve and biomass concentration reached very low levels. This run was stopped after 70 days of operation.

1.4.2.2 Challenges associated with the start-up of aerobic granular reactors for the treatment of abattoir wastewater

Two major drawbacks were identified when an aerobic granular reactor was started with floccular sludge for the treatment of abattoir wastewater. The first one was the substantial reduction in biomass before granules started to develop. Granules appeared when biomass concentration was lower than 1g MLSS/L. The second one, which is a consequence of the reduction of the biomass, is the deterioration of the nutrient removal capability of the reactor. When working with nutrient rich wastewaters, there is the additional risk of dealing with the accumulation of nutrients in the reactor. An increase of NH_4^+ to certain concentrations can have an inhibitory effect on the group of bacteria involved in the removal of this nutrient, the nitrifiers. Two of the major causes of inhibition of these bacteria are the elevated ammonium (or Free Ammonia) and nitrite (or Free Nitrous Acid) concentrations. Therefore, biological nitrogen removal should be maintained in the start up process of an aerobic granular SBR in order to avoid inhibition of the biomass.

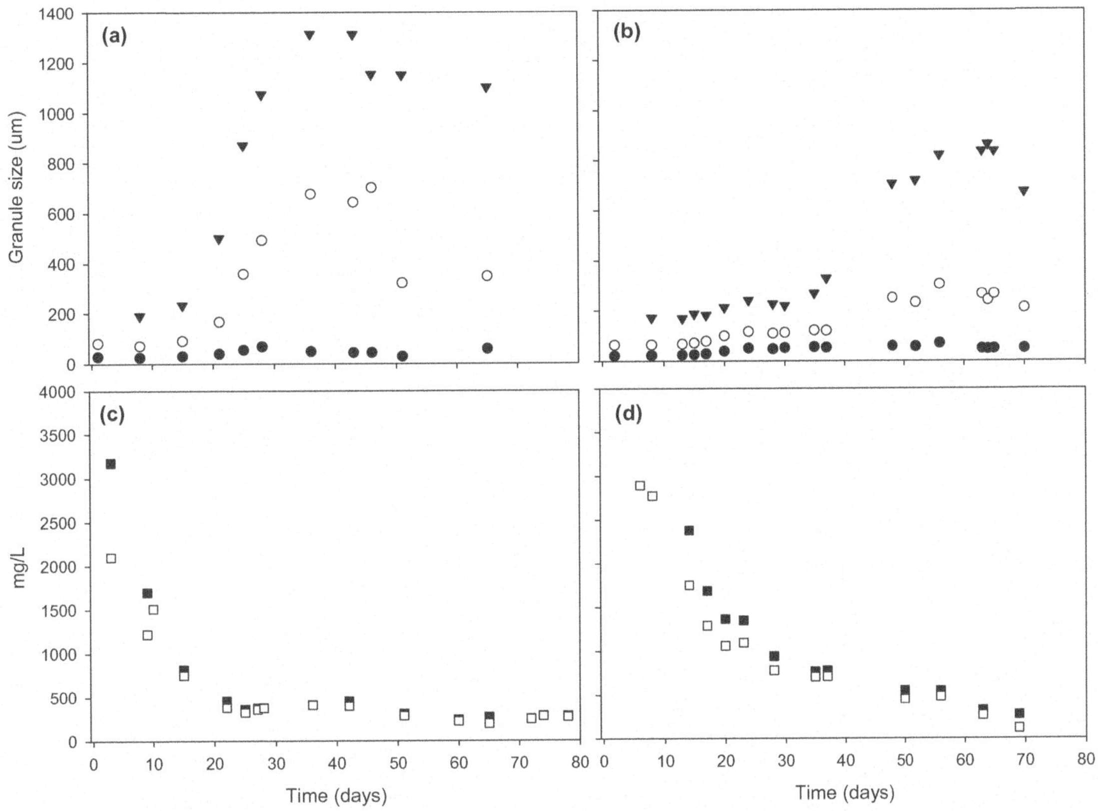

Figure 1.50 Size distribution profiles (top) and MLSS & MLVSS (bottom) of the SBRs seeded with 100% floccular sludge: a & c - 1st round; b & d - 2nd round. Percentiles: ▼ d(0.9), ○ d(0.5), ● d(0.1); □ MLVSS, ■ MLSS

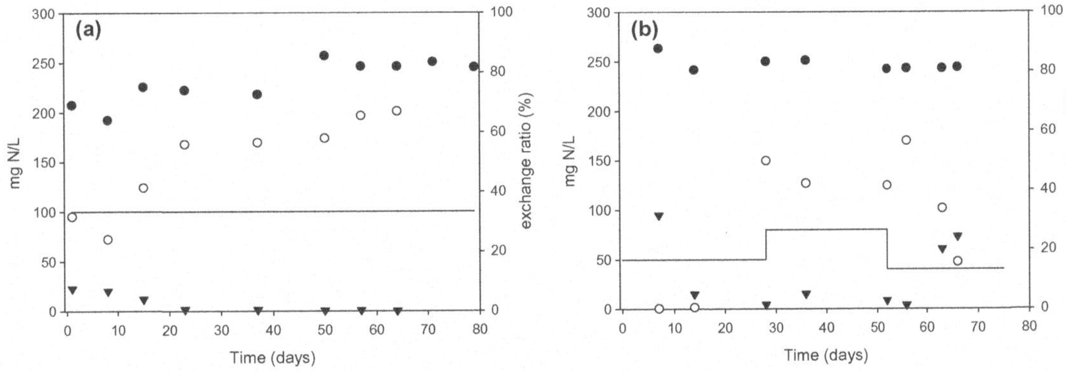

Figure 1.51 Nitrogen removal performance for SBRs seeded with 100% floccular sludge: (a) 1st round; (b) 2nd round. ● N-NH_4^+ influent, ○ N-NH_4^+, ▼ N-NOx, — exchange ratio

1.4.2.3 A novel seeding strategy: granules and floccular sludge mixture

In order to overcome the above mentioned problems a different start up strategy was applied, using a mixture of granules and floccular sludge. Different combinations were used in several SBRs and the results from five are presented.

Figure 1.52 shows the size distribution profiles for each of the 5 reactors. The 90th percentile was always substantially higher than the 50th and 10th percentiles due to the presence of these granules. For comparison purposes, we considered that complete granulation was achieved once the 10th percentile was higher than 200 µm, which is the minimum size for a particle to be considered a granule.

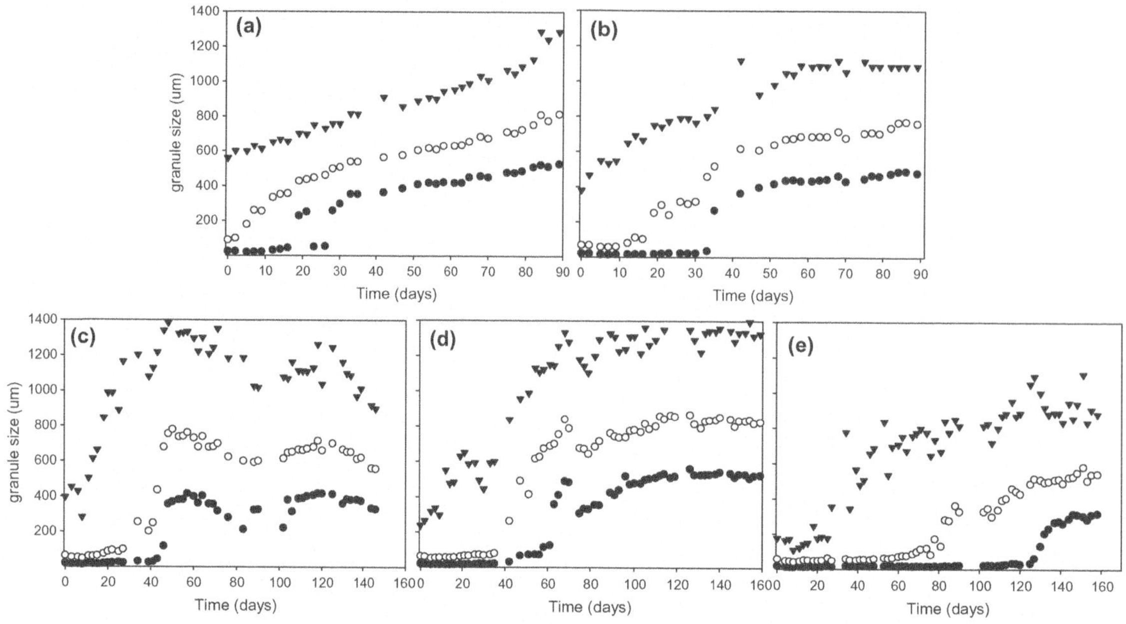

Figure 1.52 Size distribution profiles of the SBRs working with different percentages of granules: (a) 50%; (b) 25%; (c) 15%; (d) 10%; (e) 5%. Percentiles: ▾ d(0.9), ○ d(0.5), • d(0.1)

In all cases, the 90th percentile increased from the commencement of operation. After a period of time (depending on the reactor) the 50th percentile also started to increase. Eventually the 10th percentile increased to values greater than 200 μm, indicative of all biomass in the reactor to be in the form of granules.

The shortest time required for granulation was achieved with the SBR seeded with 50% granules and 50% floccular sludge (Figure 1.52a) while the longest time was found with the SBR seeded with 5% granules and 95% floccular sludge (Figure 1.52e). Biomass concentration in these reactors did not decrease but increased during granulation.

Figure 1.53 shows the correlation between the percentage of granules present in the seeding sludge and the time of granulation.

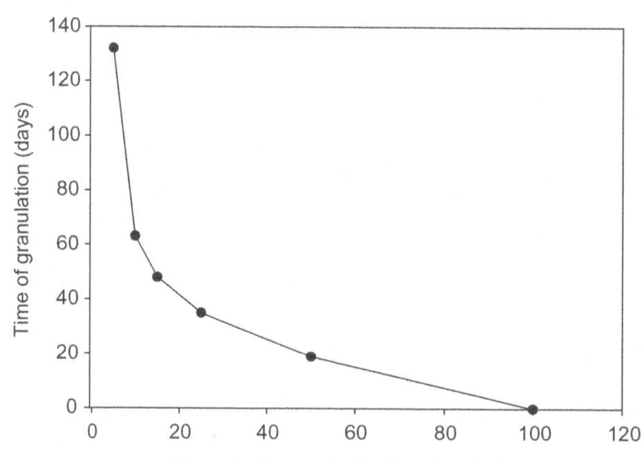

Figure 1.53 Effect of the percentage of granules in the seeding sludge on the time requirement for a reactor to become fully granulated when treating abattoir wastewater

1.4.2.4 Nutrient removal

In order to optimize the amount of carbon needed for the nutrient removal process, simultaneous nitrification, denitrification and phosphorus removal (SNDPR) was promoted in the reactors. Additionally, the nitrite pathway (ammonia oxidation to nitrite) was encouraged as it provides oxygen savings during the nitrification step and carbon savings during denitrification compared to conventional nitrogen removal treatment. Figure 1.54 shows the nitrogen present in the wastewater and in the effluent of the SBRs. High nitrogen removal efficiency was achieved in all the reactors during most of the operational period. A slight decrease in nitrogen removal was observed in some reactors during the initial days after increasing the VER but nitrogen removal efficiency was rapidly restored.

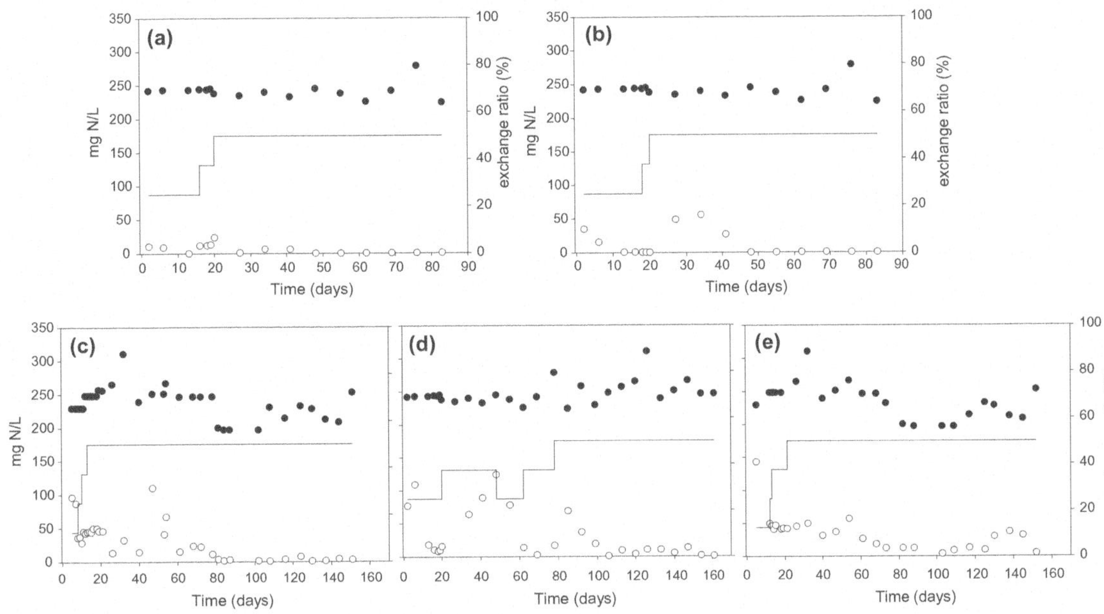

Figure 1.54 Nitrogen removal performance on SBRs working with different percentages of granules: (a) 50%; (b) 25%; (c) 15%; (d) 10%; (e) 5%. • N-NH_4^+ influent, ○ N-NH_4^++N-NOx, — exchange ratio

1.4.2.5 Conclusions

The use of a mixture of aerobic granules and floccular sludge was shown to reduce the start-up of aerobic granular sludge reactors for the treatment of abattoir wastewater. There is a positive correlation between the amount of granules used and the time required to achieve complete granulation. Additionally, 99% COD removal and 90% nitrogen removal were achieved during the operational period in all the reactors, even during the transition period.

1.4.3 Pilot-Scale Experiences – Plant Design

A previous pilot-scale investigation undertaken in Queensland, Australia successfully demonstrated the use of the Unifed feed delivery system in an SBR for achieving simultaneous N & P removal from abattoir wastewater with floccular sludge. The three main measures taken to enhance Bio-P removal in the pilot-scale SBRs were:

- The wastewater feed was added to the reactor over three feeding periods in each SBR cycle. The multi-step feeding strategy ensures the nitrite/nitrate concentrations remain low throughout the SBR cycle, allowing easier creation of anaerobic conditions in the reactor during feeding;
- The UniFed® process (in which wastewater is fed to the bottom of the reactor) was applied to ensure localised anaerobic conditions develop in the sludge blanket to enhance anaerobic phosphate release and VFA uptake;
- An anaerobic prefermenter was included to ferment the raw wastewater at an elevated temperature (due to the incoming temperature of the raw wastewater stream) to produce additional VFA. The VFA produced was added to the wastewater fed to the reactor, obtained from the effluent of an anaerobic pond. The anaerobic pond type of pre-treatment of the raw abattoir wastewater would typically be used upstream of an SBR process at full scale to reduce the high level of solids and other COD (including fat and oils) in the wastewater as it would otherwise heavily load the aerobic process and create high aeration costs.

The current project was aimed at increasing the SBR nutrient removal capacity through the use of aerobic granular sludge. Aerobic granular sludge plants have lower space requirements than conventional activated sludge systems and a higher nutrient conversion capacity due to a higher operating biomass concentration. They are also believed to generate substantially less waste sludge than conventional floccular systems.

The project aimed to characterise process parameters required for the conversion of activated sludge to aerobic granules suitable for bio-P removal from meat processing wastewater while simultaneously achieving high levels of removal for other components, notably chemical and/or biological oxygen demand (COD, BOD) and nitrogen.

More specifically, the project had the following objectives:

1. Determine key drivers for the conversion of floccular sludge to aerobic granular sludge;
2. Demonstrate aerobic granular SBR technology with a pilot plant operated under field conditions to achieve over 90% ammonia and ortho-phosphate removal from abattoir wastewater;
3. Achieve the above performance under typical load variations experienced by abattoir wastewater treatment plants through the use of on-line process control;
4. Investigate the interactions of nitrogen removal and biological phosphorus removal to ensure good, stable nutrient removal performance;
5. Investigate aerobic granule dewatering and sludge production properties.

Funding for this project was obtained from the Australian Government International Science Linkages program, the Industry Environment R&D program of Meat and Livestock Australia and the Australian Meat Processing Corporation.

1.4.3.1 Past Experiences at the Site with Floccular Sludge

The previous project used a dual SBR rig located at an abattoir for the study of biological C, N and P removal from the site's wastewater. The rig consisted of two SBRs, each with a 6 m^3 operating volume, attached to a hut which housed a PLC and other equipment. The SBRs were operated in parallel.

The operating parameters of the SBRs are indicated in Table 1.9. Each cycle consisted of multiple feed and react periods prior to settling and decant. The feed consisted of a mix of prefermenter liquor (1.5 d HRT; typically 15–20% by volume) and anaerobic pond effluent (full-scale plant; typically 80–85% by volume).

Table 1.9 Operating parameters for previous study using floccular SBR

HRT	42 hr
Cycle Length	6 hr
Feed vol./cycle	0.857 m^3
DO setpoints	1.5 $g \cdot m^{-3}$
Combined Feed Characteristics	
Av. NH_4-N	191 $g \cdot m^{-3}$
Av. NH_4-N Load	0.109 $kg \cdot m^{-3} \cdot d^{-1}$
Av. TN	234 $g \cdot m^{-3}$
Av. TN Load	0.134 $kg \cdot m^{-3} \cdot d^{-1}$
Av. PO_4-P	32 $g \cdot m^{-3}$
Av. PO_4-P Load	0.018 $kg \cdot m^{-3} \cdot d^{-1}$
Av. TP	34 $g \cdot m^{-3}$
Av. TP Load	0.019 $kg \cdot m^{-3} \cdot d^{-1}$
Av. TCOD	1773 $g \cdot m^{-3}$
Av. TCOD Load	1.013 $kg \cdot m^{-3} \cdot d^{-1}$
Av. Total VFA	184 $g \cdot m^{-3}$ as VFA 263 $g \cdot m^{-3}$ as COD
Av. TSS/VSS	771 $g \cdot m^{-3}$/701 $g \cdot m^{-3}$

The project was successful in achieving the aims of >90% TN and TP removal under the described conditions via simultaneous nitrification, denitrification and phosphorus removal (SNDPR), with denitrifying phosphorus accumulating organisms (PAOs) removing a portion of the NO_X-N. Greater than 90% TCOD removal was not achieved, largely due to some breakthrough of fat, oil and grease (FOG) from the DAF (via the prefermenter).

In the current project, the same pilot-plant was used with some modifications to allow for the development of biological granules and an increased load. The remainder of this section is a description of the pilot-plant, including modifications.

1.4.3.2 Pilot-Plant Overview

Figure 1.55 shows a flow diagram of the pilot-plant. Continuous gravity flow from the anaerobic pond flowed into an overflow tank (the pilot-plant was located approximately 300 m from the anaerobic pond). A portion of this wastewater was used for the pilot-plant feed with the remainder overflowing back to the full-scale plant.

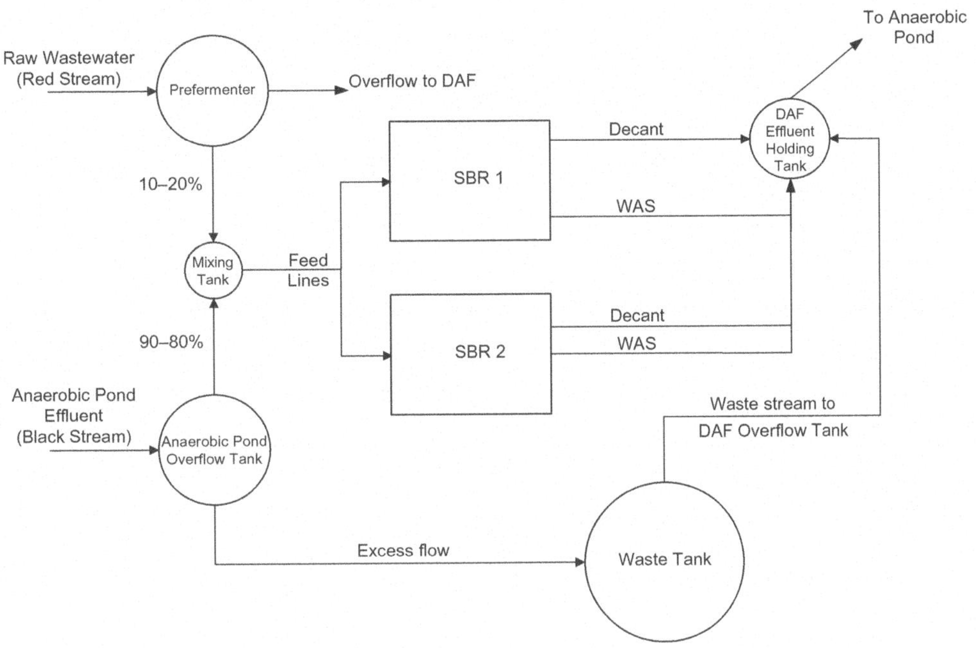

Figure 1.55 Flow diagram of pilot-plant

Due to the low VFA concentration in the anaerobic pond effluent, a fraction of the raw wastewater from the site (exiting the DAF) was pumped intermittently into a prefermenter tank (diameter 3 m, wall height 2.4 m with an overflow port for discharge to maintain a constant operating volume of 12.5 m^3; typical weekday HRT of 1-1.5 d). A portion of the prefermenter wastewater was used for the pilot-plant feed with the remainder overflowing back to the DAF.

Liquor from the anaerobic pond and prefermenter were pumped into a mixing tank periodically and mixed. This feed mixture was pumped into the two asynchronously operated SBRs during feed periods in the cycles.

The SBRs were located in an existing rig used for the previous abattoir pilot-scale study. The rig comprised of two stainless steel tanks and a control room. Each tank had dimensions of 3.9 1m (L) × 0.93 m (W) × 2.19 m (H), constructed with 3 mm thick, Grade 316 stainless steel. Both tanks were insulated and fitted with fourteen 0.9 m Aquablade diffusers (Aquatec Maxcon) for aeration. The operating volumes of each SBR were 6 m^3.

The SBRs cycled through 3 phases of feeding/non-aerated period/aerated periods, followed by settle and decant periods. Sludge was only wasted at the beginning of the project while BNR performance was increased to a base loading rate. Subsequently, all sludge wastage occurred via the decant. Effluent from the SBRs was pumped during the decant periods to the DAF effluent holding tank of the main treatment plant. Initially both SBRs had a floating decant pipe. This was later changed to a fixed, submerged decant.

Both SBRs used the UniFed feed delivery system. This system involved evenly distributing the feed through the sludge blanket on the horizontal plane. This distribution was performed near the base of the tank (above the diffusers). No additional mixing was employed during the feeding period.

1.4.4 Pilot-Scale Experiences – Results

1.4.4.1 Commissioning

Each SBR was seeded with 5 m^3 of sludge collected from a wastewater treatment plant (with biological carbon, nitrogen and phosphorus removal) on 16/4/08. Due to some mechanical issues, continuous operation was not achieved until 18/4/08. The initial startup HRT was 68.3 hr but this was rapidly decreased to 42 hr within 8 days. Sludge wastage commenced 6 days after startup and continued for 16 days while nutrient removal stabilized.

A 42 hr HRT was chosen as an initial base load starting point as this was the loading used in the previous project when excellent nutrient removal was achieved using floccular sludge. This was equivalent to a hydraulic loading rate of $0.57\ m^3 \cdot m^{-3} \cdot d^{-1}$.

Within 3 weeks of startup, >95% NH_4 and PO_4 removal had been achieved in both SBRs. On 8th May 2008 changes were made to commence granulation of the sludge.

1.4.4.2 Feed Characteristics

Both SBRs received a mixed feed consisting of fractions collected from the full-scale anaerobic pond and the pilot-scale prefermenter. The proportion (by volume) of feed streams fed to the pilot plant is shown in Figure 1.56. Typically, the combined feed contained 10–20% prefermenter volume and 90–80% anaerobic pond effluent. Throughout most of the study, both SBRs received the same proportion of feed sources. Only on few occasions were these proportions different between the SBRs and in these cases the differences were very minor.

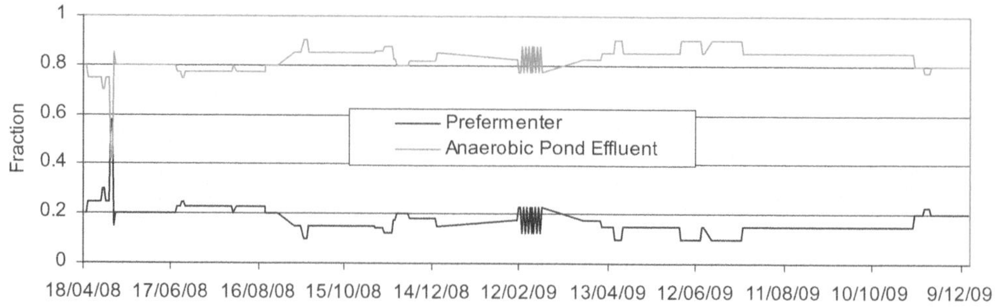

Figure 1.56 Feed proportions by volume for the SBRs

The anaerobic pond served as the primary treatment step in the mainstream abattoir wastewater treatment process, receiving both the "green stream" and the "red stream" (the latter after pre-treatment in the DAF). Despite relatively high loadings (partly due to poor operation of the DAF pre-treatment step), the pond appeared to perform very well during this study. The influent to the anaerobic pond was not measured as part of this project but based on previous studies was estimated to contain approx. 2500 to 4000 $g \cdot m^{-3}$ COD, 120 to 200 $g \cdot m^{-3}$ TKN and 20–40 $g \cdot m^{-3}$ Total P.

The data in Figure 1.57A show that the anaerobic pond effluent (overflow) contains a total COD of ~600 to 1400 $g \cdot m^{-3}$, of which the soluble component is quite stable at around 160 $g \cdot m^{-3}$. Hence the anaerobic pond probably achieved >65% COD removal, which is considered to be good for this type of system.

The total VFA concentration (Figure 1.57b), expressed as COD by appropriate conversion for each of the measured VFA species (C2 to C6, i.e. acetic to hexanoic acid), averaged 17 $g \cdot m^{-3}$ COD (range 1–69 $g \cdot m^{-3}$). On average, the acetic and propionic acids combined represented almost 90% of the C1-C6 acids produced and the acetic acid alone represented 71%.

The average pond effluent TSS concentration was 620 $g \cdot m^{-3}$ and ranged from 150–1620 $g \cdot m^{-3}$ (Figure 1.57c). The average volatile content was 80% and the average median particle size, 62 μm diameter. The solids always appeared as black particulates.

The anaerobic pond effluent ammonia and ortho-phosphate concentrations were relatively stable throughout most of the project (Figure 1.57d). This was largely due to the estimated 5 day residence time in the pond. There is an observable seasonal variation which is likely due to seasonal fluctuations of the pond temperature. The anaerobic pond effluent temperature measurements shown in Figure 1.57e are for samples collected at the pilot-plant after travelling approximately 300 m from the pond through a black pipe. Periodic measurements of the pond effluent prior to passing through the pipe show passage through the pipe typically increases the water temperature by 2–3°C, depending upon the prevailing weather conditions.

The prefermenter feed was collected from the end of the DAF which received the "red stream". Recent alterations to the full-scale pre-treatment works resulted in the removal of the "save-all" which was present during the previous project. Subsequently, all "red stream" wastewater passed directly into the DAF, eliminating the pre-removal of a portion of the fats and solids upstream of the DAF. Exacerbating this was the high temperature of the wastewater entering the DAF (typically 45–50°C), resulting in poor operational performance in terms of solids and fat removal (based on previous study measurements).

The prefermenter received an intermittent feed based on the availability of the feed source. The abattoir operates typically between 6 am and midnight on weekdays, shutting down overnight and over weekends. This resulted in the intermittent feeding of the prefermenter between 7 am and midnight on weekdays and no feeding from midnight Friday through to 7 am Monday morning. The prefermenter was sized such that sufficient volume was available for feed preparation over weekends, drawing down some of the tank volume. Occasionally the abattoir would also shutdown on weekdays and when this occurred, feeding to the prefermenter was also stopped.

Figure 1.58 shows the 7 day moving average HRT of the prefermenter. In 2008, the prefermenter was typically operated with a weekday HRT of 1.5 days, resulting in a 7-day moving average of 2.6 days. In 2009, attempts to optimize the prefermenter were undertaken, which in part involved manipulation of the HRT, hence the variability observed for this year.

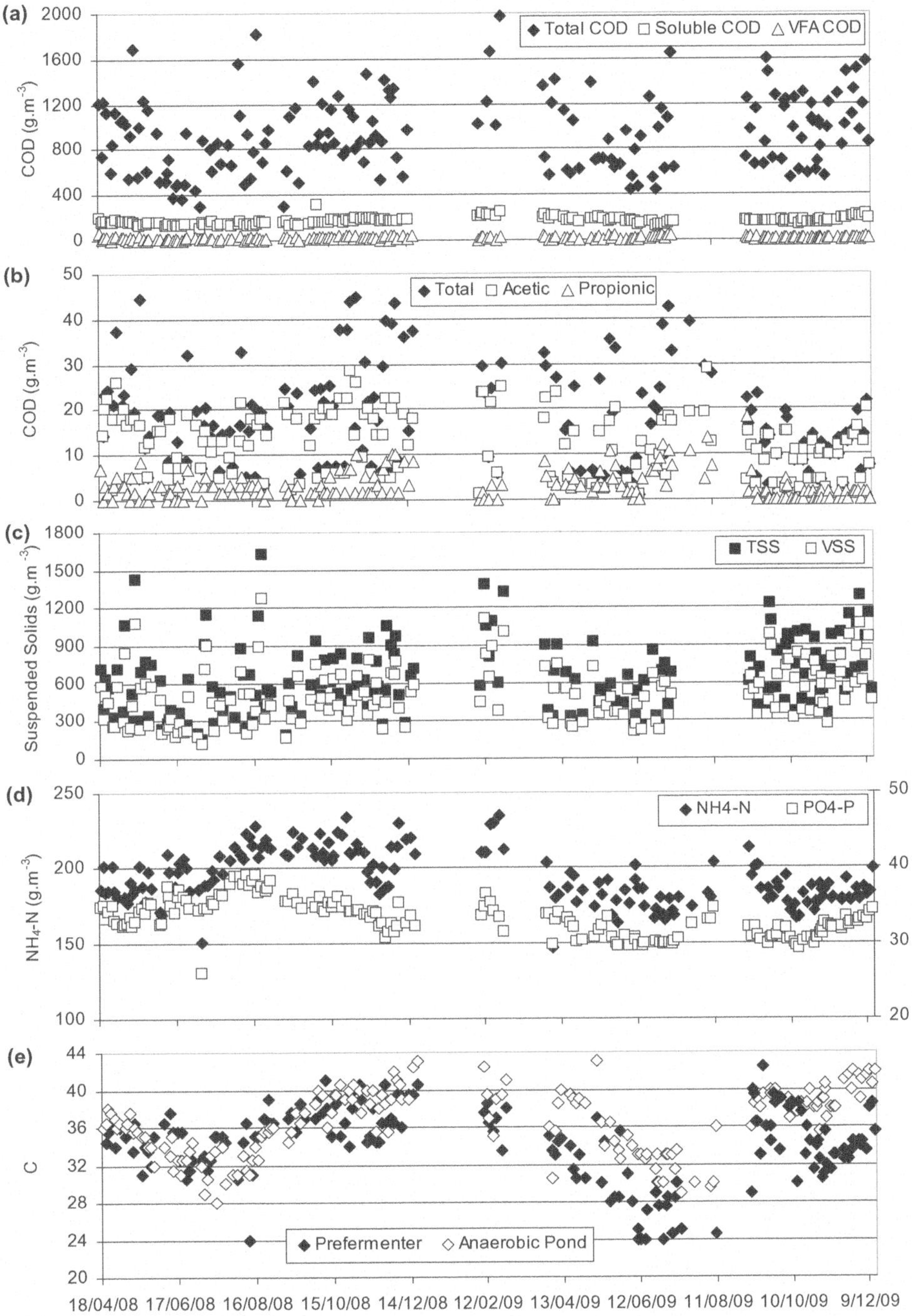

Figure 1.57 Anaerobic pond effluent trends: (a) COD fractions, (b) volatile fatty acids, (c) suspended solids, (d) soluble nutrients, (e) temperature (including prefermenter temperature trend)

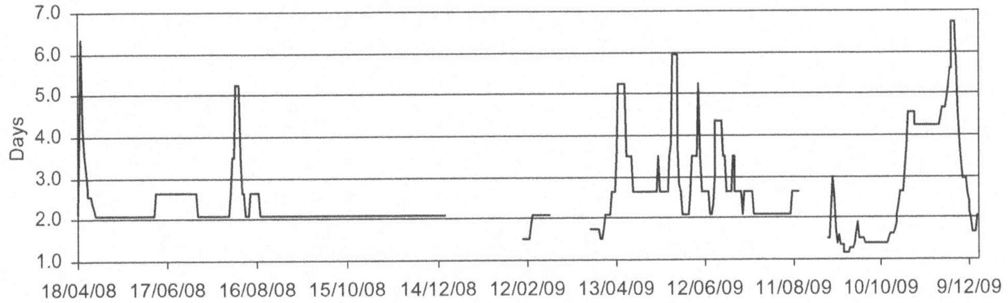

Figure 1.58 Prefermenter 7-day moving average HRT

The prefermenter effluent Total COD (TCOD) was highly variable throughout the project (Figure 1.59a). This was largely due to variability in the effluent solids concentration (Figure 1.59C) which was the result of no feeding over weekends (affecting early week grab samples) and changes to mixing regime. It is highly likely the fat content of the wastewater also affected settleability.

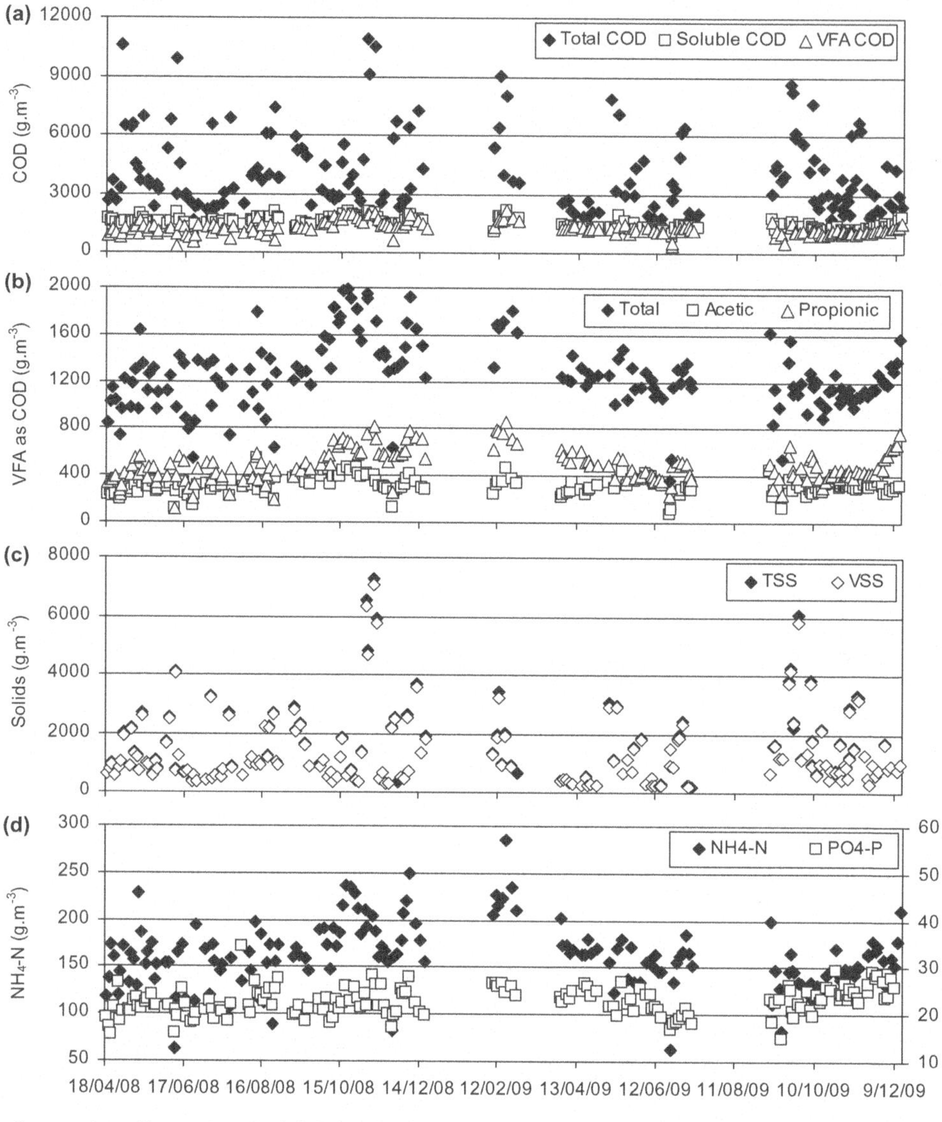

Figure 1.59 Prefermenter effluent trends: (a) COD fractions, (b) volatile fatty acids, (c) suspended solids, (d) soluble nutrients

The COD fraction attributed to VFA (COD_{VFA}) and the Soluble COD (SCOD) were very stable. Table 1.10 shows statistical analysis of the COD fractions. The COD_{VFA} on average was 83% of the SCOD fraction, indicating most of the solubilised COD was in VFA form. However, the COD_{VFA} represented only 37% of the TCOD, indicating the majority of the COD was in particulate form.

Table 1.10 Statistics of various COD parameters

Parameter		Total COD	Soluble COD	COD_{VFA}	Sol. COD/Tot. COD	COD_{VFA}/TCOD	COD_{VFA}/Sol. COD
Average	$g \cdot m^{-3}$	4209	1499	1253	44.0%	37.3%	83.0%
Std. Dev.	$g \cdot m^{-3}$	2581	248	309	17.6%	17.6%	14.3%
Min.	$g \cdot m^{-3}$	1368	722	347	7%	3.5%	17.4%
Max.	$g \cdot m^{-3}$	21885	2194	2213	88%	91.8%	100%
Count		171	169	174	169	169	167

The prefermenter preformed well in respect of VFA generation. Propionic acid was the dominant species, representing on average 39% of the total VFAs present (Figure 1.59b). Acetic acid represented 25% of total VFAs with the remainder dominated by valeric (13.6%), iso-butyric (8.4%) and butyric (7.9%).

There was a correlation between prefermenter temperatures (Figure 1.57E) and VFA concentration. Elevated prefermenter temperatures resulted in elevated VFA concentrations. Note – in September, 2009 HRT investigations resulted in elevated temperatures without elevated VFA concentrations and this can be explained by a reduced SRT of the sludge in the prefermenter.

Figure 1.59d shows the ammonia and ortho-phosphate profiles for the prefermenter effluent. There appears to be a seasonal variation in the ammonia concentration which follows the trend in temperature shift (Figure 1.57e). This is likely the result of variations in hydrolysis due to temperature shifts, with higher temperatures allowing for further hydrolysis.

The volumetric and mass loadings to the SBRs for the combined feed concentrations (calculated from the measurement on the two individual streams and the programmed feed mix ratio) are shown in Figures 1.60 and 1.61.

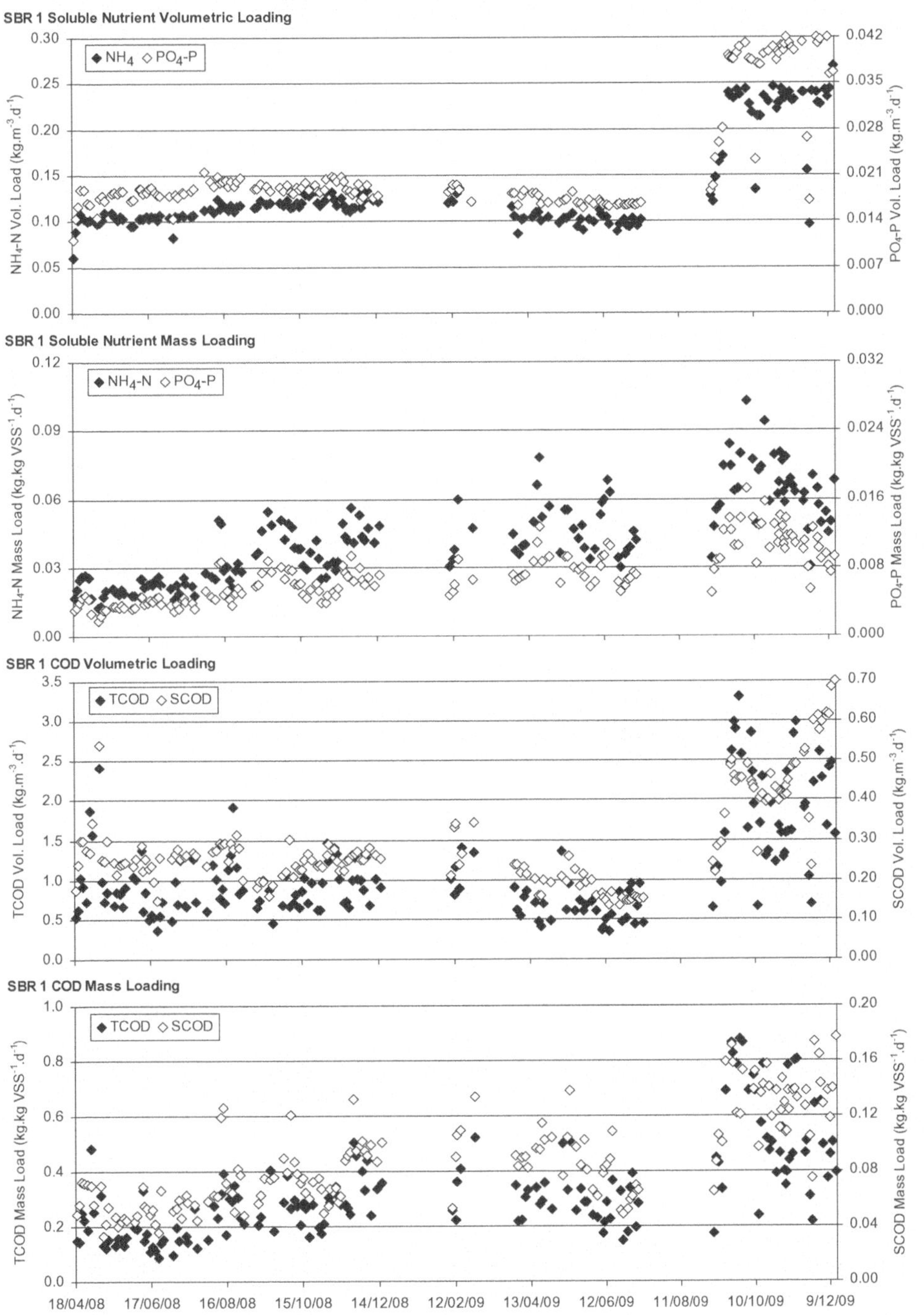

Figure 1.60 SBR 1 nutrient and COD loadings (volumetric and mass)

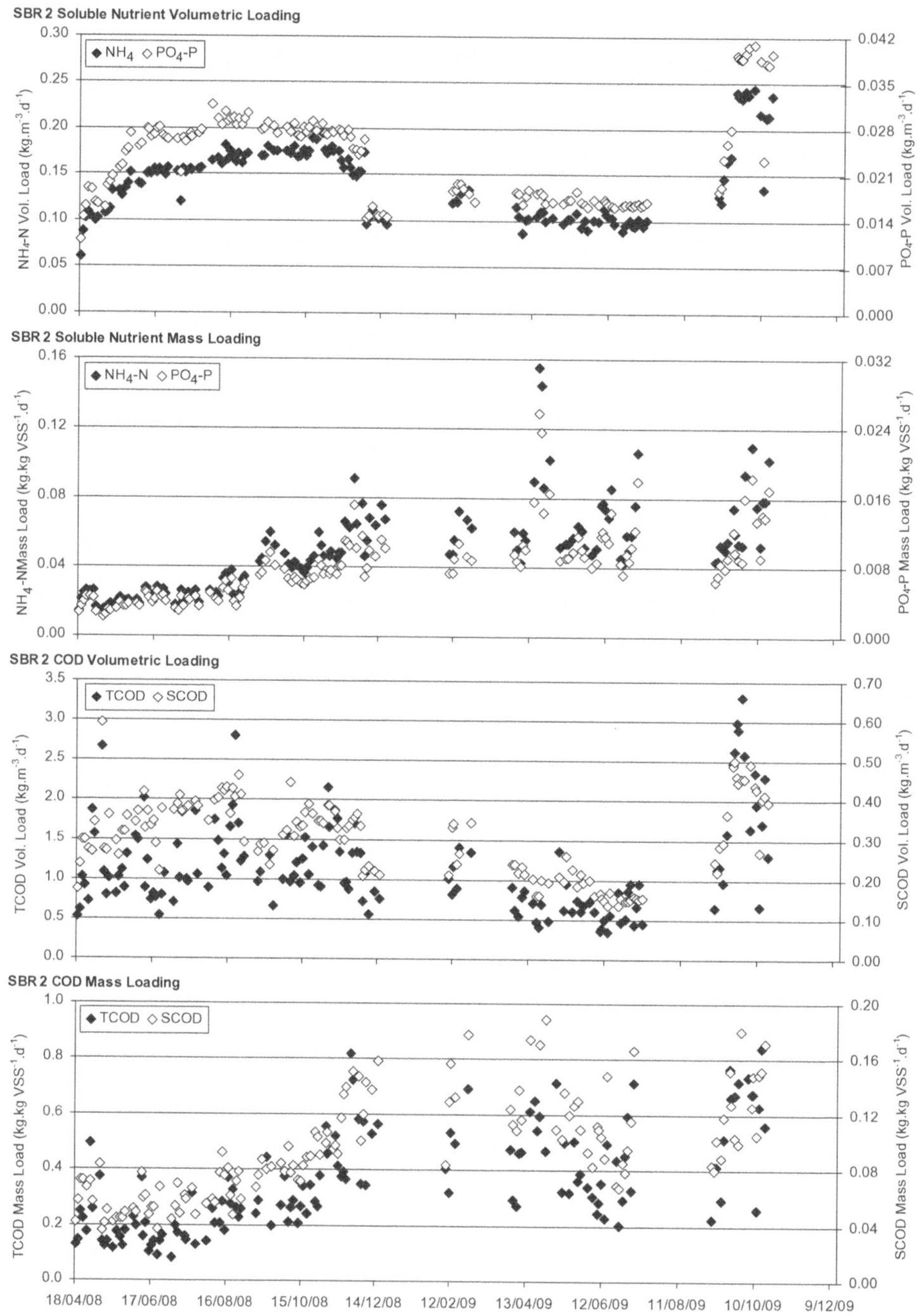

Figure 1.61 SBR 2 nutrient and COD loadings (volumetric and mass)

1.4.4.3 Experimental Phases of Operation

The pilot-plant operation can be broadly classified into a number of experimental phases or observation periods:

- Reduced settle time selective pressure
- Cycle and feeding regime manipulation
- Shock loading
- Prefermenter optimisation
- Increased hydraulic loading
- Increased Feed Event Loading

The attempts to achieve granulation and maintain the granules in the system were performed while also trying to maintain reasonably good N and P removal. At no stage was granulation attempted without maintaining N and P removal in order to avoid inhibitory affects of high ammonia, as has been observed in laboratory-scale reactors.

1.4.4.3.1 Reduced Settle Time Selective Pressure (May-Sep. '08)

SBR 1 Once the base hydraulic loading rate of $0.57\ m^3 \cdot m^{-3} \cdot d^{-1}$ was achieved, investigations into a reduced settling time as a selective pressure for granulation were undertaken. For SBR 1, the settling time was gradually reduced from 30 min (8/5/08) to 5 min (7/6/08). The cycle was maintained at 6 hr with the extra time from settle reduction added to the decant time. The plant remained with a 5 min settle without any major changes until 26th August, 2008. During this period, good NH_4-N, NO_X-N and PO_4-P removal continued (Figure 1.62). There were periods with elevated nutrient concentrations which were all related to operational issues and not loss of activity.

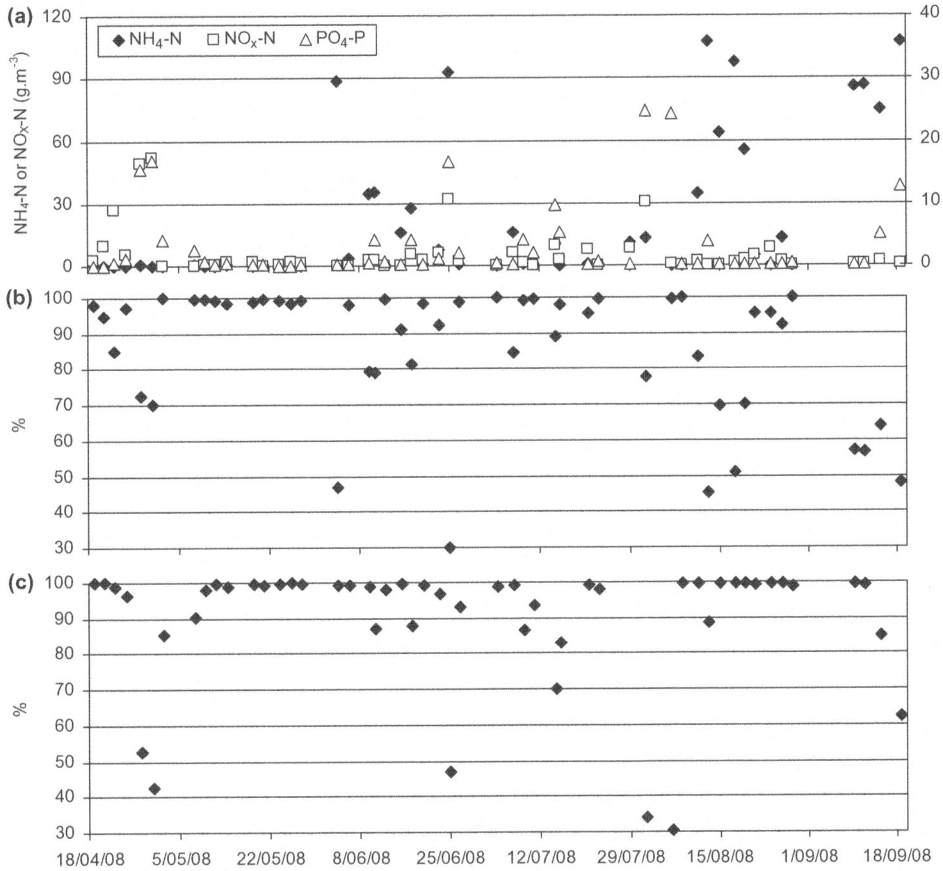

Figure 1.62 Nutrient removal in SBR 1: (a) nutrient concentrations, (b) $NH_4 + NO_X$ removal, (c) PO_4 removal (period 18/4/08–20/9/08)

The good nutrient removal continued to occur despite continued washout of sludge in the effluent. At this time, sludge wastage had ceased and so sludge turnover was due solely to washout in the effluent. Figure 1.63 shows the resulting SRT and suspended solids concentration in SBR 1 due to the washout of sludge. During this period the aerobic SRT was typical <3 days and yet the system was able to maintain excellent soluble nitrogen removal. This highlights a shortcoming with the typical SRT measurement, which does not take into account the continued addition of solids with the feed stream, which for this project was a significant load. Since most of the influent solids were small particles with limited settleability, it is expected that a substantial fraction of the effluent TSS is contributed by these incoming solids, hence the SRT does not really reflect accurately the true retention time of the biomass in the system.

Figure 1.64 shows the particle size analysis for SBR 1 suspended solids and effluent solids. Initially there was a particle size increase for the sludge, particularly for the 90%ile. However, this soon changed to a stable measurement. From 11/8/08–20/8/08 a particle size increase for the sludge was observed but then returned to prior measured values. The method of particle size loss appears to be washout of the sludge with the effluent and not solubilisation of the particles, as indicated by the particle size measurements of the effluent solids.

The SBR suspended solids concentration decreased throughout this study period at a very slow rate. This indicated that although there was significant loss of biomass in the effluent, the biomass was continually replenished. This is completely different to observations with lab-scale rigs where the settle reduction results in a significant decrease of suspended solids.

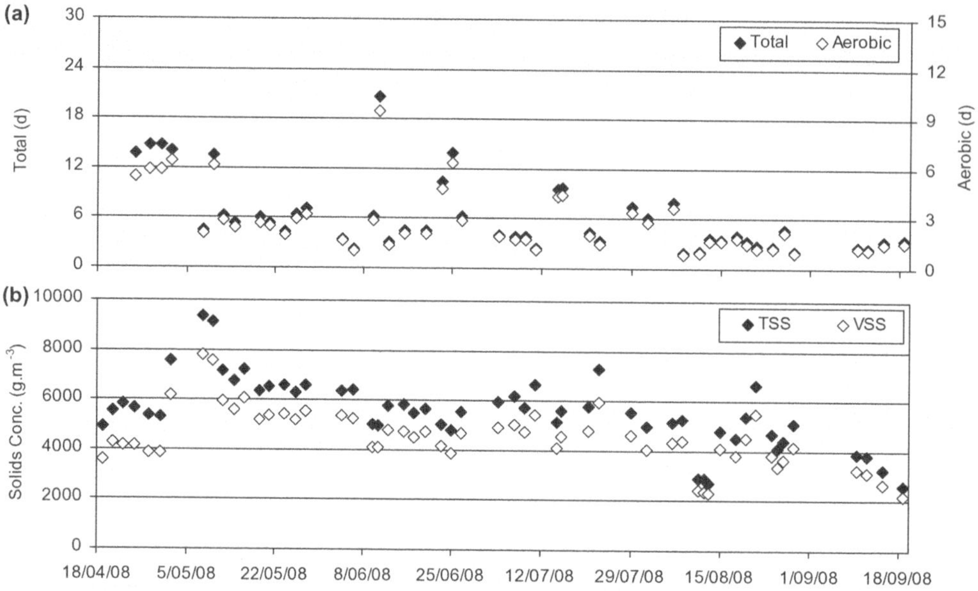

Figure 1.63 SBR 1: (a) solids retention time, (b) suspended solids concentration (period 18/4/08–20/9/08)

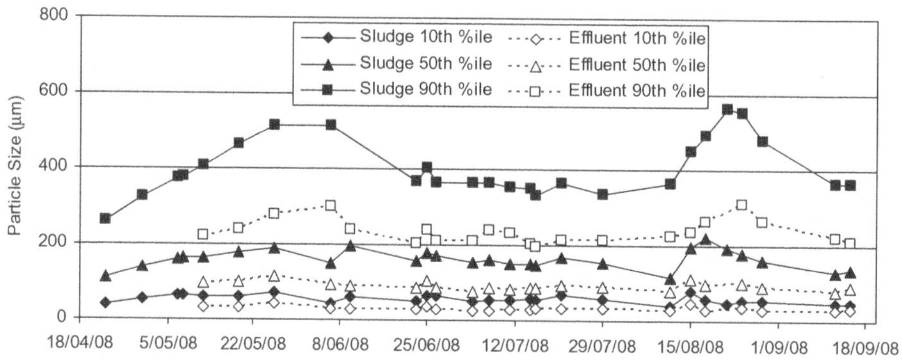

Figure 1.64 SBR 1 particle size analysis for suspended solids and effluent solids (period 18/4/08–20/9/08)

The major difference between the systems is the method of decanting. Lab-scale reactors are typically cylinders with a high aspect ratio and the decant port is located at a lower level than the top water level. In the pilot-plant, the aspect ratio is low and the decant port based on a floating system which alters as the surface level decreases. As well, the decant of lab-scale rigs is typically a fast process (within minutes) as compared to the much slower pumping process required for larger scale systems. Thus, it appeared the selective pressure of a reduced settle in the system was not sufficient to encourage granulation. With this in mind, the decant setup for SBR 1 was altered on 26/8/08. The new system involved removal of the float and connecting the flexible hose to a 2 m length of 25 mm ID pipe with 4 × 15 mm diameter holes located equidistant along its length. The pipe was fixed by hanging it from the pilot-plant upper frame and initially placed at 150 mm below bottom water level (BWL). This was later reduced to 230 mm below BWL (27/8/08). This decant apparatus change had no affect on the particle size but did appear to result in a more rapid decrease in sludge concentration.

It was concluded that settle time reduction alone, even with a submerged multiple point decant, would not cause granulation.

SBR 2 Prior to reducing the settling time for SBR 2, the load was first increased to observe if this offered any benefits. Once the base hydraulic loading rate of 0.57 $m^3 \cdot m^{-3} \cdot d^{-1}$ was achieved, it was increased from 0.57 $m^3 \cdot m^{-3} \cdot d^{-1}$ on 8/5/08 to 0.84 $m^3 \cdot m^{-3} \cdot d^{-1}$ on 27/5/08 (a 47% increase in load compared to SBR 1). The SBR operation continued with this loading and a 30 min settle time until the 20/8/08.

Nutrient removal throughout this period remained good (Figure 1.65) except for occasional spikes related to operational issues (as for SBR 1).

The SRT (Figure 1.66a) was more variable due to loss of solids in the effluent. These solids exited with the decant towards the end of the pumping period due to the decant suction port entering the settled sludge blanket (this reactor

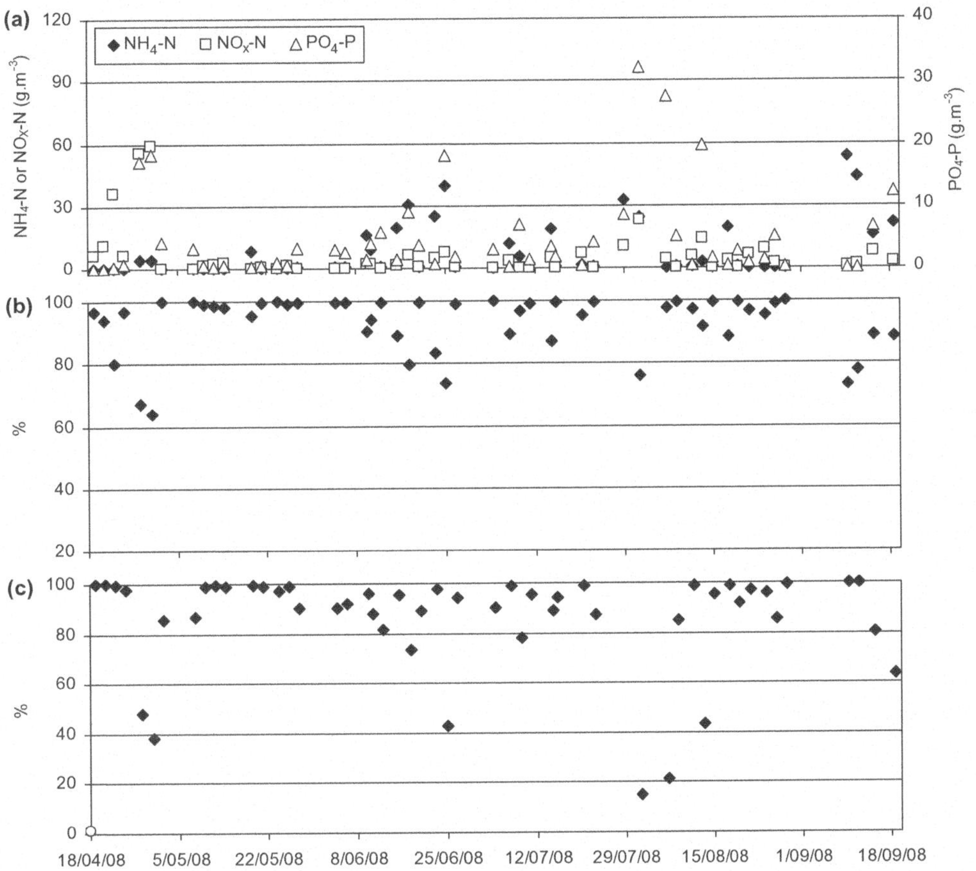

Figure 1.65 Nutrient removal in SBR 2: (a) nutrient concentrations, (b) $NH_4 + NO_X$ removal, (c) PO_4 removal (period 18/4/08–20/9/08)

continued operation with a floating decant). Despite the loss of sludge in the effluent, the reactor still maintained a high suspended solids concentration (Figure 1.66b).

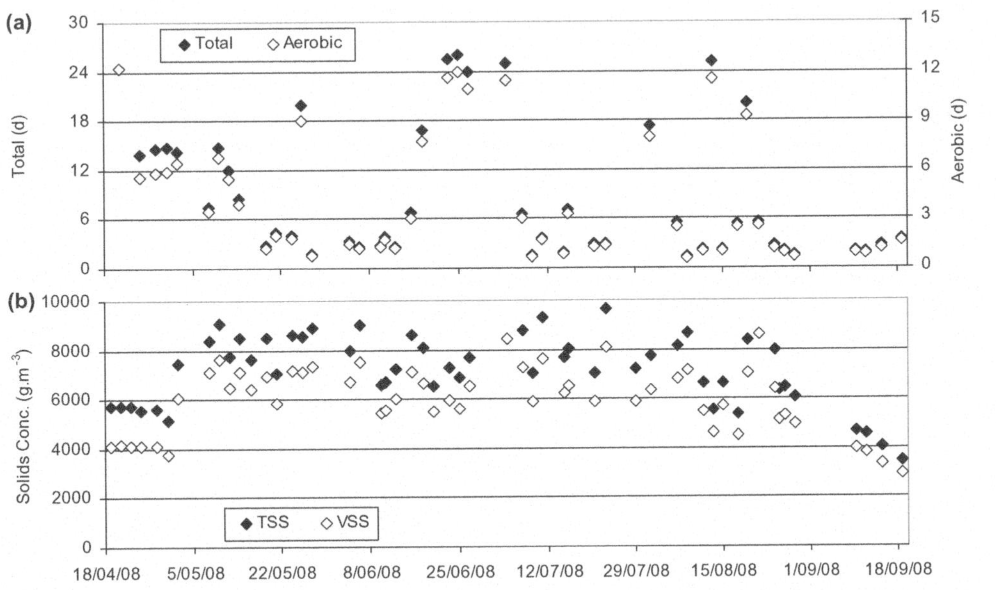

Figure 1.66 SBR 2: (a) solids retention time, (b) suspended solids concentration (period 18/4/08–20/9/08)

There was a particle size increase of the sludge towards the end of June (Figure 1.67) but this returned to previous levels shortly after. Note that this particle size increase was not observed at this time for SBR 1.

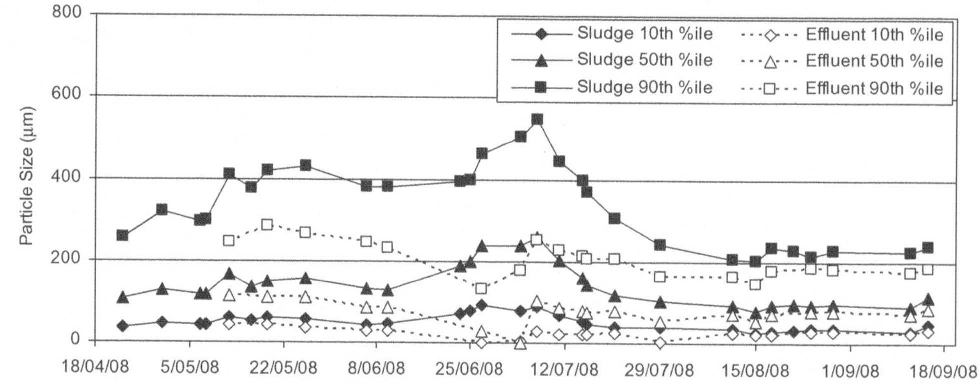

Figure 1.67 SBR 2 particle size analysis for suspended solids and effluent solids (period 18/4/08–20/9/08)

Once it was determined the additional hydraulic load would not result in granulation, the settling time was gradually reduced from 30 min on 20/8/08 to 5 min on 30/8/08. This settle reduction was performed at a considerably faster rate than was done for SBR 1. Continued operation with a 5min settle period showed no increases for the sludge particle size and it was concluded that a low settle time, even with a higher loading, would not cause granulation.

1.4.4.3.2 Cycle and Feeding Regime Manipulation (Sep.-Dec. '08)

Once it was found that a reduced settle time alone would not cause granulation in the system, major changes to the cycle timings and feed supply were undertaken. The major changes were:

- An increase in the feeding times while maintaining the same feed rate – this resulted in an additional time period after feeding where nothing happened. It was hoped further hydrolysis and fermentation of particulates from the prefermenter stream would occur within the SBRs, increasing the available VFA and decreasing the particulate COD. Implemented 20/9/08.
- Staggered feeding with feed 1 providing 50% of the total cycle load, feed 2 30% and feed 3 20%. This is the same feed strategy used in the previous study and it was hoped the higher shock loading from the first feed would help to stimulate granulation. When this was implemented, the feed preparations for the SBRs were separated so each feed batch for each SBR was prepared fresh for that SBR alone. Implemented on 8/10/08.

Following these changes, granulation was observed in SBR 1 and to a lesser extent in SBR 2 (Figure 1.68). The particle size increase for SBR 1 occurred prior to the staggered feed implementation so it can be concluded this played no role in stimulating granulation. The sludge median particle size reached a maximum of 400 µm for SBR 1 before rapidly dropping to previously measured values.

Figure 1.69a and b shows stereomicroscope images of SBR 1 sludge on 30/10/08 (peak of particle size analysis). The sludge was clearly dominated by granules. This was in direct contrast to the sludge from SBR 2 at the same time (Figure 1.69c). While some granules can be observed in SBR 2 sludge, it was largely dominated by floccular sludge.

The sludge concentration of SBR 1 began to increase during the granulating period (Figure 1.70), suggesting the granules were grown during the process rather than selectively retained over the floccular sludge during the decant washout.

The cause of granule loss from SBR 1 was likely due to a large increase in rapidly degradable solids from the prefermenter (Figure 1.59c). This increased solids loading appears to have resulted in excess floccular heterotrophic growth during the aerobic periods, which entrapped the granules during settling and they were subsequently washed out along with the floccular material. This is indicated by the delayed decrease in effluent solids particle size (Figure 1.68).

During this phase of operation, the major differences between the two SBRs was the higher loading in SBR 2 ($0.84\ m^3 \cdot m^{-3} \cdot d^{-1}$ compared to $0.57\ m^3 \cdot m^{-3} \cdot d^{-1}$ for SBR 1) and the submerged multiple point decant in SBR 1 compared to the floating one-point decant in SBR 2. At the time, it was thought the submerged decant port was critical to the extent of granule development in SBR 1, since SBR 2, which did not have this form of decant apparatus, did not develop the granules to the same extent. The same decant apparatus was incorporated into SBR 2 on 11/11/08 and both SBRs operated in the same fashion until the end of the year. However, no further granulation was subsequently observed.

1.4.4.3.3 Shock Loading (Feb.-May '09)

Prior to the granule development observed in October 2008, both SBRs were operated with a lower % volume of prefermenter feed. Initially the prefermenter feed % was 20–22.5% and this was reduced to 15% and finally 10% in the 2 weeks prior to the SBR 1 granulation period. This was then returned to a 15% load for 3 days prior to the plant shutting down for 5 days due to operational problems. Upon restarting the plant, granulation was observed in

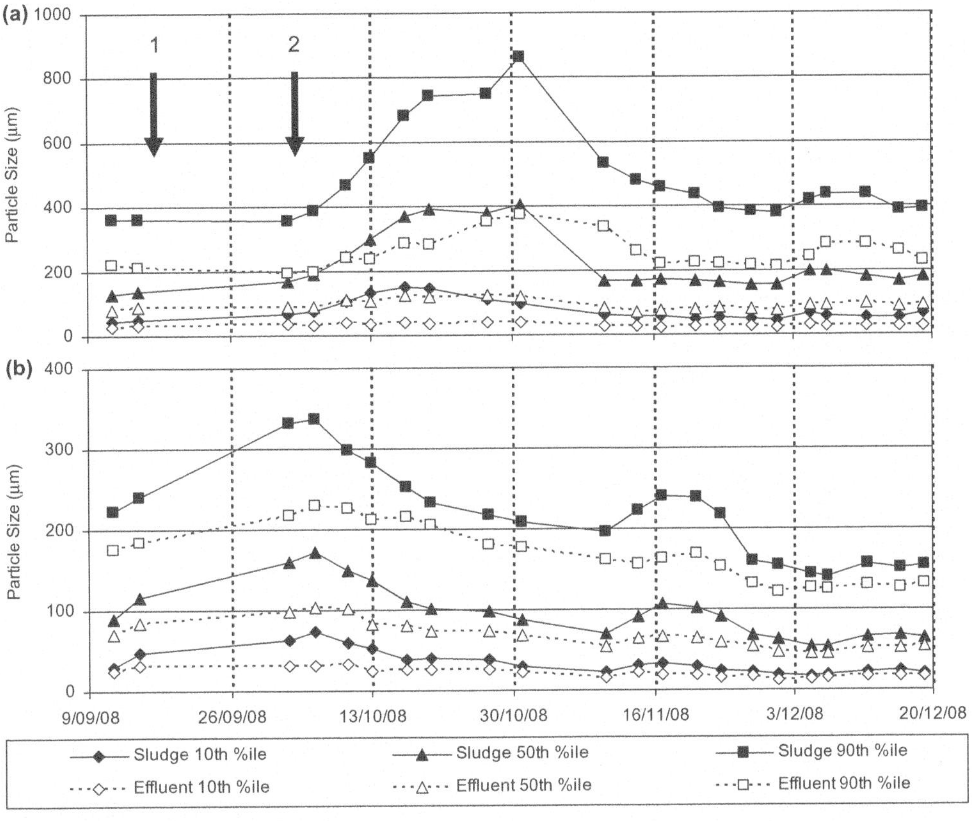

Figure 1.68 Particle size analysis for period 9/9/08–19/12/08: (a) SBR 1, (b) SBR 2, Arrow 1 indicates cycle changes; arrow 2 indicates implementation of staggered feed

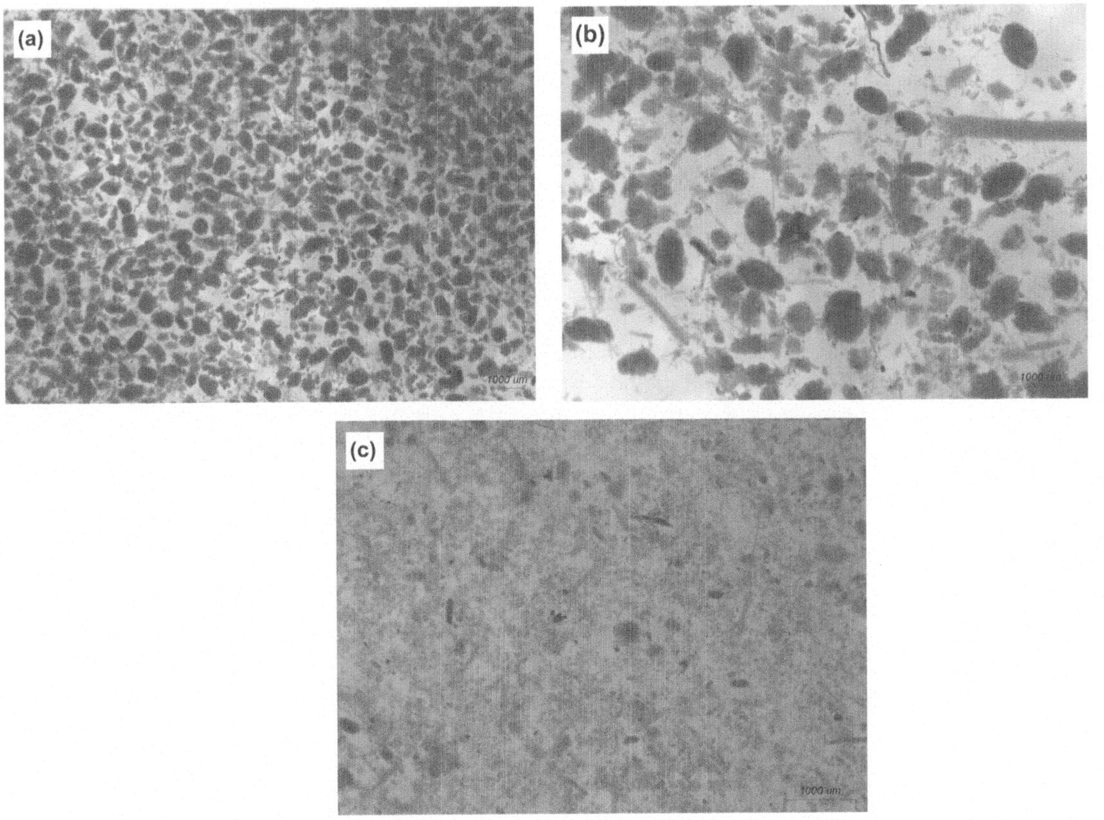

Figure 1.69 Stereomicroscope images: (a) and (b) SBR 1 sludge on 30/10/08, (c) SBR 2 sludge on 30/10/08

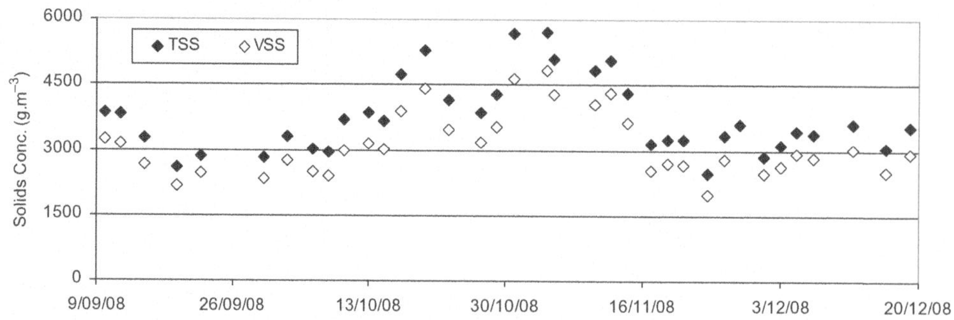

Figure 1.70 SBR 1 suspended solids concentration (period 9/9/08–19/12/08)

SBR 1. The reduction in carbon entering the SBRs resulted in reduced sludge concentrations within the reactors (approx. 2800 $g \cdot m^{-3}$ in SBR 1 and 4300 $g \cdot m^{-3}$ in SBR 2).

It was thought the rapidly increasing prefermenter loading, coupled with the reduced sludge concentration (and thus higher mass loading) may have been responsible for the initiation of SBR 1 granulation. Following the Christmas/January shutdown of the abattoir (total of 6 weeks), the pilot-plant was restarted with new cycle times which varied the length of aeration proportionally with the 50/30/20% feeding regime. The SBR 2 hydraulic loading was also changed back to the base loading of 0.57 $m^3 \cdot m^{-3} \cdot d^{-1}$ (same as for SBR 1). Once good nutrient removal was observed for both SBRs a shock loading experiment was performed to try to replicate the higher mass loadings. This was achieved by manually wasting mixed liquor from the SBRs during an aerobic period (SBR 1 was reduced to 2600 $g \cdot m^{-3}$ and SBR 2 to 2200 $g \cdot m^{-3}$ TSS). SBR 2 was then operated under the same cycle conditions as used previously but the feed for SBR 1 was alternated continuously between one day with a feed of 10% prefermenter volume and the following day of 20% prefermenter volume. This was done to determine if a shock increase in VFA loading assists in granule development.

After several weeks of operation it was observed the shock loading described had no effect on the suspended solids particle size. It simply resulted in a slightly higher effluent nutrient concentration for SBR 1. During this testing period, the prefermenter effluent COD was high and so it is not possible to determine whether the experiment failed or whether the high non-VFA COD exiting the prefermenter was responsible, preventing the accumulation of granules. The high prefermenter effluent COD appeared to be caused by high prefermenter temperatures (37–39°C) solubilising fats in the tank and reducing the settleability of the prefermenter sludge.

Subsequently the plant was shutdown in March while external temperatures decreased. Upon restart in April, the prefermenter temperature remained below 35°C and the effluent COD was also lower.

In April the conditions leading up to the previous granule formation were replicated, except for the 5 day shutdown period. This involved operating the SBRs with a 15% prefermented liquor feed, followed by a short period (several days) of 10% mix before returning to a 15% mix. Again there was no effect on the sludge particle size. It was concluded that there was another factor which was required to initiate granulation, possibly the prefermenter VFA and non-VFA COD concentrations.

1.4.4.3.4 Prefermenter Optimisation (May-Dec. '09)

By the end of 2008, it became evident that the prefermenter operation was critical to the development of granules in this system. Further examination of the data revealed that during the granulation in SBR 1 in October 2008, the prefermenter VFA concentration was higher than normal and the non-VFA COD (mainly in particulate form) was lower than normal. With this in mind attempts were made to optimize the prefermenter to reduce the non-VFA COD (mainly in particulate form) exiting the prefermenter.

The prefermenter was constructed using a simple rainwater tank without baffling. Shortly after startup in 2008, a thick fat layer developed on the top of the tank and remained throughout the year. This same fat layer was observed in the previous project as well.

Optimisation of the prefermenter included investigating:

- various depths (vertical) and positioning (horizontal) of feed delivery
- different HRTs
- different mixing regimes (frequent versus infrequent)
- incorporating an Induced Air Flotation unit
- extended settling

In May and June the feed delivery was altered in height (bottom feeding and mid-level feeding) and position (further away from the transfer pump to the mixing tank) with little affect on effluent solids or VFA concentration. After changing the inlet to trickle in at the surface (1/6/09) there was a significant reduction in effluent solids while maintaining typical VFA concentrations (Figure 1.71). However, this resulted in major accumulation of fat on the surface which eventually split the roof of the tank. Subsequently, the tank was emptied and the prefermenter restarted on 22/6/09. After restart, it was found the fat layer only redeveloped after a weekend period when no feeding occurs and the temperature typically drops. While there was no fat layer, sludge in the prefermenter could be seen to move, remaining suspended even after a 50 min non-mixing period.

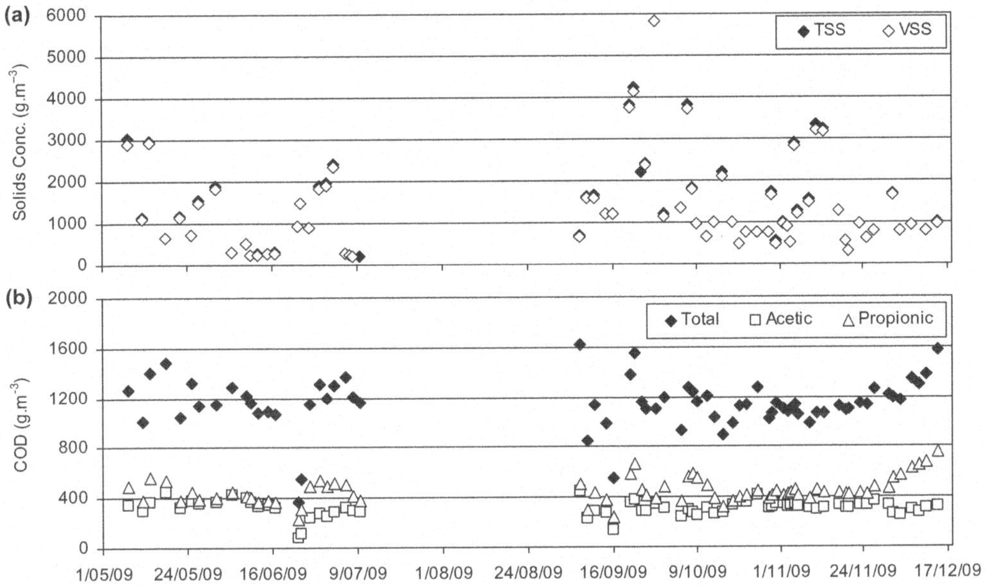

Figure 1.71 Prefermenter effluent during optimisation period: (a) solids, (b) volatile fatty acids

In July and August 2009, the plant was plagued by many operational issues, resulting in frequent shutdown of the plant. No data was collected during this period.

In September 2009, an Induced Air Flotation (IAF) unit was trialled on the prefermenter effluent stream. IAF works along the same principles as Dissolved Air Flotation, using fine bubbles to trap solids and fats and float them to the surface. It was found the IAF did remove solids but the rate of removal was considerably lower than expected and was also dependent upon the starting solids concentration. At a high TSS concentration ($4400\ g \cdot m^{-3}$) a TSS removal rate of $103\ g \cdot m^{-3} \cdot min^{-1}$ was observed compared to $3\ g \cdot m^{-3} \cdot min^{-1}$ for a low initial TSS concentration ($540\ g\ TSS \cdot m^{-3}$). An excessive amount of foam was also generated, possibly due to cleaning products used by the abattoir passing into the "red stream".

From mid-September 2009 to the end of the project, changes to the mixing regime were performed which included changing the length of the mixing event as well as the frequency of mixing. However, it was impossible to correlate any of the changes with changes to the prefermenter performance. This was likely due to the dynamic environment. Frequent mid-week shutdowns of the abattoir added to this dynamic nature.

In September and October 2009, investigations in varying HRTs was undertaken to optimise for VFA production rather than effluent suspended solids reduction. The data (Table 1.11) suggests the optimum HRT for VFA generation is between 1–3 days. However, again the study was influenced by the highly dynamic wastewater composition as well as time limitations.

Table 1.11 Effect of prefermenter HRT on VFA concentration

HRT (days)	Average VFA conc. ($g \cdot m^{-3}$)	Length of study (days)	Measurement no.
0.66d	603	9	4
1d	816	29	12
3d	771	28	14

From mid-November until the end of the project, investigations into extended settling times were undertaken. It had been observed throughout the project that the prefermenter effluent suspended solids were typically low following a period of extended non-feeding. These periods were typically weekends but also included weekdays when the abattoir was shut down. It is likely that the reason for a reduced effluent solids concentration is the solidification of entrained fats (which float to the surface) due to cooling of the liquor. The reduced soluble fat then allows for better settling of the prefermenter sludge during the non-mixed periods just prior to feed preparation.

The extended settling tests involved stopping all mixing and feeding into the prefermenter for 4-5 hours and then pumping the required prefermenter volume for 2 or 3 cycles into a separate tank. These transfers were done every 12 hours. The prefermented liquor required for feed preparation was then taken from the separate tank.

The effectiveness of the extended settling approach was highly variable, at times producing a liquor with low solids and at other times one with very high solids. The lack of success is likely due to insufficient time for cooling of the liquor necessary for the solidification of fats.

Towards the end of the project, there was an increasing VFA concentration observed. This appears directly related to the prefermenter temperature with a rising temperature resulting in either an increased VFA production rate or decreased VFA consumption rate in the prefermenter (due to methanogenesis).

1.4.4.3.5 Increased Hydraulic Loading (Sept.-Oct. '09)

To determine if an increased hydraulic loading would help to stimulate granulation, the HRT of both SBRs were reduced from 42 hr on 4/9/09 (hydraulic load of $0.57\ m^3 \cdot m^{-3} \cdot d^{-1}$) to 18 hr on 21/9/09 (hydraulic load of $1.33\ m^3 \cdot m^{-3} \cdot d^{-1}$). This was done by increasing the feed volumes per feed event (while maintaining the 50/30/20% feeding regime), decreasing the bottom water level in the SBRs (maintaining a top water level of $6\ m^3$) and also reducing the cycle times to result in a total cycle time of 4.8 hr instead of the previously used 6 hr.

SBR 1 showed a small suspended solids particle size increase in the later part of September (Figure 1.72) before decreasing to previously observed values. SBR 2 showed an even smaller increase prior to returning to previously observed values. It appears these particle size changes were not related to granulation but rather carryover of bulk liquor VFA (i.e. VFA which has not been stored intracellularly) into the aerobic phases during load increases. Such carryover would allow for rapid growth of heterotrophic bacteria and a subsequent increase in floc size. As the population of bacteria which can store VFA intracellularly (PAO and GAO) increased in number due to the higher loading, the amount of VFA carried over to the aerobic phases would decrease, slowing the floc size increase. The particle size decrease then occurred due to the continued washout of flocs during decant. Note that no VFA measurements were performed on samples collected at the start of the aerobic periods nor were microscopic examinations carried out at this time. Thus, while this explanation is plausible, it is still speculative without this supporting data.

The difference in the extent of floc size increase/decrease appears to be related to different population levels of bacteria which can accumulate VFA intracellularly within each SBR. Figure 1.73 shows that SBR 2 appeared better in handling the rapid increase in ortho-phosphate and VFA loading as compared to SBR 1, thereby reducing the VFA carryover into the aerobic phases.

SBR 2 was shutdown on 19/10/09 to allow for better manipulation of non-mixed times in the prefermenter. SBR 1 continued to operate to the end of the project with the higher loading at $1.33\ m^3 \cdot m^{-3} \cdot d^{-1}$ (18 hr HRT). Towards the end of 2009, the typical seasonal anaerobic pond effluent temperature increase was observed, particularly during November and December (Figure 1.57). However, the increased hydraulic load resulted in SBR 1 operating at relatively high temperatures for nutrient removal. Figure 1.74 indicates the ortho-phosphate removal rate was inhibited by temperatures >36°C whereas the nitrogen removal rate was inhibited by temperatures >38°C. The black pipe used for transferring the anaerobic pond liquor to the pilot-plant was known to artificially elevate the water temperature by 2–3°C on sunny days. However, when this is taken into account, it is clear that there would still be days in a sub-tropical region when the SBR temperature could adversely affect the nutrient removal process. Note that SBR 1 was shutdown on 19/12/09 (project end) and so only limited data could be collected to determine the effects of the elevated temperature.

1.4.4.3.6 Increased Feed Event Loading (Oct.–Nov. '09)

On 27/10/09, the cycle for SBR 1 was dramatically altered to only one feed, anoxic and aerobic period instead of the 3-staged feed. The same hydraulic loading was retained. Due to the limited size of the feed mixing tank, the feed preparation and delivery was altered so that the required prefermented liquor volume was first added to the mixing tank and then the tank filled to $0.8\ m^3$ with anaerobic pond effluent, thus creating a feed mix of 30% prefermented liquor. This feed batch was fed to the SBR and when the mixing tank volume had decreased to 50 L (i.e. 750 L fed), more anaerobic pond effluent was added to the mixing tank while the SBR continued to be fed. The intermittent addition of pond effluent to the mixing tank continued until the total required feed volume (as measured by the SBR pressure sensor) had been fed to the SBR. Thus, while the same overall feed fractions and total cycle loading rates were the same, the individual feed event loading rates were significantly higher.

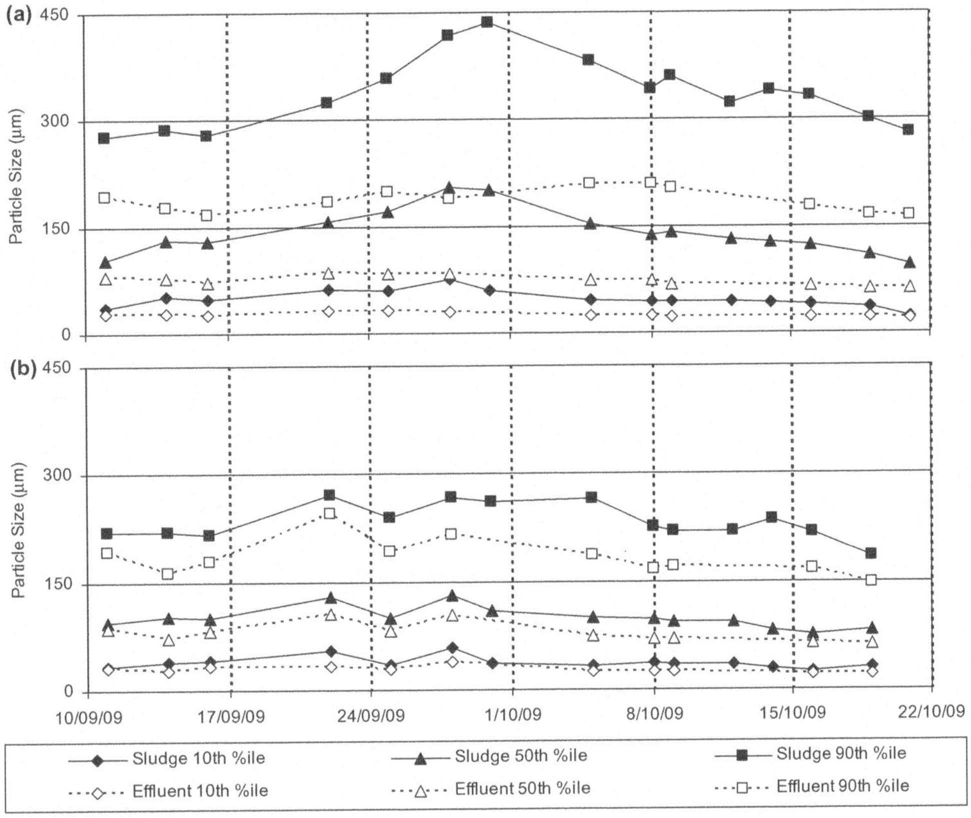

Figure 1.72 Particle size analysis during hydraulic load increase: (a) SBR 1, (b) SBR 2 (period 11/9/09–21/10/09)

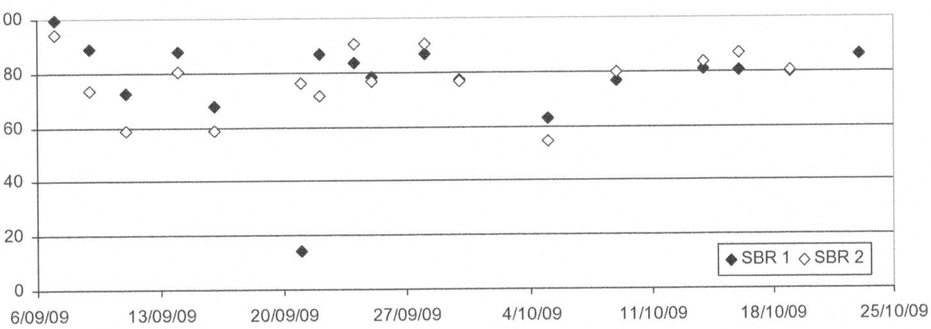

Figure 1.73 SBR 1 and 2 ortho-phosphate removal during hydraulic load increase (period 11/9/09–21/10/09)

Immediately following these changes, there was a very rapid suspended solids particle size increase (Figure 1.75a) followed by a reduction. This particle size increase again appears to have been due to carryover of VFA into the aerobic period as discussed previously. Once the bacterial population which store VFA intracellularly had increased in numbers, as indicated by the increase in ortho-phosphate removal (Figure 1.75b), VFA carryover ceased and the particle size decreased due to sludge turnover from the decant washout. On this occasion microscopic examination of the sludge was undertaken (Figure 1.76), showing that while there were some granule-like particles present, the majority of the sludge was formed from large, loose flocs.

1.4.5 Results of Pilot-Scale Experiences

1.4.5.1 Identification of Granulation Driving Factors

Granulation has been achieved at laboratory scale for a number of different systems, including removal of carbon, nitrogen or phosphorous separately and carbon, nitrogen and phosphorous removal combined. Driving factors for each system are

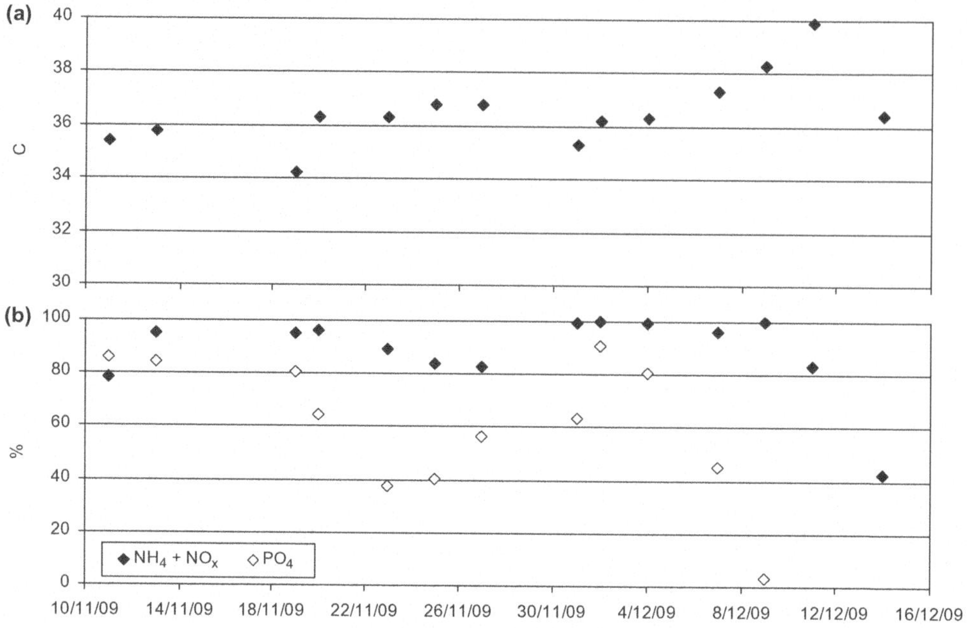

Figure 1.74 Effect of temperature (a) on $NH_4 + NO_X$ and PO_4 removal (b) in SBR 1

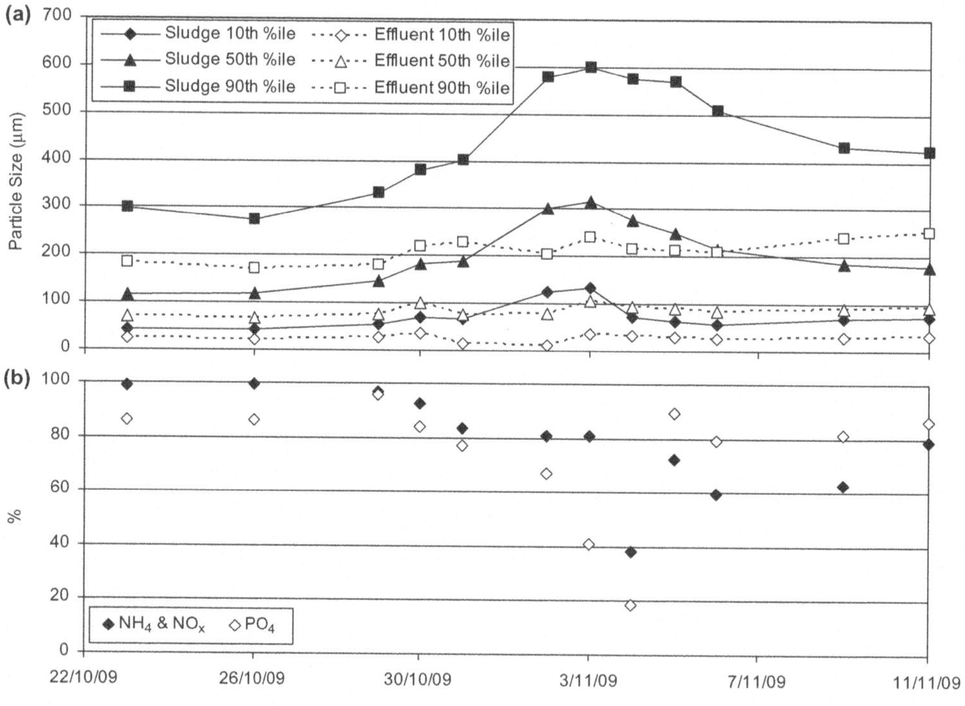

Figure 1.75 SBR 1 during the increase feed event loading trial: (a) suspended and effluent solids particle size, (b) nutrient removal

likely different due to the bacteria present. In this system, there also appear to be number of driving factors and identifying these factors was difficult due to the dynamic nature of the site.

Figure 1.77 shows graphs for SBR 1 with identified periods of granulation (or the beginning of granulation) followed by loss of forming granules. These are identified as: Granulation 1 (11/8/08–12/9/08), Granulation 2 (3/10/08–14/11/08) and Granulation 3 (1/12/08–16/12/08). All these granulation events occurred in the first year of the project. While there were periods of suspended solids particle size increase during the second year of operation, these were likely caused primarily by increasing floc size and not granule formation.

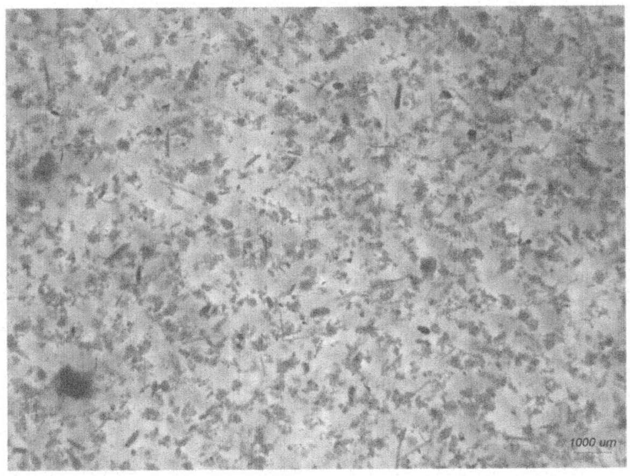

Figure 1.76 Stereomicroscope image of SBR 1 sludge on 3/11/09

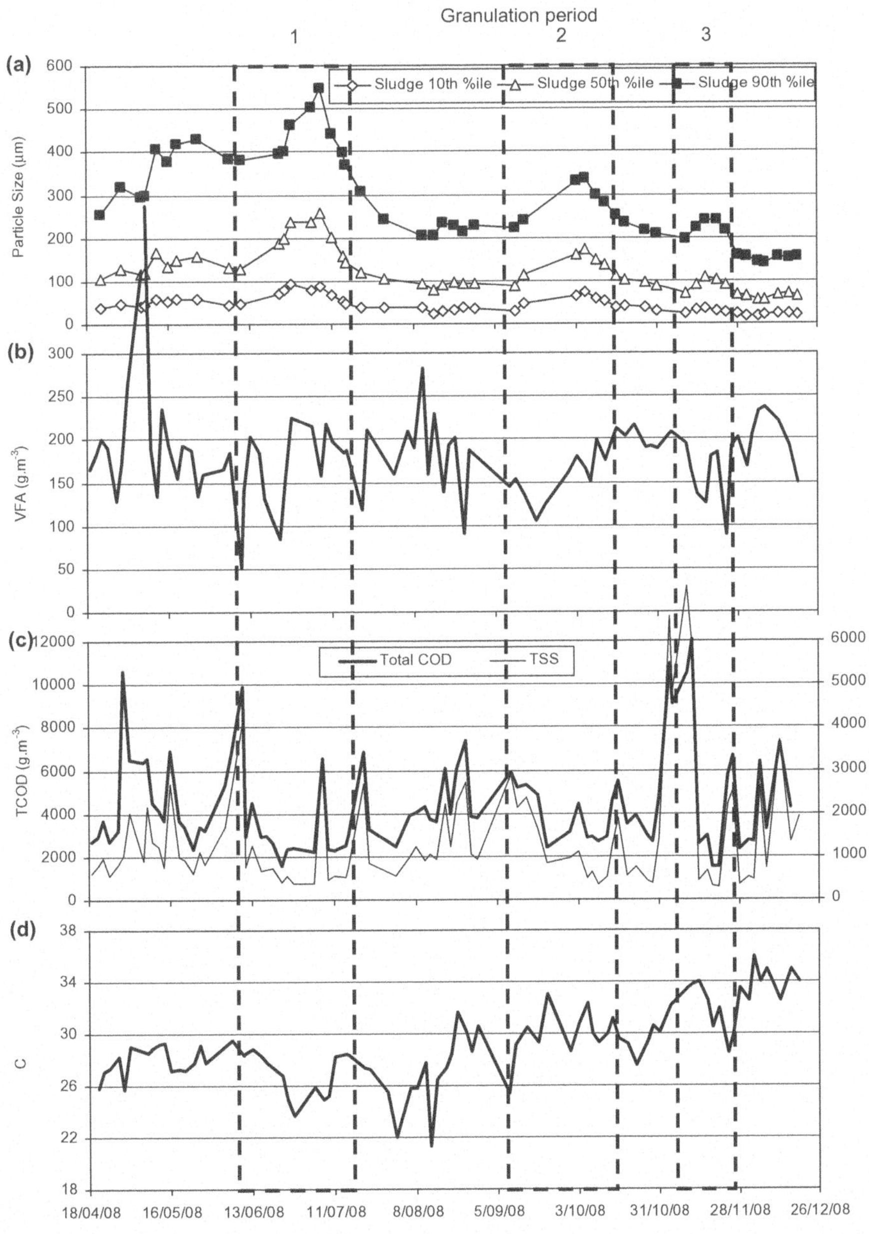

Figure 1.77 SBR 1 granulation events: (a) suspended solids particle size, (b) mixed feed total VFA concentrations, (c) prefermenter effluent TCOD and TSS levels, (d) SBR operating temperature.

The conditions present at the start of suspended solids particle size increase and decrease are shown in Table 1.12. For each period, the granulation process was accompanied by a period of high VFA concentration, low prefermenter effluent TSS and COD and medium to high SBR operating temperature (classification of low/medium/high is based on the ranges observed throughout the study).

Table 1.12 Conditions present at the start of SBR 1 sludge particle size increase and decrease

Granulation period	At start of granulation				At start of loss of granulation			
	Feed VFA conc.	**Pref. COD**	**Pref. TSS**	**SBR temp.**	**Feed VFA conc.**	**Pref. COD**	**Pref. TSS**	**SBR temp**
1 (11/8/08–12/9/08)	High	Low	Low-Medium	Medium	Low	High	Medium	Medium
2 (3/10/08–14/11/08)	High	Low	Very Low	High	High	Very high	Very high	High
3 (1/12/08–16/12/08)	High	Low	Low	High	High but dropping	High	High	High

High VFA concentrations in conjunction with low prefermenter effluent TSS and COD levels appear to be critical to the granulation process in SBR 1. Medium to high SBR temperatures also appear to play a role in this system. These higher temperatures, however, may simply allow faster granule growth due to higher bacterial growth rates and so provide a faster replenishment of granule "seed stock" to offset the granule loss due to washout in the decant.

Outside of the highlighted granulation periods, there were periods when several of the postulated "important" conditions were met, but not all and no granulation was observed. For instance prior to granulation 1, there were periods when all conditions were met except for the medium to high temperature and no median particle size increase was observed. Also, in February 2009, the feed VFA concentrations and SBR temperatures were high, but so too was the prefermenter effluent TSS and COD and subsequently no granulation was observed. Throughout most of the remainder of 2009, the VFA concentrations remained low and no granulation was observed.

These observations for the SBR 1 granulation events led to a possible hypothesis to explain how granulation may occur in this system:

- The feeding of VFA into the sludge blanket promotes the production of the exo-cellular polymers (EPS) responsible for the development of the granules (maybe as a carbon storage response to excess VFA provided). Throughout most of the study, visual observation of the sludge (by the naked eye) always showed small granule-like particles present and the sludges always exhibited excellent settling properties (SBR 1 SVI average of 59 $mL \cdot g^{-1}$; SBR 2 SVI average of 74 $mL \cdot g^{-1}$) compared to a typical floccular biomass, suggesting a mixed granular/floccular sludge.
- Under normal conditions of operation, a reactor would not normally be operated with excess VFA fed for the amount of nutrient removal required, as this would only result in unnecessary sludge production. A bottom feed, however, does provide conditions so that the VFA concentration may be in excess within the localized sludge blanket. Thus, under limiting VFA concentrations, a bottom feed is necessary to provide the "granulating" effect of excess VFA.
- The presence of non-VFA rapidly biodegradable COD (such as particulates from the prefermenter) results in rapid floccular growth due to carryover of the COD into the aerobic periods of the cycle. Even at low SRTs, washed out sludge can be quickly replenished in a single cycle if the COD is rapidly degradable (as this study has shown).
- Granule development begins with small granules which grow over time. In the early stages of granulation, the mass of the granules is insufficient to over-come entrapment within typical floccular sludge concentrations and so there is no preferential settling of the granules. If the floccular biomass concentration is sufficiently low, preferential settling of granules over flocs can occur (i.e. when less flocs are present to entrap the developing granules, preferential settling will occur) and this has been observed in laboratory-scale reactors.
- Suspended solids particle size increases observed during granulation events are the net result of granular growth (from localised excess VFA) and granular loss (from the washout of developing granules entrapped within floccular material which is washed out during decant). If the floccular growth is limited (by reducing the amount of non-VFA rapidly degradable COD fed), the net result will be an increase in measured sludge particle size due to the growth of the granules.

While this hypothesis explains the observations of SBR 1, the observations for SBR 2 were more complicated. Data examination of SBR 2 reveals 3 granulation events similar to SBR 1 (Figure 1.78, Table 1.13). However, the timing of these events was slightly different to those of SBR 1. At the start of each granulation event for SBR 2, the prefermenter effluent TCOD and TSS concentrations were always low and the SBR temperature medium to high as for SBR 1. However, the VFA concentration was not always high at the onset of granulation for SBR 2, though it was increasing for granulation 2.

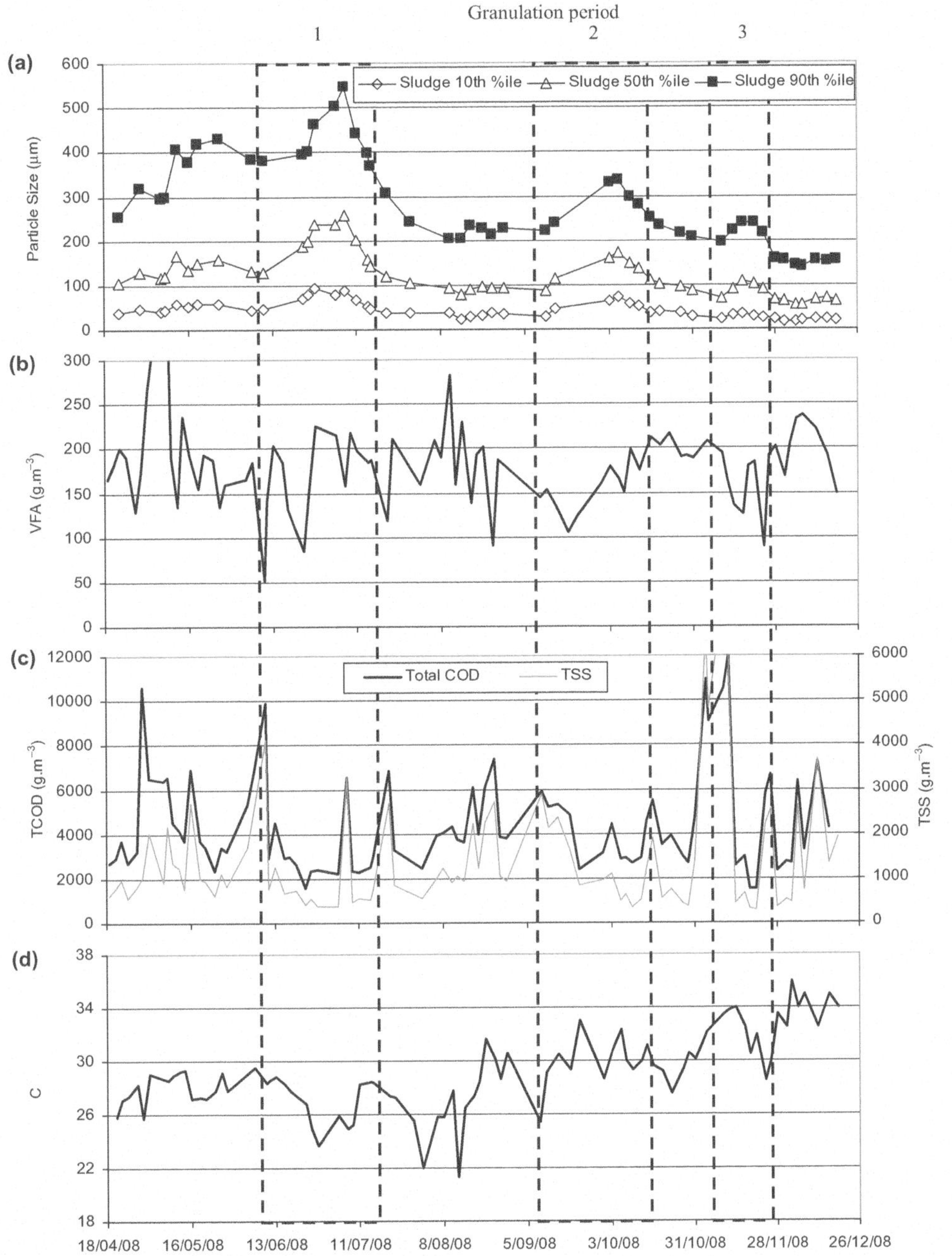

Figure 1.78 SBR 2 granulation events: (a) suspended solids particle size, (b) mixed feed total VFA concentrations, (c) prefermenter effluent TCOD and TSS levels, (d) SBR operating temperature

Similarly, at the start of loss of granulation, the prefermenter effluent TCOD and TSS concentrations were not always medium-high as was the case for SBR 1. During granulation 2, the prefermenter effluent COD and TSS levels were both low and so this does not fit well with the above mentioned hypothesis. However, observations made at the time may explain this discrepancy. Particle size reduction for granulation 2 commenced between 6/10/08 and 10/10/08. This period was preceded by observations in the SVI tests showing rising sludge in the last 10–15 minutes of the test (likely caused by denitrification during the settling test causing small N_2 gas bubbles to accumulate in the sludge and make it buoyant). Rising sludge in SVI tests were observed on 3/10/08, 6/10/08, 8/10/08 and 9/10/08 (i.e. in each test performed at this time). On 9/10/08, the DO setpoint was reduced to 1.5 $g \cdot m^{-3}$ for both SBRs and on 10/10/08 no rising sludge was observed.

Table 1.13 Conditions present at the start of SBR 2 sludge particle size increase and decrease

Granulation period	At start of granulation				At start of loss of granulation			
	Feed VFA conc.	**Pref. COD**	**Pref. TSS**	**SBR temp.**	**Feed VFA conc.**	**Pref. COD**	**Pref. TSS**	**SBR temp.**
1 (10/6/08–21/7/08)	High but variable	Low	Low	Medium	Low	High	High	Medium
2 (12/9/08–20/10/08)	Low but increasing	Low	Low	High	High	Low	Low	High
3 (10/11/08–1/12/08)	High	Low	Low	High	High	Medium, increasing	Medium, increasing	High

At the time, SBR 2 was not inspected to see if the rising sludge in the SVI translated to the tank. However, at other times in the project when rising sludge had been observed in the SVI tests, visual inspections of the SBRs also showed rising sludge in the tank, particularly during feeding. Thus, it is likely rising sludge did occur in SBR 2 feeding events just prior to the onset of granule loss in granulation 2. Under these circumstances, the biomass would no longer be exposed to the excess VFA concentrations required (since it is no longer forming a sludge blanket at the bottom of the tank) and so the granulation process would slow or stop. However, since there would still be sludge turnover in the decant, the net result is the loss of granules.

1.4.5.2 Importance of the Prefermenter

Apart from the SBR cycle timings and bottom feed arrangement, the prefermenter operation was the most critical aspect of the pilot-plant. While the prefermenter stream represented only 10–20% of the total flow to the pilot-plant, it provided, on average: 47% of the TCOD; >90% of the rapidly degradable COD; and 94% of the total VFA.

The performance of the prefermenter was critical to the success of granulation. High VFA concentrations and low particulate COD (represented in the form of TSS) in the prefermenter effluent appear to have been key drivers in the granulation process, with the converse resulting in the washout of these granules.

Achieving high VFAs and low effluent solids from the prefermenter, however, was difficult. The operating temperature of the prefermenter appears to be an important parameter, with higher temperatures (36-40°C) resulting in higher VFA production. Since the majority of the soluble COD from the prefermenter was always in VFA form, it can be concluded that such higher temperatures increase the rate of particulate hydrolysis, the preliminary step to VFA production.

However, these higher temperatures can also have the adverse effect of preventing the solidification of fats not removed upstream in the DAF. The high temperatures of the "red stream" entering the DAF (45-50°C) drastically limited the fat removal efficiency, resulting in considerable fats entering the prefermenter (as also observed in the previous project). If these fats remain liquid (and likely dispersed), as is the case at the preferred higher prefermenter temperatures, they affect the settleability of solids (influent particulates and sludge) within the prefermenter. It is likely that these liquid fat (or oil) globules will be adsorbed or otherwise associated with particulates in the water, increasing their buoyancy and hence limiting their settleability.

Investigations aimed at solids reduction during the project were largely unsuccessful. Possibly a better method of operation would be to have two prefermenters operating alternately. While one prefermenter is receiving the DAF flow and generating the VFAs, the other is intermittently mixing (and cooling) without feed and providing the feed source for the SBR, thereby drawing down the volume. These two prefermenters would alternate in their operation (probably every 2 days), thereby providing a continuous flow with likely lower effluent solids.

Another alternative is the removal of some of heat energy from the "red stream" prior to entry to the DAF. This heat energy could be transferred to the boiler feed water, thereby reducing energy requirements for heating this stream. Reducing the red stream temperature prior to the DAF should assist in the fat removal process and so reduce the fat entering the prefermenter. It could also benefit the solids removal in the DAF.

Whatever method is used, if aerobic granules are to be grown on an abattoir wastewater stream, the prefermenter operation needs to be optimised first, in terms of VFA production and solids settleability. Attempts at optimisation were limited due to time and budgetary restraints, which were not originally factored into this project.

1.4.5.3 Seeding with Aerobic Granules

While it was difficult to grow aerobic granules with the wastewater, it is likely an aerobic granule seed would be more successful, provided the seed granules have sufficient mass to overcome entrapment within floccular material. This is because the system could be started with a very low settle time at commissioning, preventing the build-up of a highly concentrated floccular sludge. If the cycle exchange ratio is sufficiently high (such as 50%) and cycle length short (3–4 hr), it could be expected that floccular material would not be retained in the system. This would most likely be the case even when high rapidly biodegradable COD is present in the feed. However, if a slowly settling particulate wastewater is fed (such as from an un-optimised prefermenter), it would be reasonable to assume such a fraction would continue in the effluent and require further treatment. Thus, while the system could be fed from an un-optimised prefermenter (in terms of effluent solids), at least some of these solids would then still need be removed after treatment in the SBR.

Currently in Australia, there are no large scale aerobic granulation plants operating to act as a suitable seed material. In the future, when such plants may exist, further investigations could be undertaken to adapt this promising technology for the very challenging abattoir wastewater situations.

1.4.5.4 SBR Temperature Issues

The relatively short HRT of the site's anaerobic pond (5 days) no doubt contributed significantly to the high pond temperatures (and subsequently the SBR temperature) towards the end of the project. The effluent temperature from a more typical abattoir anaerobic pond (with a HRT around 10 days) could be expected to be lower, relieving negative temperature effects in the SBR. Furthermore, abattoirs which experience lower temperatures than sub-tropical regions could also be expected to have lower pond effluent temperatures than experienced at this site. Thus, while this site was particularly challenging to implement this technology, particularly during the warmer summer months, other sites may be better suited for this technology.

1.4.6 Conclusions

- Aerobic granules can be developed in an SBR treating abattoir wastewater (red stream and anaerobic pond effluent) under certain conditions. These conditions include specific cycle operation, bottom feed delivery to the SBR, high feed VFA concentration ($>175\ g \cdot m^{-3}$) and low prefermenter effluent TSS levels ($300\text{–}600\ g \cdot m^{-3}$ as total feed concentration). When these conditions are maintained and a low settle time imposed, aerobic granules have been observed to develop in the reactors.
- The presence of high concentrations of non-VFA rapidly biodegradable COD during the granule development phase causes excessive floccular growth due to the carry-over of these organics from the anaerobic to the aerobic period. Such growth can entrap developing granules during settling periods and cause a washout of the immature granules during decant.
- Further optimisation of the prefermenter is critical to long-term operation of an aerobic granular system treating abattoir wastewater. In particular, maximising solids settleability and VFA production need to be addressed. Without these optimisations, the growth of aerobic granules in an SBR receiving a combined anaerobic pond effluent and prefermenter feed stream will be very difficult to achieve.
- A hypothesis for granule development in this system was proposed. This hypothesis postulates that biomass particle size increases observed during granulation events are the net result of granular growth (resulting from granule-internal growth based on the storage compounds accumulated under anaerobic conditions from the VFA provided in the feed) and granular loss (from the washout of developing granules entrapped within floccular material which is washed out during decant). This hypothesis seems to adequately reflect observations made in the study.
- Bottom feeding (Unifed system) is critical for granule development to provide localised, maximally concentrated VFA concentrations. Rising sludge during the bottom feeding process will likely result in an insufficient VFA load on the sludge in the SBR and so not result in sufficient internal carbon storage and hence good granule growth.
- Care must be taken when using particle size analysis for determining possible granular development. Breakthrough of VFA (or other rapidly degraded organics) into aerobic periods will result in a temporary biomass particle size increase. However, these increases are not due to granular growth but rather an increasing floc size, with the flocs being quite loose and weak. Microscopic examinations and nutrient removal performance must also be used for determining whether granulation is occurring.
- High anaerobic pond effluent temperatures will adversely affect the nutrient removal processes in an SBR. Loss of phosphorous removal performance occurred at SBR temperatures >36°C and nitrogen removal performance at SBR temperatures >38°C. These effects were observed only when the hydraulic loading was high ($1.33\ m^3 \cdot m^{-3} \cdot d^{-1}$) and ambient temperatures were high, resulting in high anaerobic pond effluent temperatures.

1.5 SBBGR TREATMENT OF LANDFILL LEACHATES

1.5.1 Introduction

Municipal landfill leachates, resulting from the percolation of water through solid waste, are considered one of the types of wastewater with the greatest environmental impact. Municipal landfill leachates contain pollutants that can be divided into four main groups: dissolved organic matter; inorganic compounds (such as ammonium, calcium, magnesium, sodium, potassium, iron, sulphates and chlorides); heavy metals (such as cadmium, chromium, copper, lead, nickel and zinc); and, xenobiotic organic substances (Lema *et al.*, 1988; Ehrig H.J., 1989; Andreottola *et al.*, 1990; Irene and Lo, 1996; Kjeldsen *et al.*, 2002). Organic materials found in landfill leachates are typically volatile fatty acids and humic and fulvic compounds. Xenobiotic organic compounds coming from household or industrial chemicals are present in very low concentrations (usually lower than 1 mg/L). These compounds include, among others, a variety of aromatic hydrocarbons, phenols, chlorinated aliphatics, pesticides, and plastizers (Kjeldsen *et al.*, 2002).

The composition of landfill leachates, however, varies depending on the nature of deposited wastes, soil characteristics, rainfall patterns and largely landfill age (Chen P.H., 1996; Park *et al.*, 2001). Usually, young landfill leachates contain large amounts of biodegradable organic matter (i.e. volatile fatty acids) that decrease with increasing landfill age as a result of the anaerobic decomposition that takes place in the landfill site. As volatile fatty acids content decreases, the organic matter in the leachates becomes dominated by refractory compounds, such as humic-like compounds and fulvic acid-like substances with consequent reduction of BOD/COD ratio (Harmsen, 1983). Also, the ammonia concentration increases with the increase of landfill age as a result of the fermentation of organic matter containing proteins; concentration values higher than 2 g/L are typical in old landfill leachates.

Therefore, stabilized landfill leachates are more difficult to treat compared to younger landfill sites. Firstly, the potential toxic effects of high ammonia levels makes the conventional nitrification-denitrification biological process less suitable for ammonia removal from old leachates unless very long hydraulic residence times are adopted. Moreover, the high salinity and lack of sufficient electron donors, a feature of stabilized leachates, are additional obstacles to the nitrification-denitrification process. Secondly, the significant presence of recalcitrant pollutants in old leachates makes it difficult to meet the COD discharge limits, since these compounds pass unaltered through the biological stage of a treatment plant. Therefore, the conventional biological treatments followed by classical physic-chemical methods, aimed at removing the recalcitrant substances, have been considered for many years as the most appropriate technologies for treating stabilized leachates (Renou *et al.*, 2008). Unfortunately, in the case of more stabilized leachates, this treatment scheme is typically not sufficient to meet the in force discharge limits (Renou *et al.*, 2008). Therefore, more effective treatments are required for the treatment of highly stabilized municipal landfill leachates.

An alternative process based on a pre-treatment step for nitrogen recovery as struvite, followed by biological degradation, carried out in a SBBGR system (Sequencing Batch Biofilter Granular Reactor), integrated with an ozonation step, is proposed in the present paper.

Due to its high effectiveness, reaction rate, simplicity and environmental sustainability, the chemical precipitation of ammonium as struvite (i.e. $MgNH_4PO_4 \cdot 6H_2O$; MAP – magnesium ammonium phosphate hexahydrate) is a valid alternative for the removal of high ammonium concentrations from leachates, although they are characterized by low concentrations of magnesium and phosphorus and therefore external sources of these compounds are required. Moreover, struvite is a valuable fertilizer since it is a slow release source of nitrogen, magnesium and phosphorus (Booker *et al.*, 1999; Gaterell *et al.*, 2000). This process has been widely applied for the pre-treatment of many different wastewaters such as tannery effluents (Tünay *et al.*, 1997; Zengin *et al.*, 2002; Kabdaşli *et al.*, 2003), supernatants from anaerobic sludge digestion (Maqueda *et al.*, 1994; Siegrest, 1996; Ohlinger *et al.*, 1998; Battistoni *et al.*, 2001; Munch and Barr, 2001), piggery effluents (Nelson *et al.*, 2003), human urine (Maurer *et al.*, 2006) and leachates (Günay *et al.*, 2008; Kabdaşli *et al.*, 2000; Zhang *et al.*, 2009).

SBBGR is a "time oriented" submerged biofilter in which all the steps of the biological treatment occur sequentially in the same tank instead of moving to a second tank for the continuation of the treatment as in conventional treatment systems. This technology boasts a unique feature: the presence of granular biomass (Di Iaconi *et al.*, 2005). In fact, a large fraction of the microorganisms present in the system grow as granules entrapped in the pores produced by packing the reactor with a filling material. This allows for greater biomass retention in the reactor (up to one order of magnitude higher than that recorded in conventional biological systems). As a result, a notable increase in sludge age is achieved with a consequent reduction in sludge production (Di Iaconi *et al.*, 2009; Di Iaconi *et al.*, 2010a). This technology, already applied to other kinds of wastewater, offers the following advantages (Di Iaconi *et al.*, 2010a; Di Iaconi *et al.*, 2010b): (1) high conversion capacities with consequent reduction of reaction volumes; (2) the absence of a secondary settling tank; (3) high compactness since all the steps of a biological treatment can be carried out in a single operative unit; (4) the selection of a biomass particularly effective for degrading toxic and/or recalcitrant compounds usually present in the leachates; (5) great flexibility - a feature particularly attractive for the treatment of leachates, characterised by high variable volumetric flow rate and composition and very low sludge production. Finally, SBBGR technology is effective also for treating wastewater containing recalcitrant pollutants since it allows the biological degradation to be integrated with chemical oxidation used with the aim of making recalcitrant compounds biodegradable. Moreover, due to certain features of the SBBGR (i.e. discontinuous operation conditions and separation between biomass and wastewater), the oxidant (i.e. ozone) can be used in a specific (i.e. soon after a biological degradation phase to avoid consuming the ozone for degrading biodegradable pollutants) and controlled (i.e. with the aim of rendering recalcitrant compounds biodegradable thus minimizing their mineralization) way.

This chapter reports the results of a study aimed at evaluating the effectiveness of the proposed process for treating two leachates from Italian mature municipal landfills. In order to evaluate the effect of ozone on SBBGR biomass, the structure and microbial composition of the biomass were also measured.

1.5.2 Materials and Methods

1.5.2.1 Lab-scale plant

The lab-scale plant used in the project consisted of a pre-treatment step for chemical precipitation of ammonia, followed by a biological degradation (carried out in a SBBGR) integrated with ozonation.

1.5.2.1.1 Pre-treatment for nitrogen recovery

Magnesium ammonium phosphate (MAP), commonly called struvite, forms according to the following reaction:

$$Mg^{2+} + PO_4^{3-} + NH_4^+ + 6H_2O \longrightarrow MgNH_4PO_4 \cdot 6H_2O \tag{1}$$

Looking at Eq. (1), a Mg^{2+}:NH_4^+:PO_4^{3-} stoichiometric molar ratio of 1:1:1 results. Municipal landfill leachates generally contain low concentrations of both magnesium and phosphorus so an external source of these compounds must be added. Three types of magnesium sources are usually used in the precipitation of MAP: $MgSO_4$, $MgCl_2$ and $MgO \cdot MgSO_4$ and $MgCl_2$ are more widely used than MgO, because of their lower cost, faster dissociation and shorter reaction time. However, since $MgCl_2$ and $MgSO_4$ release large amounts of salts (chlorides and sulphates, respectively), had they been used, they would have made the subsequent biological treatment impossible. Therefore, MgO was used, since it does not cause any increase in wastewater salinity. As for the phosphorus source, H_3PO_4 was used due its low cost in comparison with phosphorus salts such as Na_2HPO_4 or $Ca(H_2PO_4)_2$.

The pre-treatment unit consisted of a clarifloculator (geometric volume: 50 L) equipped with a mixer, two pumps for dosing chemicals (i.e. H_3PO_4, NaOH) and a pH controller. The clarifloculator was operated in batch mode. First the required amount of MgO powder was slowly added to 40 L of raw leachate under mixing. Successively, the required volume of H_3PO_4 was dosed by means of a peristaltic pump at a flow rate of 1 L/h; NaOH solution was added when the pH dropped to a set value. Once the whole volume of phosphoric acid had been added, the content of the clarifier was stirred slowly until equilibrium was achieved (i.e. for 30 minutes) after which the mixing was stopped and the contents were allowed to settle for 2 hours. Finally, the supernatant was ready to be pumped into the SBBGR system.

Considering that ammonia removal efficiency by struvite precipitation is greatly affected by pH and Mg:N:P ratio, preliminary tests were carried out at bench scale in order to optimise the process. In particular, MAP precipitation was carried out at pH 9.0, taking into consideration not only the high alkalinity value typical of aged leachates (which makes it difficult to vary the pH), but also the pH range required to obtain minimum struvite solubility (i.e., 9.0-10.7) (Ohlinger *et al.*, 1998; Booker *et al.*, 1999; Nelson *et al.*, 2003) and to carry out the subsequent biological treatment (i.e. 7.5-8.5). As for the $Mg:NH_4:PO_4$ molar ratio, a series of batch investigations was conducted. In particular, four molar ratios (i.e. 1.1.1, 1.5:1:1, 2:1:1 and 2.5:1:1.1) were investigated.

1.5.2.1.2 SBBGR system

The SBBGR system consisted of (Figure 1.79) a cylindrical reactor (working volume: 12 L) partly filled with biomass supporting material (wheel shaped plastic elements; features: 7 mm high, 8 mm diameter, specific area 650 m^2/m^3, 0.95 density, and bed porosity 0.74) packed between two sieves and aerated by air injection (flow rate: 250 NL/h) through porous stones placed close to the upper sieve. An external loop allowed wastewater re-circulation (at flow rate of 90 L/h) through the filling material, by means of pump P_R, in order to obtain a homogeneous distribution of substrate and oxygen. Pump P_F and motorised valve V_M provided for filling and drawing operations, respectively. A pressure meter, set at the bottom of the reactor, measured on-line biofilter head losses due to the biomass growth and captured suspended solids occurring in the leachate during the operation of the reactor. When a fixed set value (i.e. 300 cm) of head loss was reached, a washing step was carried out by compressed air until the head loss decreased to a defined value (i.e. 100 cm). Activated sludge from the municipal treatment plant of Bari (a town of Southern Italy) was used as the SBBGR inoculum.

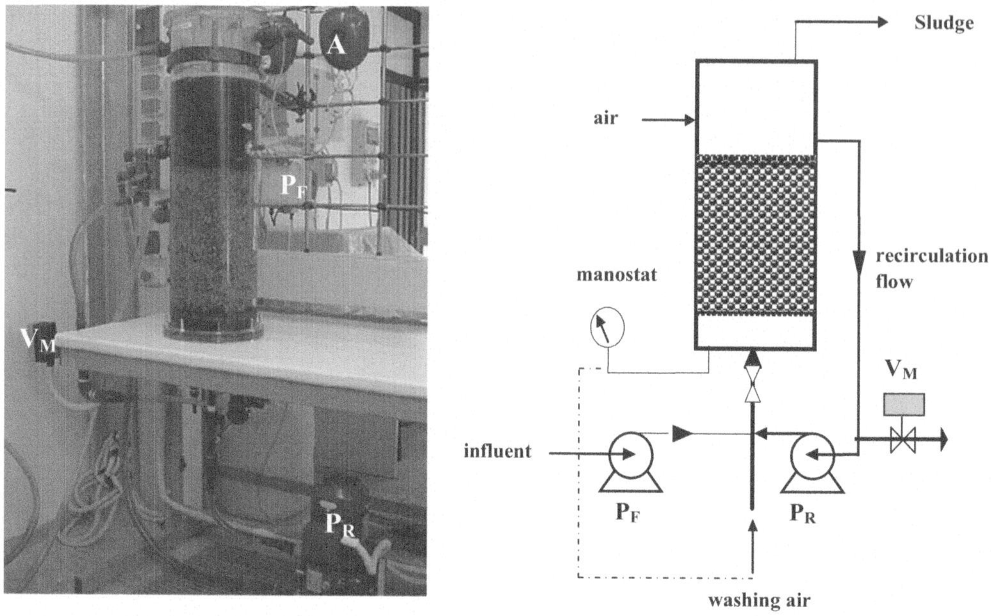

Figure 1.79 Sketch and photo of SBBGR system

1.5.2.1.3 Ozonation unit

For the chemical oxidation unit (Figure 1.80) pump P_O passed the biologically treated wastewater from the SBBGR system (70 L/h flowrate) through a glass bubble column (diameter: 85 mm; height: 90 cm; working volume: 5 L) which was supplied with a porous distributor of ozone produced from pure oxygen by means of an ozone generator (model: Modular 8HC, WEDECO, Germany). The reactor was supplied with a residual ozone destroyer and UV ozone meter (ozone analyzer BMT 964, WEDECO, Germany).

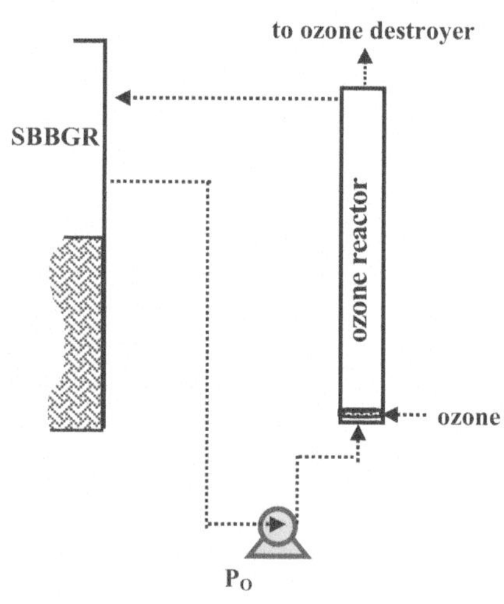

Figure 1.80 Sketch of ozonation unit.

1.5.2.2 Plant operative plan

The experimental activities were divided into four periods (Table 1.14):

- During the first period (A), clarifloculator and ozonation units were completely deactivated so that the wastewater was only treated biologically by the SBBGR system. In particular, during this period, careful attention was paid to SBBGR start-up by appropriately diluting the wastewater with tap water in order to permit the acclimatisation of microorganisms to salts and inhibitory compounds present in the leachate; this dilution ratio was progressively reduced from 25 to 0. During this period, the SBBGR system operation was cyclic (Figure 1.81), consisting of three consecutive phases of filling, reaction and drawing. In the filling phase, a fixed volume of raw leachate was added (by pump P_F, Figure 1.79) to the liquid volume retained by the biological reactor from the previous treatment cycle. During reaction phase, the liquid in the reactor was continuously aerated and recycled (by pump $P_{R,}$, Figure 1.79) through the biomass supporting material. Finally, the treated wastewater was withdrawn from the reactor by gravity using the motorised valve V_M. The system was then ready to start a new treatment cycle.

- During the second period (B), the clariflocculator was activated and the SBBGR was fed with leachate which had been pre-treated for ammonia removal by magnesium ammonium phosphate precipitation. The supernatant of the pre-treatment unit was treated by SBBGR according to the treatment cycle shown in Figure 1.81.
- During period C, the ozonation unit was also activated for rendering the recalcitrant pollutants biodegradable. Therefore, the pre-treated leachate was treated according a treatment cycle consisting of four consecutive phases (Figure 1.82): filling, biological degradation, biological degradation + ozonation, and drawing. During the filling phase, a fixed volume of pre-treated leachate was added to the liquid volume retained by SBBGR from the previous treatment cycle. In the biological degradation phase, during which the ozonation unit remained switched off in order to permit the biological removal of biodegradable pollutants, the liquid in the reactor was continuously aerated and recycled through the biomass supporting material. In the biological degradation + ozonation phase, the biological treatment was enhanced by activating the ozone unit (i.e. by switching on pump Po and the ozone generator). During this phase, two liquid flows were simultaneously present in the plant: the first (operated by pump Po) was aimed at producing biodegradable compounds by means of the partial chemical oxidation of recalcitrant pollutants in the ozone reactor, while the second (operated by pump P_R) was aimed at removing these products in the biological unit. Finally, the treated wastewater was withdrawn by gravity from the biological reactor (by motorised valve, $V_{M,}$, Figure 1.79) and the plant was ready to start a new treatment cycle. Different operative conditions of the ozone unit were trialled in order to meet the in-force discharge limits.
- In particular, period C was split into five runs (1–5), each characterised by a different operative conditions of ozone unit as reported in Table 1.14. In runs 1–3 the influence of ozonation phase time on the plant performance was evaluated for the same applied ozone dose. During runs 3–5, the influence of applied ozone dose on the plant performance was evaluated at the previously optimized ozonation phase time.
- Finally, during period D, the pre-treatment unit was deactivated and then the chemically enhanced biological treatment unit was used for both carbon and nitrogen removal according to the treatment cycle shown in Figure 1.82. Period D was split in two runs. In run 1, the influent ammonia was increased stepwise by mixing pre-treated (for ammonia removal by precipitation via struvite) and raw leachate in order to enhance the growth of nitrifying microorganisms. COD concentration of the leachate was increased stepwise as well by adding an external carbon source (glucose) in order to permit the complete denitrification of the oxidised nitrogen. In run 2 the plant was fed with raw leachate.

Table 1.14 Operative conditions of SBBGR system and ozonation unit throughout the experimentation

Run	**Period A**	**Period B**	**Period C**					**Period D**	
			1	*2*	*3*	*4*	*5*	*1*	*2*
Duration (d)	0–200	201–340	341–400	401–460	461–520	521–580	581–640	641–840	841–930
Hydraulic loading (L_{inf}/d)	4.5	4.5	4.5	4.5	4.5	4.5	4.5	4.5 up to 1.5	1.5
Daily exchange volume ratio (%)	37.5	37.5	26.5	26.5	26.5	26.5	26.5	26.5 to 8.8	8.8
Daily treatment cycles	3	3	3	3	3	3	3	3	3
Cycle time (min)	480	480	480	480	480	480	480	480	480
Filling	10	10	10	10	10	10	10	10	10
Biological degradation	455	455	335	395	425	425	425	425	425
Biological degradation+ ozonation	0	0	120	60	30	30	30	30	30
Drawing phase	15	15	15	15	15	15	15	15	15
AD (mgO_3/L_{inf})	0	0	2000	2000	2000	1000	600	500	500
AM (mgO_3/min)	0	0	25	50	100	50	30	10	10
F_{gas} (L_{gas}/h)	0	0	25	40	60	60	60	60	60
$C_{O3\ IN}$ (mgO_3/L_{gas})	0	0	60	75	100	50	30	10	10

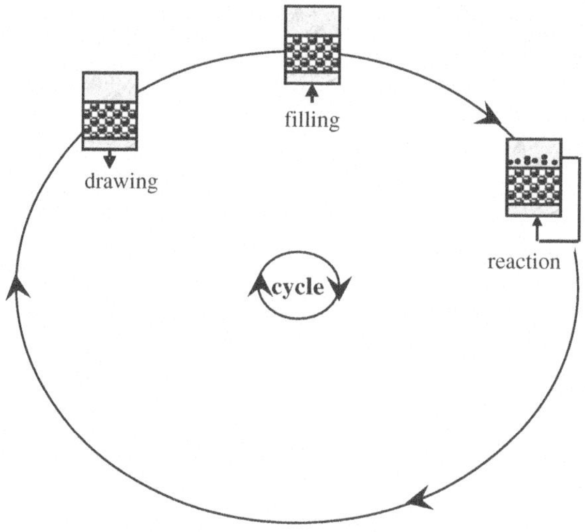

Figure 1.81 Treatment cycle of SBBGR system during periods A and B.

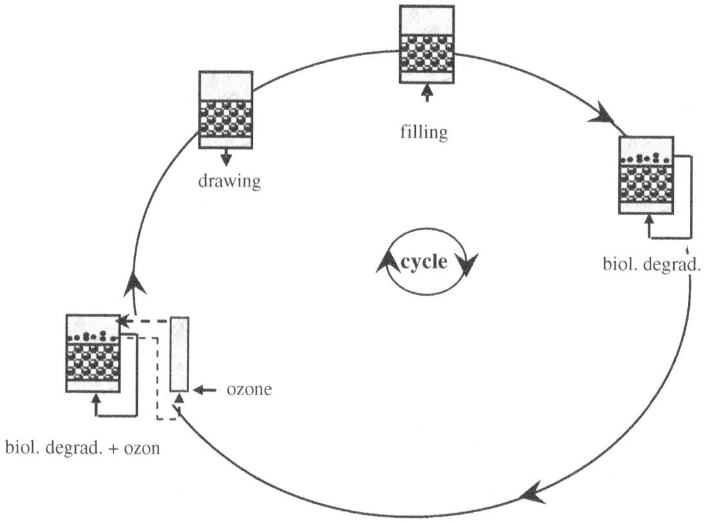

Figure 1.82 Treatment cycle of SBBGR integrated with ozonation during periods C and D.

1.5.2.3 Leachate composition

During periods A, B and C, a leachate (the composition of which is reported in Table 1.15) from a highly stabilized municipal landfill typical of low rainfall regions of southern Italy was used.

During the last period of experimentation (period D) the plant was fed with a middle stabilized municipal landfill leachate whose composition is reported in Table 1.16.

1.5.2.4 Analytical methods

COD (chemical oxygen demand), BOD_5 (biochemical oxygen demand), ammonia, TSS (total suspended solids), VSS (volatile suspended solids), P_{tot} (total phosphorous), surfactants, sulphates, and chlorides were determined according to standard methods (APHA, 1999). DOC (dissolved organic carbon) was measured by means of a carbon analyzer Shimadzu Co Japan (model 5050). The colour was measured by absorption at 660 nm. DO (dissolved oxygen) and pH were measured on-line by using selective probes. Metal concentrations were measured by inductively coupled plasma optical emission spectroscopy (ICP-OES) analysis, using an Optima 3000 (Perkin-Elmer).

Table 1.15 Composition range of the municipal landfill leachate used during period A, B and C of the experimentation

Parameter	Value range	Parameter	Value range
COD	8.0–10.8 g/L	Sulphates	4–6 g/L
BOD_5/COD	<0.15	Sodium	5–7 g/L
DOC	2.5–3.3 g/L	Potassium	1.5–2 g/L
NH_4-N	2.9–3.1 g/L	Magnesium	150–300 mg/L
pH	7.8–8.2	Pb	< 0.1 mg/L
P_{tot}	10–30 mg/L	Ni	2–4 mg/L
TSS	120–390 mg/L	Mn	0.1–0.4 mg/L
VSS	100–260 mg/L	Fe	5–7.5 mg/L
Chlorides	3.0–6.0 g/l	Zn	0.4–0.7 mg/L
Conductibility	20–30 mS/cm	Cu	0.04–0.2 mg/L

Table 1.16 Composition range of the municipal landfill leachate used during period D of the experimentation

Parameter	Value range	Parameter	Value range
COD	2.8–3.6 g/L	Sulphates	1.0–1.5 g/L
BOD_5/COD	0.15–0.2	Sodium	1.5–2.0 g/L
DOC	0.9–1.2 g/L	Potassium	1.2–1.6 g/L
NH_4-N	1.5–2.0 g/L	Magnesium	0.2–0.4 g/L
pH	7.8–8.3	Cr	<0.1 mg/L
P_{tot}	4–6 mg/L	Ni	0.5–1 mg/L
TSS	150–300 mg/L	Mn	<0.02 mg/L
VSS	120–230 mg/L	Fe	1–1.5 mg/L
Chlorides	3.0–4.0 g/l	Zn	<0.01 mg/L
Conductibility	16–22 mS/cm	Cu	0.01–0.2 mg/L

Applied (AM) and transferred (TM) ozone mass flow rates were calculated by the following equations:

$$AM = F_{gas} \times C_{O_3 IN} \tag{2}$$

$$TM(t) = F_{gas} \times (C_{O_3 IN} - C_{O_3 OUT}) \tag{3}$$

where F_{gas} (L_{gas}/min), $C_{O3\ IN}$ (mgO_3/L_{gas}) and $C_{O3\ OUT}$ (mgO_3/L_{gas}) are the flow rate and ozone concentration at the entry and exit of the ozone reactor at time (t) of ozonation phase, respectively. The inlet ozone concentration ($C_{O3\ IN}$) remains constant over time of the "biological and chemical degradation" phase for each run while $C_{O3\ OUT}$ changes. F_{gas} was measured by means of the ozone generator flow meter while $C_{O3\ IN}$ and $C_{O3\ OUT}$ by an UV analyzer (ozone analyzer BMT 964, WEDECO, Germany).

Applied (AD) and transferred (TD) ozone dose (i.e. the quantity of ozone dosed and transferred per litre of influent, respectively) was calculated by the following equations:

$$AD = \frac{AM \times t_{O_3}}{L_{inf}} \tag{4}$$

$$TD = \frac{\int_0^t TM(t)\,dt}{L_{inf}} \tag{5}$$

where t_{O3} (min) is the duration of ozonation phase (i.e. the duration of "biological degradation+ozonation" phase of Figure 1.80).

Ozone transfer efficiency (TE) was calculated by the following equation:

$$TE = \frac{TD}{AD} \times 100 \tag{6}$$

Gas holdup in the ozone reactor was calculated by measuring the increase in liquid height after introducing the ozone gas.

Ozone factor (f_{O3}) was calculated by means of the following equation:

$$f_{O_3} = \frac{TD}{(COD_{eff\,of\,period\,B} - COD_{eff\,of\,period\,C})} \tag{7}$$

SBBGR head loss was measured by means of a manostat on the bottom of SBBGR bed.

The specific sludge production (SSP) was calculated by dividing the TSS leaving the system (i.e. TSS discharged with the effluen + TSS removed during the washing operation) by the amount of COD removed in the period between two washing operations.

Biomass concentration was evaluated in terms of TSS and VSS, on representative samples of biomass detached from a bed volume previously determined by counting the carrier elements and relating them to carrier number in 1 L of bed (i.e. 1023). In particular, a bed volume was withdrawn from the reactor by a particular ladle and suspended in tap water. The biomass was completely separated from the carrier material by shaking with glass rod many times and then measured as TSS and VSS.

Sludge age (θ_c) was calculated by the following equation:

$$\theta_c = \frac{\text{biomass concentration } (gTSS/L_{bed}) \times \text{bed volume } (L_{bed})}{\text{SSP } (gTSS/gCOD_{removed}) \times \text{hydraulic loading } (L/d) \times COD_{removed} \ (gCOD_{removed}/L)} \tag{8}$$

The microbial composition and structure of the SBBGR biomass was evaluated by Fluorescence *In Situ* Hybridization (FISH). Biomass samples were periodically taken during the whole operation of SBBGR system and fixed in paraformaldehyde according to the procedure described by Amann (1995). Oligonucleotide probes specific for *Alpha*, *Beta*- and *Gamma-proteobacteria* (ALF968, Bet-42a and Gam-42a probes, respectively), most *Flavobacteria* (CF319a probe), *Actinobacteria* (HGC69A probe), *Firmicutes* (LGC354mix probes), *Chloroflexi* (CFX1223 probe), *Deltaproteobacteria* (Delta495mix probes), *Planctomycetales* (Pla46 probe), *Thauera/Azoarcus* group (THAU646 and AZO644 probes) and for *Archaea* domain (ARC915 probe) were used. Ammonia (Nso1225 probe) and nitrite oxidisers (Ntspa662, NSR1156 and NIT3 probes) were also monitored by FISH. All the hybridizations with group specific probes were carried out simultaneously with probes EUB338, EUB338-II and EUB338-III combined in a mixture (EUB338mix) for the detection of most bacteria and with DAPI staining for quantifying the total number of cells. To estimate the amount of bacteria out of the total number of cells, DAPI was directly added to the hybridization buffer at a final concentration of 1 mg/ml. Twenty different fields of view were selected randomly for each reported population measurements. Area measurements with all probes were reported as a proportion of the area of total cells in each field according to DAPI staining. Area measurements were performed on TIFF images using ImageJ software package (National Institutes of Health, USA). The area of all pixels above a manually determined minimum pixel intensity was measured from the greyscale image of each probe in a field. The upper greyscale pixel intensity value remained constant at the maximum greyscale value of 255, thus allowing for the varying fluorescent signals from different populations of cells to be measured above the highest possible background and non-specific levels of fluorescence.

Details on oligonucleotide probes are available at probeBase (Loy *et al.*, 2007). All the probes were synthesised with 5′-FITC and -Cy3 labels and purchased from MWG AG Biotech (Germany). Slides were examined under an epifluorescence microscope (Olympus BX51) and CLSM (Leica SP5). Three different EPS staining methods for the main EPS components were tested on SBBGR biomass: carbohydrate-containing EPS staining (mainly a-glucans) according to the method described in Strathmann *et al.* (2002); carbohydrate-containing EPS staining (mainly b-glucans) according the method described in De Beer *et al.* (1996); amino-sugars and proteins containing EPS using a modified form of the method described in Schmid *et al.* (2003).

Activated sludge from a municipal wastewater treatment plant was used as inoculum for the start-up of the SBBGR reactor. *Betaproteobacteria* were estimated by FISH as 25% of the total DAPI stained cells (18% of *Betaproteobacteria* were ammonia oxidisers) while *alphaproteobacteria* and *gammaproteobacteria* were 12.5% and 7%, respectively. Additionally, cells belonging to *Actinomycetes* (5.5%, HGC69a probe) and to CFB group (5.7%, CF319 probe) as well as nitrite oxidisers (2%, Ntspa662, NSR1156 and NIT3 probes) and Archaea (2%) were detected by FISH.

1.5.3 Results

1.5.3.1 Leachate treatment by SBBGR (Period A)

As reported in the Section 1.5.2.2, during the first period (period A), pre-treatment and ozonation units were not used so the wastewater was treated biologically only by the SBBGR system. In particular, during the SBBGR start-up, the leachate was appropriately diluted with tap water in order to permit the acclimatisation of microorganisms to the inhibitory compounds present in the leachate. The dilution ratio was progressively reduced to zero at the end of the start-up when acclimatisation had been achieved.

Figures 1.83, 1.84 and 1.85 show the concentration/time profiles of COD, TSS and NH_4-N, respectively, as well as their relative removal efficiencies recorded during the start-up period. Regarding COD, the data in Figures 1.83a and b show an increase in removal efficiency (from 30 to 50%) with the decrease of dilution ratio, up to day 60, at least. This result can be ascribed to acclimatisation improvement of the micro-organisms to inhibitory compounds present in the leachate during the start-up period. Furthermore, it can be observed that from day 60 the removal efficiency continuously decreased, reaching a value of 30% by the end of the start-up period. This reduction could have been due to the increase in salinity as the dilution ratio was decreased (since above a certain value, salinity negatively influences the metabolism of microorganisms) or to the toxic effect of ammonia (see below). However, Figure 1.83b indicates that when undiluted leachate was fed into the SBBGR, the concentration in the effluent was still too high (about 6 g/L).

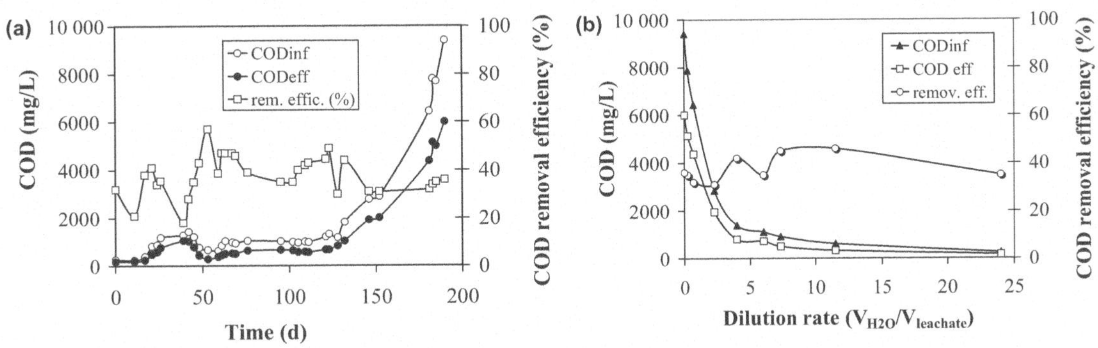

Figure 1.83 COD concentrations in the influent and effluent as well as its removal efficiency as functions of time (a) and dilution ratio (b).

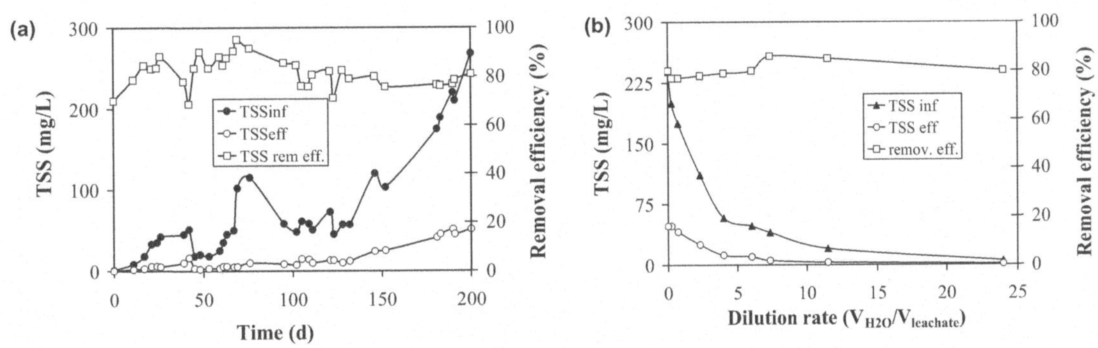

Figure 1.84 TSS concentrations in the influent and effluent as well as its removal efficiency as functions of time (a) and dilution ratio (b).

As for total suspended solids, the profiles in Figures 1.84a and b confirm the results obtained for COD. In fact, an increase of TSS removal efficiency (up to 90%) was obtained up to day 60–70 (i.e. up to a dilution ratio of 7). From day 70, the removal efficiency continuously decreased, reaching a value of 80% (with a residual concentration in the effluent of 60 mg/L) when the undiluted leachate was fed into the SBBGR.

The ammonia removal was almost complete up to a dilution rate of 12 on day 20 (Figures 1.85a and b). When the dilution ratio was decreased from 12 down to 6, a reduction in ammonia removal efficiency was recorded. In order to confirm this result, the dilution ratio was again increased from 6 to 12 and the complete removal of ammonia was recovered. As this decrease in ammonia removal could have been due to the rapid change of the dilution rate, from day 60–90 the dilution rate was changed slowly. Nevertheless, the ammonia removal efficiency still decreased.

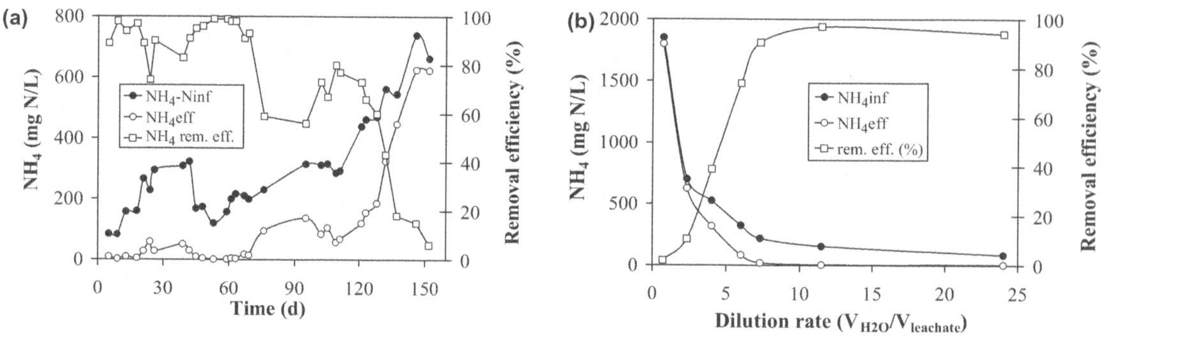

Figure 1.85 NH_4 concentrations in the influent and effluent as well as its removal efficiency as functions of time (a) and dilution ratio (b).

In order to enhance the ammonia removal efficiency, from day 90–110, the hydraulic residence time of the reactor was doubled by reducing the hydraulic load to the SBBGR. This resulted only in a slight increase in the ammonia removal efficiency. However, the ammonia removal efficiency decreased when the dilution ratio was reduced again.

These results indicate the inhibition of the nitrification process was probably due to the presence of inhibitory compounds in the leachates, such as ammonia. Therefore, the inhibition effect of ammonia concentration on its removal efficiency was investigated. Figure 1.86 shows ammonia concentration profiles in the reactor at the end of filling phase (i.e. the maximum concentration of ammonia in the reactor) and in the effluent as well as its removal efficiency as a function of time. The ammonia concentration at the end of filling phase greatly influenced the ammonia removal efficiency. When the ammonia concentration in the reactor exceeded 150–200 mg/L, the inhibition of nitrification took place. Taking into account the concentration of ammonia in the leachate (i.e. around 3 g/L, Table 1.15), a hydraulic residence time as high as 15–20 days was then required with consequent large reaction volumes. Therefore, a physic-chemical treatment of leachate for removing ammonia would appear to be a good solution, at least for this type of leachate.

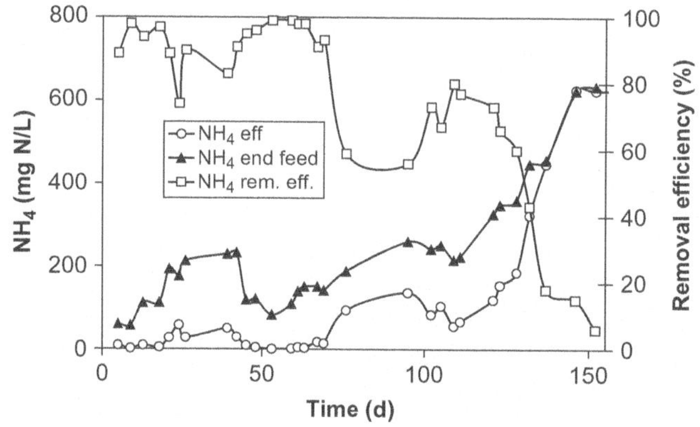

Figure 1.86 End filling phase and effluent ammonia concentrations together its removal efficiency as functions of time.

Biomass samples were collected during the reactor operation at different values of leachate dilution ratio (0, 2 and 5 VH_2O/Vleachate, respectively). Figure 1.87 shows the results of microbial characterization of the SBBGR biomass within period A by applying a large set of FISH probes for detecting the main phyla within the Bacteria domain. The initial higher dilution ratio allowed the acclimatisation of microorganisms to inhibitory compounds present in the leachate. Nitrifiers (mainly ammonia-oxidizers identified by Nso1225 and Nso190 probes) were initially enriched in the biomass with respect to the activated sludge inoculum (from 4.5% up to 15% ammonia oxidising bacteria/total cells retrieved during reactor operation at dilution ratio of 5) due most likely to the high ammonia/COD ratio present in the influent.

A change in biomass composition was observed in the sample collected with a lower dilution ratio. The strong reduction of *Betaproteobacteria* was mainly due to the los of ammonia-oxidizers in the system. Furthermore, Gram positive bacteria became the dominant microbial components (mainly *Actinomycetes* and *Firmicutes* identified by HGC69A and LGC69A probes, respectively) able to grow on a wastewater with high salinity and the presence of inhibitory compounds such as high ammonia.

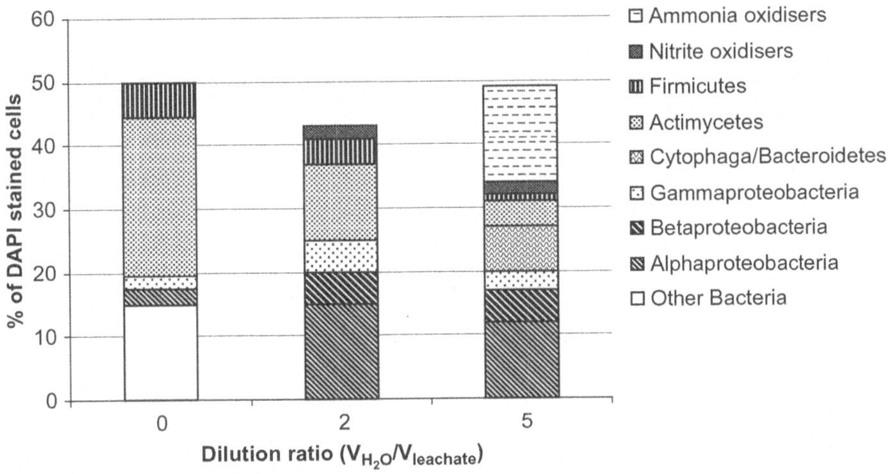

Figure 1.87 Main bacterial components of SBBGR biomass at different values of leachate dilution ratio during period A.

However, despite the observed reduction of bacterial biodiversity, highlighted by the analysis with phylum specific probes, the SBBGR biomass maintained a high *Bacteria*/total cell ratio (EUB338mix/DAPI stained cells), meaning only a change in bacteria composition occurred whereas the active fraction of the biomass was stable in the system during the reactor operation at lower dilution ratios. When the system was fed with undiluted leachate (i.e. $V_{H_2O}/V_{leachate} = 0$), an increase in the number of total *Bacteria* detected by FISH was observed (up to 50% of the total cells, Figure 1.87).

1.5.3.2 Leachate treatment by struvite precipitation followed by SBBGR (period B)

As reported in Section 1.5.2.2, during Period B, the pre-treatment unit was activated and the SBBGR was fed with leachate which had been pre-treated for ammonia removal by magnesium ammonium phosphate precipitation. Preliminary tests were carried out at bench scale in order to optimize the precipitation process, taking into consideration the successive biological treatment. In particular, four $Mg:NH_4:PO_4$ molar ratios were investigated (1.1.1, 1.5:1:1, 2:1:1 and 2.5:1:1.1). Table 1.17 reports the values of the selected parameters for each molar ratio investigated. Using the stoichiometric molar ratio, an ammonia removal efficiency of 66.5% was obtained, with a residual concentration in the supernatant of 870 mg/L. This limited removal efficiency value, however, could be ascribed to the low dissolution of MgO in the leachate leading to low MAP precipitation efficiency. This was confirmed by the Mg data (also reported in Table 1.17) which showed that at the stoichiometric molar ratio, the magnesium concentration measured in the supernatant was very low (<3 mg/L), while the phosphorus concentration was as high as 2274 mg/L, indicating a deficiency of magnesium only. By doubling the magnesium dose (by changing the molar ratio from 1:1:1 to 2:1:1), an increase in ammonia removal efficiency was achieved, increasing from 66.5% to 98.6% efficiency. Moreover, at this molar ratio range, the residual magnesium concentration in the supernatant remained very low, indicating that the efficiency of the struvite precipitation depended on the amount of MgO added, at least up to a molar ratio of 2:1:1. It is interesting to observe that at this molar ratio, all the MgO and H_3PO_4 which had been added was consumed during the precipitation process. This was very important in cases where enhanced phosphorus removal in the subsequent biological treatment was not foreseen.

Table 1.17 also compares the theoretical and experimental amounts of dried struvite produced per m^3 of leachate at each investigated molar ratio value. Theoretical struvite is the amount of precipitate that one would expect on the basis of ammonia removals, whereas the experimental struvite is the amount of precipitate actually weighed during the tests. From Table 1.17, the measured amount of struvite produced increased with an increasing molar ratio (from 31.4 kg at the stoichiometric molar ratio to 60.6 kg/m^3 of leachate at a molar ratio of 2.5:1:1.1) because of the increased quantity of removed ammonia. The difference between the theoretical and experimental struvite values can be ascribed to the presence in the latter of the magnesium oxide overdosed. As for the amount of chemicals consumed, the data shows opposite trends for the MgO and NaOH consumptions. In particular, an increase in the magnesium oxide consumed results in a decrease in NaOH consumed due to the alkaline properties of magnesium oxide, which consequently reduced the NaOH needed for fixing the pH at 9.0. However, at molar ratio values higher than 2:1:1, the saving in NaOH dosing becomes smaller in proportion compared to the increase in the costly MgO added.

On the basis of the data reported in Table 1.17, and considering that the residual ammonia concentration in the supernatant at a molar ratio of 2:1:1 (i.e. 35 mg/L) is compatible with a successive biological treatment, this molar ratio was chosen for MAP precipitation. In order to evaluate its possible utilisation as a fertilizer, the chemical composition of the struvite obtained at the 2:1:1 molar ratio was performed (Table 1.18). The struvite obtained had a composition not far from the theoretical one (i.e. 9.9% Mg, 5.7% N and 12.6 P). The higher magnesium content confirms the presence of MgO in the precipitate due to the limited solubility in water of this compound. The struvite formed had a low metal content, an important aspect if the struvite obtained were to be used as a fertilizer. Moreover, by comparing these values with those typical of Italian soils (Leita *et al.*, 1998), it can be observed that the struvite produced in this study was characterized by a metal content much lower than that of raw soil. Therefore, there would be no need for concern regarding heavy metal contamination if the struvite was applied to the soil.

Table 1.17 Precipitation process performance as a function of the $Mg:NH_4:PO_4$ molar ratio.

Parameter	$Mg:NH_4:PO_4$ molar ratio			
	1:1:1	**1.5:1:1**	**2:1:1**	**2.5:1:1.1**
Residual NH_4^+ concentration (mg N/L)	870	295	35	20
NH_4^+ removal efficiency (%)	66.5	88.6	98.6	99.2
Residual Mg^{2+} concentration (mg/L)	2.9	3.5	3.8	528
Residual PO_4^{3-} concentration (mgP/L)	2274	640	8	24
Theoretical amount of struvite (kg/$m^3_{leachate}$)	31	39.9	43.3	45.5
Experimental amount of struvite (kg/$m^3_{leachate}$)	31.4	48.1	53.6	60.6
MgO consumption (kg/$m^3_{leachate}$)	7.5	11.2	15	18.7
H_3PO_4 consumption (L/$m^3_{leachate}$)	15.4	15.4	15.4	16.9
NaOH consumption (L/$m^3_{leachate}$)	29.3	20.6	13.9	11.8

Table 1.18 Chemical composition of struvite produced at a molar ratio of 2:1:1.

Parameter	Value (mg/kg)	Parameter	Value (mg/kg)
NH_4^+-N	46 200	Cadmium	<dl
PO_4^{3-}-P	107 200	Copper	<dl
Magnesium	168 300	Zinc	7
Potassium	1100	Nickel	45
Calcium	550	Lead	<dl
Chromium	15	Manganese	80

dl, *detection limit*

On the basis of the results reported above, ammonia precipitation at a $Mg:NH_4:PO_4$ molar ratio of 2:1:1 was transferred to the clarifloculator of 50 L. The data in Table 1.19, reporting the composition range of the supernatant, confirmed the high ammonia removal efficiencies obtained at bench scale (i.e. 98–99%) with residual concentrations lower than 45 mg/L. In addition, by comparing the COD ranges of Tables 15 (raw leachate composition) and 19, it is worth noting that COD was not significantly removed during MAP precipitation. Following MAP precipitation, the pH of supernatant was adjusted to a value compatible for the biological treatment (i.e. 7.5–8.0) by adding a solution of concentrated sulphuric acid.

Figure 1.88 shows the effluent COD concentration profile of the SBBGR during period B. When the SBBGR was fed with the pre-treated leachate, the COD concentration in the effluent progressively decreased from around 6 g/L (i.e. the concentration when the SBBGR was fed with raw leachate) down to 4.5 g/L. Therefore, the ammonia concentration in the leachate greatly affected not only nitrifying bacteria but also the heterotrophic biomass.

The performance of the SBBGR (average values) with an organic loading rate of 2.0–3.0 g COD/$L_{bed} \cdot d$ (which corresponds to HRT of 2.7 days) is summarised in Table 1.20. In particular, this table reports COD, DOC, BOD_5, TSS, ammonia and surfactants concentrations in the SBBGR influent and effluent, as well as their removal efficiencies. The COD and DOC removal efficiencies were somewhat low (48% and 35%, respectively) with very high residual concentrations in the effluent (4500 and 1750 mg/L, respectively). This was due to the presence in the leachate of a fraction of recalcitrant pollutants, as confirmed by BOD_5 values (the removal of biodegradable matter was nearly complete). The presence of recalcitrant organics was also confirmed by the COD profile shown in Figure 1.89, clearly indicating the residual COD in a typical treatment cycle did not decrease any further during the last part of the cycle.

Table 1.19 Composition range of the pre-treated leachate for ammonia removal by struvite

Parameter	Value range	Parameter	Value range
COD	7.5–9.8 g/L	Sulphates	3.5–5.0 g/L
BOD_5/COD	<0.15	Sodium	4.5–5.8 g/L
DOC	2.2–2.9 g/L	Potassium	1.5–2 g/L
NH_4-N	25–45 mg/L	Magnesium	0–15 mg/L
pH	7.5–8.0	Pb	<0.1 mg/L
P_{tot}	6–10 mg/L	Ni	2–4 mg/L
TSS	110–290 mg/L	Mn	0.1–0.2 mg/L
VSS	80–250 mg/L	Fe	2.0–4.0 mg/L
Surfactants	60–80 mg/L	Zn	0.2–0.5 mg/L
Chlorides	3.0–6.0 g/l	Cu	0.04–0.2 mg/L
Conductibility	25–35 mS/cm	Cr	0.1–0.6 mg/L

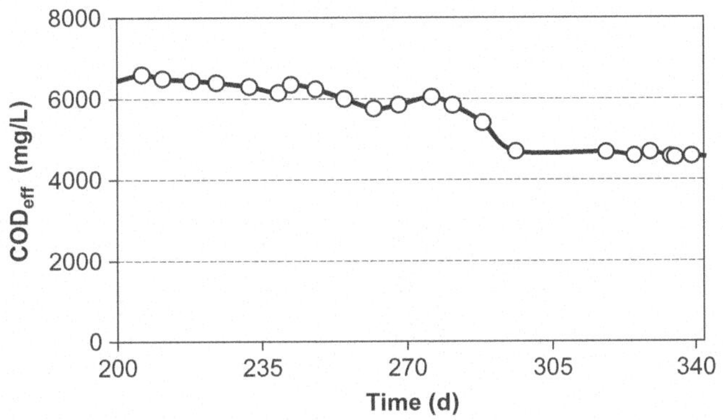

Figure 1.88 COD concentration in the effluent of SBBGR during period B

Table 1.20 Average SBBGR performances during period B of experimentation

Parameter		Value
COD	effluent [mg/L]	4500±300
	removal efficiency [%]	48±5
DOC	effluent [mg/L]	1750±125
	removal efficiency [%]	35±4
BOD5	effluent [mg/L]	15±12
	removal efficiency [%]	98±1
TSS	effluent [mg/L]	65±20
	removal efficiency [%]	85±4
NH4	effluent [mg N/L]	15±5
	removal efficiency [%]	60±2
Surfactants	effluent [mg/L]	10±5
	removal efficiency [%]	86±3
Colour removal	[%]	4±2

The microscopic analysis of the SBBGR biomass after the introduction of the struvite pre-treatment showed a strong change of the microbial composition (Figure 1.90). With the removal of the toxic ammonia concentrations, the previous Gram positive bacteria disappeared and a shift to loosely aggregated bacilli belonging to Gammaproteobacteria was observed. The latter group increased up to 52% of the total cells in the last sample analysed within period B (sample taken at day 340), with only a marginal presence of a few other bacterial taxa. As expected, due to the low ammonia concentration in pre-treated leachate, no ammonia oxidisers were retrieved by FISH in the biomass. The application of staining techniques for analysing the presence and the quality of the EPS showed a slight presence of EPS material, mainly constituted by amino-sugars. In accordance with this finding, a biomass aggregation was microscopically visible for the first time since the start up of the SBBGR operation.

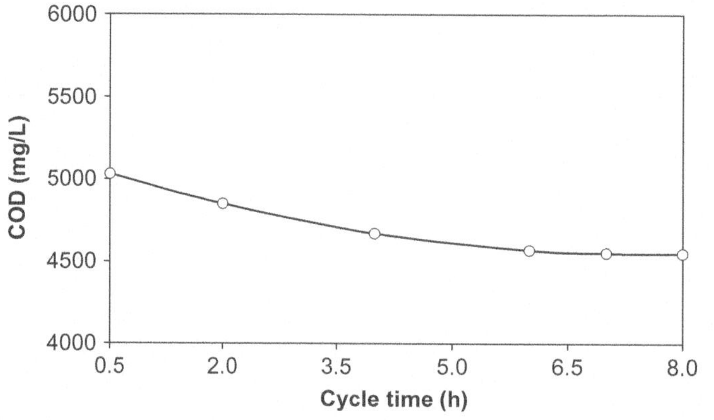

Figure 1.89 COD concentration profile during a typical treatment cycle of the SBBGR in period B (the value 0.5 on the x-axis corresponds to the end of the filling phase)

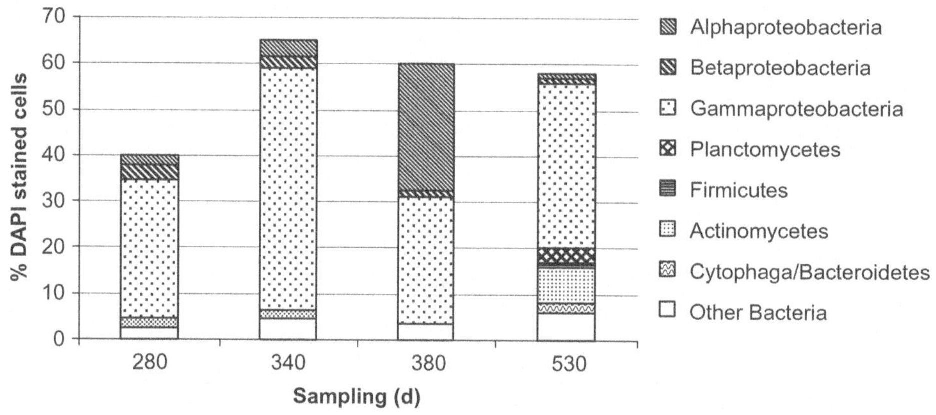

Figure 1.90 Microbial composition of the SBBGR biomass within operation period B (samples: 280 and 340 days) and period C (samples: 380 and 530 d). Cells hybridised with FISH probes are estimated as a fraction of total DAPI stained cells

1.5.3.3 Leachate treatment by struvite precipitation followed by SBBGR integrated with ozonation (period C)

In order to reduce the residual COD, DOC, surfactant concentration and colour, during Period C the biological treatment was chemically enhanced by activating the ozonation unit so that the supernatant from the MAP precipitation was treated according to the treatment cycle shown in Figure 1.16.

Two operative parameters must be selected when the ozonation is used as in Figure 1.16. The first is the point of the treatment cycle where the ozonation phase is inserted. Basically, an ozonation phase should be inserted after the complete removal of the biodegradable organic content in order to save the costly oxidant. Therefore, on the basis of the COD profile shown in Figure 1.89, the ozone unit should be activated not before the sixth hour of the treatment cycle.

The second parameter is the quantity of ozone to be used (AD). Equation (4) shows that the same amount of ozone can be dosed over a long or short time, t_{O3}, by using a low or high ozone mass flow rate, AM, respectively. Therefore, the ozonation phase time should be optimized before testing different AD values. In runs 1, 2 and 3, the influence of the ozonation phase time ("biological degradation + ozonation" phase; see Figure 1.82) on the plant performance was evaluated by using the same AD. The results (in terms of average values) are reported in Table 1.21.

The effluent COD and DOC concentrations decreased with a shortening the ozonation phase time (achieved with an increase of the AM). The low values of ozone factor (<1) indicate the ozone was mainly used to render the recalcitrant compounds biodegradable which were then removed biologically. Furthermore, it is worth noting that by using the same AD (i.e. 2,000 mgO_3/L_{inf}), the effluent COD concentration can be reduced from 545 mg/L (run 1) to 160 mg/L (run 3) by increasing the AM by four times (i.e. from 25 mgO_3/min in run 1 to 100 mgO_3/min in run 3; see

Table 1.14). The ozone factor values reported in Table 1.21, ranging from 0.3–0.4, are much lower than those found by Ried *et al.* (2007) for a combined biological and ozonation processes (falling in the range 2.5–1.1), thus showing a greater synergistic effect between ozone and biological degradation in the SBBGR system.

Table 1.21 Average plant performances during run 1, 2 and 3 of period C

Parameter		**Run**		
		1	**2**	**3**
AD	[mgO_3/L_{inf}]	2000	2000	2000
"Biol. degr. + O_3" phase time	[min]	120	60	30
COD	effluent [mg/L]	545±25	460±20	160±10
	removal efficiency [%]	91±1	94±1	98±1
DOC	effluent [mg/L]	395±12	370±10	141±8
	removal efficiency [%]	82±1	85±1	94±1
TSS	effluent [mg/L]	40±10	35±10	20±10
	removal efficiency [%]	88±2	90±2	94±2
Surfactants	effluent [mg/L]	4±1	3±1	1±0.5
	removal efficiency [%]	94±1	95±1	98±1
Colour removal	[%]	92±2	93±2	98±1
Ozone factor (f_{O3})		0.31	0.35	0.38

The improvement of plant performance in the runs 1-3 can be explained by considering the ozone concentration profiles at the entry and exit of ozone reactor, measured during the ozonation time (Figure 1.91). Applying equation (3) to these profiles it is possible to calculate for the three runs the TM (i.e. the transferred ozone mass flow rate) profiles shown in Figure 1.92. By applying equation (5) to the TM profiles, the TD values (i.e. the transferred ozone dose) can be calculated, which were 1211, 1420 and 1633 mg/L_{inf} for runs 1, 2 and 3, respectively. Finally, applying equation (6) to the TD values, the TE (i.e. the ozone transfer efficiency) values obtained for runs 1, 2 and 3 were 61%, 71% and 81%, respectively. Therefore, these results clearly indicate that the decrease of COD and DOC concentration in the effluent of runs 1 to 3 was due to the higher ozone transfer efficiency. In fact, it is well known that the rate-limiting step of the ozonation process is the low gas–liquid mass transfer rate due to the low solubility of ozone (Zhou and Smith, 2000). Therefore, the process performance can be greatly affected by the ozone transferred from gas to liquid phase.

The increase of TD (and thus TE) recorded in the runs can be ascribed to the higher gas flow rate used, F_{gas} (25 L/h in run 1 to 60 L/h in run 3; see Table 1.14), which leads to an increase of the gas/liquid interfacial surface area, as confirmed by the gas holdup measured in the ozone reactor, a parameter related to bubble surface (Finch *et al.*, 2000; Zhou and Smith, 2000). The gas holdup increased linearly as the gas flow rate increased (35 mm at 25 L/h to 100 mm at 60 L/h) in agreement with the results reported by Roustan *et al.* (1996).

The benefits produced by the ozone unit can be appreciated by looking at the COD and DOC profiles recorded during a typical treatment cycle of run 1, 2 and 3 (Figure 1.93). These profiles clearly show a sharp reduction of COD and DOC during the "biological degradation + ozonation" phase. This reduction must be ascribed to the ozone effect since the concentration of these parameters did not decrease any further during the last part of the previous phase (i.e. during "biological degradation" phase). Furthermore, this reduction increases as ozonation time decreases in accordance with the TD and TE trends.

Therefore, on the basis of the results reported above, at the same applied ozone dose, the plant performance can be improved by shortening the ozonation phase time so as to use a high ozone mass flow rate. In particular, the data reported in Table 1.21 indicates that the plant was able to meet the COD discharge limit of 160 mg/L (i.e. the Italian limit for discharging into water bodies) when an AD and ozonation phase time of 2,000 mgO_3/L_{inf} and 30 min, respectively, were used (run 3).

Figure 1.94, showing effluent samples from periods B and C, which clearly reflects visually the benefit yielded by the ozonation process. The sample from period C appeared more like a tap water sample.

As reported in Section 1.5.2.2 (see also Table 1.14), during runs 3, 4 and 5, the influence of applied ozone dose (AD) on the plant performance was evaluated by using the same previously optimized ozonation phase time (30 min). The results obtained during these runs are reported in Table 1.22. The data clearly indicates that the plant performance deteriorates as the AD decreases. The ozone factor also proportionally decreases with a decreasing AD, indicating an improvement of ozone efficiency. This can be partially attributed to the higher ozone transfer efficiency.

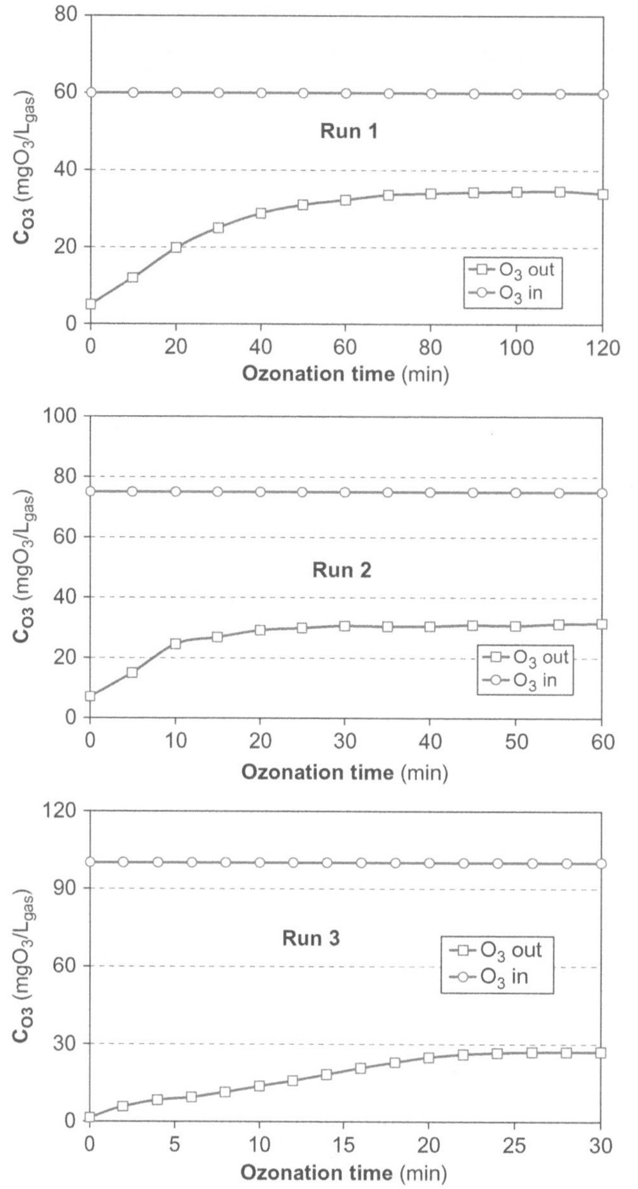

Figure 1.91 Profiles of ozone concentration at the entry and exit of the ozone reactor, measured during the "biological degradation + ozonation" phase of run 1, 2 and 3 of period C

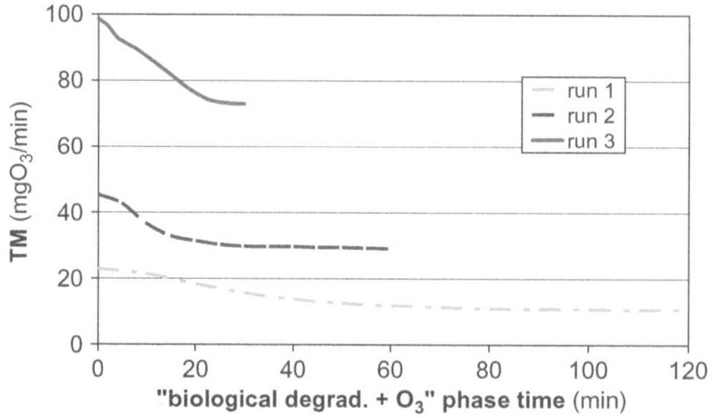

Figure 1.92 Transferred ozone mass flow (TM) profiles recorded during "biological degradation + ozonation" phase in run 1, 2 and 3 of period C

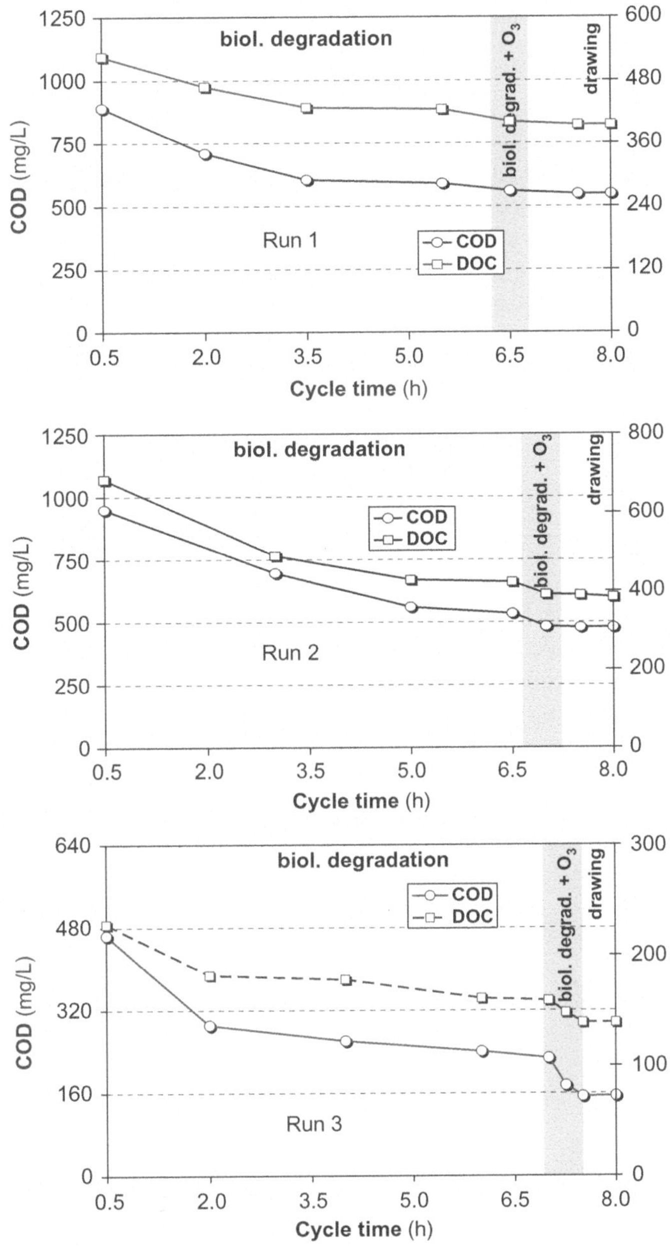

Figure 1.93 COD and DOC profiles during a typical treatment cycle in run 1, 2 and 3 of period C (the value 0.5 on the x-axis corresponds to the end of the filling phase)

Figure 1.94 Photograph of an effluent sample from period B and C (run 3)

Table 1.22 Average plant performances during run 3, 4 and 5 of period C

Parameter		Run		
		3	4	5
AD	[mgO_3/L_{inf}]	2000	1000	600
COD	effluent [mg/L]	160 ± 10	260 ± 10	500 ± 10
	removal efficiency [%]	98 ± 1	97 ± 1	93 ± 1
DOC	effluent [mg/L]	141 ± 8	195 ± 8	380 ± 5
	removal efficiency [%]	94 ± 1	92 ± 1	85 ± 1
TSS	effluent [mg/L]	20 ± 10	25 ± 7	30 ± 10
	removal efficiency [%]	94 ± 2	93 ± 1	91 ± 2
Surfactants	effluent [mg/L]	1 ± 0.5	1.5 ± 0.5	2 ± 0.5
	removal efficiency [%]	98 ± 1	97 ± 1	96 ± 1
Colour removal	[%]	98 ± 1	97 ± 1	95 ± 3
Ozone factor (f_{O3})		0.38	0.20	0.14

Applying equation (3) to the ozone concentration profiles at the entry and exit points of ozone reactor, the TM profiles (transferred ozone mass flow rate) can be calculated for runs 3, 4 and 5 (Figure 1.95). Then, by inserting the TM profiles into equation (5), the TD values (transferred ozone dose) obtained were 1 633 847 and 567 mgO_3/L_{inf} for runs 3, 4 and 5, respectively. Finally, applying equation (6), the TE (ozone transfer efficiency) values obtained for runs 3, 4 and 5 were 81%, 85% and 94%, respectively. These results clearly indicate that TE decreases as applied ozone dose AD increases. The ozone factor values reported in Table 1.22 (as well as those of Table 1.21) show an increasing synergistic effect between ozone and biological treatment as the effluent COD concentration decreases. In particular, it is interesting to note that in run 5, the ozone factor was 7 times lower than the lowest value reported by Ried *et al.* (2007).

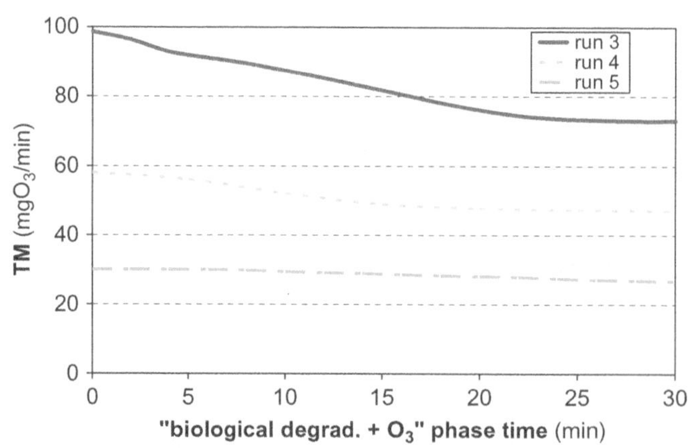

Figure 1.95 Transferred ozone mass flow (TM) profiles recorded during "biological degradation+ozonation" phase of a typical treatment cycle in run 3, 4 and 5 of period C

The results show that even at the minimum investigated AD (600 mgO_3/L_{inf} in run 5), the plant was able to achieve high removal efficiencies for all parameters, with residual concentrations lower than the Italian limits for discharging into the sewer system. The effects of the ozone unit on plant performance are evident in Figure 1.96, which shows COD and DOC profile during a typical treatment cycle of run 5.

The comparison of the plant performance recorded in run 1 (when AD and ozonation time of 2,000 mgO_3/L_{inf} and 120 min, respectively, were used) and 5 (when AD and ozonation time of 900 mg and 30 min, respectively, were used) highlights the role played by the operative conditions of the ozone unit. The performance of the plant in run 5 was better than that of run 1, although in the latter a considerably higher AD was used (>3 times higher).

Finally, the process was characterised by a very low sludge production (<0.05 g TSS/g $COD_{removed}$), ascribable to the long sludge age (θ_c >200 d) and sludge concentration (22–35 g TSS/L_{bed}; VSS/TSS: 0.4-0.5) of the biomass in the system. Figure 1.97 shows a photograph of a biomass sample drawn from the SBBGR bed. It is possible to see the presence of the granular biomass.

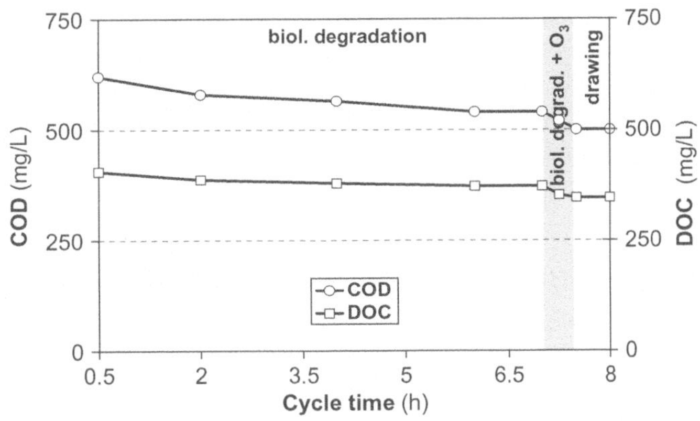

Figure 1.96 COD and DOC profiles during a typical treatment cycle in run 5 of period C (the value 0.5 on the x-axis corresponds to the end of the filling phase)

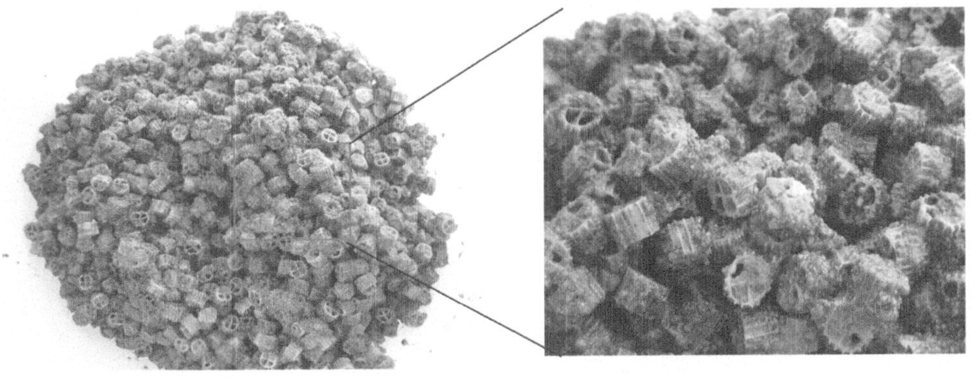

Figure 1.97 Photograph of the biomass sample drawn from the bed of the SBBGR system

After the activation of ozone unit (Period C), the microscopic analysis showed an increase of biomass aggregation due to granular biomass generation (granule average diameter of 2–3 mm, Figure 1.98a). This led to a substantial increase in the biomass concentration (from 18 to 30 g TSS/L.bed) and EPS content. In accordance with previous experiences (Di Iaconi *et al.*, 2010; De Sanctis *et al.*, 2010), the granules were generated by the biofilm particles detached from the surface of the filling material, following the thickening of the biofilm. No differences were observed by FISH in granule and biofilm compositions.

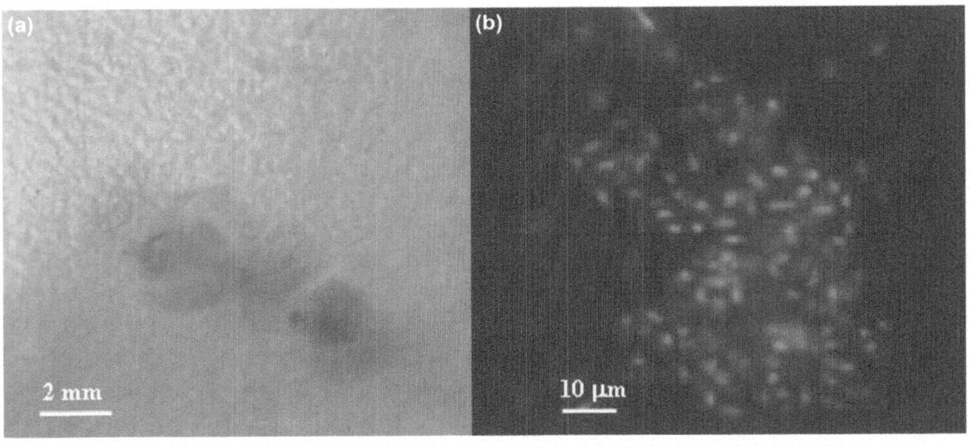

Figure 1.98 (a) Granules growing in SBBGR reactor during Period C; (b) Cells belonging to *Gammaproteobacteria* detected by FISH (GAM42A probe) in SBBGR granular biomass

As observed in Period B, the SBBGR biomass in this phase was also dominated by bacteria belonging to *Gammaproteobacteria* (Figure 1.98b) with the presence of a wider number of minor bacterial components mainly detected in the sample taken under steady state operating conditions (day 530 in Figure 1.90). Approximately 60% of the total DAPI stained cells were detectable by FISH in both SBBGR samples with bacterial probes (EUB338mix) and almost completely identified by the group specific probes utilized in this study. This value is similar to those previously estimated in a SBBGR granular biomass treating municipal and synthetic wastewater (Di Iaconi *et al.*, 2010; De Sanctis *et al.*, 2010) and to those typically recorded on activated sludge samples (Eschenhagen *et al.*, 2003) which are characterized by lower biomass concentrations (3–4 g TSS/L) and sludge ages (<30 d).

1.5.3.4 Leachate treatment by SBBGR integrated with ozonation (period D)

As reported in the Sections 1.5.2.2 and 1.5.2.3, during the last period of experimentation (period D), the plant was fed with a different municipal landfill leachate, characterised by a composition reported in Table 1.16. Moreover, the pre-treatment unit was deactivated and the chemically enhanced biological treatment unit was used for both carbon and nitrogen removal. Period D was split in two runs (1–2): the first referred to the biomass acclimatisation to the new leachate whereas the second to steady state conditions. In particular, in run 1, the influent ammonia concentration was increased stepwise by mixing pre-treated (for ammonia removal by precipitation via struvite) and raw leachate in order to enhance the growth of nitrifying microorganisms. COD concentration of the leachate was increased stepwise as well during this run by adding an external carbon source (glucose) in order to permit the complete denitrification of the oxidised nitrogen. Furthermore, the hydraulic retention time (HRT) of the plant was progressively increased from 3.7 d (the HRT in Period C) to 12 d, by reducing the hydraulic load (4.5 L/d down to 1.5 L/d) in accordance with the increase of ammonia concentration in the influent. The profiles of ammonia concentration in the influent and effluent of the plant as well as effluent oxidised nitrogen concentration throughout run 1 are shown in Figure 1.99.

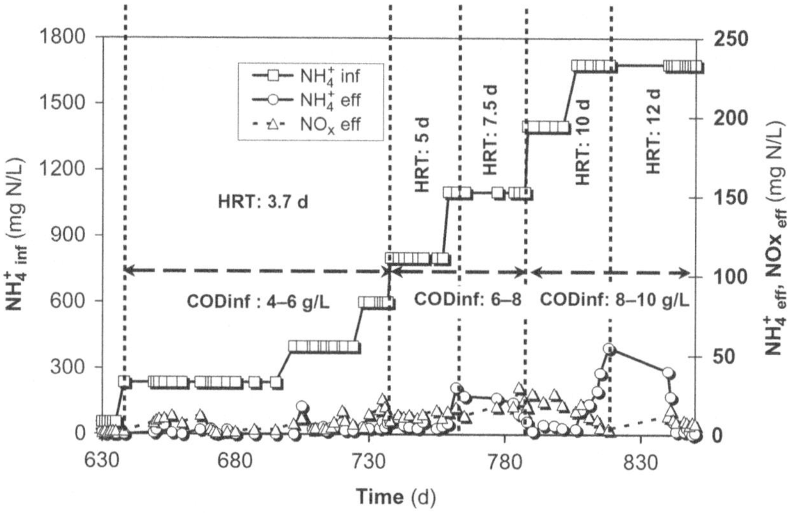

Figure 1.99 Ammonia and oxidised concentration profiles in the influent and effluent of the plant during run 1 of Period D of experimentation

The HRT of the plant greatly affected ammonia removal efficiency. When raw leachate was fed to the plant (at day 800), a sharp increase in effluent ammonia concentration was observed (>50 mg/L) with a HRT of 10 days. By increasing the HRT from 10 to 12 d, the effluent ammonia concentration decreased to less than 10 mg/L. Oxidised nitrogen was always lower than 20 mg/L, indicating the denitrification process was occurring.

Following the raw leachate feeding, the plant was operated for a further 3 months (run 2) in order to monitor the process effectiveness. The plant performance (in terms of average values) recorded during this run is reported in Table 1.23. In this run high removal efficiencies for COD, DOC, NH_4, NO_x, surfactants and colour (i.e., 95%, 93%, 86%, 99.5%, 99.5% and 98%, respectively) were achieved with residual concentrations lower than Italian limits for discharging into sewer system.

Figure 1.100 shows the COD and DOC profiles recorded during a typical treatment cycle of run 2. It is possible to distinguish 3 consecutive trends: the first of these is typical of biodegradable pollutant removal; the second trend shows that the biological treatment is complete, and the third, which coincides with the ozonation phase, highlights the benefits produced by using ozone. The COD balance of the "biological degradation + O_3" phase demonstrates that the COD removal (40 mg/L COD) was due mostly to biological degradation of the biodegradable organic products formed by ozonation. The TM profile was calculated by applying equation (3) to the ozone concentration profiles at the entry and exit points of ozone reactor during "biological degradation + O_3" phase. Then, inserting this profile into equation (5) provided the TD value of 330 mg of ozone transferred (and then consumed) per treatment cycle. Taking into account the working volume of the plant (17 L), the ozone consumed was 19.4 mgO_3/L. Therefore, this balance clearly shows that ozone was used also to render the recalcitrant compounds biodegradable that were then biologically removed.

Table 1.23 Average plant performance during run 2 of period D

Parameter		Value
COD	effluent [mg/L]	485 ± 20
	removal efficiency [%]	95 ± 2*
DOC	effluent [mg/L]	290 ± 12
	removal efficiency [%]	93 ± 1*
BOD_5	effluent [mg/L]	5 ± 5
	removal efficiency [%]	99 ± 1
TSS	effluent [mg/L]	30 ± 5
	removal efficiency [%]	86 ± 1
NH_4	effluent [mg N/L]	4 ± 3
	removal efficiency [%]	99.5 ± 0.5
NO_x	effluent [mg N/L]	6 ± 2
	removal efficiency [%]	99.5 ± 0.5
Surfactants	effluent [mg/L]	1.5 ± 0.5
	removal efficiency [%]	92 ± 2
Colour removal	[%]	98 ± 1

It includes also the removal of the external carbon source (i.e., glucose) added to the leachate.

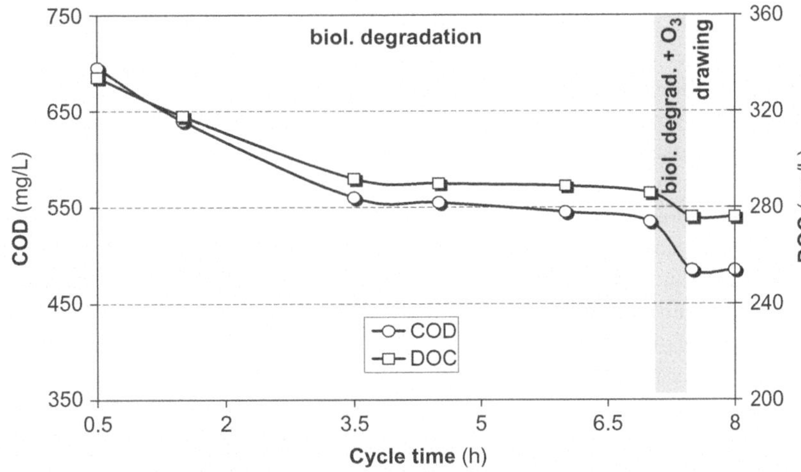

Figure 1.100 COD and DOC profiles during a typical treatment cycle of run 2 of Period D (the value 0.5 on the x-axis corresponds to the end of the filling phase)

Ammonia removal was observed to occur simultaneously with the COD removal and the removal rate was almost constant (Figure 1.101). The oxidised nitrogen remained low throughout the cycle ($<$10 mg/L), including during the aerobic phase, indicating simultaneous nitrification and denitrification occurred (the nitrogen consumption for biomass growth can be neglected because of the very low excess biomass production as reported below). This can be ascribed to the high biomass concentration (around 30 kg TSS/m^3), to the transient conditions (typical of sequential reactors) and to granule size, which generated contiguous anoxic and aerobic biomass layers (it has been reported in previous studies on granular biomass that the ammonium oxidizers situated in the biomass outer layers carried out the oxidation of ammonium to nitrite/nitrate while denitrifying bacteria located in deeper layers, where oxygen cannot penetrate, reduced these compounds to nitrogen gas by using carbon sources coming from storage or hydrolysis products of particulate organic matter present in the feed (Jang *et al.*, 2003; De Kreuk *et al.*, 2005b; De Sanctis *et al.*, 2010)).

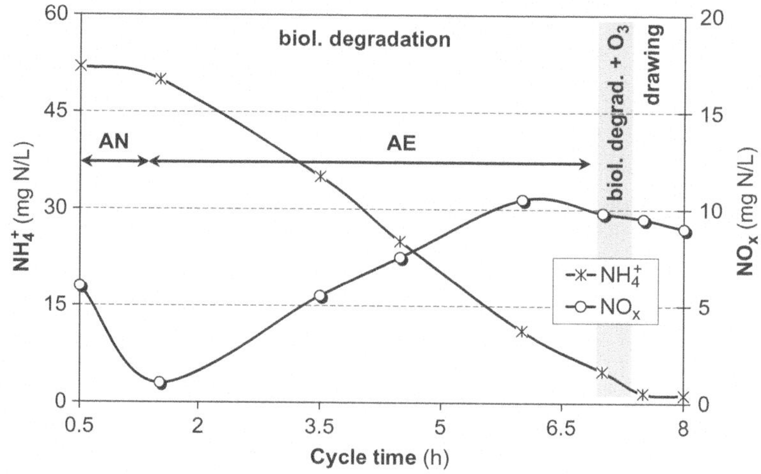

Figure 1.101 Ammonia and oxidised nitrogen profiles during a typical treatment cycle of run 2 of Period D (the value 0.5 on the x-axis corresponds to the end of the filling phase)

The head loss in the reactor ranged from a minimum of 100 cm (the value just after the washing operation) to a maximum of 300 cm (the value set for carrying out the washing operation). The observed head loss increase recorded in the bed of the SBBGR was much slower than that recorded in continuous conventional biofilters (Boller *et al.*, 1997) where washing operations are usually carried out every 1–2 days, resulting in an excess sludge production as high as 0.5–0.6 kg TSS/kg $COD_{removed}$ (Pujol *et al.*, 1992; Canler and Perret, 1994; Le Tallec *et al.*, 1999; Barjenbruch, 2007). For the SBBGR, the washing operation was only necessary every 15-20 days and the sludge production rate less than 0.05 kg TSS/kg $COD_{removed}$ was achieved.

During period D, the SBBGR was started with the granular biomass already selected during the previous experiments. Samplings of the granules for the microbiological characterisation were performed at different bed heights (mainly at the top, middle and bottom of the reactor bed). Multiple samplings within the reactor bed were performed due to previous findings of asymmetrical biomass distribution through the reactor bed after granulation, caused by a biomass concentration gradient (Di Iaconi *et al.*, 2010). The analysed granules were ~3 mm and both crushed granules and granule sections were utilised for the molecular characterisation.

As for previous samples, the ratio of bacteria identified by FISH (EUB338mix probes) with respect to total cells (DAPI stained) was utilised as a gross parameter for estimating the active fraction of the mixed microbial population. The values of EUB338mix/DAPI ratios estimated in the screened samples is reported in Figure 1.102 as the sum of the cells binding the group specific probes and those binding only the bacterial probes (EUB338mix, cited as "other Bacteria" in Figure 1.102). As expected, a decrease in bacterial activity from the top to the bottom of the bed was observed; ranging between 45–35% in the first round of samples (day 625) and 60-40% in the second sampling round (day 710). Taking into account the very high sludge age of the system (>100 d) and biomass concentration (>30 g/l), these values are satisfactory and close to those reported in activated sludge systems (characterised by significantly lower values of both parameters). The increase in Bacteria/total biomass ratio, observed in the sample taken under steady state operating conditions (day 710), was most likely caused by the increase of readily biodegradable COD (RBCOD) supplied to the system for optimising the nitrogen removal (by promoting simultaneous nitrification and denitrification processes).

As for the microbial composition of granules, the change in the influent composition (leachate and RBCOD addition) caused a modification of the bacterial populations present in the system (Figure 1.102). In particular, three bacterial phyla showed an increase in abundance compared with the microbial composition of the granular biomass at the end of period C (see sample day 530 in Figure 1.90). They were comprised of members of *Actinobacteria* (HGC69A probe), *Flavobacteria-Bacteroidetes-Sphingobacteria* (CF319A probe) and *Betaproteobacteria* (Bet42A). *Actinobacteria* were the dominant microbial group and mainly morphologically represented by tetrad forming organisms. These bacteria are commonly observed in sequential reactors (i.e. SBRs) and are capable of producing intracellular storage compounds (glycogen or PHA) in the presence of high concentrations of RBCOD. Another important phylum was the *Betaproteobacteria*. A deeper analysis of this phylum revealed that it was almost composed completely by two bacterial groups: ammonia-oxidizers (identified by Nso1225 probe) and denitrifiers (represented by *Thauera-Azoarcus* group and identified by THAU-646 and Azo-644 probes) (Figure 1.103). The simultaneous presence of nitrifying and denitrifying biomass has already been observed in a SBBGR operating under different conditions (De Sanctis *et al.*, 2010).

A remarkable peculiarity of the granular biomass observed in this study was the high recovery ability of the SBBGR system after exposure to prolonged stress conditions. The SBBGR recovered the nitrification capability after a long operating period (around 1.5 years) without ammonia oxidisation (most likely caused by the presence of high ammonia concentrations and salinity during the feeding with the first landfill leachate). The EPS analysis showed a similarity among these new samples and the granular biomass used as the inoculum. The high presence of EPS was maintained with α-glucans being the main EPS component while only low binding was observed with staining specific for β-glucans and amino-sugars.

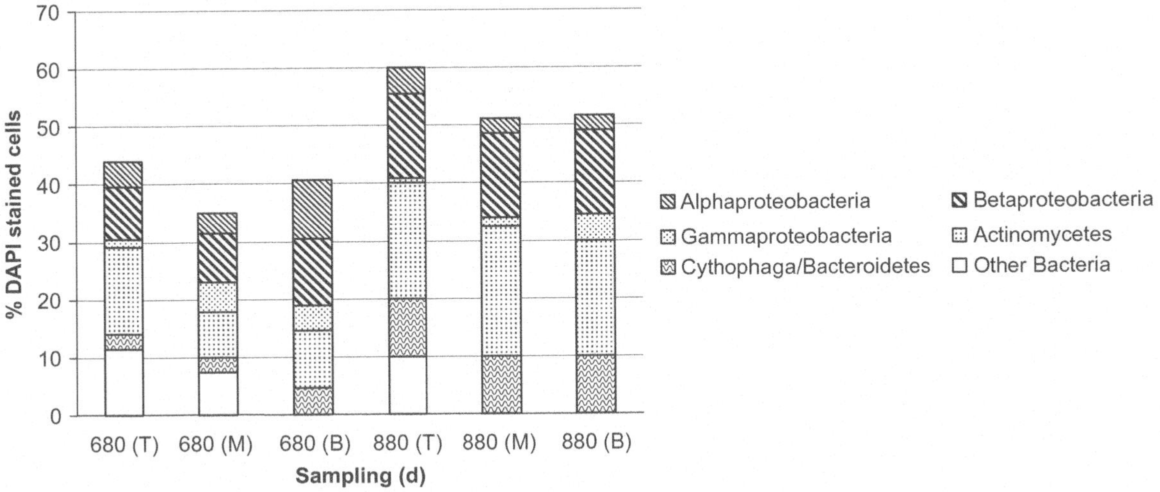

Figure 1.102 Microbial composition of the SBBGR biomass during period D (680 and 880 days) at three different bed's heights (T = top; M = middle; B = bottom)

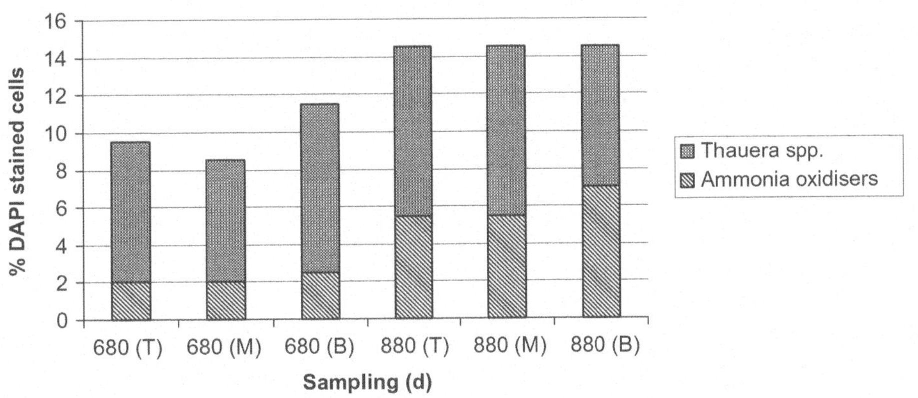

Figure 1.103 Main bacterial functional groups observed in the SBBGR biomass during period D (680 and 880 days) at three different bed's heights (T = top; M = middle; B = bottom)

1.5.4. Conclusions

The treatment based on nitrogen recovery as struvite followed by ozone enhanced biological degradation, carried out in an aerobic granular biomass system (SBBGR – Sequencing Batch Biofilter Granular Reactor), was able to meet the current discharge limits even in the case of highly stabilized municipal landfill leachates. The treatment required a specific ozone dose of 2000 and 600 mgO_3/L-influent for discharging into water bodies and sewer systems, respectively. A strong synergy between ozone and biological degradation was noticed with a O_3consumed/CODremoved ratio lower than 0.3. The process was characterized by a negligible excess biological sludge production (one magnitude order lower than that recorded in conventional biological systems) ascribable to the long sludge age ($\theta_c = 200$ d) and sludge concentration (22–35 g TSS/Lbed; VSS/TSS: 0.4–0.5) of the biomass in the system. In situ hybridization and staining techniques allowed the microbial composition and structure of the biomass growing under the different operating conditions of the SBBGR system to be defined. Microscopic analysis showed that ozone improves the biomass aggregation due to granular biomass generation (granule average diameter of 2–3 mm) which leads to a substantial increase in biomass concentration (i.e. from 18 to 30 g TSS/Lbed) and EPS content; SBBGR biomass was dominated by Gammaproteobacteria. Finally, the ozone enhanced biological degradation was also effective for removing nitrogen from a typical medium-age landfill leachate.

REFERENCES

Amann R. I., Ludwig W. and Schleifer K.-H. (1995). Phylogenetic identification and in situ detection of individual microbial cells without cultivation. *Microbiology and Molecular Biology Reviews*, **59**, 143–169.

Amor L., Eiroa E., Kennes C. and Veiga M. C. (2005). Phenol biodegradation and its effect on the nitrification process. *Wat. Res.*, **39**, 2915–2920.

Andreottola G., Canavas P. and Cossu R. (1990). Overview on landfill leachate quality. CISA, Environmental Sanitary Engineering Centre, Technical note n. 3.

Antoniou P., Hamilton J., Koopman B., Jain R., Holloway B., Lyberatos G. and Svoronos S. A. (1990). Effect of temperature and pH on the effective maximum specific growth rate of nitrifying bacteria. *Wat. Res.* **24**(1), 97–101.

APHA, AWWA, WPCF. (1999). 20th ed., Standards Methods for the Examination of Water and Wastewater. American Public Health Association, Washington, DC.

Arnz P. (2001). Enhanced Biological Phosphorus Removal. Institute of Water Quality Control and Waste Management. Technical University of Munich. PhD, Munich.

Arrojo B., Mosquera-Corral A., Garrido J. M. and Méndez R. (2004). Aerobic granulation with industrial wastewater in sequencing batch reactors. *Water Research*, **38**(14–15), 3389–3399.

Barjenbruch M. (2007). Benchmarking of BAF plants: operational experience on 40 full-scale installations in Germany. *Water Science and Technology*, **55**(8–9), 91–98.

Battistoni P., De Angelis A., Pavan P., Prisciandaro M. and Cecchi F. (2001). Phosphorus removal from a real anaerobic supernatant by struvite crystallization. *Water Research*, **35**, 2167–2178.

Beun J. J. (2001). Phb Metabolism and n-removal in Sequencing Batch Granular Sludge Reactors. Kluyver laboratory of Biotechnology. Delft, Technical University, The Netherlands.

Beun J. J., Hendriks A., van Loosdrecht M. C. M., Morgenroth M., Wilderer P. A. and Heijnen J. J. (1999). Aerobic granulation in a sequencing batch reactor. *Wat. Res.*, **33**(10), 2283–2290.

Beun J. J., van Loosdrecht M. C. M. and Heijnen J. J. (2000). Aerobic granulation. *Wat. Sci. Technol.*, **41**(4–5), 41–48.

Beun J. J., Van Loosdrecht M. C. M. and Heijnen J. J. (2001). N-removal in a granular sludge sequencing batch airlift reactor. *Biotechnol. Bioeng.*, **75**(1), 82–92.

Beun J. J., van Loosdrecht M. C. M. and Heijnen J. J. (2002). Aerobic granulation in a sequencing batch airlift reactor. *Wat. Res.*, **36**(3), 702–712.

Boller M., Kobler D. and Koch G. (1997). Particle separation, solids budget and headloss development in different biofilters. *Water Science and Technology*, **36**(4), 239–247.

Bond P. L., Keller J. and Blackall L. L. (1998). Characterisation of enhanced biological phosphorus removal activated sludges with dissimilar phosphorus removal performances. *Water Sci. Technol.*, **37**(4–5), 567–571.

Booker N. A., Priestley A. J. and Fraser I. H. (1999). Struvite formation in wastewater treatment plants: opportunities for nutrient recovery. *Environmental Technology*, **20**, 777–782.

Brothers A. M. and Wyatt R. D. (2000). The antifungal activity of natamycin toward molds isolated from commercially manufactured poultry feed. *Avian Diseases*. **44**(3), 490–497.

Brothers A. M. and Wyatt R. D. (2000). The antifungal activity of natamycin toward molds isolated from commercially manufactured poultry feed. *Avian Diseases*, **44**(3), 490–497.

Campos J. L., Mosquera-Corral A., Sánchez M., Méndez R. and Lema J. M. (2002). Nitrification in saline wastewater with high ammonia concentration in an activated sludge unit. *Wat. Res.*, **36**, 2555–2560.

Canler J. P. and Perret J. M. (1994). Biological aerated filters: assessment of the process based on 12 sewage treatment plants. *Water Science and Technology*, **29**, 13–22.

Crocetti G. R., Banfielda J. F., Keller J., Bond P. L. and Blackall L. L. (2002). Glycogen-accumulating organisms in laboratory-scale and full-scale wastewater treatment processes. *Microbiology*, **148**, 353–3364.

Crocetti G. R., Hugenholtz P., Bond P. L., Schuler A., Keller J., Jenkins D. and Blackall L. L. (2000). Identification of polyphosphate-accumulating organisms and design of 16S rRNA-directed probes for their detection and quantitation. *Appl. Env. Micrbiol.*, **66**(3), 1175–1182.

Chen P. H. (1996). Assessment of leachates from sanitary landfills: impact of age, rainfall and treatment. *Environment International*, **22**, 225–237.

Daims H., Bruhl A., Amann R., Schleifer K. H. and Wagner M. (1999). The domain-specific probe EUB338 is insufficient for the detection of all Bacteria: development and evaluation of a more comprehensive probe set. *Syst. Appl. Microbiol.*, **22**(3), 434–444.

Dangcong P., Bernet N., Delgenes J.-P. and Moletta R. (1999). Aerobic granular sludge–a case report. *Wat. Res.*, **33**(3), 890–893.

De Beer D., O'Flaharty V., Thaveesri J. and Lens P. (1996). Distribution of extracellular polysaccharides and flotation of anaerobic sludge. *Appl. Microbiol. Biotechnol.*, **46**, 197–201.

De Beer D. and Stoodley P. (1995). Relation between the structure of an aerobic biofilm and transport phenomena. *Wat. Sci. Technol.*, **32** (8), 11–18.

De Bruin L. M. M., Van der Roest H. F., de Kreuk M. K. and Van Loosdrecht M. C. M. (2005). Promising results Pilot research aerobic granular sludge technology at WWTP Ede. In: Aerobic Granular Sludge. S. Bathe, M. K. de Kreuk, B. S. McSwain and N. Schwarzenbeck (eds), IWA Publishing, London, UK, pp. 135–142.

De Bruin L. M. M., De Kreuk M. K., van der Roest H. F. R., Uijterlinde C. and Van Loosdrecht M. C. M. (2004). Aerobic granular sludge technology, alternative for activated sludge technology? *Water Science and Technology*, **49**(11–12), 1–9.

De Kreuk M. K. and De Bruin L. M. M. (2004). Aerobic Granule Reactor Technology. London, IWA Publishing.

De Kreuk M. K. and Van Loosdrecht M. C. M. (2004). Selection of slow growing organisms as a means for improving aerobic granular sludge stability. *Wat. Sci. Technol.*, **49**(11–12), 9–19.

De Kreuk M. K. and Van Loosdrecht M. C. M. (2006). Formation of aerobic granules with domestic sewage. *J. Environ. Eng. - ASCE*, **132**(6).

De Kreuk M. K., Picioreanu C., Hosseini M., Xavier J. B. and Van Loosdrecht M. C. M. (2007). Kinetic model of a granular sludge sbr: Influences on nutrient removal. *Biotechnol. Bioeng.*, **97**(4), 801–815.

De Kreuk M. K., Heijnen J. J. and van Loosdrecht M. C. M. (2005a). Simultaneous COD, Nitrogen and Phosphate removal by aerobic granular sludge. *Biotechnol. Bioeng.*, **90**(6), 761–769.

De Kreuk M. K., Pronk M. and Van Loosdrecht M. C. M. (2005b). Formation of aerobic granules and conversion processes in an aerobic granular sludge reactor at moderate and low temperatures. *Wat. Res.*, **39**(18), 4476–4484.

De Kreuk M. K., Kishida N. and Van Loosdrecht M. C. M. (In preparation). Conversion pf particulate and colloidal cod with aerobic granular sludge and their influence on morphology.

De Sanctis M., Di Iaconi C., Lopez A. and Rossetti S. (2010). Granular biomass structure and population dynamics in Sequencing Batch Biofilter Granular Reactors (SBBGR). *Bioresource Technology*, **101**, 2152–2158.

Di Iaconi C., De Sanctis M., Rossetti S. and Ramadori R. (2010a). SBBGR technology for minimising excess sludge production in biological processes. *Water Research*, **44**, 1825–1832.

Di Iaconi C., Del Moro G., De Sanctis M. and Rossetti S. (2010b). A chemically enhanced biological process for lowering operative costs and solid residues of industrial recalcitrant wastewater treatment. *Water Research*, **44**, 3635–3644.

Di Iaconi C., Del Moro G., Lopez A. and Ramadori R. (2009). The essential role of filling material in aerobic granular biomass generation in a periodic submerged biofilter. *World Review of Science, Technology and Sustainable Development*, **6** (2/3/4), 144–155.

Di Iaconi C., Ramadori R. and Lopez A. (2005). Aerobic granulation during the start up period of a periodic biofilter. In: Aerobic Granular Sludge Water and Envir. Manag. Series (WEMS), London, UK, pp. 25–34.

Di Iaconi C., Ramadori R., Lopez A. and Passino R. (2005). Hydraulic shear stress calculation in a sequencing batch biofilm reactor with granular biomass. *Env. Sci. Technol.*, **39**(3), 889–894.

Dinçer A. R. and Kargi F. (2001). Salt inhibition kinetics in nitrification of synthetic saline wastewater. *Enzyme Microb. Technol.*, **28**, 661–665.

Dulekgurgen E., Ovez S., Artan N. and Orhon D. (2003). Enhanced biological phosphate removal by granular sludge in a sequencing batch reactor. *Biotechnol. Lett.*, **25**(9), 687–693.

Ehrig H. J. (1989). Leachate quality. In: Sanitary landfilling: process technology environmental impact, T. H. Christensen, R. Cossu, R. Stegmann (eds), Academic Press, London.

Eikelboom D. H. and van Buijsen H. J. J. (1983). 2nd ed, Microscopic sludge investigation maual, Delft, TNO Research Institute for Environmental Hygiene, Water and Soil Division, 81.

Fang H. H. P. (2000). Microbial distribution in UASB granules and its resulting effects. *Water Sci. Technol.*, **42**, 201–208.

Finch J. A., Xiao J., Hardie C. and Gomez C. O. (2000). Gas dispersion properties: bubble surface area flux and gas holdup. *Minerals Engineering*, **13**(4), 365–372.

Garrido J. M., Van Benthum W. A. J., Van Loosdrecht M. C. M. and Heijnen J. J. (1997). Influence of dissolved oxygen concentration on nitrite accumulation in a Biofilm airlift suspension reactor. *Biotechnology and Bioengineering*, **53**(2), 168–178.

Gaterell M. R., Gay R., Wilson R., Gochin R. J. and Lester J. N. (2000). An economic and environmental evaluation of the opportunities for substituting phosphorus recovered from wastewater treatment works in existing UK fertiliser markets. *Environmental Technology*, **21**, 1067–1084.

Goel R., Mino T., Satoh H. and Matsuo T. (1998a). Comparison of hydrolytic enzyme systems in pure culture and activated sludge under different electron acceptor conditions. *Water Science and Technology*, **37**(4–5), 335–343.

Goel R., Mino T., Satoh H. and Matsuo T. (1998b). Enzyme activities under anaerobic and aerobic conditions in activated sludge sequencing batch reactor. *Water Research*, **32**(7), 2081–2088.

Goel R., Mino T., Satoh H. and Matsuo T. (1998c). Intracellular storage compounds, oxygen uptake rates and biomass yield with readily and slowly degradable substrates. *Water Science and Technology*, **38**(8–9), 85–93.

Guellil A., Boualam M., Quiquampoix H., Ginestet P., Audic J. M. and Block J. C. (2001). Hydrolysis of wastewater colloidal organic matter by extracellular enzymes extracted from activated sludge flocs. *Wat. Sci. Technol.*, **43**(6), 33–40.

Günay A., Karadag D., Tosun I. and Ozturk M. (2008). Use of magnesit as a magnesium source for ammonium removal from leachate. *Journal Hazardous Material*, **156**, 619–623.

Hall I. R. (1974). Some studies on nitrification in the activated-sludge process. *Wat. Pollut. Control*, **73**, 538–547.

Hao X. D., Joseph J., Heijnen J. J. and Van Loosdrecht M. M. C. (2002). Sensitivity analysis of a biofilm model describing a one-stage completely autotrophic nitrogen removal (CANON) process. *Biotechnology and Bioengineering*, **77**(3), 266–277.

Harmsen J. (1983). Identification of organic compounds in leachate from a waste tip. *Water Research*, **17**, 699–705.

Heijnen J. J. and Van Loosdrecht M. C. M. (1998). Method for acquiring grain-shaped growth of a microorganism in a reactor. Patent Cooperation Treaty (PCT). US, European patent, Technische Universiteit Delft: 13.

Heijnen J. J., Loosdrecht M. C. M., Mulder R., Weltevrede R. and Mulder A. (1993). Development and scale-up of an aerobic biofilm air-lift suspension reactor. *Wat. Sci. Technol.*, **27**(5–6), 253–261.

Hellinga C., Schellen A. A. J. C., Mulder J. W., van Loosdrecht M. C. M. and Heijnen J. J. (1998). The sharon process: an innovative method for nitrogen removal from ammonium-rich waste water. *Wat. Sci. Technol.*, **37**(9), 135–142.

Inizan M., Freval A., Cigana J. and Meinhold J. (2004). Aerobic granulation in sequencing batch reactor (sbr) for industrial wastewater treatment. World Water Congress & Exhibition 2004, IWA Publishing, Marrakech, Marocco.

Irene M. and Lo C. (1996). Characteristics and treatment of leachates from domestic landfills. *Environment International*, **22**, 433–442.

Jang A., Yoon Y. H., Kim I. S., Kim K. S. and Bishop P. L. (2003). Characterization and evaluation of aerobic granules in sequencing batch reactor. *Journal of Biotechnology*, **105**, 71–82.

Jiang H. L., Tay J. H. and Tay S. T. L. (2002). Aggregation of immobilized activated sludge cells into aerobically grown microbial granules for the aerobic biodegradation of phenol. *Letters Appl. Microbiol.*, **35**(5), 439–445.

Jiang H. L., Tay J. H. and Tay S. T. L. (2003). Changes in structure, activity and metabolism of aerobic granules as a microbial response to high phenol loading. *Appl. Microbiol. Biotechnol. on-line publication (12 june 2003).*

Jimenez B., Noyola A., Capteville B., Roustan M. and Faup G. (1988). Dextran blue colorant as a reliable tracer in submerged filters. *Water Research*, **22**(10), 1253–1257.

Kabdaşli I., Ölmez T. and Tünay O. (2003). Nitrogen removal from tannery wastewater by protein recovery. *Water Science and Technology*, **48**, 215–223.

Kabdaşli I., Öztürk Í., Tünay O., Yilmaz S. and Arikan O. (2000). Ammonia removal from young landfill leachate by magnesium ammonium phosphate precipitation and air stripping. *Water Science and Technology*, **41**, 237–240.

Karahan Ö., Martins A., Orhon D. and Van Loosdrecht M. C. M. (2006). Experimental evaluation of starch utilization mechanism by activated sludge. *Biotechnol. Bioeng.*, **93**(5), 964–970.

Kincannon D. F. and Gaudy A. F. (1966). Some effect of high salt concentrations on activated sludge. *J. Wat. Poll. Cont. Fed.*, **38**, 1148–1159.

Kjeldsen P., Barlaz M. A., Rooker A. P., Baun A., Ledin A. and Christensen T. H. (2002). Present and long-term composition of MSW landfill leachate: a review. *Critical Reviews in. Env. Sci. & Tech.*, **32**, 297–336.

Lawton G. W. and Eggert E. (1957). Effect of high sodium chloride concentrations on tricking filter slimes. *Sew. Ind. Wastes*, **29**, 1228–1237.

Le Tallec V., Vidal A. and Thornberg D. (1999). Upflow biological filter: modelling and simulation of filtration. *Water Science and Technology*, **39**(4), 79–84.

Lema L. M., Mendez R. and Blazquez R. (1988). Characteristics of landfill leachates and alternatives for their treatment: a review. *Water air soil pollution*, **40**, 223–250.

Levine A. D., Tchobanoglous G. and Asano T. (1985). Characterization of the size distribution of contaminants in wastewater treatment and reuse implications. *Journal Water Pollution Control Federation*, **57**(7), 805–816.

Liu Q. S., Tay J. H. and Liu Y. (2003a). Substrate concentration-independent aerobic granulation in sequential aerobic sludge blanket reactor. *Environ. Technol.*, **24**(10), 1235–1242.

Liu Q. S., Liu Y., Show K. Y. and Tay J. H. (2009). Toxicity effect of phenol on aerobic granules. *Environ. Technol.*, **30**(1), 69–74.

Liu Y. and Tay J.-H. (2004a). State of the art of biogranulation technology for wastewater treatment. *Biotechnol. Adv.*, **22**(7), 533–563.

Liu Y., Yang S.-F. and Tay J.-H. (2004b). Improved stability of aerobic granules by selecting slow-growing nitrifying bacteria. *J. Biotechnol.*, **108**(2), 161–169.

Liu Y. Q., Tay J. H., Ivanov V., Moy B. Y. P., Yu L. and Tay S. T. L. (2005a). Influence of phenol on nitrification by microbial granules. *Proc. Biochem.*, **40**, 3285–3289.

Liu Y., Liu Y.-Q., Wang Z.-W., Yang S.-F. and Tay J.-H. (2005b). Influence of substrate surface loading on the kinetic behaviour of aerobic granules. *Appl. Microbiol. Biotechnol.*, **67**(4), 484–488.

Liu Y., Yang S. F. and Tay J.-H. (2003). Elemental compositions and characteristics of aerobic granules cultivated at different substrate n/c ratios. *Appl. Microbiol. Biotechnol.*, **61**(5–6), 556–561.

Loveless J. E. and Painter H. A. (1968). The influence of metal ions and pH value on the growth of a Nitrosomonas strain isolated from activated sludge. *J. Gen. Microbiol.*, **52**, 1–14.

Loy A., Maixner F., Wagner M. and Horn M. (2007). ProbeBase an online resource for rRNA-targeted oligonucleotide probes: new features 2007. *Nucleic Acids Research*, **35**, D800–D804.

Ludzack F. J. and Noran P. K. (1965). Tolerance of high salinities by conventional wastewater treatment process. *J. Water Pollut. Control Fed.*, **37**(10), 1404–1416.

Maqueda C., Perez Rodriguez J. L. and Lebrato J. (1994). Study of struvite precipitation in anaerobic digesters. *Water Research*, **28**, 411–416.

Martins A. M. P. (2004). Bulking Sludge Control. Kinetics, Substrate Storage, and Process Design Aspects. Applied Sciences. PhD, Delft University of Technology, Delft.

Matsuo Y. and Hosobora K. (1988). An experiment study on anaerobic aerobic sludge process characteristic of the phosphorus uptake reaction The Third WPCF/JSWA Joint Technical Seminar on Sewage Treatment Technology, Tokyo.

Maurer M., Pronk W. and Larsen T. A. (2006). Review: treatment processes for source-separated urine. *Water Research*, **40**, 3151–3166.

McSwain B. S., Irvine R. L. and Wilderer P. A. (2004a). The effect of intermittent feeding on aerobic granule structure. *Wat. Sci. Technol.*, **49**(11), 19–25.

McSwain B. S., Irvine R. L. and Wilderer P. A. (2004b). The influence of-settling time on the formation of aerobic granules. *Wat. Sci. Technol.*, **50**(10), 195–202.

Metcalf and Eddy I. (2003). Wastewater Engineering, Treatment and Reuse. McGraw-Hill, New York.

Mishima K. and Nakamura M. (1991). Self-immobilization of aerobic activated sludge - a pilot study of the aerobic upflow sludge blanket process in municipal sewage treatment. *Wat. Sci. Technol.*, **23**(4–6), 981–990.

Mobarry B. K., Wagner M., Urbain V., Rittmann B. E. and Stahl D. A. (1996). Phylogenetic probes for analyzing abundance and spatial organization of nitrifying bacteria. *Appl. Env. Micrbiol.*, **62**(6), 2156–2162.

Moon B. H., Seo G. T., Lee T. S., Kim S. S. and Yoon C. H. (2003). Effects of salt concentration on floc characteristics and pollutants removal efficiencies in treatment of seafood wastewater by SBR. *Water. Sci. Technol.*, **47**(1), 65–70.

Morgenroth E., Kommedal R. P. and Harremoe?s (2002). Processes and modeling of hydrolysis of particulate organic matter in aerobic wastewater treatment - a review. *Wat. Sci. Technol.*, **45**(6), 25–40.

Morgenroth E., Sherden T., Van Loosdrecht M. C. M., Heijnen J. J. and Wilderer A. P. (1997). Aerobic granular sludge in a sequencing batch reactor. *Water Research*, **31**(12), 3191–3194.

Mosquera-Corral A., Montras A., Heijnen J. J. and van Loosdrecht M. C. M. (2003). Degradation of polymers in a biofilm airlift suspension reactor. *Wat. Res.*, **37**(3), 485–492.

Mosquera-Corral A., de Kreuk M. K., Heijnen J. J. and van Loosdrecht M. C. M. (2005). Effects of oxygen concentration on N-removal in an aerobic granular sludge reactor. *Wat. Res.*, **39**(12), 2676–2686.

Moussa M. S., Sumanasekera D. U., Ibrahim S. H., Lubberding H. J., Hooijmans C. M., Gijzen H. J. and van Loosdrecht M. C. M. (2006). Long term effects of salt on activity, population structure and floc characteristics in enriched bacterial cultures of nitrifiers. *Wat. Res.*, **40**, 1377–1388.

Moy B. Y. P., Tay J. H., Toh S. K., Liu Y. and Tay S. T. L. (2002). High organic loading influences the physical characteristics of aerobic sludge granules. *Letters Appl. Microbiol.*, **34**(6), 407–412.

Munch E. V. and Barr K. (2001). Controlled struvite crystallization for removing phosphorus from anaerobic digester sidestream. *Water Research*, **35**(1), 151–159.

Nelson N. O., Mikkelsen R. E. and Hesterberg D. L. (2003). Struvite precipitation in anaerobic swine lagoon liquid: effect of pH and Mg: P ratio and determination of rate constant. *Bioresource Technology*, **89**, 229–236.

Ng H. Y., Ong S. and Ng W. J. (2005). Effects of sodium chloride on the performance of a sequencing batch reactor. *J. Environ. Eng. ASCE*, **131**, 1557–1564.

Nicolella C., van Loosdrecht M. C. M. and Heijnen J. J. (2000). Waste water treatment with particulate biofilm reactors. *J. Biotechnol.*, **80**, 1–33.

Nicollela C., van Loosdrecht M. C. M. and Heijnen J. J. (2000). Particle-based biofilm reactor technology. *TIBTECH*, **18**(July), 312–320.

O'leary P. R., Tchobanoglous G. and Kreith F. (eds), 2002. Handbook of Solid Waste Management, McGraw-Hill.

Ohlinger K. N., Young T. M. and Schroeder E. D. (1998). Predicting struvite formation in digestion. *Water Research*, **32**, 3607–3614.

Oostendorp J. (1981). Natamycin®. *Antonie van Leeuwenhoek*, **47**(2), 170–171.

Orhon D. and Çokgör E. U. (1997). Cod fractionation in wastewater characterization - the state of the art. *Journal of Chemical Technology & Biotechnology*, **68**(3), 283–293.

Painter H. A. and Loveless J. E. (1983). Effect of temperature and pH value on the growth-rate constants of nitrifying bacteria in the activated-sludge process. *Wat. Res.*, **17**, 237–248.

Panswad T. and Anan C. (1997). Salt tolerance of carbon and nitrogen bacteria in an anerobic/anoxic/aerobic process. Proceedings of the Third Australian Conference on Biological Nutrient Removal from Wastewater, Australia Water and Wastewater Association, Victoria.

Panswad T. and Anan C. (1999). Impact of high chloride wastewater on anaerobic/anoxic/aerobic process with and without inoculation of chloride acclimated bacteria. *Wat. Res.*, **35**(5), 1165–1172.

Park S., Choi K. S., Joe K. S., Kim W. H. and Kim H. S. (2001). Variations of landfill leachate's properties in conjunction with the treatment process. *Environmental Technology*, **22**, 639–45.

Pujol R., Canler J. P. and Iwema A. (1992). Biological aerated filters: an attractive and alternative biological process. *Water Science and Technology*, **26**, 693–702.

Qin L., Liu Y. and Tay J.-H. (2004). Effect of settling time on aerobic granulation in sequencing batch reactor. *Biochem. Eng. J.*, **21**(1), 47–52.

Renou S., Givaudan J. G., Poulain S., Dirassouyan F. and Moulin P. (2008). Landfill leachate treatment: review and opportunity. *Journal Hazardous Material*, **150**, 468–493.

Ried A., Mielcke J., Wieland A., Schaefer S. and Sievers M. (2007). An overview of the integration of ozone systems in biological treatment steps. *Water Science and Technology*, **55**(12), 253–258.

Roessink R. and Eikelboom D. H. (1997). Characterization of suspended solids in/out airlift biofilm-reactors. *Wat. Sci. Technol.*, **36**(1), 237–245.

Rogalla F. and Sibony J. (1992). Biocarbone aerated filters - 10 years after - past, present, and plenty of potential. *Wat. Sci. Technol.*, **26** (9–11), 2043–2048.

Roustan M., Wang R. Y. and Wolbert D. (1996). Modelling hydrodynamics and mass transfer parameters in a continuous ozone bubble column. *Ozone Science & Engineering*, **18**(2), 99–115.

San Pedro D. C., Mino T. and Matsuo T. (1994). Evaluation of the rate of hydrolysis of slowly biodegradable cod (sbcod) using starch as substrate under anaerobic, anoxic and aerobic conditions. *Wat. Sci. Technol.*, **30**(11), 191–199.

Schmid M., Thill A., Purkhold U., Walcher M., Bottero J. Y., Ginestet P., Nielsen P. H., Wuertz S. and Wagner M. (2003). Characterization of activated sludge flocs by confocal laser scanning microscopy and image analysis. *Water Res.*, **37**, 2043–2052.

Schwarzenbeck N., Borges J. M. and Wilderer P. A. (2005). Treatment of dairy effluents in an aerobic granular sludge sequencing batch reactor. *Appl. Microbiol. Biotechnol.*, **66**(6), 711–718.

Schwarzenbeck N., Erley R. and Wilderer P. A. (2004a). Aerobic granular sludge in an sbr-system treating wastewater rich in particulate matter. *Wat. Sci. Technol.*, **49**(11–12), 41–46.

Schwarzenbeck N., Erley R., Mc Swain B. S., Wilderer P. A. and Irvine R. L. (2004b). Treatment of malting wastewater in a granular sludge sequencing batch reactor (SBR). *Acta. Hydrochim. Hydrobiol.*, **32**(1), 16–24.

Shammas N. K. (1986). Interactions of temperature, pH and biomass on the nitrification process. *J. WPCF.*, **58**, 52–59.

Sharma B. and Ahlert R. C. (1977). Nitrification and nitrogen removal. *Wat. Res.*, **11**, 897–925.

Sheng G.-P., Yu H.-Q. and Li X.-Y. (2006). Stability of sludge flocs under shear conditions: roles of extracellular polymeric substances (eps). *Biotechnol. Bioeng.*, **93**(6), 1095–1102.

Siegrist H. (1996). Nitrogen removal from digester supernatant – comparison of chemical and biological methods. *Water Science and Technology*, **34**, 399–406.

Smolders G. J. F., van der Meij J., Van Loosdrecht M. C. M. and Heijnen J. J. (1994). Stoichiometric model of the aerobic metabolism of the biological phosphorus removal process. *Biotechnol. Bioeng.*, **44**, 837–848.

Smolders G. J. F., van Loosdrecht M. C. M. and Heijnen J. J. (1994). pH: key factor in the biological phosphorus removal process. *Wat. Sci. Technol.*, **29**(7), 71–74.

Sophonsiri C. and Morgenroth E. (2004). Chemical composition associated with different particle size fractions in municipal, industrial, and agricultural wastewaters. *Chemosphere*, **55**(5), 691–703.

Stewart M. J., Ludwig H. F. and Kearns W. H. (1962). Effect of varying salinity on extended aeration process. *J. Wat. Poll. Cont. Fed.*, **34**, 1161–1177.

Strathmann M., Wingender J. and Flemming H. C. (2002). Application of fluorescently labelled lectins for the visualizatin and biochemical characterization of polysaccharides in biofilms of Pseudomonas aeruginosa. *J. Microbiol. Methods.*, **50**, 237–248.

Tay J. H., Liu Q. S. and Liu Y. (2001a). The effects of shear force on the formation, structure and metabolism of aerobic granules. *Appl. Microbiol. Biotechnol.*, **57**, 227–233.

Tay J. H., Liu Q. S. and Liu Y. (2001b). The role of cellular polysaccharides in the formation and stability of aerobic granules. *Letters Appl. Microbiol.*, **33**(3), 222–226.

Tay J. H., Liu Q. S. and Liu Y. (2002). Characteristics of aerobic granules grown on glucose and acetate in sequential aerobic sludge blanket reactors. *Environ. Technol.*, **23**(8), 931–936.

Tijhuis L. (1994). The biofilm airlift suspension reactor; biofilm formation, detachment and heterogenity. Department of Biotechnology. Technical University Delft, Delft, 296.

Tijhuis L., Van Loosdrecht M. C. M. and Heijnen J. J. (1994). Formation and growth of heterotrophic aerobic biofilms on small suspended particles in airlift reactors. *Biotechnol. Bioeng.*, **44**(5), 595–608.

Tünay O., Kabdaşli I., Orhon D. and Kolçak S. (1997). Ammonia removal by magnesium ammonium phosphate in industrial wastewaters. *Water Science and Technology*, **36**, 225–228.

Uygur A. (2006). Specific nutrient removal rates in saline wastewater treatment using sequencing batch reactor. *Proc. Biochem.*, **41**, 61–66.

Uygur A. and Kargi F. (2004). Salt inhibition on biological nutrient removal from saline wastewater in a sequencing batch reactor. *Enzyme Microb. Technol.*, **34**, 313–318.

Van Benthum W. A. J., Garrido-Fernandez J. M., Tijhuis L., van Loosdrecht M. C. M. and Heijnen J. J. (1996). Formation and detachment of biofilms and granules in a nitrifying biofilm airlift suspension reactor. *Biotechnel. Prog.*, **1996**(12), 764–772.

Van Kempen R., Mulder J. W., Uijterlinde C. A. and van Loosdrecht M. C. M. (2001). Overview: full scale experience of the SHARON® process for treatment of rejection water of digested sludge dewatering. *Water Science and Technology*, **44**(1), 145–152.

Van Loosdrecht M. C. M. and De Kreuk M. K. (2004). Method for the treatment of waste water with sludge granules. *Bureau voor Industriele Eigendom.*, The Netherlands, International, Technische Universiteit Delft.

Van't Riet K. and Tramper J. (1991). Basic Bioreactor Design. Marcel Dekker, New York.

Yang S.-F., Tay J.-H. and Liu Y. (2005). Effect of substrate nitrogen/chemical oxygen demand ratio on the formation of aerobic granules. *Journal of Environmental Engineering*, **131**(1), 86–92.

Zengin G., Ölmez T., Dogruel S., Kabdaşli I. and Tünay O. (2002). Assessment of source-based nitrogen removal alternatives in leather tanning industry wastewater. *Water Science and Technology*, **45**, 205–215.

Zhang T., Ding L. and Ren H. (2009). Pretreatment of ammonium removal from landfill leachate by chemical precipitation. *Journal Hazardous Material*, **166**, 911–915.

Zhou H. and Smith D. W. (2000). Ozone mass transfer in water and wastewater treatment: experimental observations using a 2d laser particle dynamics analyzer. *Water Research*, **34**(3), 909–921.

Chapter 2

Integrated processes

Isabel Oller, Sixto Malato, Ana Zapata, Carla Sirtori, Félicien Mazille, Cesar Pulgarin, Daniela Cassano, Sara Diomede, Piera Ielpo, Giuseppe Laera, Giuseppe Mascolo, Natascha Michel, Isabel Pariente, Angela Pinto, Alfieri Pollice, Carlo Salerno, Benedek Gy. Plósz and Christian Vogelsang

2.1 BACKGROUND

One of the major threats to water quality is chemical pollution from heavy metals, solvents, dyes, pesticides, etc. Chemicals enter the aquatic medium in several different ways, either dumped directly, such as industrial effluents, or from wastewater treatment plants (WWTP) that do not fulfil their obligations. They may also enter the water indirectly through the use of plant health products, such as biocides and fertilizers, in agriculture. In general, very water-soluble substances can be transported and distributed more easily in the water cycle. Discharge resulting from lax enforcement of the rules, illegal use and inappropriate application of substances may be considerable.

In the past, the focus was on detecting the severe, direct effects of individual pollutants and the short-term negative impact on ecosystems. But as scientific understanding has advanced, and the more concentrated emissions have been lowered, environmental evaluation reveals a considerable number of chronic effects that can usually only be detected after a long period of time. Furthermore, larger and larger quantities of persistent substances are being found at long distances from their sources of discharge (Meyer and Wania, 2007). Evaluation also used to concentrate mostly on the effect of individual substances, whereas we are now beginning to study and understand interactions in mixtures of these substances (Hernando *et al.*, 2005; Hildebrant *et al.*, 2006). The main routes for destroying toxic compounds in natural water are biodegradation and photodegradation. Photodegradation may be by direct or indirect photolysis. In indirect photolysis, a photosensitizer (as nitrate or humic acids) absorbs the light and transfers the energy to the pollutants, which otherwise would not react, since they do not absorb light in the wavelength interval of the solar photons that arrive on the Earth's surface (i.e. >300 nm). Biological degradation of a chemical refers to the elimination of the pollutant by the metabolic activity of living organisms, usually microorganisms and in particular bacteria and fungi that live in natural water and soil.

In this context, conventional biological processes do not always provide satisfactory results, especially for industrial wastewater treatment, since many of the organic substances produced by the chemical industry are toxic or resistant to biological treatment (Pulgarín and Kiwi, 1996; García *et al.*, 2001; Muñoz and Guieysee, 2006; Lapertot *et al.*, 2006). Conventional methods of water decontamination can address many of these problems. However, these treatment methods are often chemically, energetically and operationally intensive, focused on large systems, and thus require considerable infusion of capital, engineering expertise and infrastructure, all of which precludes their use in much of the world. Furthermore, intensive chemical treatments (such as those involving ammonia, chlorine compounds, hydrochloric acid, sodium hydroxide, permanganate, alum and ferric salts, coagulation and filtration aids, anti-scalants, corrosion control chemicals, and ion exchange resins and regenerants) and residuals resulting from treatment (sludge, brines, toxic waste) can add to the problems of contamination and salting of freshwater sources. Air stripping and adsorption, merely transferring toxic materials from one medium to another, is not a long-term solution. Incineration is capable of converting toxics to carbon dioxide, water and inorganic acids, but negative public perception has very often prevented its implementation. Therefore, a feasible option for such biologically persistent wastewater is the use of advanced technologies based on chemical oxidation, such as the advanced oxidation processes (AOPs), widely recognized as highly efficient treatments for recalcitrant wastewater. These methods rely on the formation of highly reactive chemical species which degrade even the most recalcitrant molecules into biodegradable compounds. Although there are different reacting systems (see http://www.jaots.net/), all of them are characterized by the same chemical feature: production of hydroxyl radicals ($^{\bullet}OH$), which are able to oxidize and mineralize almost any organic molecule, yielding CO_2 and inorganic ions. Rate constants (k_{OH}, $r = k_{OH}$ [$^{\bullet}OH$] C) for most reactions involving hydroxyl radicals in aqueous solution are usually on the order of 10^6 to 10^9 M^{-1} s^{-1}. They are also characterized by their non-selective attack, which is a useful attribute for wastewater treatment and solution of pollution problems. The versatility of the AOPs is also enhanced by the fact that there are different ways of producing hydroxyl radicals, facilitating compliance with the specific treatment requirements. Methods based on UV, H_2O_2/UV, O_3/UV and H_2O_2/O_3/UV combinations use photolysis of H_2O_2 and ozone to produce the hydroxyl radicals. Other methods, like heterogeneous photocatalysis and homogeneous photo-Fenton, are based on the use of a wide-band-gap semiconductor and addition of H_2O_2 to dissolved iron salts, respectively, and irradiation with UV-VIS light (Pignatello *et al.*, 2006; Comninellis *et al.*, 2008; Shannon *et al.*, 2008).

Chemical oxidation for complete mineralization is generally expensive because the oxidation intermediates formed during treatment tend to be more and more resistant to their complete chemical degradation, and furthermore, they all consume energy (radiation, ozone, etc.) and chemical reagents (catalysts and oxidizers) which increase with treatment time (Muñoz *et al.*, 2008). One attractive potential alternative is to apply these chemical oxidation processes in a pre-treatment to convert the initially persistent organic compounds into more biodegradable intermediates, which would then be treated in a biological oxidation process with a considerably lower cost. Therefore, the main role of the chemical pre-treatment is partial oxidation of the biologically persistent part to produce biodegradable reaction intermediates. The percentage of mineralization should be minimised during the pre-treatment stage in order to avoid unnecessary expenditure of chemicals and energy, thereby lowering the operating cost. Experimental examples of sequential chemical and biological oxidation treatment have been previously reviewed by (Mantzavinos and Psillakis, 2004; Augugliaro *et al.*, 2006; Ikehata and Gamal El-Din, 2006; Gaya and Abdullah, 2008).

Selection of the best treatment option for remediation of a specific industrial wastewater is a highly complex task. The choice of one or more processes to be combined in a certain situation depends on the quality standards to be met and the most effective treatment with the lowest reasonable cost. Therefore, the main factors which must be considered in the decision on the wastewater treatment technologies to be applied are:

1. The quality of the original wastewater.
2. Removal of parent contaminants.
3. Conventional treatment options.
4. Final wastewater treatment system efficiency.
5. Economic studies.
6. Life Cycle Assessment to determine environmental compatibility of the wastewater treatment technology.

In general, the possibilities and capabilities of the conventional treatments available are widely known. However, information on the efficiency of the new technologies (AOPs) for eliminating certain specific pollutants present in wastewater compared to conventional options is necessary. For this reason, bench-scale and pilot plant studies are required to develop the technologies and generate information on new industrial wastewater treatment processes. Such scaled studies are even more decisive when combining several technologies for decontamination of a specific industrial wastewater, as stated in this chapter.

When preliminary chemical oxidation is applied in a combination treatment line, sometimes its effect is insignificant or even harmful to the properties of the original effluent, even though it is conceptually advantageous. There are several reasons for this, the most common of which are:

- Formation of stable intermediates which are less biodegradable than the original molecules.
- Lack of selectivity for preferential attack on the more bioresistant fractions of the wastewater during chemical treatment.
- Poor selection of treatment conditions as, for example, excessive pre-oxidation can lead to generation of an effluent with too little metabolic value for the microorganisms.
- Too much oxidant and/or catalyst used for oxidation remained in the pre-treated wastewater. Compounds such as ozone and hydrogen peroxide (both known as biocides), metals, metal oxides and metal salts (catalysts in many processes), are normally toxic to microorganisms.

These limitations underline the need to establish a step-by-step research methodology which takes these effects into account, because operating conditions effect on the original properties of the pre-treatment stream (contact time, oxidant and/or catalyst type, dose and toxicity, temperature, etc.) must be known. Such studies must employ analytical tools to infer the reaction mechanisms, pathway and kinetics, evaluate the effect of the chemical pre-treatment on toxicity and biodegradability, the effect of cations and anions in the wastewater matrix, and the application of various techniques for determining biodegradability and toxicity (Sarria *et al.*, 2002; Lapertot and Pulgarin, 2006; Amat *et al.*, 2009; Malato *et al.*, 2009).

Figure 2.1 summarizes the different steps necessary to evaluate the feasibility of applying a combined AOP/biological treatment strategy for industrial wastewater (toxic and/or non-biodegradable) treatment. This general diagram depicts not only the necessary chemical and biological analyses that must be performed in a certain industrial wastewater treatment line, but also the different situations which could appear depending on the special characteristics of the particular wastewater. As observed in this figure, the variables usually used to describe chemical oxidation are the dissolved organic carbon concentration (DOC) and the chemical oxygen demand (COD). Apart from these measurements, another series of parameters related to the process conditions should be monitored during the AOP selected for each treatment, which, in general, include the catalyst concentration (Fe(II), Fe(III), TiO_2, etc.), the hydrogen peroxide or ozone concentration, the radiation intensity, pH and temperature.

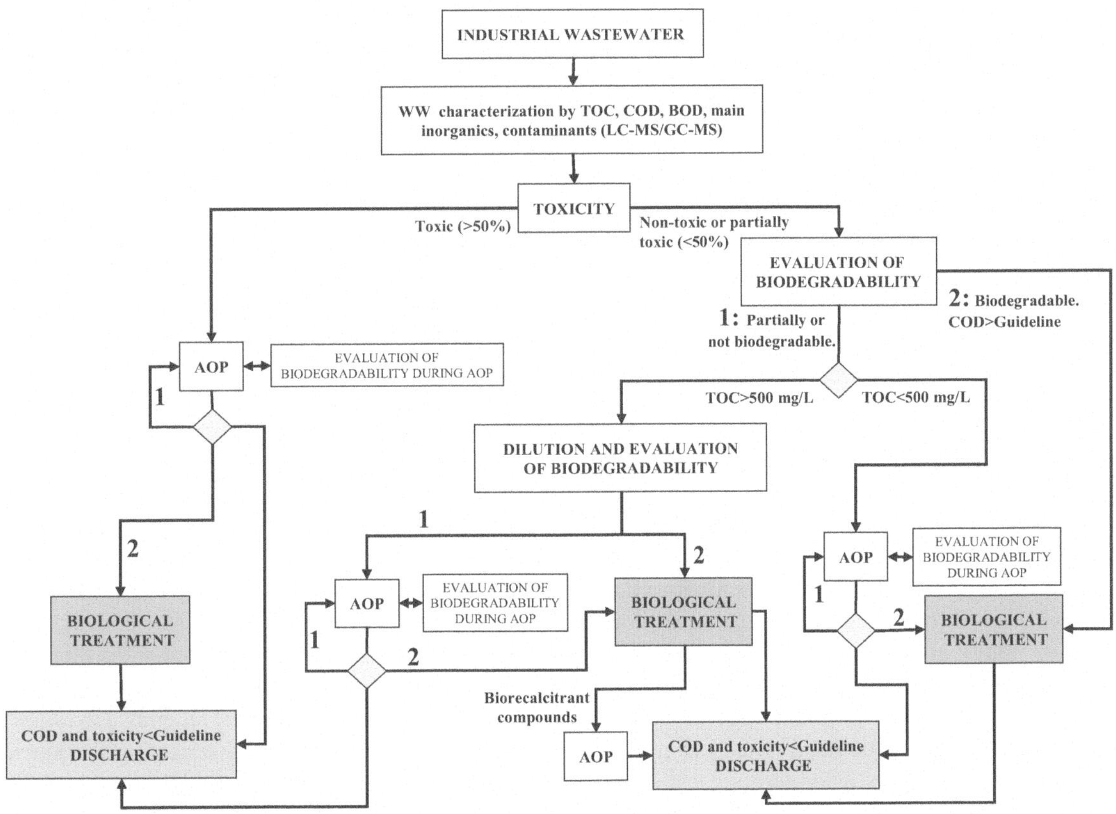

Figure 2.1 Strategy for the selection of treatment options for toxic and/or non biodegradable industrial wastewater

Appropriate techniques must be combined to provide technically and economically feasible options. The performance of an AOP treatment could be enhanced in several ways. The first possibility is to position the AOP in a sequence of physical, chemical and biological treatments. Such a treatment approach often involves at least one AOP step and one biological treatment step. Whether the AOP or the biological process is first in the treatment line, the overall purpose of reducing costs will be nearly the same as minimizing AOP treatment and maximizing the biological stage, because of the wide difference in the cost of the two treatments. In combined chemical and biological wastewater treatment, to keep in mind how the characteristics of each individual treatment, such as the chemical oxidant to be used (photo-Fenton or Fenton reagent, O_3/H_2O_2, H_2O_2/UV, etc.) can improve the destruction of a persistent contaminant is very important. Measurement of the combined process efficiency depends on the purpose of the treatment, but normally requires the independent optimization of each chemical and biological step. For example, the extent of mineralization of the organic compounds may be a measure of efficiency if highly pure water is needed or an effluent with a specific dissolved organic carbon limit. The main purpose of other treatments may be reduction of toxicity or elimination of a specific pollutant. Determining the target is an essential step in combination studies since it helps define process efficiency and provides a basis for comparing the different operating conditions and optimizing the process.

Calculation of the individual biological and chemical oxidation efficiencies is important for finding the optimal operating conditions for the combined process. This task involves profound knowledge of both biological and chemical processes. Therefore, several analytical parameters must be monitored during each step of the treatment line. Chemical parameters normally measured are the total organic carbon (and/or chemical oxygen demand), the concentration of specific pollutants which could be present in the target wastewater and complete degradation of heteroatoms released (Cl, N, P...) as inorganic species (Cl^-, NO_3^-, PO_4^{3-}...) into the media from contaminants completely degraded during the AOP treatment. Regarding biological assays, it is very important to perform toxicity analyses (with organisms like *Vibrio fischeri, Daphnia magna*, activated sludge by respirometric assays, etc.) and biodegradability tests (using activated sludge) to ensure the conditions of the AOP effluent to be subsequently treated by a conventional biodegradation process. In the biological system itself, and apart from daily control analyses the measurement of inorganics present in the biological media is also essential since, on one hand, nutrients are vital to the micoorganism populations in the activated sludge, and on the other hand, monitoring the nitrogen species provides much information related to nitrification and denitrification. If further understanding of the underlying processes is sought, additional analytical methods may have to be applied, which allow the identification of unknown intermediate degradation products. Such studies involve a considerable effort, dedication, and sophisticated analytical equipment (based on chromatography and mass spectrometry), but this is sometimes the only way to explain why acute toxicity arises during treatment, for example, by pinpointing a single intermediate degradation product that is more toxic than the original pollutant (Malato *et al.*, 2009).

This chapter highlights not only efforts in applying AOPs as a pre-treatment, but also real cases in which the combination strategy is in the opposite direction, first eliminating the highly biodegradable part of the wastewater and then degrading the recalcitrant contaminants (non toxic) by a post-treatment AOP. Industrial wastewaters are much more diverse than domestic wastewater, which is usually qualitatively and quantitatively similar in its composition. On the contrary, treatment of industrial wastewater is a complex problem due to the wide variety of compounds and concentrations it may contain. Although studies with model effluents provide useful information for optimising the chemical pre-treatment, this information cannot be applied directly to real wastewater. Therefore, the following sections will be mainly focused on showing different applications of integrated processes for treating pesticide, pharmaceuticals wastewaters and landfill leachate. Among the priority substances polluting water are soluble pesticides, which are a serious threat to surface and ground water, since their high solubility in water makes their propagation in the environment extremely easy. Most pesticides are resistant to chemical and/or photochemical degradation under typical environmental conditions. Pharmaceutical residues are another group of compounds of particular interest and unknown fate. For instance, during monitoring in Italy, France, Greece and Sweden, carbamazepine, clofibrate, phenazone and aminopyrine, clofibri acid, diclofenac, fenofibrate, fenoprofen, flurbiprofen, gemfibrozil, ibuprofen, ketoprofen and naproxen, all belonging to different therapeutic pharmaceuticals classes, were found in the effluents of sewage treatment plants (Andreozzi *et al.*, 2003). In general, the presence of residual pharmaceuticals in the environment and in aquatic systems in particular constitutes a serious environmental problem, as these compounds are extremely resistant to biological degradation and usually escape intact from conventional treatment plants. Leachates are defined as the aqueous effluent generated by rainwater percolation through wastes, biochemical processes in waste cells and the inherent water content of the waste itself. Leachates may contain large amounts of organic matter (biodegradable, but also refractory to biodegradation), where humic-type components comprise an important group, as well as ammonia-nitrogen, heavy metals, chlorinated organic and inorganic salts. The removal of ammonium and organic material verified by chemical oxygen demand (COD) and biological oxygen demand (BOD) from leachate is usually a prerequisite for discharging the leachates into natural water. Toxicity analysis using various test organisms have confirmed the potential dangers of landfill leachates (Marttinen *et al.*, 2002) and the need to treat them to meet the standards for discharge into receiving water. However, biological treatment is hampered by specific toxic substances (such as PAHs-polyaromatic hydrocarbons, AOXs-adsorbable organic halogens, PCBs-polychlorinated biphenyls) and/or by the presence of bio-refractory organics (such as humic acids or surfactants).

2.2 RECALCITRANT WASTEWATER TREATMENT BY INTEGRATING SOLAR ADVANCED OXIDATION PROCESSES AND IMMOBILISED BIOMASS REACTOR

2.2.1 Introduction

It is clear that the study of real wastewater is inevitably compromised by the amount and quality of information available. If the exact composition of the original effluent is hard to find, the identification of the reaction intermediates is impracticable, and therefore, the kinetics must be based on dissolved organic carbon or chemical oxygen demand measurements. Although studies with model effluents provide useful information for optimising the chemical pre-treatment, this information cannot be applied directly to real wastewater. The combination of these two extremes requires a chain of experimental studies beginning with model wastewaters before undertaking the more complex real effluents. In this context, it must also be considered that the real wastewater might contain substances that could compete with the target pollutant for the oxidizing agent (•OH), such as other organic and inorganic compounds and natural organic matter. This usually causes a decrease in process efficiency. Even in effluents in which the target pollutants are predominant, oxidation may lead to the quick formation of a multitude of intermediates that could lower overall process efficiency, since these species could be less reactive with the hydroxyl radicals. In this sense, Scott and Ollis (1995) have identified four types of wastewater as potentially treatable by combined AOPs/biological degradation: wastewater containing biorecalcitrant compounds such as large macromolecules like soluble polymers that are not easily biodegradable due to their large size and lack of active centers; highly biodegradable industrial wastewater which still requires chemical post treatment as it contains a large amount of biodegradable organic compounds in addition to small concentrations of recalcitrant compounds; wastewater containing inhibiting compounds which are somewhat toxic to a certain percentage of some biological cultures; wastewater containing inert intermediates such as specific metabolites which must be effectively degraded or, they would accumulate in the medium and inhibit growth of the microorganisms. This section will focus on wastewaters related with the last three types.

But if the original wastewater contains a considerable amount of biodegradable compounds, the pre-oxidation step obviously will not lead to a significant improvement of biodegradability; rather, it will only cause unnecessary consumption of chemicals. In such cases, a biological pre-treatment (removing biodegradable compounds) followed by an AOP may prove to be more useful. Such integrated systems are particularly favourable for effluents which initially contain some biodegradable fractions (i.e. sugars and proteins) which could easily be removed first and so, not compete for the chemical oxidant. It is important to take into account that combination of chemical and biological processes presents an additional difficulty for efficient operation of a large-scale plant, as the chemical oxidant and biological culture must not come in undue contact with each other. For example, high concentrations of hydrogen peroxide cause an adverse effect on the microorganisms, although relatively low concentrations are not a problem. The chemical oxidation reaction time must also be taken into account, because when a biological process receives a chemically over degraded effluent, the highly oxidised products metabolic value for the microorganisms is minimal. Furthermore, large doses of oxidant may be spent in the degradation of easily biodegradable intermediates, reducing overall system efficiency. This section will focus also in those aspects.

2.2.1.1 Fenton and photo-Fenton

For the treatment of industrial wastewater Fenton and Fenton-like processes are probably among the most applied advanced oxidation processes (Legrini *et al.*, 1993; Suty *et al.*, 2004). The first proposals for wastewater treatment applications were reported in the 1960's. Yet, it was not until the early 1990's when first works of the application of the photo-Fenton process for the treatment of wastewater were published by the groups of Pignatello, Lipcznska-Kochany, Kiwi, Pulgarín and Bauer (Pignatello *et al.*, 2006). Much of the literature treating the photo-Fenton process takes into account the possibility of driving the process with solar radiation. This fact is due to that a priori the photo-Fenton process seems to be the most apt of all AOPs to be driven by sunlight, because soluble iron-hydroxy and especially iron-organic acid complexes absorb even part of the visible light spectrum, not only ultraviolet radiation (Malato *et al.*, 2009). Consequently, photo-Fenton is a very efficient reaction system for the generation of radicals available for oxidative processes. Similar as in literature on other AOPs, originally toxic waste water has been proven to lose its toxicity upon treatment by photo-Fenton process before total mineralisation has been achieved. Therefore, photo-Fenton process and AOPs in general have been proposed as a pre-treatment to biological treatment for toxic wastewater streams (Mantzavinos and Psillakis, 2004). Though several excellent and comprehensive reviews on the process exist (Neyens and Baeyens, 2003; Pignatello *et al.*, 2006) for the sake of completeness and clarity of the following discussion we will give a short summary of the principles of reactions occurring in the photo-Fenton system.

Hydrogen peroxide is decomposed to water and oxygen in the presence of iron ions in aqueous solution in the Fenton reaction, Eq. (1), which was first reported by H. J. H. Fenton (1894). Mixtures of ferrous iron and hydrogen peroxide are called Fenton reagent. If ferrous is replaced by ferric iron it is called Fenton-like reagent. Eqs. 1–7 show the reactions of ferrous iron, ferric iron and hydrogen peroxide in the absence of other interfering ions and organic substances. The regeneration of ferrous iron from ferric iron by Eqs. 4–6), is the rate limiting step in the catalytic iron cycle, if iron is added in small amounts.

$$Fe^{2+} + H_2O_2 \longrightarrow Fe^{3+} + OH^- + OH^\bullet \tag{1}$$

$$Fe^{2+} + OH^\bullet \longrightarrow Fe^{3+} + OH^- \tag{2}$$

$$Fe^{2+} + HO_2^\bullet \longrightarrow Fe^{3+} + HO_2^- \tag{3}$$

$$Fe^{3+} + H_2O_2 \longrightarrow Fe^{2+} + HO_2^\bullet + H^+ \tag{4}$$

$$Fe^{3+} + HO_2^\bullet \longrightarrow Fe^{2+} + O_2 + H^+ \tag{5}$$

$$Fe^{3+} + O_2^{\bullet -} \longrightarrow Fe^{2+} + O_2 \quad (6)$$

$$OH^{\bullet} + H_2O_2 \longrightarrow H_2O + HO_2^{\bullet} \quad (7)$$

Furthermore, radical-radical reactions have to be taken into account:

$$2OH^{\bullet} \longrightarrow H_2O_2 \quad (8)$$

$$2HO_2^{\bullet} \longrightarrow H_2O_2 + O_2 \quad (9)$$

$$HO_2^{\bullet} + OH^{\bullet} \longrightarrow H_2O + O_2 \quad (10)$$

If organic substances (quenchers, scavengers or in the case of wastewater treatment pollutants) are present in the system $Fe^{2+}/ Fe^{3+}/ H_2O_2$, they react in many ways with the generated hydroxyl radicals. Yet, in all cases the oxidative attack is electrophilic and the rate constants are close to the diffusion-controlled limit. The following reactions with organic substrates have been reported (Legrini *et al.*, 1993): Hydrogen abstraction from aliphatic carbon atoms, Eq. (11); electrophilic addition to double bonds or aromatic rings, Eq. (12); electron transfer reactions, Eq. (13).

$$OH^{\bullet} + RH \longrightarrow R^{\bullet} + H_2O \quad (11)$$

$$R{-}CH{=}CH_2 + OH^{\bullet} \longrightarrow R{-}C^{\bullet}H{-}CH_2OH \quad (12)$$

$$OH^{\bullet} + RX \longrightarrow RX^{\bullet} + OH^{-} \quad (13)$$

The generated organic radicals continue reacting prolonging the chain reaction and thereby contribute to reduce the consumption of oxidant in wastewater treatment by Fenton and photo-Fenton method. In the case of aromatic pollutants the ring system usually is hydroxylated before it is broken up during the oxidation process. Substances containing quinone and hydroquinone structures are typical intermediate degradation products. Anyway, sooner or later ring opening reactions occur, which further carry on the mineralisation of the molecule (Chen and Pignatello, 1997). But there is one great setback of the Fenton method. Especially when the treatment goal is the total mineralisation of organic pollutants, carboxylic intermediates cannot be further degraded. Carboxylic and dicarboxylic (L: Mono- and Dicarboxylic acids) acids are known to form stable iron complexes, which inhibit the reaction with peroxide (Kavitha and Palanivelu, 2004). Hence, the catalytic iron cycle reaches a standstill before total mineralisation is accomplished, Eq. (14).

$$Fe^{3+} + nL \longrightarrow [FeL_n]^{x+} \xrightarrow{H_2O_2,\ dark} \text{no further reaction} \quad (14)$$

The primary step of the photoreduction of dissolved ferric iron is a ligand-to-metal charge-transfer (LMCT) reaction. Subsequently, intermediate complexes dissociate as shown in reaction 15. The ligand can be any Lewis base able to form a complex with ferric iron (OH^-, H_2O, HO_2^-, Cl^-, R-COO^-, R-OH, R-NH_2, etc.). Depending on the reacting ligand, the product may be a hydroxyl radical such as in Eq. 16 and 17 or another radical derivated from the ligand. The direct oxidation of an organic ligand is possible as well as shown for carboxylic acids in Eq. 18.

$$[Fe^{3+}L] + h\nu \longrightarrow [Fe^{3+}L]^* \longrightarrow Fe^{2+} + L^{\bullet} \quad (15)$$

$$[Fe(H_2O)]^{3+} + h\nu \longrightarrow Fe^{2+} + OH^{\bullet} + H^+ \quad (16)$$

$$[Fe(OH)]^{2+} + h\nu \longrightarrow Fe^{2+} + OH^{\bullet} \quad (17)$$

$$[Fe(OOC{-}R)]^{2+} + h\nu \longrightarrow Fe^{2+} + CO_2 + R^{\bullet} \quad (18)$$

Depending on the ligand the ferric iron complex has different light absorption properties and reaction 15 takes place with different quantum yields and also at different wavelengths. Consequently, the pH plays a crucial role in the efficiency of the photo-Fenton reaction, because it strongly influences which complexes are formed. Thus, pH 2.8 was frequently postulated as an optimum pH for photo-Fenton treatment (e.g. Pignatello, 1992), because at this pH precipitation does not take place yet and the dominant iron species in solution is $[Fe(OH)]^{2+}$, the most photoactive ferric iron–water complex. In fact, as shown in its general form in Eq. 15, ferric iron can form complexes with many substances and undergo photoreduction. Of special importance are carboxylic acids because they are frequent intermediate products in an oxidative treatment. Such ferric iron–carboxylate complexes can have much higher quantum yields than ferric iron–water complexes. It is therefore a typical observation that a reaction shows an initial lag phase, until intermediates are formed, which can regenerate more efficiently ferrous iron from ferric iron accelerating the process. This behaviour is observed in most of the degradation results shown in the following sections.

Fe(III) complexes present in mildly acidic solution, such as $Fe(OH)^{2+}$ absorb light appreciably in the UV and into the visible region. The quantum yield for Fe^{2+} formation in Eq. 17 is wavelength dependent. It is 0.14–0.19 at 313 nm and 0.017 at 360 nm (Faust and Hoigne, 1990). Fe(III) may complex with certain target compounds or their by-products. These complexes typically have higher molar absorption coefficients in the near-UV and visible regions that the aquo complexes. Polychromatic quantum efficiencies in the UV/visible range from 0.05 to 0.95 are common (Pignatello *et al.*, 2006). This is why the photo-Fenton process is apt to be driven by sunlight.

2.2.1.2 Solar photocatalysis hardware

For many of the solar detoxification system components (Blanco and Malato, 2003), the equipment is identical to that used for other types of water treatment and construction materials are commercially available. Most piping may be made of polyethylene or polypropylene, avoiding the use of metallic or composite materials that could be degraded by the oxidant conditions of the process. Neither must materials be reactive, interfering with the photocatalytic process. All materials used must be inert to degradation by UV solar light in order to be compatible with the minimum required lifetime of the system. Photocatalytic reactors must transmit UV light efficiently because of the process requirements. With regard to the reflecting/concentrating materials, aluminium is the best option due to its low cost and high reflectivity in the solar UV spectrum on earth surface. The reflectivity (reflected radiation/incident radiation) between 300 and 400 nm of traditional silver-coated mirrors is very low and aluminium-coated mirrors is the best option in this case. Aluminium is the only metal surface that is highly reflective throughout the ultraviolet spectrum. Reflectivitiy range from 92.3% at 280 nm to 92.5% at 385 nm. Comparable values for silver are 25.2% and 92.8%, respectively.

The photocatalytic reactor must be transparent to UV radiation. The choice of materials that are both transmissive to UV light and resistant to its destructive effects is limited. Common materials that meet these requirements are fluoropolymers, acrylic polymers and several types of glass. Quartz has excellent UV transmission as well as good temperature and chemical resistance, but high cost makes it completely unfeasible for photocatalytic applications. Fluoropolymers are a good choice of plastic for photoreactors due to their good UV transmittance, excellent ultraviolet stability and chemical inertness. One of their greatest disadvantages is that, in order to achieve a desired minimum pressure resistance, the wall thickness of a fluoropolymer tube has to be increased, which in turn will lower its UV transmittance. Acrylics could also potentially be used but they are very brittle. Other low cost polymeric materials are significantly more susceptible to be attacked by •OH radicals. Standard glass, used as protective surface, is not satisfactory because it absorbs part of the UV radiation that reaches it, due to its iron content. Borosilicate glass has good transmissive properties in the solar range with a cut-off of about 285 nm (Blanco *et al.*, 2000). Therefore, such a low-iron-content glass would seem to be the most adequate. Therefore, both fluoropolymers and glass are valid photoreactor materials. On this basis, if a large field is being designed, large collector area means also a considerable number of photoreactors and, as consequence, high system pressure drop. Thus, fluoropolymer tubes are not the best choice of material in these cases.

The original solar photoreactor designs (Goswami, 1995). for photochemical applications were based on line-focus parabolic-trough concentrators (PTCs). In part, this was a logical extension of the historical emphasis on trough units for solar thermal applications. Furthermore, PTC technology was relatively mature and existing hardware could be easily modified for photochemical processes. The main disadvantages are that the collectors (i) use only direct radiation (ii) are expensive and (iii) have low optical efficiencies. One-sun (non-concentrating) collectors have no moving parts or solar tracking devices. They do not concentrate radiation, so efficiency is not reduced by factors associated with concentration and solar tracking. As there is no concentrating system (with its inherent reflectivity), the optical efficiency is higher than for PTCs. They are able to utilize the diffuse as well as the direct portion of the solar UV-A. An extensive effort in the design of small non-tracking collectors has resulted in the testing of several different non-concentrating solar reactors (Blanco *et al.*, 2007). Although one-sun collector designs possess important advantages, the design of a robust one-sun photoreactor is not trivial, due to the need for weather-resistant and chemically inert ultraviolet-transmitting reactors. In addition, non-concentrating systems require significantly more photoreactor area than concentrating photoreactors and, as a consequence, full-scale systems must be designed to withstand the operating pressures for fluid circulation.

To design a solar collector for photocatalytic purposes, there is a group of main constraints for performing the optimization: (1) the collection of UV radiation, (2) working temperatures as close as possible to ambient temperature, and (3) quantum efficiency. Finally, its construction must be economical and should be efficient, with a low pressure drop. As a consequence, the use of tubular photoreactors has a decisive advantage because of the inherent structural efficiency of tubing. Tubing is also available in a large variety of materials and sizes and is a natural choice for a pressurized fluid system. There is a category of low concentration collectors, called Compound Parabolic Concentrators (CPCs), that are used in thermal applications. They are an interesting option, between parabolic concentrators and static flat systems. Thus they also constitute a good option for solar photochemical applications (Ajona and Vidal, 2000). CPCs are static collectors with a reflective surface designed to be ideal in the sense of Non-Imaging Optics and can be designed for any given reactor shape. They do so illuminating the complete perimeter of the receiver, rather than just the "front" of it, as in conventional flat plates. These concentrating devices have ideal optics, thus maintaining both the advantages of the PTC and static systems (Colina *et al.*, 2009). The concentration factor (R_C) of a two dimensional CPC collector is given by Eq. 19 and defined in Figure 2.2.

$$R_{C,CPC} = \frac{1}{\sin \theta_a} = \frac{A}{2\pi r} \quad (19)$$

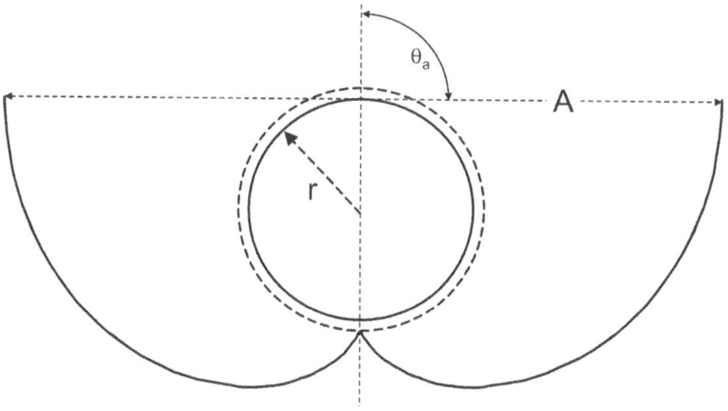

Figure 2.2 Schematic drawing of Compound Parabolic Concentrators (CPC) with a semi-angle of acceptance of 90°

The normal values for the semi-angle of acceptance (θ_a), for photocatalytic applications, are between 60 and 90 degrees. A special case is the one in which $\theta_a = 90°$, where by $R_C = 1$ (non-concentrating solar system). When this occurs, all the UV radiation that reaches the aperture area of the CPC (direct and diffuse) can be collected and redirected to the reactor. If the CPC is designed for an acceptance angle of +90° to −90°, all incident solar diffuse radiation can be collected (Figure 2.2). The light reflected by the CPC is distributed all around the tubular receiver so that almost the entire circumference of the receiver tube is illuminated. CPCs have the advantages of both technologies (PTCs and non concentrating collectors) and none of the disadvantages so they seem to be the best option for photocatalytic processes based on the use of solar radiation. They can make highly efficient use of both direct and diffuse solar radiation, without the need for solar tracking.

One important factor related to the photoreactor design is its diameter. It seems obvious that uniform flow must be maintained at all times in the reactor, since non-uniform flow causes non-uniform residence times that can lower efficiency compared to the ideal. As already commented, the Fenton reactant consists of an aqueous solution of hydrogen peroxide and ferrous ions providing hydroxyl radicals. When the process is complemented with UV/visible radiation, it is called photo-Fenton. In this case, the process becomes catalytic. Fe^{3+} (related species and organic complexes) absorbs solar photons as a function of its absorptivity. This effect must be considered when determining the optimum load as a function of light-path length in the photoreactor. The optimum concentrations of 0.2–0.5 mM Fe as function of photoreactor diameter have been proposed after many experiments with different photoreactors under sunlight at the PSA (Malato *et al.*, 2009). In the results shown in this chapter, and attending to the diameter of the photoreactor used, 20 mg/L of Fe have been used.

2.2.2 Wastewater containing inhibiting/toxic compounds (pesticides)

The enormous growth of intensive agriculture along the Mediterranean coast in recent years has polluted water with pesticides. These substances are often toxic and recalcitrant, and conventional biological treatments are ineffective for their elimination. As a result, they have been designated as priority substances (PSs) in EU legislation (Directive 2008/105/EC of the European parliament of 16 December 2008. Although several combined systems have been proposed as a solution for the treatment of different kinds of industrial wastewater, they are still not in general commercial or industrial use. This may be because most of the research on these systems is limited to evaluation of overall parameters such as BOD_5, COD and DOC and use of activated sludge from treatment plants in small laboratory devices, very often working with model wastewaters prepared by dissolving a single compound in demineralised water. The work presented here intends to go a step forward in this direction; by proposing a design strategy for an industrial combined solar photo Fenton/aerobic biological system for the decontamination of wastewater polluted with commercial pesticides. For this purpose, a mixture of commercial pesticides in natural water was selected as the model waste water because it is close to real biorecalcitrant wastewater containing different compounds (pesticides + formulation components). After preliminary laboratory testing, the feasibility of the two possibilities (photo-Fenton/biotreatment and biotreatment/photo-Fenton) of the combined system was evaluated in pilot plant, being the most successful applied at industrial scale to upgrade the treatment in a commercial solar photo-Fenton installation. Such an industrial solar photo-Fenton system has been in operation since 2004 at ALBAIDA RESIDUOS S.L., a company that selectively collects empty plastic pesticide containers from the greenhouses in El Ejido (southeastern Spain) for recycling. These containers are shredded and washed, contaminating the rinse water with pesticide residues. The rinse water is then treated by solar photo-Fenton. This photocatalytic plant has recently been upgraded to include the photo-Fenton and biotreatment combination, using the strategy included in this paper.

2.2.2.1 Preliminary, pilot plant and industrial scale tests using model wastewater

Commercial formulations of Vydate® (10% w/v oxamyl, $C_7H_{13}N_3O_3S$), Metomur®, (20% w/v methomyl, $C_5H_{10}N_2O_2S$), Couraze® (20% w/v imidacloprid, $C_{16}H_{22}ClN_3O$), Ditimur-40® (40% w/v dimethoate, $C_5H_{12}NO_3PS_2$) and Scala® (40% w/v pyrimethanil, $C_{12}H_{13}N_3$) were used as model contaminants for preliminary studies. Demineralised water was used in the pilot plant and the photo-Fenton experiments were performed using $FeSO_4 \cdot 7H_2O$, reagent-grade hydrogen peroxide (30% w/v) and sulfuric acid. For industrial scale experiments, the model compounds were dissolved in natural water (2 mS/cm; 350 mg/L Cl^-, 80 mg/L SO_4^{2-}, 3 mg/L of dissolved organic carbon, DOC). High-performance liquid chromatography HPLC with a UV-DAD detector and a C-18 column was used to monitor the pesticide concentrations during degradation. Mineralization was followed by measuring the dissolved organic carbon (DOC) by direct injection of filtered samples into a Shimadzu-5050A TOC analyzer. Hydrogen peroxide, chemical oxygen demand (COD) and iron were measured by standard methods.

A commercial assay marketed as Biofix®Lumi-10 was employed to evaluate the toxicity at different stages of the treatment using the marine bacterium *Vibrio fischeri*. Toxicity was also analyzed by respirometry using a BMT Respirometer (SURCIS, S.L). The biodegradability of the pesticide mixture was evaluated by an adaptation of the Zahn-Wellens (Z-W) test (Directive 88/303/EEC). The biodegradation process was monitored at appropriate regular time intervals, by DOC analysis of the filtered solution. The ratio of DOC eliminated after each interval to initial DOC compared to the DOC in a blank reference is expressed as the percentage of biodegradability. When this percentage reaches 70%, the test sample is considered biodegradable.

2.2.2.1.1 Pilot and industrial scale plants

Photo-Fenton at pilot scale was carried out under sunlight in a pilot plant specially developed for photo-Fenton applications installed at the Plataforma Solar de Almeria (PSA). A diagram of this system has been published elsewhere (Lapertot *et al.*, 2006). It consists of four compound parabolic collector (CPC) modules (1.02 m^2/module). Solar ultraviolet radiation (UV) was measured by a global UV radiometer (KIPP & ZONEN, model CUV 3) mounted on a platform tilted 37° (the same as the CPCs). With Eq. 20, combination of the data from several days' experiments and their comparison with other photocatalytic experiments is possible, where t_n is the experimental time for each sample, UV is the average solar ultraviolet radiation measured during Δt_n, and t_{30W} is a "normalized illumination time". In this case, time refers to a constant solar UV power of 30 W/m^2 (typical solar UV power on a perfectly sunny day around noon). V_T is the total volume of the water loaded in the pilot plant (75 L), V_i is the total irradiated volume (44.6 L). All experiments with model wastewaters were carried out at an initial DOC of 500 mg/L (100 mg/L of DOC coming from each commercial pesticide). The pesticide mixture was added directly to the plant and homogenized by turbulent recirculation for half an hour. With the collectors covered, the pH was adjusted to 2.7–2.9 and iron salt was added (20 mg/L of Fe^{2+}). Then the hydrogen peroxide was added and the collectors were uncovered, which is when photo-Fenton began.

$$t_{30W,n} = t_{30W,n-1} + \Delta t_n \frac{UV}{30}\frac{V_i}{V_T}; \quad \Delta t_n = t_n - t_{n-1} \tag{20}$$

The industrial photoreactor configuration is similar to the pilot plant described, but larger. It has 4 parallel rows of 14 photocatalytic reactor modules (20 tubes/module, 2.7 m^2/module) mounted on a 37°-tilted platform (local latitude). The total collector surface is 150 m^2 and total photoreactor volume is 1060 L. The industrial-scale experiments were also performed with 500 mg/L of initial DOC, 20 mg/L of Fe^{2+} and pH adjusted to 2.7–2.9. In this case, a matrix of raw water was employed to approximate real wastewater conditions and the operating temperature was not controlled. A diagram of the plant has been previously published (12). Figure 2.3 shows a photograph of the industrial CPC plant including a detail of the solar collector.

The biological pilot plant consists of a 160-L flat-bottom tank filled with 90–95 L of polypropylene Pall Ring® supports colonized by activated sludge (IBR-PP). The system is also provided with a 100-L conditioner tank with a pH control system (between 6.5 and 7.5), which is connected to the IBR by a recirculation pump. The system was operated in batch mode with a recirculation flow of 500 L/h. Dissolved oxygen; pH and temperature were automatically measured and registered. Furthermore, dissolved oxygen was also automatically kept between 4 and 6 mg/L.

Figure 2.3 Industrial CPC photoreactor for treating wastewater containing pesticides by solar photo-Fenton

The biological industrial plant (IBR-IP), designed taking into account the pilot plant results, consists of a neutralization tank (NT, 5000 L), a conditioner tank (CT, 1290 L), two immobilized biomass reactors (IBR1 and IBR2) filled with Pall®Ring supports colonized by activated sludge (1230 L each tank), centrifugal pumps, a blower, two dosing pumps (H_2SO_4 and NaOH), tubing and valves (see Figure 2.4). In the biological system, dissolved oxygen and pH were automatically kept in a range of 4–6 mg/L and 6.5–7.5, respectively. The industrial biological system was operated also in batch mode with a recirculation flow of 1500 L/h.

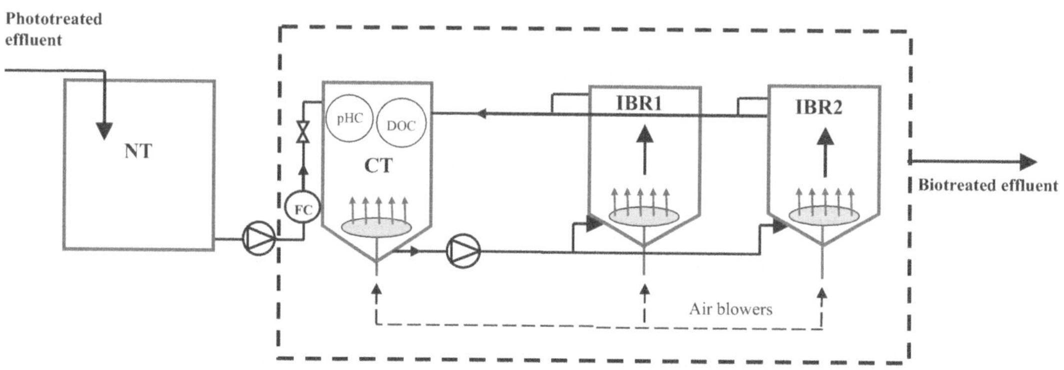

Figure 2.4 Biological industrial plant (IBR-IP), designed within Innowatech project and operated in combination with Industrial CPC photoreactor

2.2.2.1.2 Pilot scale results

The photo-Fenton degradation of the mixture of commercial pesticides was studied at different initial DOC (100, 200 and 500 mg/L) to observe the effect of the initial organic content on the decontamination rate at 20 mg/L Fe^{2+}. In Table 2.1 and subsequent results, initial concentration of DOC included was that really measured, very often not exactly the same that intended due to the size of the pilot plant and to the excipient of the commercial pesticides. H_2O_2 concentration was always maintained between 100-300 mg/L, and the DOC from each commercial pesticide was 1/5 of the total mixture. Initial DOC of the mixture and initial concentration of the active ingredients as measured by liquid chromatography are shown in Table 2.1.

It may be observed in Figure 2.5 that, as expected, overall H_2O_2 consumption and illumination time required for achieving the total elimination of all the active ingredients and mineralization increased with higher initial DOC, since the higher organic load requires more hydroxyl radicals, and consequently, degradation of the pesticide mixture takes longer. However, the active ingredients (considered as a whole) were all eliminated after a certain amount of mineralization of the initial DOC and approximately at the same rate (zero order kinetics), which means that the reaction mechanism does not depend on initial organic content. The illumination time necessary for 65% mineralization was also proportional to initial DOC, which reinforces the above. Also worthy of mention is that under the same operating conditions, less H_2O_2 was required for mineralizing DOC as the initial concentration increased (H_2O_2 consumed/mg DOC mineralized). As H_2O_2 was used more efficiently as DOC increased, this means that there was always excess H_2O_2 and the concentration of $OH^{\bullet}$ radicals was constant and governed by reactions 16 and 17. Therefore, the reaction rate measured for pesticide degradation or degradation products mineralization did not depend on organics concentration. It is well-known that excess H_2O_2 could be consumed by inefficient reactions and therefore, when the organic load in wastewater is low, fewer $OH^{\bullet}$ radicals are required and H_2O_2 is not used efficiently. This information was so useful for H_2O_2 dosing in a the industrial plant for treating real wastewater, as the H_2O_2 concentration should be kept lower as pesticide concentrations in wastewater are lower. Moreover, the process would also be more efficient from the point of view of H_2O_2 consumption, at higher concentrations of pollutants in wastewater. Therefore, the rest of the study was done at the highest concentration.

Table 2.1 Initial concentration of each active ingredient at different initial DOC of the total mixture

DOC_0 (mg/L)	Oxamyl (mg/L)	Methomyl (mg/L)	Imidacloprid (mg/L)	Dimethoate (mg/L)	Pyrimethanil (mg/L)
107	42	10	8.5	15	19
230	79	22	21	32	47
550	210	48	48	69	82

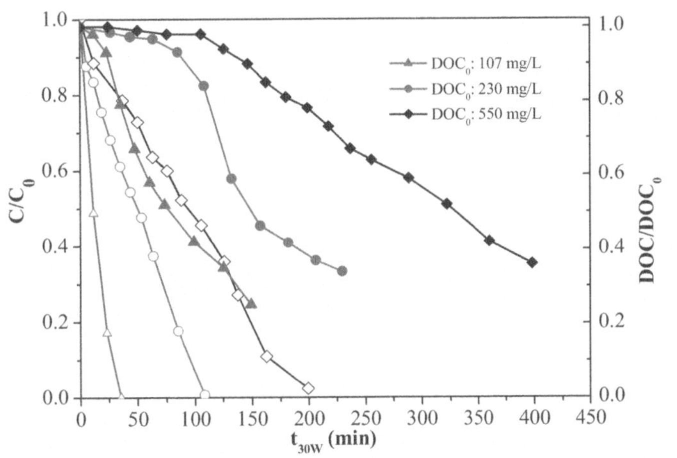

Figure 2.5 Mineralization (solid symbols) and degradation (open symbols) of the pesticide mixture at different initial DOC concentrations during the photo-Fenton experiments

The first step was to demonstrate that photo-Fenton could be employed as a pretreatment to enhance the biodegradability of the mixture of five commercial pesticides. The original DOC of 500 mg/L (100 mg/L DOC from each pesticide) corresponded to pesticides and other unknown components of the commercial formulations. During the process, partly photodegraded samples with different DOCs (from 500 to 150 mg/L) were collected. At the end of the experiment (after 430 minutes of illumination time) the total mineralization was 70% and H_2O_2 consumption was 48 mM. The active ingredients were completely eliminated after 23 mM of H_2O_2 consumed (DOC = 345 mg/L). The relation between the hydrogen peroxide consumption and the illumination time during the photo-Fenton process was 0.1 mM H_2O_2/minute, approximately. In addition to DOC, COD was also measured during treatment to evaluate the average oxidation state (AOS) of the solution according to Eq. 21 (Parra *et al.*, 2000), where DOC and COD are expressed in molar concentrations. AOS is between +4 for CO_2, the most oxidized state of C, and −4 for CH_4, the most reduced state of C. AOS is a valuable parameter that can be used to estimate oxidation in a complex solution consisting of the initial substances and their oxidation by-products. It can also provide indirect information on its biodegradability, as it indicates variations in the composition of the effluent that could result in changes in biodegradability/toxicity of the solution.

$$AOS = \frac{4 \cdot (DOC - COD)}{DOC} \tag{21}$$

Figure 2.6 shows the progress of COD, DOC and AOS during pesticide mixture degradation by photo-Fenton. AOS increased gradually throughout the experiment. These results suggest that more oxidized organic intermediates were formed during the photo-Fenton process, changing the average chemical nature of the mixture (note degradation rate for COD was higher at the beginning of the test than for DOC). After a certain amount of H_2O_2 consumption, COD and DOC behave similarly, stabilizing the AOS. Formation of more oxidized intermediates is an indirect indication of the ability of the treatment to improve biodegradability. Taking these changes in AOS into account, toxicity and biodegradability studies were performed until 75% mineralization.

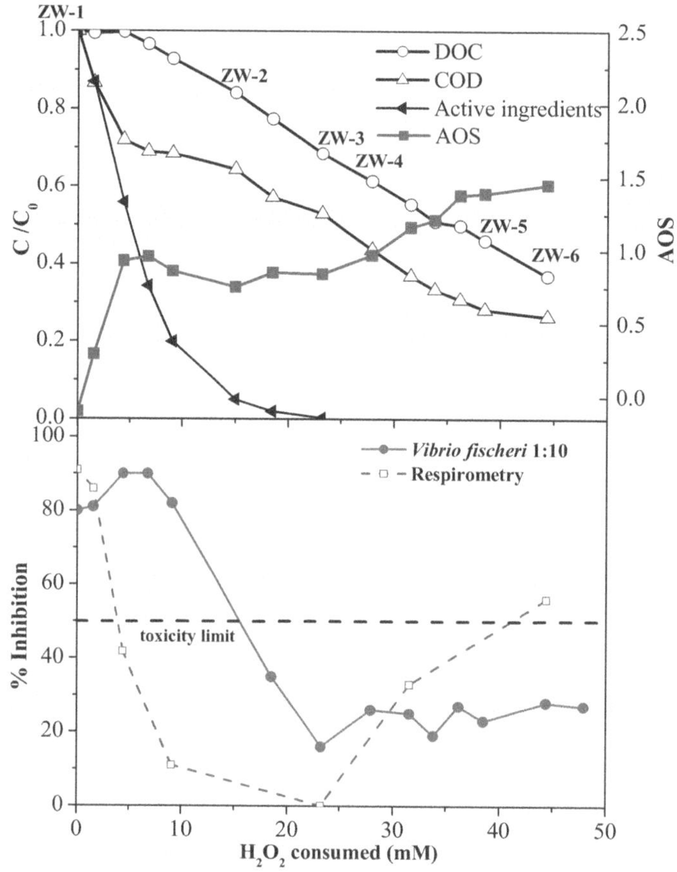

Figure 2.6 Up: DOC, COD and AOS during photo-Fenton degradation of the pesticide mixture as a function of H_2O_2 dose. Active ingredients refer to all five pesticides. Down: Toxicity expressed as percentage of inhibition (evaluated with *Vibrio fischeri* and activated sludge)

Toxicity was monitored using *Vibrio fischeri* and respirometry tests, as shown also in Figure 2.6. The *Vibrio fischeri* toxicity was evaluated in 1:10 dilutions knowing that the sample was especially toxic, dropping from 80% to 20%. These results confirm that photo-Fenton degradation of the pesticide mixture was able to gradually reduce the toxicity of the initial solution. Toxicity was also studied with a respirometric assay. In general, the combination of different microorganism populations makes activated sludge more resistant to toxic effluents than *Vibrio fischeri*. As also shown in Figure 2.6, as soon as the active ingredients were eliminated by photo-Fenton process, the activated sludge did not show toxicity. However the final inhibition of 56% correlates with the inhibition measured in the *Vibrio fischeri* assay. Considering the results above, the total elimination of the active ingredients might be a key point of the treatment. Afterwards, respirometric and *Vibrio fischeri* tests show that toxicity increases, which could be (or not) related to biodegradability, as discussed below. Biodegradability of the mixture during phototreatment was also evaluated in different samples by the Zahn-Wellens (Z-W) test along 28 days. Among the different samples taken during photo-Fenton, two of them (ZW1 and ZW2) contained varying amounts of the active ingredients of the commercial pesticides. The results are summarized in Figure 2.7. ZW1 (highest concentration of the active ingredients) was non-biodegradable (around 50% of biodegradability in 28 days), whereas the rest of the samples were biodegradable according to the Z-W test. ZW2 (containing 5% of initial active ingredients) was biodegradable after a long period of 9 days. However, once the active ingredients had been eliminated (ZW3: 345 mg/L of DOC), all the samples became biodegradable in a short time (5 days or less). ZW5 and ZW6 had the highest biodegradation percentages. The evolution of the five active ingredients concentration during the 28 days of test was monitored in ZW1, remaining constant in the solution, confirming that they are non biodegradable. This demonstrates that the most suitable point for combining photo-Fenton with the biological treatment was after the total elimination of the five active ingredients, somewhere between ZW3 and ZW4.

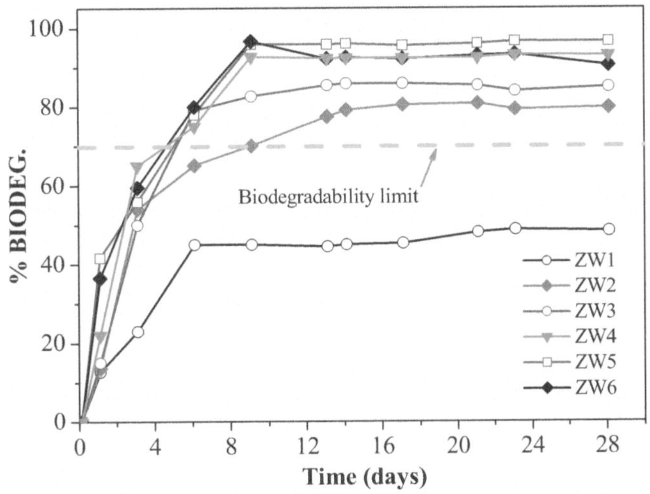

Figure 2.7 Zahn-Wellens (Z-W) biodegradability analysis of samples taken during photo-Fenton. The H_2O_2 doses corresponding to S1, S2, S3, S4, S5, S6, S7 and S8 are 0, 6.3, 11.3, 16.4, 23, 27.7, 36.6 and 39.1 mM, respectively

Therefore, the next step was to test the degradation of the commercial pesticide mixture in the combined photo-Fenton/biological system. To do this, the five commercial pesticides were degraded by photo-Fenton until total elimination of the active ingredients, according to the above biodegradability results. This partly oxidized effluent was then discharged into the IBR-PP system after adjusting the pH to around 7. After three days of biotreatment, the DOC had dropped to 28 mg/L. The results found in the IBR were more favorable than in Z-W test (higher percentage of biodegradation in less treatment time). Overall mineralization efficiency of the combined system was 94%, of which 35.5% corresponds to the AOP and 58.5% to the aerobic biological treatment.

2.2.2.2 Industrial scale results

After these preliminary tests, the combined photo-Fenton/IBR system was reproduced in an up-scaled system designed from the pilot plant results to prove the feasibility of the technology. The commercial pesticide mixture and the experimental conditions (500 mg/L of initial DOC and 20 mg/L of Fe^{2+}) were maintained. The photo-Fenton treatment was also extended until complete degradation of the active ingredients (34% mineralization, 600 minutes illumination time and 20 mM of H_2O_2 consumed). Figure 2.8 shows the degradation of the five active compounds, the mineralization of the mixture (reduction in COD and DOC) during photo degradation over hydrogen peroxide consumption. As observed, the H_2O_2 required for complete elimination of the active ingredients is in the same range as in the pilot plant (20–25 mM). This confirmed our previous statements (Gernjak *et al.*, 2006): H_2O_2 consumption of a specific wastewater could be used as a reference parameter for controlling the most important factor of a combined photo-Fenton/biological system, the moment when the wastewater reaches the desired purification (defined by certain DOC mineralisation and disappearance of the key compounds) for transferring from the AOP to the biotreatment. It should be mentioned here that H_2O_2 analyses are quick and easy and can be applied during normal operation of treatment plants. This effluent was discharged into the IBR-IP (after adjusting the pH to around 7). After 5 days of biotreatment, DOC was reduced 71% (80 mg/L of final DOC). The efficiency of the industrial-scale combined system was 84%, 35% corresponding to the photo-Fenton treatment and 49% to the biological stage.

Although the pesticide mixture was successfully degraded in both combined systems, two points deserve special mention in the industrial plant experiments. First, photo-Fenton degradation was much slower (600 minutes compared to 250 minutes illumination time for the elimination of the active ingredients at pre-industrial and pilot-plant scale, respectively). This significant increase in the illumination time can be attributed to: (i) water in the industrial plant was salty (around 2 mS/cm) groundwater, and the effect of salts on the photo-Fenton reaction rate is well known (De Laat *et al.*, 2004; Bacardit *et al.*, 2007); (ii) temperature in the photoreactor of the industrial plant was around 20°C, in winter in South-Spain, which was much lower than experiments at pilot-plant scale (operated at 35°C).

Secondly, IBR-IP efficiency in terms of DOC reduced per volume of supports colonized by sludge and time was considerably lower than in the pilot plant (50 versus 160 mg DOC reduced/day L of Pall Ring supports respectively).

This may be due to the difference in the bioconsortia formed in the two bioreactors, and also the higher salinity of the water in the industrial-scale plant. However, the biological treatment in the normal operation of the industrial plant would have been enhanced by specific biomass adaptation to this particular wastewater, which takes several months. Both aspects were taken into account when treating real wastewaters, as commented below.

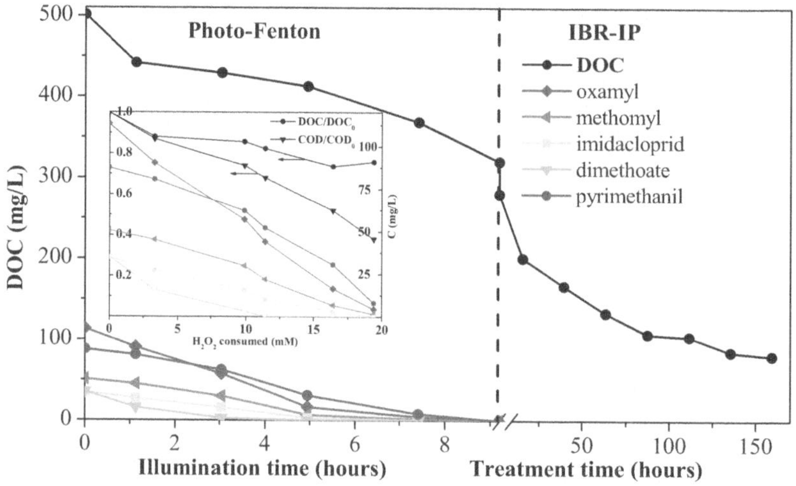

Figure 2.8 Pesticide degradation and mineralization of the mixture by the industrial- scale combined photo-Fenton/IBR-IP system (DOC_0: 500 mg/L, 20 mg/L Fe^{2+}). The inset shows degradation and mineralization over H_2O_2 consumed during the photo-Fenton process

The real industrial wastewater generated in the washing of commercial pesticide containers has a heterogeneous nature (variable DOC, ionic concentration and pesticide content). The wastewater was collected and stored (approximately 1200 L of wastewater per 1000 kg of plastic containers, corresponding to an initial rate of DOC between 200 and 500 mg/L). For the decontamination of this wastewater, batches of 2000 L were treated by photo-Fenton in the solar reactor. The photo treated effluent was then discharged in a tank of 5000 litres for neutralizing its pH before the biological treatment. This tank received the water from several batches of the photocatalytic plant and it was used as a water source for the biological reactor. Regarding the former results with model wastewater, different considerations were taken into account: (i) biodegradability was enhanced when active ingredients of the pesticides were removed but the remaining DOC exhibited a low biodegradation rate, (ii) the composition of the real wastewater is so variable that it is very difficult to adjust the initial dose of H_2O_2 to carry out the photo-Fenton process until complete degradation of the active ingredient is attained. Consequently, it was decided to adopt a compromise solution and to extend photo-Fenton treatment to a larger mineralization degree in order to ensure pesticide degradation and to increase the biodegradability of the remaining DOC. Therefore, the proposed strategy for the combined system was: extending the photo-Fenton process until 40–50 % of mineralization before discharging the partially oxidized solution in the IBR-IP.

DOC_0 was reduced from 480 mg/L to 300 mg/L in the photocatalytic step (37.5 % mineralization) with 15 mM of H_2O_2 consumed (see Figure 2.9). It is interesting to note that using a similar H_2O_2 dose, in the experiment with simulated wastewater containing five pesticides, lower mineralization was attained. Therefore, it means that the organic content of the real wastewater contained less persistent contaminants than the model wastewater. The neutralized mixture was treated in the IBR-IP, and after only 1.5 days the DOC was 75 mg/L. Continuing the biotreatment did not improve the efficiency outstandingly. In this case, the overall efficiency of the coupled system, in terms of initial DOC elimination, was 84% (37.5% photo-Fenton and 46.5% IBR) after 2 hours of photo-Fenton and 1.5 days of biotreatment, respectively.

The identification and quantification of the pesticides present in real wastewater was performed, during the combined system, by liquid chromatography with a time-of-flight mass spectrometry detector (LC-TOF-MS) based on the use of an accurate-mass database. We used a rapid automated screening method developed by Mezcua *et al.* (2009) to verify the presence of 300 compounds in less than 20 minutes. Twelve active ingredients in diverse concentration ranges were identified in the initial sample of the real wastewater (see Figure 2.9). Three pesticides identified by LC-TOF-MS in real wastewater were contained in the model mixture previously studied (imidacloprid, dimethoate and pyrimethanil). After the biological treatment, the majority were totally removed, except for two of them (pyrimethanil and thiacloprid), that were found in the range of μg/L. Figure 2.9 also shows the reduction of each pesticide (in percentage) in the integrated system.

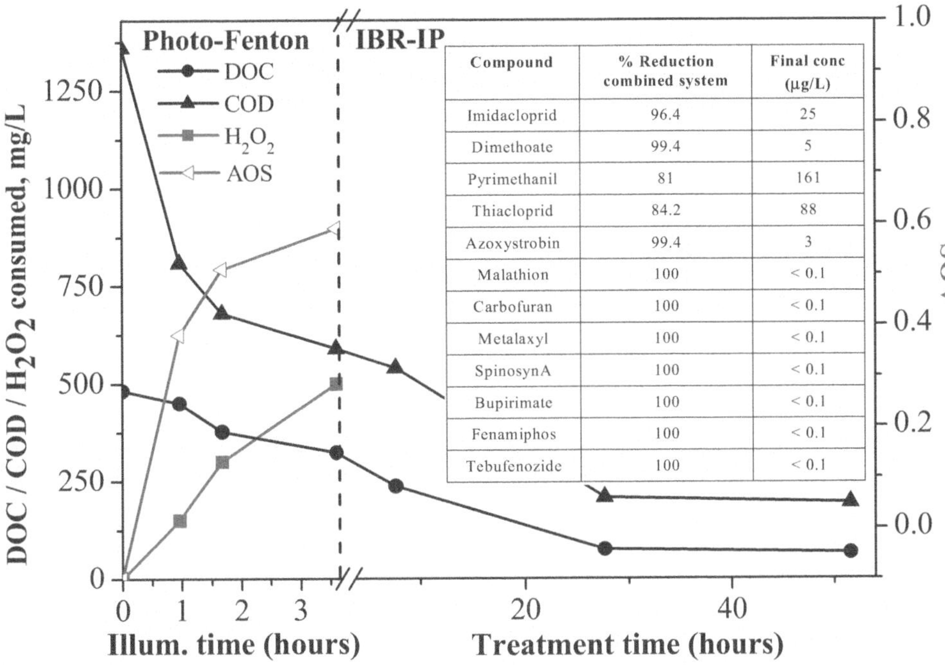

Compound	% Reduction combined system	Final conc (μg/L)
Imidacloprid	96.4	25
Dimethoate	99.4	5
Pyrimethanil	81	161
Thiacloprid	84.2	88
Azoxystrobin	99.4	3
Malathion	100	< 0.1
Carbofuran	100	< 0.1
Metalaxyl	100	< 0.1
SpinosynA	100	< 0.1
Bupirimate	100	< 0.1
Fenamiphos	100	< 0.1
Tebufenozide	100	< 0.1

Figure 2.9 Mineralization of the real wastewater (DOC and COD), H_2O_2 consumed, AOS evolution in the combined system photo-Fenton/IBR-IP. The inset shows the concentration in the IBR-IP effluent of the pesticides determined by LC-TOF-MS

2.2.3 Wastewater containing a large amount of biodegradable organic compounds in addition to small concentrations of recalcitrant compounds (pharmaceuticals)

2.2.3.1 Introduction

ECs (Emerging Contaminants) have been defined in several ways, but essentially they are either naturally occurring or synthetic substances that interfere with the functioning of endocrine systems resulting in unnatural responses. Pharmaceuticals, personal-care products, steroid sex hormones, illicit drugs, flame retardants and perfluorinated compounds, considered environmental emerging contaminants of particular concern because of their endocrine-disrupting properties, enter the wastewater network after their use in households and industry. In view of the widespread occurrence and potential impact of ECs, they must be removed before discharge or reuse (Heberer, 2002; Zhang and Zhou, 2008). They may have serious toxic and other effects on humans and other living organisms, and they are present at minute concentrations, thus requiring more sophisticated and laborious analytical tools for their accurate determination. Therefore, the ability to increase biodegradability and detoxify effluent streams containing polar and hydrophilic chemicals by alternative treatment with AOPs has been studied in detail during the last years and reviewed by Kosjek and Heath (2008), Gültekin and Ince (2007) and Klavarioti *et al.* (2009).

Application of AOPs to wastewater containing pharmaceuticals or ECs has been widely studied, but not many cost-effective chemical and biological treatment combinations are available. In this sense, ozonation has been employed as a pre-oxidation step in a combined treatment line. However, no mineralization was usually observed during the ozonation pre-treatment. It has been demonstrated that the use of advanced oxidation processes (AOPs) can break down a wide range of pharmaceutical molecules due to the on-site generation of sufficient hydroxyl radicals ($^{\bullet}OH$). The efficacy of AOPs in treating a wastewater is strongly dependent on (i) the composition and concentration of the wastewater in question and (ii) the treatment target, so the heavier the pollutant load and the more removal is required, the stronger the treatment conditions must be. From this point of view, treatment performance can be enhanced by the application of AOPs as a pre-treatment stage to enhance biodegradability and reduce toxicity followed by biological post-treatment. This concept, which has attracted much attention over the past several years as commented before, is relatively straightforward and based on (i) biological treatment being less expensive and more environmentally friendly than any other destructive treatment, and (ii) complete mineralisation by AOPs leads to excessive treatment cost since the highly oxidized end-products tend to be resistant to total chemical oxidation. But studies involving real wastewater treatment are scarce, and still scarcer those taking into account that wastewater often contains a significant number of biodegradable compounds along with those that are biorecalcitrant. Nevertheless, there are very few studies that combine the information of chemical analysis, toxicity analysis and biodegradability analysis to study the viability of the combination of photo-Fenton and biological treatment on actual industrial wastewater, not only model wastewater.

Hence, there is still a major need for a scientific rationale on which an "a priori" choice of the most appropriate treatment can be based and additional case-studies like the present one are required to enhance the common knowledge database.

The aim of this section was to explain how to find the best way of treating real wastewater containing nalidixic acid (a synthetic antibacterial agent frequently used in the treatment of urinary tract infections involving Gram-negative organisms), by combining an AOP and a biotreatment. This is an example of wastewater containing a considerable amount of biodegradable compounds, in which the pre-oxidation step would not lead to a significant improvement of biodegradability; rather, it will only cause unnecessary consumption of chemicals. In such cases, a biological pre-treatment (removing biodegradable compounds) followed by solar photo-Fenton may prove to be more useful. Both combinations AOP/Bio and Bio/AOP will be discussed and compared. This section will focus also in those aspects. The work also involved a detailed liquid chromatography-time of flight-mass spectrometry (LC-TOF-MS) study of intermediates formed during both treatments.

2.2.3.2 Experimental details

Photo-Fenton conditions (Fe, H_2O_2, pH, etc.) and analyses were performed as commented in Section 2.2.2. The pharmaceutical wastewater had a pH of 3.98, 725 mg/L of DOC, 3400 mg/L of COD, 0.4 g/L of TSS, 40 mg/L of nalidixic acid (NXA) and inorganic ions ($Cl^- = 2.7$ g/L, $Na^+ = 1.8$ g/L, $PO_4^{3-} = 0.018$ g/L and <0.1 mg/L of NH_4^+ and NO_3^-). These data varied slightly between different experiments as real wastewater composition was not constant in all the tested samples. Anyway, experimental details are stated in any case. Other organics present in the pharmaceutical wastewater were analysed by LC-TOF-MS. The wastewater was treated directly, without dilution. The actual treatment procedure in the industrial facility includes blending the wastewater with other water and the combined stream is disposed of in a central biotreatment plant. NXA was found to cause serious problems in such a centralized treatment plant. Solid-phase extraction using OasisTM HLB cartridges was applied first to reduce the salt content in the matrix before LC-TOF-MS, and to improve intermediate detectability. NXA degradation intermediates were studied by Liquid Chromatography-Time of Flight-Mass spectrometry using positive-ion mode electrospray ionization in positive mode, LC-TOF-MS (ESI+). Photo-Fenton at pilot scale was carried out in the pilot plant installed at the Plataforma Solar de Almeria (PSA), consisting of four compound parabolic collector (CPC) modules ($1.02\ m^2$/module) and described in Section 2.2.2. The biological reactor was IBR-PP described also in a former section, colonized by activated sludge from a conventional wastewater treatment plant (WWTP) in Almeria (southeast Spain). The biological reactor was started up and adapted to high NaCl content (approximately 5 g/L). Bioassays included toxicity by *Vibrio fischeri* (Biofix®Lumi-10) and biodegradability by Zahn-Wellens (Z-W) test (an adaptation of the EC protocol, Directive 88/303/EEC).

2.2.3.3 Wastewater treatment by AOP/Biotreatment

The main parameters of the industrial pharmaceutical wastewater have been described in the former paragraph and one relevant point was the high conductivity, associated to the presence of large amounts of inorganic ions, such as chloride and sodium, found in grams per litre. The sample further contained a significant concentration of suspended solids. The most important organic compound studied in the matrix was nalidixic acid (Figure 2.9). The concentration of this compound in the wastewater was 45 mg/L. Together with NXA, which was the target constituent of the wastewater effluent, a second compound was also importantly present (P1, see Figure 2.10), which was identified as l-ethyl-l,4 dihydro-7-methyl-4-oxo-1,8-naphthyridine, a decarboxylated form of NXA, which has also been described as a product of NXA photolysis and thermolysis (Fernández and Cárdenas, 1990). An isobaric compound (P14) was also identified at a different retention time, which could correspond with a positional isomer of P1, probably generated during the synthesis of NXA. Other lower intensity peaks were present in the chromatogram (see Figure 2.10), showing the presence of a complex mixture of NXA derivatives. The most relevant were identified and their structures are shown in Figure 2.10. They may correspond to previous stages of NXA synthesis or to degradation products generated in the wastewater during storage. The compound labelled P3 was identified as 7-hydroxymethyl-NXA, which has been described as the primary NXA metabolite. This compound exhibits at least in vitro anti-bacterial activity identical in spectrum and potency to that of the parent compound (Guiberteau *et al.*, 2007). The absence of commercial standards makes their quantitative evaluation impossible and the material balance could therefore not be closed but take into intensity scale to be compared with the results obtained after treatment. Nevertheless, NXA was quantified during all the process.

An initial biodegradability test was performed at two different dilutions (1:2 and 1:8) of the original wastewater. The 1:2 dilution was selected according to the initial DOC recommended in Z-W standard methodology. The other dilution (1:8) was studied in order to understand the behaviour of sludge in contact with small concentrations of nalidixic acid. As shown after in Figure 2.14, both samples were biodegradable. However, the concentration of NXA remained constant in both cases. Adaptation of biomass also took much longer (around 15 days) for the 1:2-diluted sample, mainly due to the higher NXA concentration. This pharmaceutical may therefore be considered recalcitrant and possibly inhibitory to sludge metabolism, though not very acutely toxic. NXA inhibition of sludge metabolism was also confirmed during photo-Fenton tests, as discussed below.

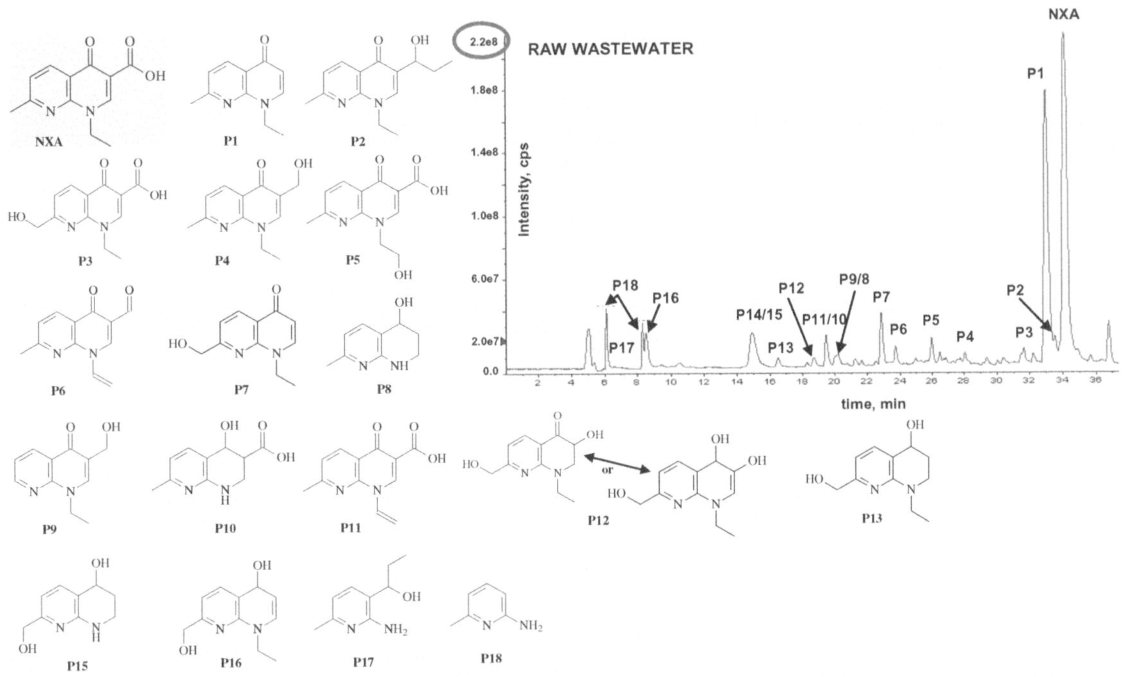

Figure 2.10 LC-TOF-MS chromatogram obtained by the analysis of a wastewater sample before treatments and chemical structures proposed

Figure 2.11 shows photo-Fenton treatment of the real wastewater. The experimental conditions were 20 mg/L of total iron and H_2O_2 concentration was maintained between 200 and 400 mg/L. 90% of the initial DOC was removed in 400 minutes of illumination and NXA had completely disappeared after 190 minutes. Since the biodegradability of the initial wastewater was relatively high, NXA degradation was the main goal of the treatment. The important issue was to determine the behavior of biodegradability and toxicity of wastewater during the photo-Fenton treatment, mainly during the degradation of NXA and shortly afterwards, in order to find the best treatment time with the minimum hydrogen peroxide consumption. For this aim, it was necessary to reproduce the previous experiment maintaining all the parameters except H_2O_2 dosing. A small quantity of H_2O_2 (in the order of a few mM) was added to the photoreactor, and consumption monitored. After the peroxide was consumed, a sample was taken for bioassay, ensuring that no H_2O_2 remained (which could affect the bioassays). H_2O_2 was again added, and another sample was taken after all the peroxide had been consumed. As NXA and DOC were also determined in these experiments, results can be included with the kinetic results shown in Figure 2.11. Different groups of samples were selected for the *Vibrio fischeri* toxicity bioassay and COD analysis.

All the toxicity results were quite similar, and it can therefore be concluded that photo-Fenton treatment did not decrease *Vibrio Fischeri* toxicity. Therefore, toxicity bioassays show that photo-Fenton was unsuccessful. But this result is not surprising for real wastewater containing hundreds of mg/L of different organic compounds (for example, carboxylic acids) and their degradation intermediates as this bioassay has been described as very sensitive (Hernando *et al.*, 2007). Toxicity tests (usually a quick method) can sometimes help to select the stage of an AOP treatment at which the water becomes nontoxic and, presumably, biodegradable but in view of the results shown in Figure 2.11, such information was not found from the toxicity assays.

Variation in the COD during the experiment was also determined along with DOC. The considerable decrease in this parameter agrees with the strong oxidation of organic matter. The efficiency of the oxidative process is more clearly shown by the AOS parameter (see Eq. 21). The maximum AOS of −0.3 was reached after approximately 25 mM of H_2O_2 consumed and remained almost constant until the end of the test. AOS usually increases with treatment time until almost reaching a plateau. At the moment that AOS stabilizes, the chemical treatment is only mineralizing organic contaminants, but with no partial oxidation. The changes in AOS and toxicity were taken into account in selecting the treatment stage at which the water might presumably be biodegradable. Z-W tests (usually a time-consuming method) were only performed on samples taken after sample 4. Before this treatment stage the concentration of NXA and toxicity was rather high and AOS increased. Z-W tests (Figure 2.12) were performed on six undiluted samples at different stages during the photo-Fenton process. All samples were at least 90% biodegradable at the end of the Z-W assay. In three of them (ZW4, ZW6 and ZW7), in which NXA concentrations were 20.7, 8.5 and 4.6 mg/L, respectively, 70% biodegradability was attained only after 10 days of biotreatment (Samples ZW6 and ZW7 after 8 days). On the other hand, samples with very low concentrations (<1 mg/L) or without NXA (ZW9, ZW11 and ZW13) were biodegradable after 3 days. Results show that untreated samples need much longer adaptation periods than treated samples and that NXA was also detrimental for biodegradation efficiency. These results demonstrated that photo-Fenton should be performed until total degradation of NXA before coupling a biological treatment.

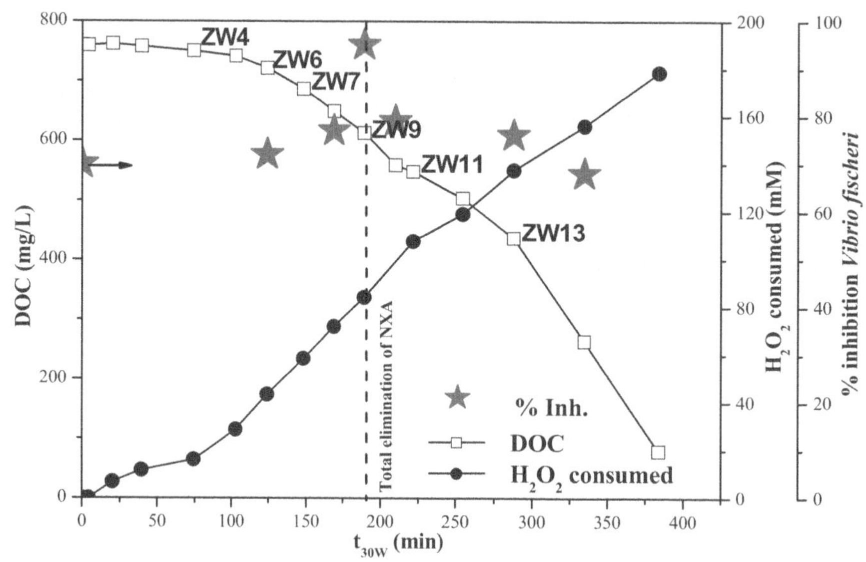

Figure 2.11 Mineralization of industrial pharmaceutical wastewater, H_2O_2 consumed and toxicity bioassay (*Vibrio fischeri*) during photo-Fenton. ZWn (from ZW1 to ZW113) were selected samples for biodegradability studies

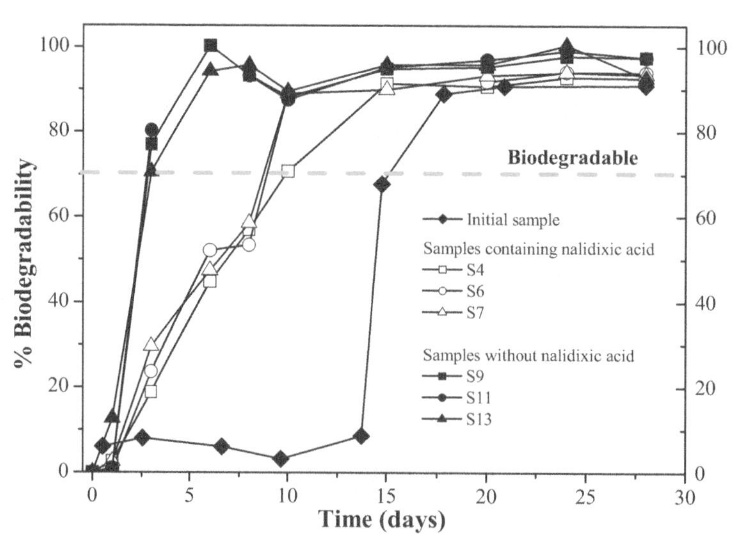

Figure 2.12 Zahn-Wellens test for selected samples during the photo-Fenton process. Initial sample is also shown

When the best photo-Fenton wastewater pretreatment for biodegradability had been determined, the combined photo-Fenton/biological treatment was carried out in a pilot bioreactor. Before performing the experiment in the combined system, the IBR was inoculated with 150 L of concentrated activated sludge from a municipal wastewater treatment plant. Then, recirculation was maintained between the conditioner tank and the IBR-PP in order to ensure optimum fixation of the sludge on the propylene Pall Ring supports. Total suspended solids, DOC and inorganic ion concentration (mainly ammonia and nitrate) were measured daily. The system was maintained with controlled addition of glucose and ammonium chloride, keeping the carbon/nitrogen ratio at 100/20. The next step was adapting the sludge to high salinity, as the wastewater contains large quantities of NaCl. During the whole adaptation process, NaCl concentration was increased gradually in five cycles from 1 to 5 g/L. The photo-Fenton runs were accumulated in a large volume tank (neutralization tank) connected to the bioreactor where the pH was adjusted roughly to 7, as automatic pH control is in the bioreactor itself. In the coupled system, wastewater containing NXA (45 mg/L) was pre-treated by photo-Fenton until its total elimination during 190 minutes of illumination time. The chemical characterization of the photo-Fenton effluent was 530 mg/L of DOC, 6.5 g/L of NaCl and absence of NXA (<1 mg/L). The pre-treated effluent was pumped from the neutralization tank to the conditioner tank connected to the IBR-PP.

The system was operated in batch mode, with a recirculation flow rate of 500 L/h between the conditioner tank and the IBR-PP until the effluent was bio-mineralized (final DOC 35 mg/L). The DOC and evolution of nitrogen (as NH_4^+ and NO_3^-) during biological treatment are shown in Figure 2.13. DOC decreased 495 mg/L in 4 days, similar to the result obtained in the Z-W test. The consumption of ammonia and the generation of nitrates demonstrated nitrification and nitrogen assimilation through biomass growth as overall nitrogen content decreased. Overall mineralisation efficiency of the combined photo-Fenton and biological treatment in batch mode for the degradation of the real pharmaceutical wastewater was over 95%, of which 33% corresponds to the solar photochemical process and 62% to the biological treatment. *Vibrio Fischeri* bioassays were also performed on the biotreated effluent showing below-threshold toxicity. Therefore, the combined treatment was also successful from the viewpoint of biotoxicity removal.

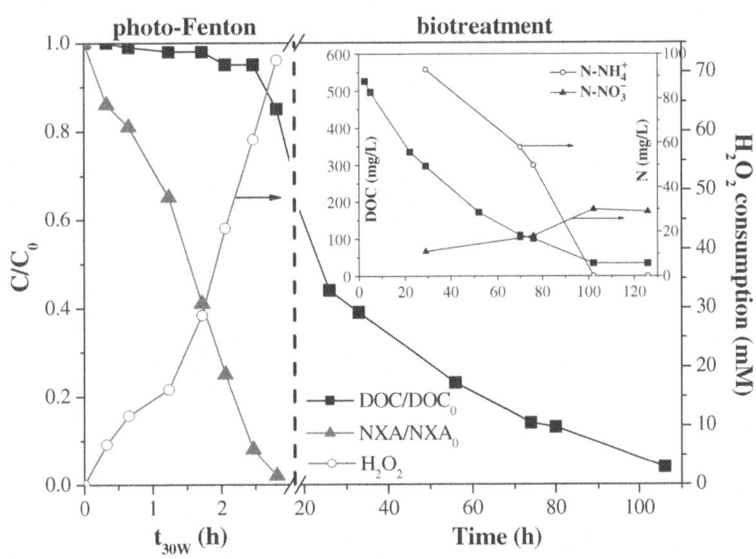

Figure 2.13 Wastewater treatment during photo-Fenton process coupled with biological system. The inset shows mineralisation of the pre-treated wastewater in the IBR-PP

2.2.3.4 Wastewater treatment by Biotreatment/AOP

It has been demonstrated that an industrial wastewater containing a biorecalcitrant compound can be successfully treated by photo-Fenton enhancing the wastewater biodegradability. For maximizing the efficiency of the process, a suitable selection of the photo-Fenton treatment time and hydrogen peroxide dose is required. However, as it was previously demonstrated by Z-W tests (shown in Figures 2.12 and 2.14) despite the most of the organic content of the wastewater could be biodegraded, the biomass required a shorter adaptation period as NXA concentration was lower, and NXA persisted in all cases. This demonstrated that a biological system could treat the matrix studied, but not NXA, which is not biodegradable and should be removed afterwards by an AOP. Therefore, it would be pertinent to test if a bioreactor could give better results than Z-W after an adaption period.

The IBR-PP was filled with untreated pharmaceutical wastewater with pH previously adjusted to around pH = 7, and maintained between 7 and 8 by a pH control system. DOC and NXA concentrations were monitored throughout the biological treatment. After 4 days, 96% of the initial DOC was removed, demonstrating that biomass immobilized in the IBR-PP was more efficient (and not inhibited by NXA) than activated sludge utilized in Z-W tests. On the first day, NXA concentration was reduced to 14 mg/L and remained constant until the end of the process. This sharp drop in NXA could only be related to the adsorption of the compound on the IBR-PP (biomass or 15 mm polypropylene

Pall Rings). After the biological treatment, the partially biotreated wastewater was treated by photo-Fenton. Two degradation studies were carried out. In the first, the biotreatment effluent (NXA 13 mg/L and DOC 22 mg/L) was treated directly. In the second study, the biotreated wastewater was enriched by adding NXA standard until initial concentration, simulating absence of IBR-PP adsorption of NXA after biomass saturation in normal plant operation. The other parameters, common to both studies, were 20 mg/L of iron, pH = 2.7 and H_2O_2 concentration kept between 200 and 400 mg/L. In both cases, NXA was completely degraded at around 25 minutes of illumination time and 12 mM of H_2O_2 consumed, reaching a DOC of around 20 mg/L, which is low enough for disposal. The overall efficiency of the combined IBR + solar photo-Fenton treatment in terms of elimination of DOC was over 97%. Biorefractory compounds are very often not so toxic to activated sludge, and biodegradation prior to AOP may be successful, avoiding long treatment times and high chemical consumption, the main drawbacks of AOP treatments. In these cases, AOP treatment is more suitable as a polishing step than as a pretreatment.

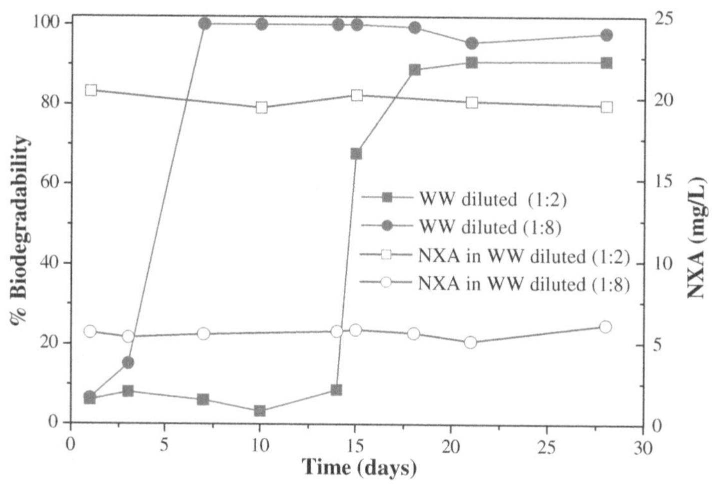

Figure 2.14 Zahn-Wellens biodegradability test of the pharmaceutical wastewater at 1:2 and 1:8 dilutions

It should be mentioned (as stated in former section, see Figure 2.13) that direct photo-Fenton treatment of this wastewater required around 3 hours of treatment time and 80 mM of H_2O_2 to completely degrade NXA, and almost 600 mg/L of DOC still remained after NXA had completely disappeared.

Toxicity was monitored using *Vibrio fischeri*. It showed high toxicity in the first wastewater sample, but did not show any after biological treatment, and at the end of photo-Fenton, inhibition was 59%. Toxicity could be related to the presence of organic acids characteristic of the last stages of the oxidation process. As NXA is an antibacterial agent, chronic toxicity could develop in the IBR-PP after long exposure. Therefore, biomass stability and adsorption studies were performed during continuous exposure to NXA (see Figure 2.15). The reactor was emptied between runs. The 5th run was maintained for seven days to observe the progress of the parameters studied during longer periods and determine any change in NXA concentration with low DOC or biomass death (DOC should increase) due to chronic toxicity, etc. After that, the system was emptied and filled with biodegradable wastewater without NXA (last cycle) to evaluate possible desorption of NXA or biomass inhibition. In all of the steps, significant DOC reduction was observed (>90% mineralisation), while the pH remained between 6.8–8.0.

NXA concentration was only reduced in the first runs due to adsorption on the biomass, after which it returns to the original concentration. However, NXA concentration increased slightly in the following runs. This means that the biomass adsorption rate decreased over exposure time to NXA, and part of the absorbed NXA during the first runs was later desorbed. NXA is a very weak organic acid that may be present in both neutral and anionic forms in the neutral pH range. Therefore, the adsorption capacity of this compound may also vary with pH. As pH during the biotreatment varied between 6.8 and 8 and the composition of the water also varied due to DOC biodegradation, NXA adsorption/desorption phenomena did not follow any predictable pattern.

The disappearance of the initial compounds (see Figure 2.10) and formation of new compounds (intermediates) during the different degradation steps were studied by LC-TOF-MS (ESI+). The elemental composition proposed for the molecules of interest was based on its accurate mass measurements. In this case, 23 different compounds could be identified 18 in the wastewater (see Figure 2.10) and 5 more during biotreatment (see Figure 2.16).

After biotreatment, compounds originally present in the wastewater were completely degraded (P3-P10, P13, P17), or their concentrations dropped as observed in Figure 2.16 (see intensity scale). NXA and P1 still remained the main components of the mixture, and a new group of compounds (P19-P23) were identified. The absence of commercial standards for the intermediates identified makes their quantitative evaluation impossible.

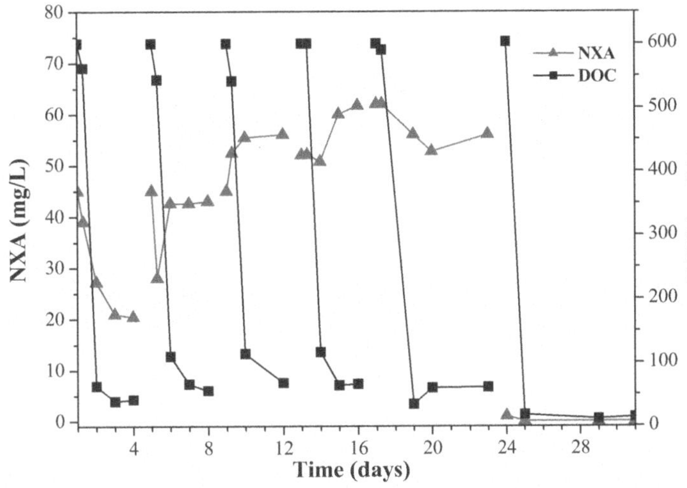

Figure 2.15 Long exposure of IBR-PP to NXA wastewater in batch mode

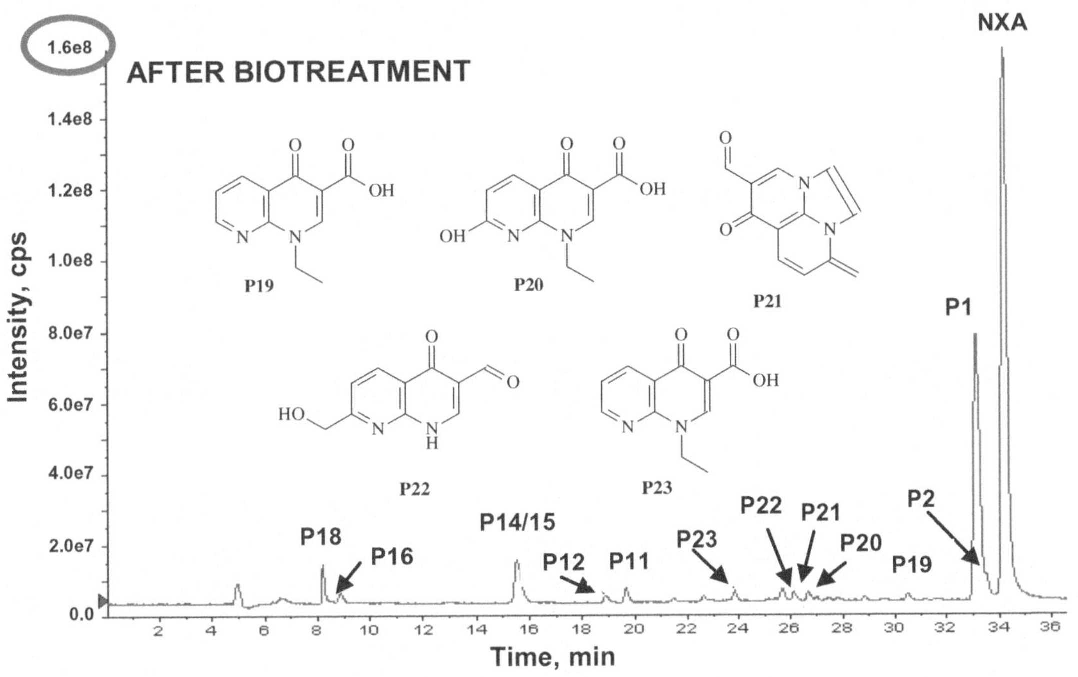

Figure 2.16 LC-TOF-MS chromatogram obtained by the analysis of a wastewater sample after biotreatment and chemical structures proposed for the new 5 compounds found

Only after photo-Fenton treatment, were NXA and analogous compounds totally eliminated, no further intermediates detected (Figure 2.17). Residual DOC measured is thus assigned to the presence of organic acids characteristic of the last stages of the oxidative process. This biotreatment, coupled with photo-Fenton, degraded the recalcitrant antibiotic (NXA) originally present in the matrix, and eliminated the by-products generated during biotreatment, demonstrating that photo-Fenton applied to finish off biological treatment of pharmaceutical wastewater is a useful approach.

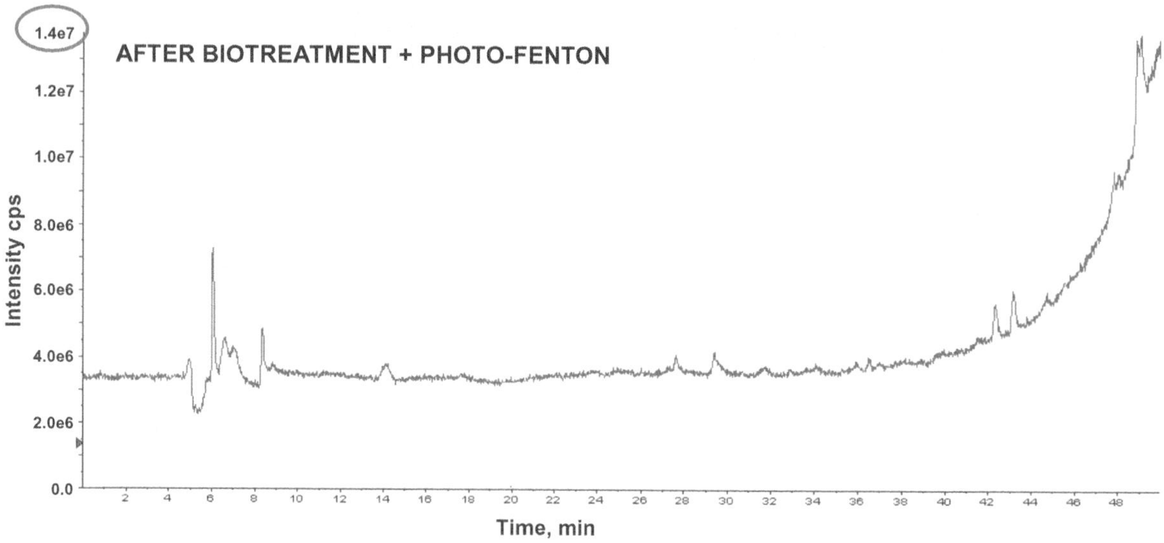

Figure 2.17 LC-TOF-MS chromatogram obtained by the analysis of a wastewater sample after biotreatment+ photo-Fenton. Compare intensity scale with Figure 2.10 and 2.16

2.2.4. Conclusions

It has been demonstrated that photo-Fenton is able to enhance biodegradability of real wastewaters containing high concentration of biorecalcitrant compounds but it has been also pointed out that the process should be optimized by a proper selection of treatment time and H_2O_2 dose. Operational conditions should be selected in a suitable range (longer treatment time and higher H_2O_2 dose than the minimum pre-determined at lab scale or with model wastewaters) ensuring proper managing of real wastewaters of unpredictable composition. This coupled technology can be applied in industrial plants with difficult access to the municipal sanitation and sewage system, making possible the direct discharge of the effluent or its reutilization in industrial facilities. On the other hand, the correct operation of the solar photo-Fenton systems could very often permit the disposal of the pretreated wastewater in conventional sewage systems.

It has been also shown an example of wastewater containing a considerable amount of biodegradable compounds jointly with a low concentration of biorecalcitrant compounds. In this case, a biological pre-treatment followed by solar photo-Fenton has been the best choice. Such integrated systems are particularly favourable for effluents which initially contain biodegradable fractions which could easily be removed first and so, not consume H_2O_2 in the photo-Fenton process. Biorefractory compounds are very often not so toxic to activated sludge, and biodegradation prior to AOP may be successful. In these cases, AOP treatment is more suitable as a polishing step than as a pretreatment.

This chapter also highlighting the importance of applying a consistent methodology for evaluating these systems, as each wastewater should be studied on a case by case basis.

2.3 FUNDAMENTAL STUDIES ON IMMOBILIZED PHOTO FENTON CATALYST

2.3.1 Introduction

The immobilization of photo-Fenton active species (iron complexes or oxides) on a suitable support to prepare heterogeneous catalysts for wastewater decontamination is an emergent research field as such systems present several advantages if compared with conventional homogeneous photo-Fenton process. By nature, with supported systems the final separation of catalyst is facilitated. Besides, heterogeneous photo-Fenton catalysts are in most of cases efficient to decontaminate wastewaters at pH close to neutrality.

Different approaches were proposed to design supported photo-Fenton catalyst: the use of iron-rich porous minerals such as Fe^{3+} exchanged zeolites, Noorjahan *et al.* (2008), clays Feng *et al.* (2003) etc.; or the deposition of iron complexes or oxides on different supports like silica fabric Bozzi *et al.* (2004), ionic exchange resins Lv *et al.* (2005), Nafion membranes Dhananjeyan *et al.* (2001). In the frame of INNOWATECH project, the use of transparent polymer films as iron oxide support was investigated as such supports let the light pass more efficiently to the reactor.

However, the use of polymer films as substrate is a challenge as it induces several practical drawbacks that have to be overtaken: (i) commercial polymer films surfaces are slick and intrinsically not very adhesionable (Mittal, 1996), limiting iron oxide crystal nucleation and its adhesion to the substrate; (ii) polymer films are destroyed under high temperature which prevents from the proceeding of high temperature thermal treatment in preparation procedure. Although commercial available polymers have surface roughness, electron donor groups, relatively high melting points and a high stability under solar light, many commonly used polymers films such as polyethylene (PE), polyethylene terephtalate (PET) or fluorinated polymers etc. do not show such properties. When latter polymers are used as photocatalyst supports, a functionalization of their surface is required.

Radio frequency plasma (RF-P) and vacuum ultraviolet (V-UV) treatment have been applied for polymer surface functionalization Chan *et al.* 1996. RF-P and V-UV treatments were also applied to textiles to promote the deposition of silver Yuranova *et al.* (2003) for antibacterial applications.

Photocatalytic treatments such as TiO_2 photocatalysis (Ti-PC) and photo-Fenton oxidation (P-FO) are attractive alternatives for polymer surface modification because of the mild conditions needed (solar light, low temperature and atmospheric pressure). Besides, Ti-PC treatment of polymer films was studied recently for increasing adhesion strength before electroless Cu plating by Kim *et al.* 2006 and iron oxide deposition Mazille *et al.* 2009a.

This investigation is directed toward the fixation and photocatalytic performance of iron oxide on different polymer films previously functionalized by different methods. To compare the photo-activity as a function of preparation procedure (different functionalization methods, iron oxide coating conditions, or polymer substrates etc.) hydroquinone was selected as a model compound. An optimized photocatalyst was thus obtained. The most efficient photocatalysts were adapted to compound parabolic collectors (CPC) photoreactors at larger scale for the degradation of real pollutants and their long-term stability was tested.

2.3.2 Optimisation of photocatalyst preparation and lab scale experiments

2.3.2.1 Experimental

2.3.2.1.1 Chemicals

Hydroquinone (HQ), Nalidixic acid (NA), NaOH, NaCl HNO_3, $FeCl_3$ anhydrous, $FeSO_4 \cdot 7H_2O$, ferrozine, hydroxylamine hydrochloride, acetate buffer (pH = 4.65) were Fluka p.a. reagents (Buchs, Switzerland) and used as received. Hydrogen peroxide (35%) was supplied by Merck AG (Darmstadt, Germany). TiO_2 P25 powder (anastase to rutile ratio between 70:30 and 80:20) was supplied by Degussa. The polyvinyl fluoride (PVF), polyethylene (PE) and polyethylene terephtalate films were 72 μm thick and were supplied by Goodfellow (Cambridge Ltd. United Kingdom).

2.3.2.1.2 Catalyst characterization

The UV-Visible spectra of different photocatalysts were recorded on a Varian Cary 5 equipped with an integration sphere.

X-ray photoelectron spectroscopy data were collected by Axis Ultra system (Kratos analytical, Manchester, UK) under ultra-high vacuum condition ($<10^{-8}$ Torr), using a monochromatic Al K_α X-ray source (1486.6 eV) at the laboratory of Chemical Metallurgy at EPFL. The source power was maintained at 150 W (10 mA, 15 kV). The emitted photoelectrons were sampled from a square area of 750 * 350 μm. The gold (Au $4f_{7/2}$) and copper (Cu $2p_{3/2}$) lines at 84.0 and 932.6 eV respectively, were used for calibration, and the adventitious carbon 1s peak at 284.6 eV was used as an internal standard to compensate for charging effects.

The surface morphology of the catalysts were investigated using a scanning electron microscope (Phillips XL30 SFEG) equipped with X-ray detector.

2.3.2.1.3 Catalyst preparation

Photoactive iron oxide was deposited on commercial polymer film substrates following the strategy presented in Figure 2.18. Before use, the films were washed in a diethyl ether: ethanol (1:1) mixture and in MilliQ water in order to eliminate surface contaminants. Then polymer films surfaces were functionalized by different methods described in the next paragraphs.

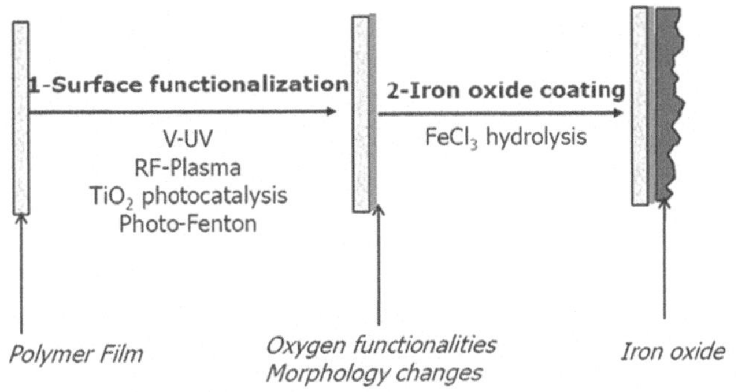

Figure 2.18 Scheme of catalyst preparation strategy (not to the scale). Reprinted from Mazille *et al.* (2009a) with permission of Editor.

Functionalization with radio frequency plasma (RF-P). The polymers films were treated in a plasma cavity (Harricks Corp, 13.56 MHz, power 100 W), at a pressure of 1.0 mbar. The gas used for the plasma generation was air. The functionalization process occurs in the topmost layers and increases with longer treatment time, but remains constant with treatment time >30 min, Yuranova *et al.* (2003). For this reason the latter time has been chosen for RF-P treatment. A previous report, Golub *et al.* (1989) presented a detail electron spectroscopy for chemical analysis (ESCA) study of the effect of oxygen RF-P treatment on PVF surface which was shown to produce C=O, fluorine loss and a slight loss of carbon atoms (by etching).

Functionalization with V-UV. The polymer film surfaces were also functionalized by V-UV irradiation using the 185 nm (6 W) line from a 25 W (254 + 185 nm) low pressure mercury lamp (Ebra Corp, Tokyo, Japan). The lamp was a synthetic silica tube. The polymer film was attached around the lamp tube. The gas used was air (0.8 mbar) and the treatment time was 30 minutes (condition selected from (Yuranova, 2003)). A detail investigation on the effect of V-UV radiation on PVF films was reported recently (Everett, 2006). It was found that with short treatment times (i.e. <24 hours) the main effect induced by this treatment was a washing of surface contamination a decrease of fluorine surface concentration and an increase of carbon surface concentration.

Functionalization with photo-Fenton oxidation (P-FO). The polymer films were attached on a cylindrical steal support and immersed in solutions of Fe^{2+} (50 mg/L) containing H_2O_2 (30 mM) at pH 3. The batch reactor was irradiated in a solar simulator CPS Suntest system (ATLAS). The polymer surface undergoes the photo-Fenton functionalization during two hours under magnetic stirring.

Functionalization with TiO_2 photocatalysis. The polymer films were attached on a cylindrical steal support and immersed in well dispersed solutions of TiO_2 0.2–1.7 g/L at initial pHs 3, 7, natural, 11. The batch reactor was irradiated in a solar simulator CPS Suntest system (ATLAS). The polymer surface underwent the TiO_2 photocatalytic functionalization under magnetic stirring during 2 hours. The species generated by Ti PC treatment are radicals such as hydroxyl and super-oxide, as well as holes and electrons. These reactive species react with polymer surface generating a variable density of negative groups such as COO^{2-}, and can induce fluorine elimination in the case of fluoro-polymers. TiO_2 particles also bind to the freshly functionalized polymer surface generating roughness and supplementary photocatalytic activity, Mazille *et al.* 2009a.

Figure 2.19 shows the SEM pictures of PVF before and after the Ti-PC functionalization. The non-modified PVF surface, Figure 2.19a, is predominantly smooth. After a TiO_2 PC treatment, aggregated TiO_2 particles cover a significant part of the surface exposed to the light irradiation (Figure 2.19b).

Figure 2.19 Scanning electron microscopic images of (a) PVF, (b) PVF^{TiPC}. Reprinted from Mazille *et al.* (2009a) with permission of Editor.

Iron oxide coating. To immobilize iron oxide, functionalized polymer films (P^f) were immersed in a solution of $FeCl_3$ (anhydrous) 5 g/L and heated under stirring at 70, 80 or 90°C during one hour. Optionally, the composite material was heated in an oven at 100°C for one hour. The final films showed a brownish iron oxide coat.

Figure 2.20 shows the XPS spectum representative of iron oxide on modified PVF films. The Fe (2p) spectrums of as prepared $PVF^{Ti\text{-}PC}$-Fe oxide (trace a), of $PVF^{Ti\text{-}PC}$-Fe oxide and $PVF^{RF\text{-}P}$-Fe oxide samples after use are pretty similar and unambiguously characteristic of Fe^{3+} in an oxide (Fe $2p_{1/2}$, Fe $2p_{3/2}$ and satellite peaks are at a binding energy of 725, 711 and 719 eV, respectively; the satellite peak is a distinct peak, Yashamita and Hayes (2008), which suggest that no significant changes in the iron oxidation state occurred after use. Akaganeite, Deliyanni *et al.* (2006), is probably the oxide present due to the significant presence of chlorine in the oxide (see Table 2.2). Various reports confirm this hypothesis, as the thermal hydrolysis of $FeCl_3$ solutions leads mainly to the formation of akaganeite, Music *et al.* (2003). Figure 2.21 shows the SEM pictures typical of iron oxide-coated polymer surfaces loaded with iron oxide aggregates of about 1 μm. These aggregates are composed of nanoparticles of approximately 50 nm, which seem to cover the polymer surface (Figure 2.21b).

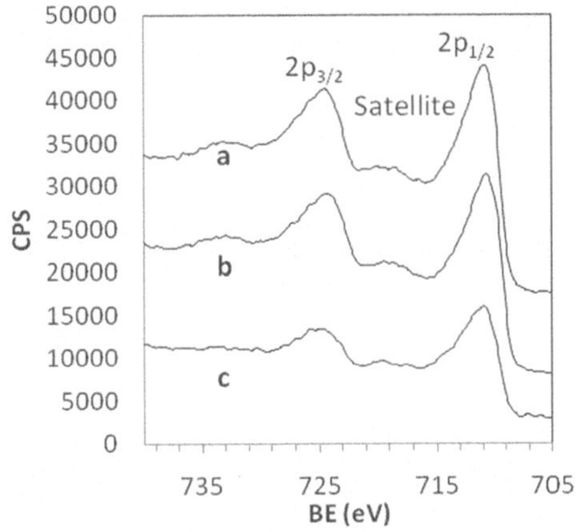

Figure 2.20 Fe (2p) core level photoelectron spectrum of (a) as-prepared PVF$^{Ti\text{-}PC}$-Fe oxide, (b) and (c), PVF$^{Ti\text{-}PC}$-Fe oxide and PVF^{RF-P}-Fe oxide, respectively, after three runs. Reprinted from Mazille *et al.* (2009a) with permission of Editor.

Table 2.2 Composition of PVF^f-TiO_2-Fe oxide surface (atomic percent) as prepared and after three runs

Samples	C	O	F	Cl	Ti	Fe	Fe/Ti
PVF^f-TiO_2-Fe oxide	22.7	46.0	2.0	6.4	0	23.8	–
PVF^f-TiO_2-Fe oxide after use	35.1	45.1	1.5	1.7	0.3	16.8	56

Figure 2.21 Scanning electron microscopic images of as prepared PVF^{TiPC}-Fe oxide. Reprinted from Mazille *et al.* (2009a) with permission of Editor.

2.3.2.1.4 Photo-reactor and irradiation procedure

All the degradation experiments were carried out under simulated solar light, using thin film Pyrex glass reactors (Figure 2.22).

A peristaltic pump allows circulation of water (flow rate of 100 mL/min) using an Erlenmeyer flask as a recirculation tank (Figure 2.23). The total volume of the system (110 mL) consists of two parts: 25 mL irradiated volume and the dead volume in connecting tubes and Erlenmeyer tank. Internal PVC supports were placed in the reactor to carry the

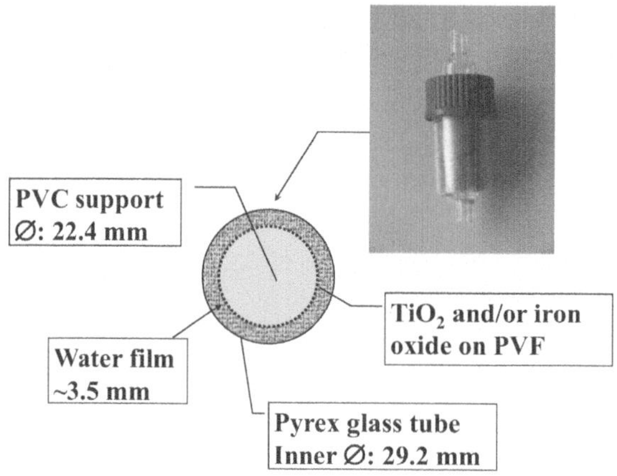

Figure 2.22 Transversal cut and characteristics of the photo-reactor. Reprinted from Mazille *et al.* (2009a) with permission of Editor.

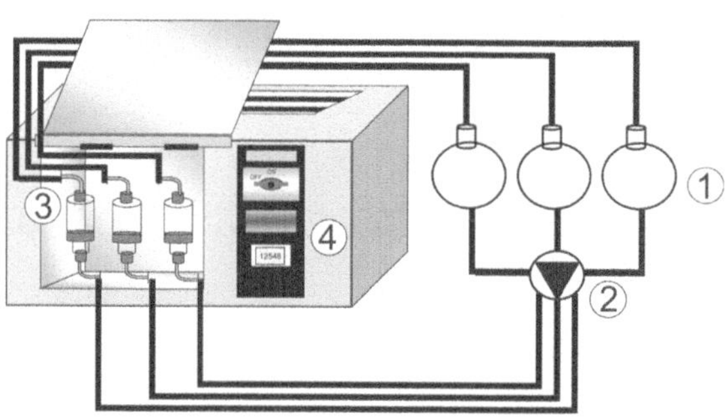

Figure 2.23 Photocatalytic setup: (1) tanks; (2) peristaltic pump; (3) photo-reactors; (4) solar simulator. Reprinted from Mazille *et al.* (2009a) with permission of Editor.

photocatalyst films. The reactor was illuminated inside a solar simulator CPS Suntest system (Atlas GmbH). This solar box has a spectral distribution of about 0.5% of the emitted photons at wavelengths shorter than 300 nm, and about 7% within 300 and 400 nm. The emission spectrum between 400 and 800 nm follows the solar spectrum The irradiation experiments were started at room temperature (20°C) and progressively the temperature increased up to approximately 30°C. Three reactors were placed in parallel inside the solar box. The degradation experiments were performed using solutions containing HQ or NA (0.18 mM) and H_2O_2 (1.6 mM).

2.3.2.1.5 Analysis of the irradiated solutions

The quantitative determination of organic compounds was carried out by HPLC chromatography using a LC system HPLC–UV: Shimatzu LC-2010A equipped with a UV detector. Samples were injected via an auto-sampler. Samples of the treated HQ solution were eluted at a flow rate of 1 mL min^{-1} through a column (nucleosil C18 Marcherey Nagel) and using as mobile phase HPLC grade acetonitrile–acetic acid solution (10%) in a 40–60%. Samples of the treated NA solution were eluted at a flow rate of 1 mL min^{-1} through a column (nucleosil C8 HPLC) and using as mobile phase HPLC grade acetonitrile–formic acid solution (25 mM) in a 50–50%. Total organic carbon (TOC) was monitored via an instrument (Shimadzu 500) equipped with an automatic sample injector (ASI). The peroxide concentrations were assessed by Merkoquant1paper at levels between 0.5 and 25 mg/L. The analyses were performed as fast as possible after sampling in order to minimize the effect of dark Fenton reaction. The total iron concentration in the irradiated solutions was measured by complexation with Ferrozine® (Aldrich 16.060–1) in the presence of hydroxylamine hydrochloride and acetate buffer (pH = 4.65) (Gibbs *et al.* 1979).

2.3.2.2 Results

2.3.2.2.1 Effect of functionalization on photocatalytic activity of PVFf-Fe oxide

Figure 2.24 shows (i) HQ, (ii) TOC and (iii) dissolved iron concentrations ($[Fe]_d$) over time during degradation experiments mediated by different photocatalysts in presence of H_2O_2 and light. The pH initially 5.7 decreased to 4.5 after 30 min mostly due to the formation of short acidic degradation intermediates.

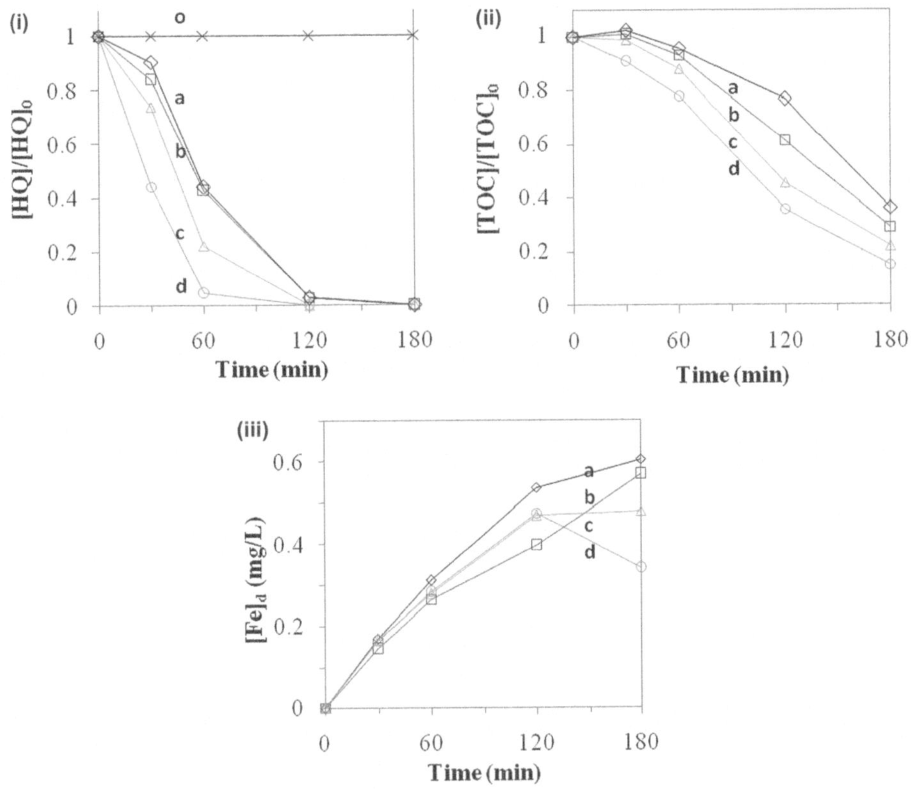

Figure 2.24 Evolution of (i) HQ concentration, (ii) TOC and (iii) $[Fe]_d$ during degradation of 0.18 mM of HQ at initial pH 5.7, in presence of 1.6 mM H_2O_2, 75 cm^2 of catalyst under solar simulation (light): (o) photolysis; (a) PVF-Fe-oxide; (b) PVF^{V-UV}-Fe-oxide and PVF^{P-FO}-Fe oxide*; (c) PVF^{RF-P}-Fe oxide; (d) PVF^{Ti-PC}-Fe-oxide. The traces represent an average over three runs. *Due to similar result obtained for PVF^{V-UV}-Fe-oxide and PVF^{P-FO}-Fe oxide trace (b) represent the results relative to these two catalysts. Reprinted from Mazille *et al.* (2009a) with permission of Editor.

Figure 2.24 (i) shows that HQ was resistant to direct photolysis by simulated solar light alone (trace o). During the degradation of HQ mediated by PVF-Fe oxide (trace a Figure 2.24 (i)), PVF^{P-FO}-Fe oxide and PVF^{V-UV}-Fe oxide (both represented by trace b Figure 2.24 (i)), the HQ degradation was low during the first 30 min of treatment but was complete at 120 min. The initial phase is slow and then fast degradation sets seems to be due to homogeneous photo-Fenton oxidation, since after 60 min the concentration of dissolved iron ($[Fe]_d$) detected is around 0.3 mg/L (Figure 2.24 (iii)). For the photo-assisted degradations (Figure 2.24 (i)) mediated by PVF^{RF-P}-Fe oxide (Trace c) and PVF^{Ti-PC}-Fe oxide (trace d), the initial degradation (first 30 min) was important with about 30 and 60% of HQ degradation, respectively.

Figure 2.24 (ii) shows that during the first 30 min of irradiation, the TOC slightly increased for PVF-Fe oxide (trace a), PVF^{V-UV}-Fe oxide and PVF^{P-FO}-Fe oxide (both in trace b), due to a degradation of the polymer substrate, but remains constant for PVF^{RF-P}-Fe oxide. After that the TOC begins to decrease and about 25, 40 and 55% of mineralization was observed for PVF-Fe oxide, both PVF^{V-UV}-Fe oxide and PVF^{P-FO}-Fe oxide, and, PVF^{RF-P}-Fe oxide, respectively, after 120 min of treatment. In contrast, a significant TOC decrease was observed from the beginning, when PVF^{Ti-PC}-Fe oxide was used and 65% of TOC was mineralized after 120 min of treatment.

Figure 2.24 (iii) shows that the leaching of iron was similar for the five tested catalysts. Only for the PVF^{Ti-PC}-Fe oxide, the $[Fe]_d$ decreased slightly at the end of the treatment. This fact suggests that dissolved iron species were re-adsorbed on the catalyst surface possibly on TiO_2 particles. Thus the heterogeneous contribution seems to be involved in the observed differences as follows: no treatment $\leq$ P-FO $\approx$ V-UV $<$ RF-P $\ll$ Ti-PC. The possible heterogeneous reactions involved in the degradation and mineralization of HQ are photo-Fenton oxidation (Eq. 1–4 where $=$Fe(II) and $=$Fe(III) represent the surface Fe(II)/Fe(III) species) and semiconductor photocatalysis (Eq. 5–9 where SC represent FeOOH or/and TiO_2 in the case of Ti-PC). Besides synergistic effects between Fenton and TiO_2 photocatalysis is likely to occur (Eq. 7 where SC $=$ TiO_2) which was discussed in more details elsewhere, Mazille *et al.* (2009b).

$$=\mathrm{Fe(II)} + H_2O_2 \longrightarrow =\mathrm{Fe(III)} + OH^{\bullet} + OH^- \tag{1}$$

$$=\mathrm{Fe(III)} + H_2O_2 \longrightarrow =\mathrm{Fe(II)} + HO_2^{\bullet} + H^+ \tag{2}$$

$$=\mathrm{Fe(III)(OH)}^{2+} + h\nu \longrightarrow =\mathrm{Fe(II)} + OH^{\bullet} \tag{3}$$

$$\equiv Fe(III)L_n + h\nu \longrightarrow \equiv Fe(II)L_{n-1} + L_{ox}^{\bullet +} \quad (4)$$

$$SC + h\nu \longrightarrow SC(e^- + h^+) \quad (5)$$

$$SC(e^-) + H_2O_2 \longrightarrow SC + OH^\bullet + OH^- \quad (6)$$

$$SC(e^-) + \equiv Fe(III) \longrightarrow SC + \equiv Fe(II) \quad (7)$$

$$SC(h^+) + R_{ad} \longrightarrow SC + R_{ad}^{\bullet +} \quad (8)$$

$$SC(h^+) + H_2O \longrightarrow SC + OH^\bullet + H^+ \quad (9)$$

Figure 2.25 shows the UV-Vis absorption spectrum of several photocatalysts between 360 and 700 nm before their first use. The absorbance of PVF film in this region was negligible. The spectra of PVF-Fe oxide, $PVF^{V\text{-}UV}$-Fe oxide and $PVF^{P\text{-}FO}$-Fe oxide were similar and are thus represented by trace a. In contrast, $PVF^{RF\text{-}P}$-Fe oxide (trace b), and more markedly $PVF^{Ti\text{-}PC}$-Fe oxide (trace c) spectra show that the absorption was higher. The iron oxide on the polymer surface is responsible for the absorption of light. Hence the absorbance can be correlated with the thickness of iron oxide coating. These results show that RF-P and Ti-PC treatments favour iron oxide nucleation and/or deposition on the functionalized film. On the contrary, P-FO and V-UV treatments did not improve significantly iron oxide deposition on polymers if compared to not treated ones.

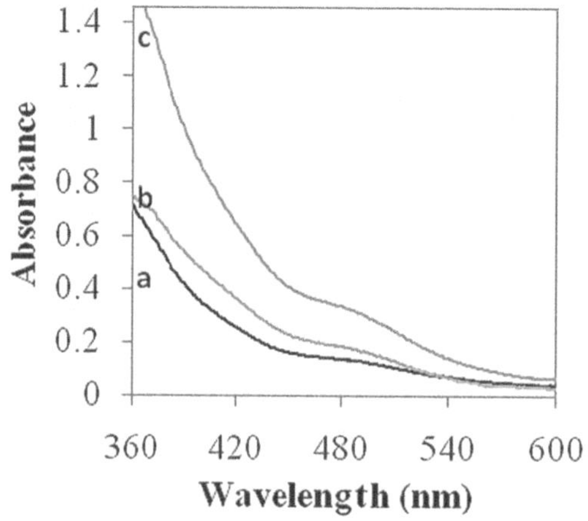

Figure 2.25 UV-vis absorption spectrum of (a) PVF-Fe oxide, $PVF^{V\text{-}UV}$-Fe oxide, $PVF^{P\text{-}FO}$-Fe oxide; (b) $PVF^{RF\text{-}P}$-Fe oxide; (c) $PVF^{Ti\text{-}PC}$-Fe oxide. Reprinted from Mazille *et al.* (2009a) with permission of Editor.

Several causes could explain the beneficial effects induced by the functionalization of PVF surface: (i) a chemical functionalization where oxygen surface functionalities increase the surface polarity; (ii) the modification of polymer surface morphology induced by RF-P (etching) and by Ti-PC (deposition of particles); and (iii) the additional photocatalytic activity induced by the deposition of TiO_2 particles. The fact that V-UV and P-FO pre-treatments did not increase the photocatalytic activity of as prepared material (and by extension iron oxide deposition) can be explained by: (i) Short V-UV treatment does not allow the production of polar groups on polymer surface and thus will not enhance iron oxide deposition; (ii) P-FO treatment may produce polar groups susceptible to bind iron oxide particles, however iron aqua-complexes might immediately bind with these freshly formed polar moieties during the pre-treatment process.

2.3.2.2.2 Effect of polymer film nature

The different polymeric substrates used to support iron oxide are represented in Figure 2.26.

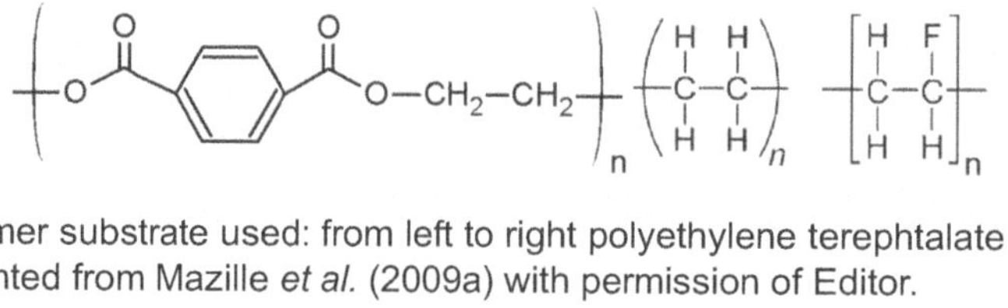

Figure 2.26 Structure of polymer substrate used: from left to right polyethylene terephtalate (PET), polyethylene (PE) and polyvinyl fluoride (PVF). Reprinted from Mazille *et al.* (2009a) with permission of Editor.

Figure 2.27 shows (i) TOC, (ii) HQ and (iii) $[Fe]_d$ over time during the HQ degradations experiments mediated by photocatalysts composed of three types of polymeric substrates (PVF, PE, PET) prepared under conditions slightly different than previously (i.e. TiO_2 1 g/L in the Ti-PC treatment and $FeCl_3$ 2 g/L in the hydrolysis step).

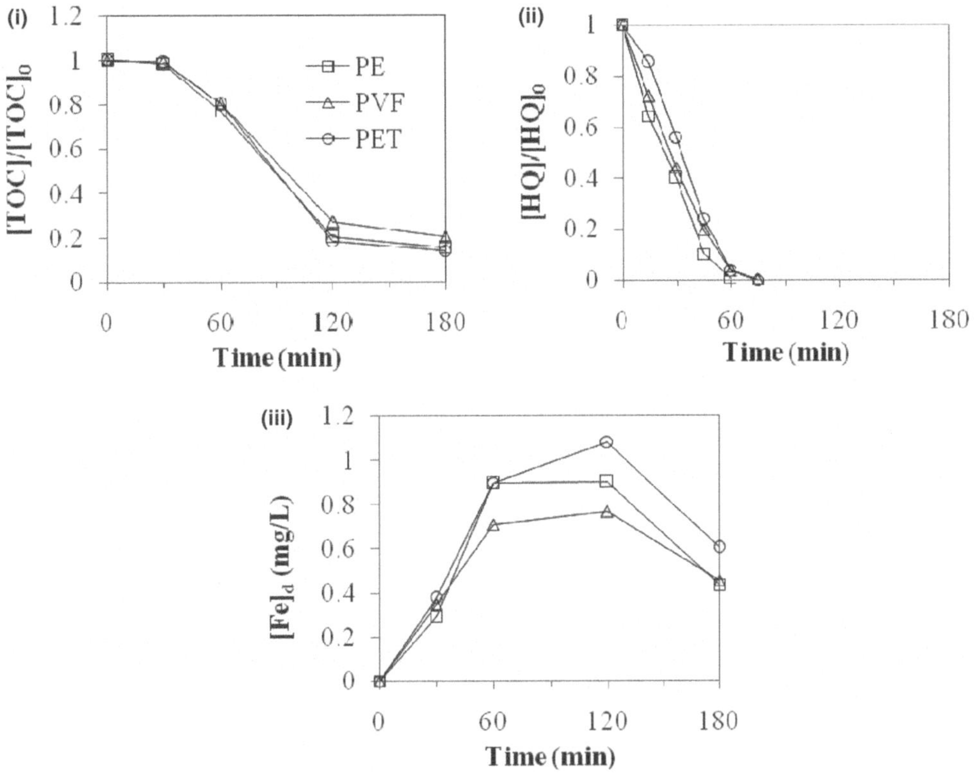

Figure 2.27 Evolution of (i) TOC, (ii) HQ concentration and (iii) $[Fe]_d$ during degradation of 0.18 mM of HQ at initial pH 5.7, in presence of 1.6 mM H_2O_2, 75 cm^2 of catalyst ($PE^{Ti\text{-}PC}$-Fe oxide, $PVF^{Ti\text{-}PC}$-Fe oxide, $PET^{Ti\text{-}PC}$-Fe oxide) under solar simulation for different polymeric substrates. The traces represent an average over three runs. Reprinted from Mazille *et al.* (2009a) with permission of Editor.

During the photocatalytic experiment mediated by the supported iron oxide on Ti-PC treated PE, PET and PVF, HQ and TOC reduction were similar, reaching HQ total degradation between 60–90 minutes, and around 80% TOC decrease in 120 min. This fact indicates the negligible effect of the nature of polymer substrate on catalyst performance.

In contrast, Figure 2.23 (iii) shows that the amount of leached iron ions was dependant on the polymer nature with a maximum around 1.1, 0.9 and 0.7 mg/L after 120 min of treatment for PET, PVF and PE, respectively. A larger iron leaching could be linked to larger iron oxides quantities adsorbed on the substrate and, by extension, to the achievement of more surface functionalization in the pre-treatment, and to the own nature of the substrate (higher polarized surface).

The latter results suggest that different polymer substrate can be used depending on the application required. PET bottles could be used as catalyst support for drinking water production by modified solar water disinfection (SODIS).

2.3.2.2.3 Effect of the pH during Ti-PC treatment

The initial pH of the aqueous medium during the Ti-PC treatment determines the amount of oxygenated groups formed and the superficial charge of TiO_2 nanoparticles. The extent of TiO_2 photo-deposition depends on the pH as well Mazille *et al.* (2009a). At acidic pH, the TiO_2 being positively charged can be stably deposited on electronegative oxygenated groups on the polymer surface (Figure 2.28). At basic pH, the TiO_2 and polymer are both negatively charged and the interaction is repulsive, limiting the polymer functionalization and the TiO_2 deposition. Consequently, since iron oxide crystallization is favored by TiO_2 presence, less iron oxide is deposited on its surface leading to a less efficient $PVF^{Ti\text{-}PC}$-Fe oxide photocatalyst. HQ mineralization rates by $PVF^{Ti\text{-}PC}$-Fe oxide/H_2O_2/light decrease as the pH for Ti-PC treatment increased with 73, 68 and 46% of HQ mineralized after 120 minutes for pH 3, 7 and 11 respectively.

Figure 2.29 represents TOC removal by different photocatalyst during HQ degradation in presence of H_2O_2 under light. The disparity between PVF-Fe oxide (trace a) and $PVF^{Ti\text{-}PC}$-Fe oxide prepared at pH 11 and 3 (trace b-c) shows the advantage of applying the Ti-PC treatment before iron oxide coating. Indeed at pH 3, more TiO_2 fixation induces as well more iron oxide coating. The Ti-PC treatment also avoids the initial TOC increase by the PVF-Fe oxide/H_2O_2/light system as the polymer already was degraded to a stable state under this treatment. In general, TOC removal is faster for the Ti-PC treated polymers particularly when the Ti-PC treatment was carried out at acidic pHs.

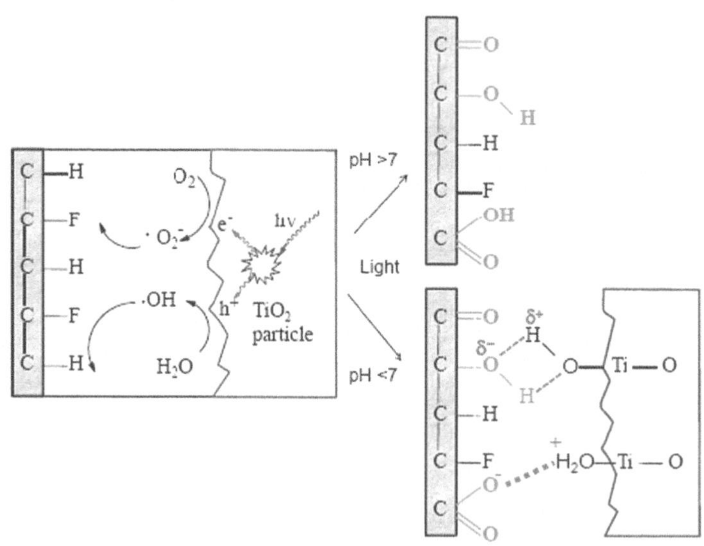

Figure 2.28 Mechanistic proposition for TiO_2 photocatalytic surface functionalization deposition as a function of pH. Reprinted from Mazille *et al.* (2009a) with permission of Editor.

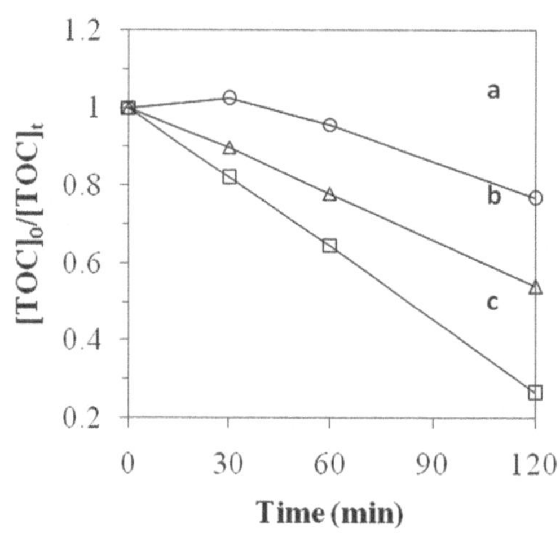

Figure 2.29 TOC removal of 0.18 mM of HQ, initial pH: 5.7, 75 cm^2 of heterogeneous photocatalyst 1.6 mM H_2O_2 under solar simulation: (a) PVF-Fe-oxide/H_2O_2/light; (b) $PVF^{Ti\text{-}PC}$–Fe oxide (Ti-PC treatment at pH 11)/H_2O_2/light; (c) $PVF^{Ti\text{-}PC}$-Fe oxide (Ti-PC treatment at pH 3)/H_2O_2/light. Reprinted from Mazille *et al.* (2009a) with permission of Editor.

2.3.2.2.4 Effect of TiO_2 concentration used for the Ti-PC treatment on HQ mineralization by the system $PVF^{Ti\text{-}PC}$-Fe oxide/H_2O_2/light

Figure 2.30 shows that maximum HQ mineralization under light is reached when TiO_2 concentration used during Ti-PC treatment is ~1 g/L. The TiO_2 concentration influences both the extent of polymer surface functionalization and the amount of deposited TiO_2. Figure 2.30 shows that during Ti-PC treatment, the first run (curve a) of photocatalytic mineralization is less efficient than following one (curve b). Thus, the first photocatalytic run enhances the photocatalyst performance for subsequent runs.

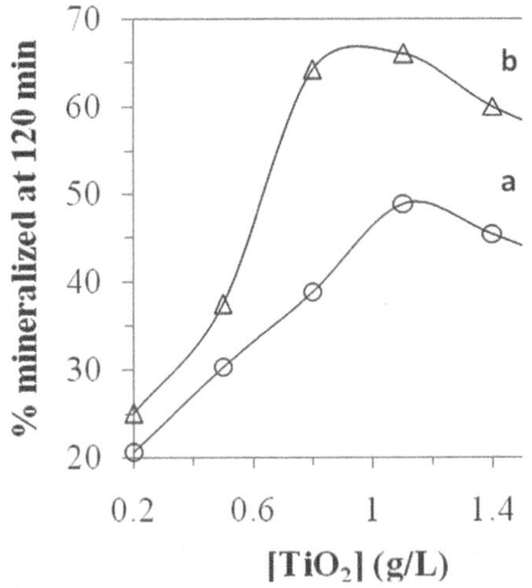

Figure 2.30 Effect of TiO_2 concentrations (used for previous Ti-PC treatment of PVF) on HQ mineralization (of 0.18 mM HQ, initial pH = 5.7, $PVF^{Ti\text{-}PC}$-TiO_2-Fe oxide 75 cm^2, H_2O_2 1.6 mM under solar simulation): (a) first; (b) second consecutive HQ degradation. Reprinted from Mazille *et al.* (2009a) with permission of Editor.

2.3.2.2.5 Effect of iron oxide coating conditions

$FeCl_3$ hydrolysis was carried out at different temperatures (70, 80 and 90°C) for the iron oxide generation and deposition. Applied temperatures during hydrolysis were found to play a significant role only for mineralization efficiency during the first degradation run: as temperature during hydrolysis was increased the mineralization rate during the first run increased. During the following runs, HQ mineralization was similar for photocatalysts prepared at 70, 80 or 90°C. This may be due to the formation of oxide during the hydrolysis at low temperatures, which is stabilized during the first run.

An additional thermal treatment at 100°C during one hour after hydrolysis was applied to $PVF^{Ti\text{-}PC}$-Fe oxide. The results obtained concerning HQ degradation suggested that this additional treatment was unnecessary. Actually a temperature of 100°C is not sufficiently high to induce phase changes of iron oxide or to melt PVF substrate.

The optimisation of preparation procedure led to an efficient photocatalyst composed of TiO_2 and iron oxide supported on functionalized PVF films (so called PVF^f-TiO_2-Fe oxide. The preparation conditions are the following: $[TiO_2] = 1$ g/L, pH 5 during the Ti-PC functionalization step and $[FeCl_3] = 5$ g/L, T = 80°C for iron oxide coating.

2.3.2.2.6 Nalidixic acid degradation: effect of salt content

Figure 2.31(i) presents the evolution of normalized NA and TOC concentrations within 240 min under different conditions. During homogeneous photo-Fenton (Fe^{2+}(0.3 mg/L)/H_2O_2/light) oxidation (trace a), mineralization of NA was very slow (15%). For the system PVF^f-TiO_2-Fe oxide/H_2O_2/light (trace b) the mineralization was higher (35%) and NA degradation (trace c) was fast with a half-time around 25 min. During NA photocatalytic degradation by the system PVF^f-TiO_2-Fe oxide/H_2O_2/light, Figure 2.31(ii) shows that $[Fe]_d$ increased during the first 120 min of treatment up to 0.3 mg/L, then this concentration was chosen for the homogeneous photo-Fenton treatment. These results show the low but significant contribution of homogeneous photo-Fenton reaction to the degradation process mediated by PVF^f-TiO_2-Fe oxide/H_2O_2/light system.

The effect induced by the addition of NaCl (5 g/L) on NA degradation and mineralization mediated by the systems PVF^f-TiO_2-Fe oxide/H_2O_2/light (i) and Fe^{2+}(2 mg/L)/H_2O_2/light (ii) is illustrated in Figure 2.32. The salt adjunction was detrimental for the mineralization process of both homogeneous and heterogeneous systems. The initial mineralization rate (first 60 min) was around 2 times lower in presence (trace a) than in absence (trace b) of salt. For NA degradation in absence of salt, the rate relative to homogeneous system (trace d Figure 2.27 (ii)) was 3 times higher than those of heterogeneous system (trace d Figure 2.27 (i)). However, in presence of salt (traces c), the rates observed were similar for homogeneous and heterogeneous systems.

This comparison suggests that the heterogeneous system is advantageous for the treatment of saline wastewater since the degradation process is slightly faster in presence of salt. It is known that the inhibitory effect of Cl^- on Fenton reaction is due to the scavenging of $^{\bullet}OH$ by this ion (Pignatello, 2006). Thus for PVF^f-TiO_2-Fe oxide/H_2O_2/light system: (i) mineralization process seems largely controlled by photo-Fenton reaction initially induced by $^{\bullet}OH$ attacks and then by charge transfers occurring on iron oxide surface or dissolved complexes with intermediates; (ii) degradation process is heterogeneous and involves charge transfers from catalyst surface to pollutant rather than the participation of $^{\bullet}OH$.

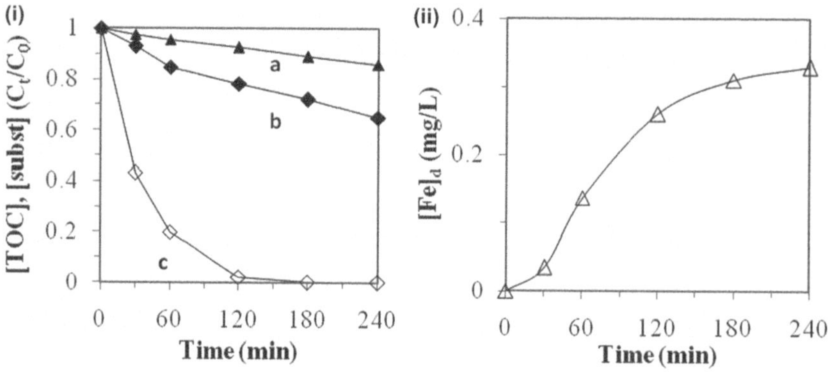

Figure 2.31 (i) Degradation of 0.18 mM of NA, initial pH 6.5, 30°C, 3.2 mM H_2O_2, 75 cm^2 of PVF^f-TiO_2-Fe oxide under solar simulation (a) TOC removal by 0.3 mg/L Fe^{2+}/H_2O_2/light; (b) TOC removal by PVF^f-TiO_2-Fe oxide/H_2O_2/light; (c) NA degradation by PVF^f-TiO_2-Fe oxide/H_2O_2/light; (ii) The traces represent average over 3 runs. Reprinted from Mazille *et al.* (2009b) with permission of Editor.

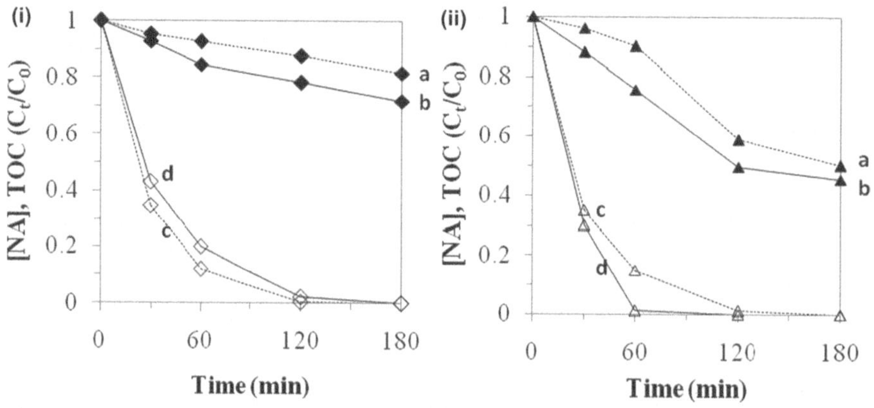

Figure 2.32 Variation of HQ and TOC normalized concentrations during degradation of 0.18 mM NA by (i) 75 cm^2 of PVF^f-TiO_2-Fe oxide, (ii) 2 mg/L Fe^{2+}: TOC removal (a) in presence and (b) without NaCl; NA degradation (c) in presence and (d) without NaCl (experimental conditions at initial pH 6.5, 30°C, 3.2 mM H_2O_2, 0.86 M NaCl under solar simulation). Reprinted from Mazille *et al.* (2009b) with permission of Editor.

2.3.2.2.7 Long-term stability

The long term stability of PVF^f-TiO_2-Fe oxide was studied by evaluating catalytic performances during repetitive nalidixic acid degradations. Figure 2.33 represents the results of this stability showing NA evolution as a function of time. These results show that the PVF^f-TiO_2-Fe oxide is extremely stable during 47 consecutive runs corresponding to about 150 hours of operation. The degradation rate relative to the first run, an average of runs 2–6, 10–30 and 45–47 were calculated. Surprisingly, instead of a slow decrease of the degradation rates, a slow increase was observed. The rate undergoes a 1.5-fold increase between runs 2–6 and runs 45–47. These results suggest that the presence of TiO_2 particles deposited on PVF substrate not only increased iron oxide formation and deposition on polymer surface but also other beneficial effects: (i) as long as the PVF^f-TiO_2-Fe oxide is used, the TiO_2 uncovering and exposition to light, induced by iron photo-dissolution, increases the Ti/Fe surface concentration (Table 2.2) which increases the synergistic effects and the degradation rates and (ii) The catalyst was protected from self-degradation since TiO_2 diminished the contact between radicals and polymer surface and can trap charges which otherwise would degrade the polymeric substrate.

2.3.3 Applications: Adaptation to compound parabolic collector photoreactors

2.3.3.1 Experimental

2.3.3.1.1 Chemicals

P.A. reagent phenol ($C_0 = 0.7$ mM) and nalidixic acid ($C_0 = 0.13$ mM) were supplied by Panreac and Fluka, respectively. Pesticide mixtures were prepared from commercial formulations of Vydate (10%, w/v oxamyl), Metomur, (20%, w/v methomyl), Couraze (20%, w/v imidacloprid), Ditimur-40 (40%, w/v dimethoate) and Scala (40%, w/v pyrimethanil).

The mixture contained 20% of each commercial formulation and had an original TOC of 50 mg L^{-1}. The amounts of active ingredients in the mixture were 21 mg L^{-1} of oxamyl, 5.1 mg L^{-1} of methomyl, 5 mg L^{-1} of imidacloprid, 8 mg L^{-1} of dimethoate, and 13.5 mg L^{-1} of pyrimethanil. Analytical standards (>98%) for chromatographic analysis were purchased from Sigma-Aldrich. The emerging contaminants were atenolol, metroprolol, sulfamethoxazole, flumequine, ketorolac and diclofenac, acetaminophen and progesterone p.a. reagents from Sigma-Aldrich. Antipyrine, caffeine, hydroxybiphenyl and triclosan were Fluka p.a. reagents; atrazine and isoproturon were p.a. reagents from Riedel-De-Haën. The mixture contained a 100 µg L^{-1} concentration of each emerging contaminants. The structures of all the organic compounds used were shown in Figure 2.24–2.26 NaOH, HNO_3, $FeCl_3 \cdot 6H_2O$ were Fluka p.a. reagents (Buchs, Switzerland) and used as received. TiO_2 P25 (anastase to rutile ratio between 70:30 and 80:20) was supplied by Degussa. The polyvinyl fluoride (PVF) films were 72 µm thick and were supplied by Goodfellow (Cambridge Ltd. United Kingdom). Distilled water used in the pilot plant was supplied by the plataforma solar de Almería (PSA) distillation plant (conductivity <10 microScm^{-1}, Cl^- = 0.2–0.3 mg/L, NO_3^- < 0.2 mg/L, organic carbon < 0.5 mg/L). The experiments were performed using reagent-grade hydrogen peroxide (30% w/v).

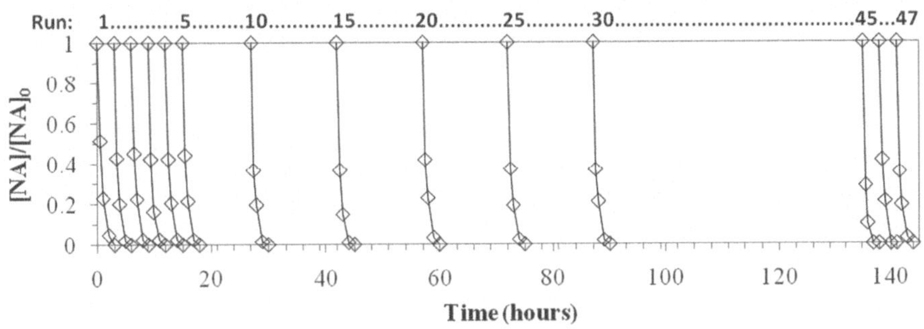

Figure 2.33 Repetitive NA degradation (experimental conditions: 0.18 mM NA, 30°C, solar simulation, initial pH = 6.5, PVF^f-TiO_2-Fe oxide 75 cm^2, H_2O_2 3.2 mM). Reprinted from Mazille *et al.* (2009b) with permission of Editor.

Nalidixic acid, a quinolone antibiotic (which structure is shown in Figure 2.34) was selected as a real pollutant for pilot scale experiments since this compound is present in the wastewater of an INNOWATECH partner.

The degradation of many real pollutants was also performed at pilot scale in solar compound parabolic collector photo-reactors. Figure 2.35 shows the name and structure of the pesticides present in the mixture. Figure 2.36 shows the name and structure of emerging contaminants studied and their properties. Many of these compounds are common pharmaceutical, pesticides and personal care products.

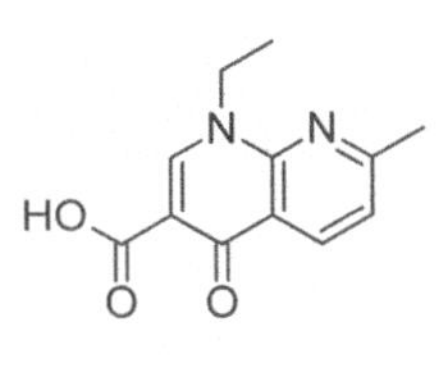

Figure 2.34 Chemical structure of nalidixic acid

2.3.3.1.2 Catalyst preparation

The optimisation of preparation condition was reported Section 2.3.2. Here, optimized material was prepared: a 3-hour Ti-PC treatment was applied to PVF film using aqueous TiO_2 P25 suspensions (1 g/L) at natural pH (≈ 5) and under simulated sunlight irradiation (leading to so called PVF^f-TiO_2) . For the deposition of an iron oxide coat, PVF^f-TiO_2 was immersed in an aqueous solution of $FeCl_3$ (5 g/L) and heated at 80°C while stirring for one hour. The resulting materials (PVF^f-TiO_2-Fe oxide) were dried in an oven at 120°C for one hour.

2.3.3.1.3 Photoreactor and irradiation procedure

The solar degradation experiments were performed under natural sunlight in a pilot plant installed at the Plataforma Solar de Almeria in Spain, consisting of compound parabolic collectors (CPC), a reservoir tank, a recirculation pump and connecting tubing. The configuration of this pilot plant was adapted for these experiments to insert the PVF^f-TiO_2-Fe oxide photocatalyst in the CPC glass tubes. Solar ultraviolet radiation (UV) was measured by a global UV radiometer (KIPP & ZONEN, model CUV 3) mounted on a platform 37°-tilted (the same as the CPC). With Eq. (14), combination of the data from several days' experiments and their comparison with other photocatalytic experiments is possible (Malato, 2003).

$$t_{30w,n} = t_{30w,n-1} + \Delta t_n \frac{UV}{30} \frac{V_i}{V_T}; \quad \Delta t_n = t_n - t_{n-1} \tag{14}$$

where t_n is the experimental time for each sample, UV is the average solar ultraviolet radiation measured during Δt_n, and t_{30W} is a "normalized illumination time". In this case, time refers to a constant solar UV power of 30 W m^{-2} (typical solar UV power on a perfectly sunny day around noon). V_T is the total volume of the water loaded in the pilot plant (8.5 L), V_i is the total irradiated volume (1.8 L). The flow rate was 4.5 L/min. The operation temperature of the pilot plant was around 35 ($\pm$5)°C.

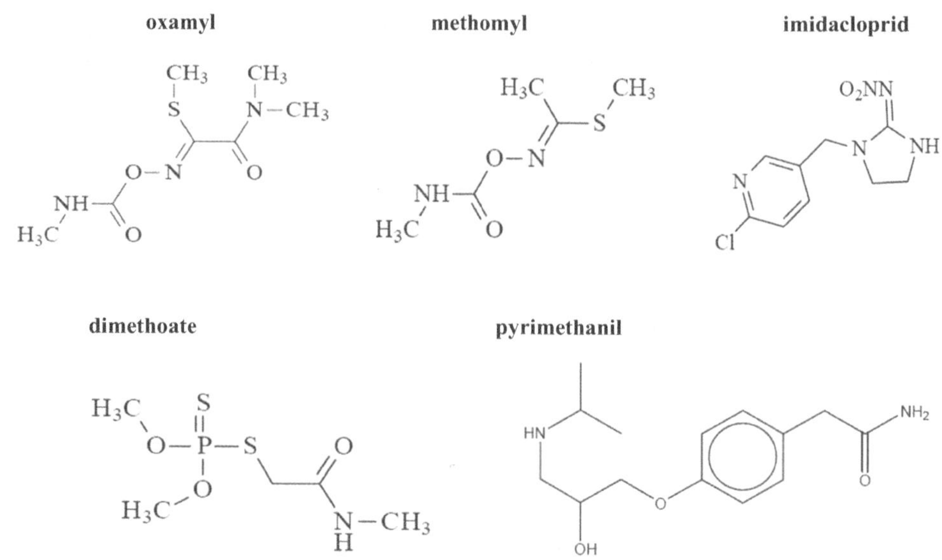

Figure 2.35 Chemical structure and name of the studied pesticides

Several PVF^f-TiO_2-Fe oxide films (26 × 12 cm) were inserted concentrically inside two CPC tubes (1.8 L) in the pilot plant and were used for all the following experiments (Figure 2.37). Before experiments started, the CPC reactor was covered to prevent any photochemical reaction, and filled with distilled water. The chemicals were added directly into the pilot-plant and homogenized by recirculation for 15 min. Then hydrogen peroxide was added and homogenized for 15 min, the collectors were uncovered and the photocatalytic reaction began. Samples were taken every 15 minutes for the first hour after uncovering the reactor, then every 30 minutes for the rest of the experiment.

2.3.3.1.4 Analysis of the irradiated solutions

High-performance liquid chromatography HPLC (Agilent Technologies, series 1100) was used to monitor the compounds using a reverse-phase C-18 analytical column (LUNA 5 mm, 3 mm × 150 mm from Phenomenex) with a UV-DAD detector. The UV signal was recorded at a different wavelength for each compound depending on its maximum light absorption wavelength. The mobile phases H_2O/methanol (60/40), methanol/formic acid 0.25 mM (50:50), acetonitrile/H_2O (15/85), acetonitile/formic acid 0.25 mM (1:10) to 100% acetonitrile (in 40 min), were used for phenol, nalidixic acid, pesticide mixture, and, emerging contaminant mixture, respectively. Mineralization was followed by measuring the dissolved organic carbon (DOC) by direct injection of filtered samples (PTFE 0.22 μm, Millipore Millex1 GN) into a Shimadzu 5050A TOC analyzer with an NDIR detector and calibrated with standard solutions of potassium phthalate. Spectrophotometry for the determination of iron and hydrogen peroxide was performed with a UNICAM 2 spectrophotometer. Total iron was determined with 1,10-phenanthroline, at 510 nm, following ISO 6332, hydrogen peroxide was analyzed using ammonium metavanadate, at 450 nm. The peroxide concentrations are calculated from absorption measurements by a ratio found by Nogueira *et al.* (2005).

2.3.3.2 Results

Experiments were performed on consecutive days separated by a dark period of 17 hours. The illumination time (t_{30W}) included in the discussion was calculated by Eq. 14, which takes this into consideration. The same PVF^f-TiO_2-Fe oxide photocatalyst films were used for all the consecutive experiments included in this chapter.

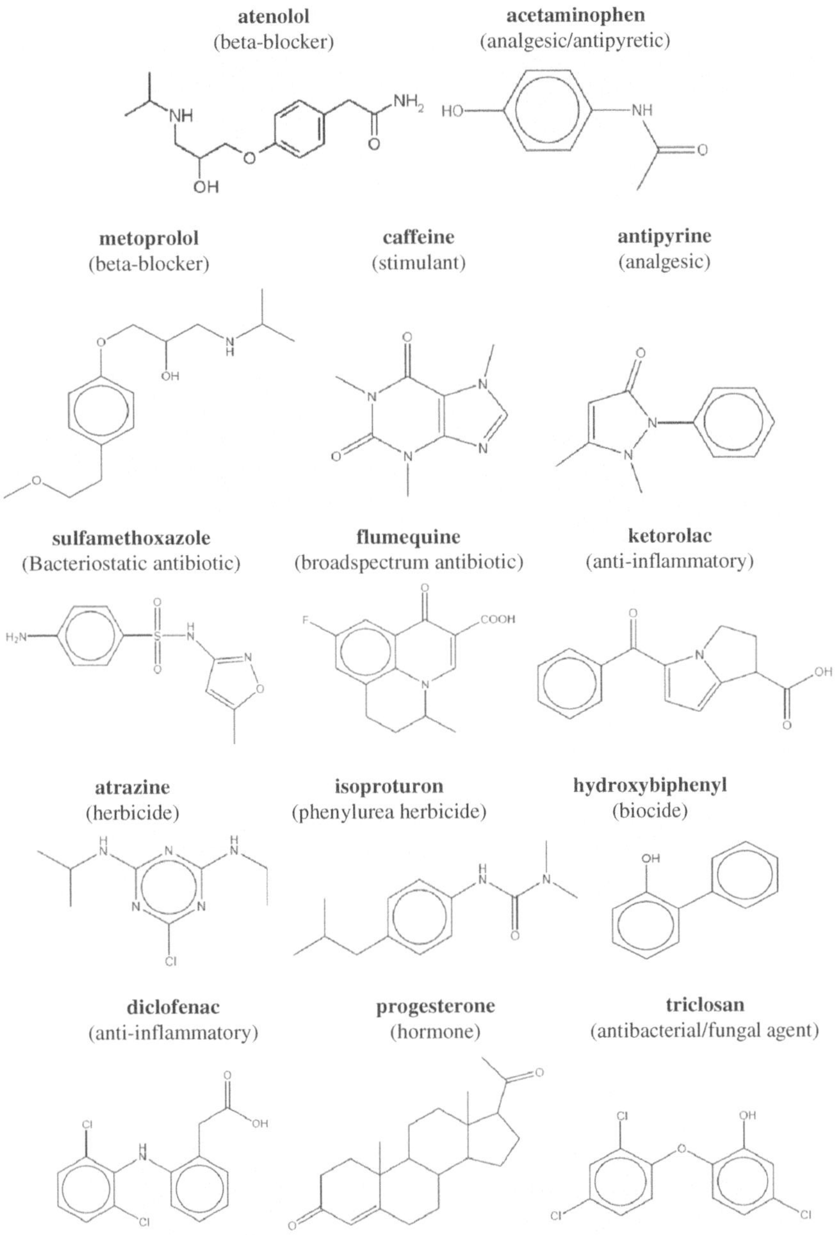

Figure 2.36 Names, structures and properties of emergent organic pollutants chosen for pilot scale study

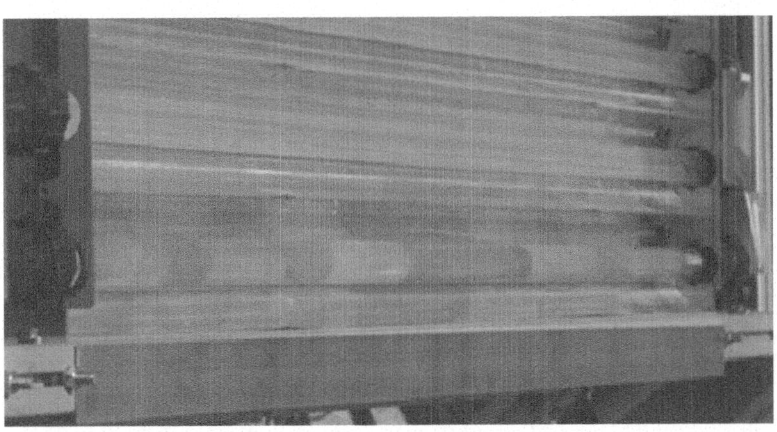

Figure 2.37 Picture of PVF^f-TiO_2-Fe oxide films in the CPC photo-reactor. Reprinted from Mazille *et al.* (2009a) with permission of Editor.

2.3.3.2.1 Photocatalytic degradation of single compounds

Figure 2.38 (i) shows phenol concentration, DOC and pH over t_{30W} [min] during the first degradation experiments mediated by PVF^f-TiO_2-Fe oxide/H_2O_2/light. Only one of two replications is shown, as there was no significant difference between the two consecutive runs. Figure 2.38 (i) shows that the phenol concentration remained almost constant during the first 30 min of treatment and then decreased rapidly until the end of the treatment (60 min). During phenol degradation, pH fell from 6.5–5 from 0 to 25 min to 5–4 after 25 min due to the formation of carboxylic acids. Iron leaching from the catalyst to the solution took place likely by iron oxide photo-reductive dissolution Borer *et al.* (2009). Dissolved iron concentration was always below 0.2 mg/L, suggesting that phenol degradation is mainly induced by a heterogeneous reaction at the catalyst surface although a significant participation of dissolved iron ions in a homogeneous photo-Fenton reaction cannot be totally excluded (dissolved iron concentration is not reported in any of the following experiments, because the quantities of iron ions detected were always below the method detection limit (<0.2 mg/L)).

The observed enhancement of phenol degradation rate after 25 min treatment can be attributed to (i) the formation of degradation intermediates such as hydroquinone which exhibit significant catalytic action in photo-Fenton reaction (Eq. 15–18, Chen *et al.* 2002) and (ii) the pH value drops below 5 which may simultaneously increase iron oxide reductive dissolution (Huang *et al.* 2009) and the catalytic action of iron ions (adsorbed at the interface solid-liquid or dissolved in solution) in photo-Fenton reaction.

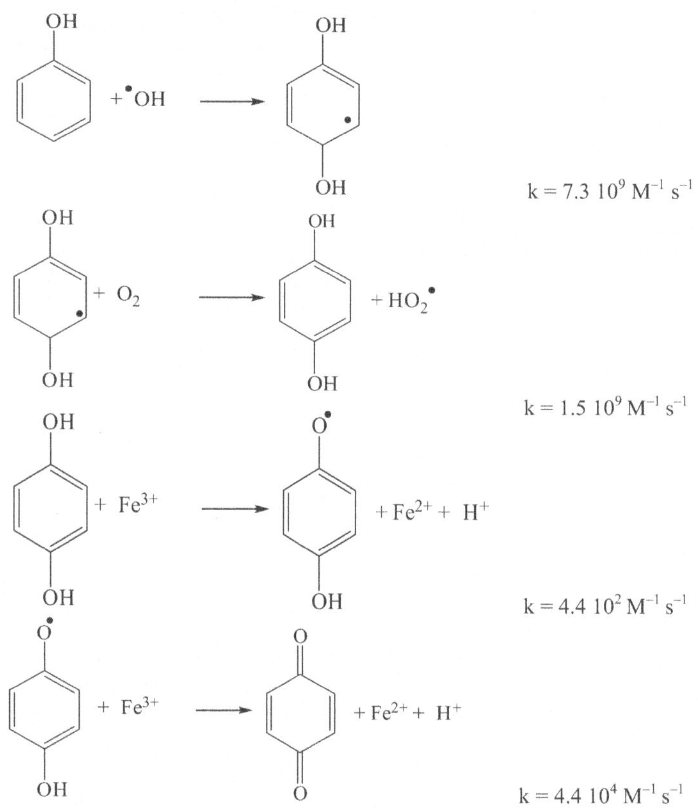

(15)

(16)

(17)

(18)

Phenol degradation was repeated after 7 runs using PVF^f-TiO_2-Fe oxide (called "old" photocatalyst) to test long-term stability of the catalyst. The results are shown in Figure 2.38 (ii). Mineralization was almost complete (>90%) after 135 min of irradiation and during the 17-hour night phase a significant amount of DOC (20%) was removed by dark Fenton reaction. With the old photocatalyst, degradation was slow at the beginning (35%) until t_{30w} = 60 min, then the rate increased and total abatement was observed after 120 min. This sharp enhancement of degradation rates occurred when pH fell below 5. A significant difference in initial phenol degradation rate between new (0.10 mg/(L min)) and old (0.27 mg/(L min)) PVF^f-TiO_2-Fe oxide was observed when pHs > 5. This difference can be attributed to changes occurring on the PVF^f-TiO_2-Fe oxide surface during use, that is, TiO_2 was uncovered, allowing the photocatalyst to be efficient even at high pH. In contrast, in the freshly prepared photocatalyst, only iron oxide is present on the surface (Table 2.2) explaining the low degradation rates observed close to pH 7. When the operating pH is below 5, smaller differences in degradation rate between new (1.0 mg/(L min)) and old (1.5 mg/(L min)) PVF^f-TiO_2-Fe oxide were observed suggesting that under acidic conditions, photo-Fenton oxidation plays a stronger role in degradation.

The degradation of nalidixic acid mediated by PVF^f-TiO_2-Fe oxide/H_2O_2/light is shown in Figure 2.38 (ii). The pH was kept above 7 by appropriate addition of NaOH because the degradation was efficient in neutral solutions. Nalidixic acid, completely degraded rapidly, had a half-life of only 3 minutes under these conditions. This result demonstrates that PVF[f]-TiO_2-Fe oxide is highly active for nalidixic acid, even at neutral operating pH. Mineralization was about 35% after 100 min which suggests that a biological coupling could be performed, Sarria *et al.* (2001).

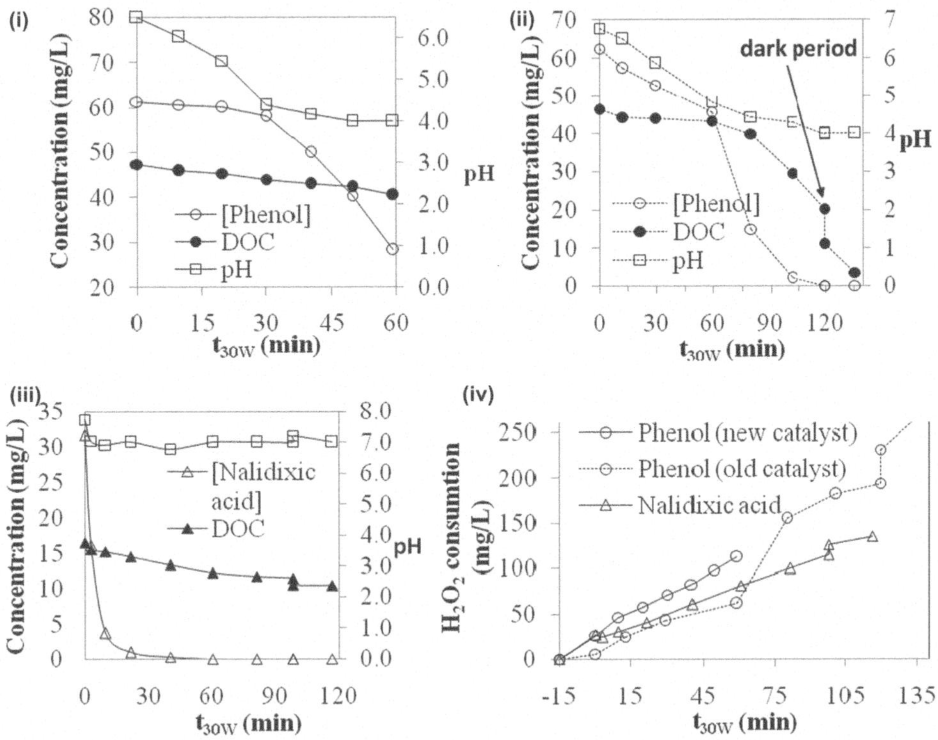

Figure 2.38 DOC and pH during phenol degradation (0.7 mM phenol, 6.2 mM H_2O_2) (i) first and (ii) second set of experiments, (iii) DOC and pH during nalidixic acid degradation (0.13 mM nalidixic acid 3.1 mM H_2O_2), and (iv) evolution of H_2O_2 consumption. The degradations were mediated by PVF[f]-TiO_2-Fe oxide under solar light. Reprinted from Mazille *et al.* (2009a) with permission of Editor.

The differences of reactivity between nalidixic acid and phenol can be assigned to the fact that these two molecules have different abilities to bind with photocatalyst. Nalidixic acid (see Figure 2.34) has a chelating ability and a carboxylate group susceptible to bind with TiO_2, iron oxide surface or iron ions. In contrast phenol has only an alcohol function and is not charged. A recent article (Xue, 2009) claims that organic compound degradation mediated by Fenton heterogeneous oxidation is significantly improved by using chelating agents at neutral pH.

Figure 2.29 shows the evolution of hydrogen peroxide consumption during single compounds degradation. For phenol degradation the initial hydrogen peroxide concentration used was 6.2 mM. Initial H_2O_2 consumption rates (first 30 min) were higher for the new catalyst than for the old one although the degradation rates were higher for the old catalyst. This result suggests that, initially, in contrast with new catalyst, phenol degradation mediated by the old catalyst, involves mechanisms efficient without H_2O_2, such as TiO_2 photocatalysis and synergistic effects. Between 60 and 80 min of treatment, the H_2O_2 consumption rate relative to old catalyst increased significantly but then decreased due to the shortage of H_2O_2. Thus at the end of the first day, H_2O_2 was added until a concentration of 2 mM and during the night phase a significant amount of H_2O_2 was consumed (40 mg) by Fenton process. For nalidixic acid degradation, the initial H_2O_2 concentration used was 3.1 mM and hydrogen peroxide was added when the concentration was too low (60% of consumption).

2.3.4 Photocatalytic degradation of compound mixtures

2.3.4.1 Pesticide mixture

Figure 2.39 (i) shows oxamyl, methomyl, imidacloprid, dimethoate and pyrimethanyl concentrations over t_{30w} during the degradation experiment mediated by PVF[f]-TiO_2-Fe oxide/H_2O_2/light.

For all compounds (except pyrimethanil), initial degradation rates were low and at the end of the treatment (150 min), only about 50% of the substances were degraded. Degradation of compounds during the 17-hour dark phase was negligible. This slow degradation of the pesticides could be due to the use of commercial pesticide formulations, in which some coadjutants might compete with the pesticides in the degradation processes. Pyrimethanyl appeared to be rapidly degraded between 30 and 60 min treatment. However, this fast depletion could be due to an adsorption of the substance to the discovered PVF surface until saturation after 60 min. Pyrimethanil is the less water soluble of the tested pesticides (0.1 g/L, 25°C) and could have high affinity with PVF film.

Figure 2.39 (ii) shows that 30% of the total DOC was removed during the treatment. Most of the DOC decayed the first day under illumination (20%) and during the dark phase (10%). Afterwards, mineralization stopped, possibly because of the formation of resistant intermediates complexing Fe with carboxylic acids produced during degradation and adsorption of part of the organics on the PVF^f-TiO_2-Fe oxide, or because the coadjutants easily mineralized had been completely removed.

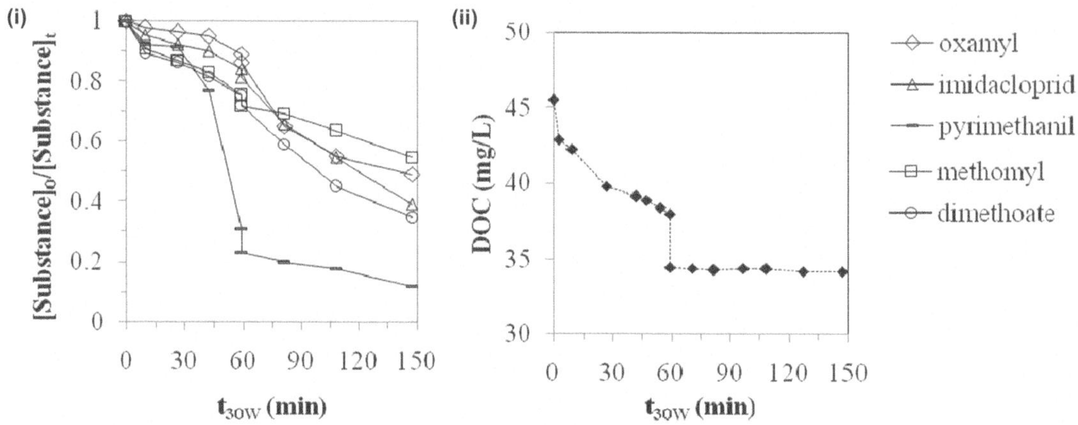

Figure 2.39 Normalized pesticide concentrations (i), DOC (ii) during degradation mediated by PVF^f-TiO_2-Fe oxide in presence of 3.1 mM H_2O_2 (constant) under sunlight ($DOC_0 = 50$ mg/L). Reprinted from Mazille *et al.* (2009a) with permission of Editor.

Table 2.3 shows the initial degradation rates (R_0) relative to each pesticide for the process mediated by PVF^f-TiO_2-Fe oxide/H_2O_2/light and the constants for homogeneous photo-Fenton oxidation (5 mg/L Fe^{3+}/H_2O_2/light) reported by Zapata *et al.* (2009). This comparison shows that homogeneous rates were between 5 to 10 times higher than heterogeneous rates. However, in the homogeneous process, pH was close to 3.

Table 2.3 Initial degradation rates with PVF^f-TiO_2-Fe oxide /H_2O_2/light and zero order rate with Fe^{3+}/H_2O_2/light (in μg/(L.min)) for each compound in the pesticide mixture

Compound	R_0^a	k (Fe^{3+} 5 mg/L)^b	k/R_0
Oxamyl	27	290	11
Methomyl	21	100	5
Imidacloprid	15	100	7
Dimethoate	28	240	9
Pyrimethanyl	73	390	5

[a]*Calculated from Figure 2.39 (first 60 min).*
[b]*Reproduced from Zapata et al. (2009).*

2.3.4.2 Emerging contaminants mixture

The 14-compound mixture used in this study is a reconstruction of pollution typical of a municipal wastewater treatment plant effluent. In such wastewater, there is a large number of low-concentration (<0.1 mg/L) pollutants (Martínez Bueno *et al.*, 2007). Concentration of each substance in the two consecutive experiments is shown over t_{30W} in Figure 2.40. Eight of the fourteen compounds (atenolol, acetaminophen, metropolol, flumequine, hydroxybiphenyl, diclofenac, progesterone and triclosan) were efficiently degraded by PVF^f-TiO_2-Fe oxide/H_2O_2/light after 150 min and after 100 min of treatment in two consecutive experiments (Figure 2.40a_1, b_1 and a_2, b_2, respectively). The other six compounds, caffeine, antipyrine, sulfamethoxazole, ketorolac, atrazine and isoproturon were more resistant, with around 40-70% (Figure 2.40c_1) and 50–80% (Figure 2.40c_2) degradation after 180 min for the first and second consecutive runs, respectively. Caffeine and antipyrine were degraded the least (50%) under illumination and only caffeine and isoproturon showed significant degradation (≈10% decay) during dark Fenton.

In the two successive experiments, substances showed a rather similar trend, although degradation rates were 1 to 5 times higher in the second experiment. In Figure 2.40, degradation rates are dependent on the nature of the substance, from 0.2

to 2 µg/(L min) for antipyrine (minimum) and for flumequine (maximum), respectively. Thus PVF[f]-TiO_2-Fe oxide/H_2O_2/light degraded pollutants selectively, which was not the case for homogeneous photo-Fenton degradation of the emerging contaminant mixture (Klamerth *et al.*, 2009). The differences observed in degradation rates between the emergent compounds can be assigned to several possible mechanisms: (i) the electrostatic attraction or repulsion between photocatalyst surface and substance (ii) the intrinsic reactivity of the substance (presence of weak bonds or electron donor groups, possibility to form complex with iron surface species).

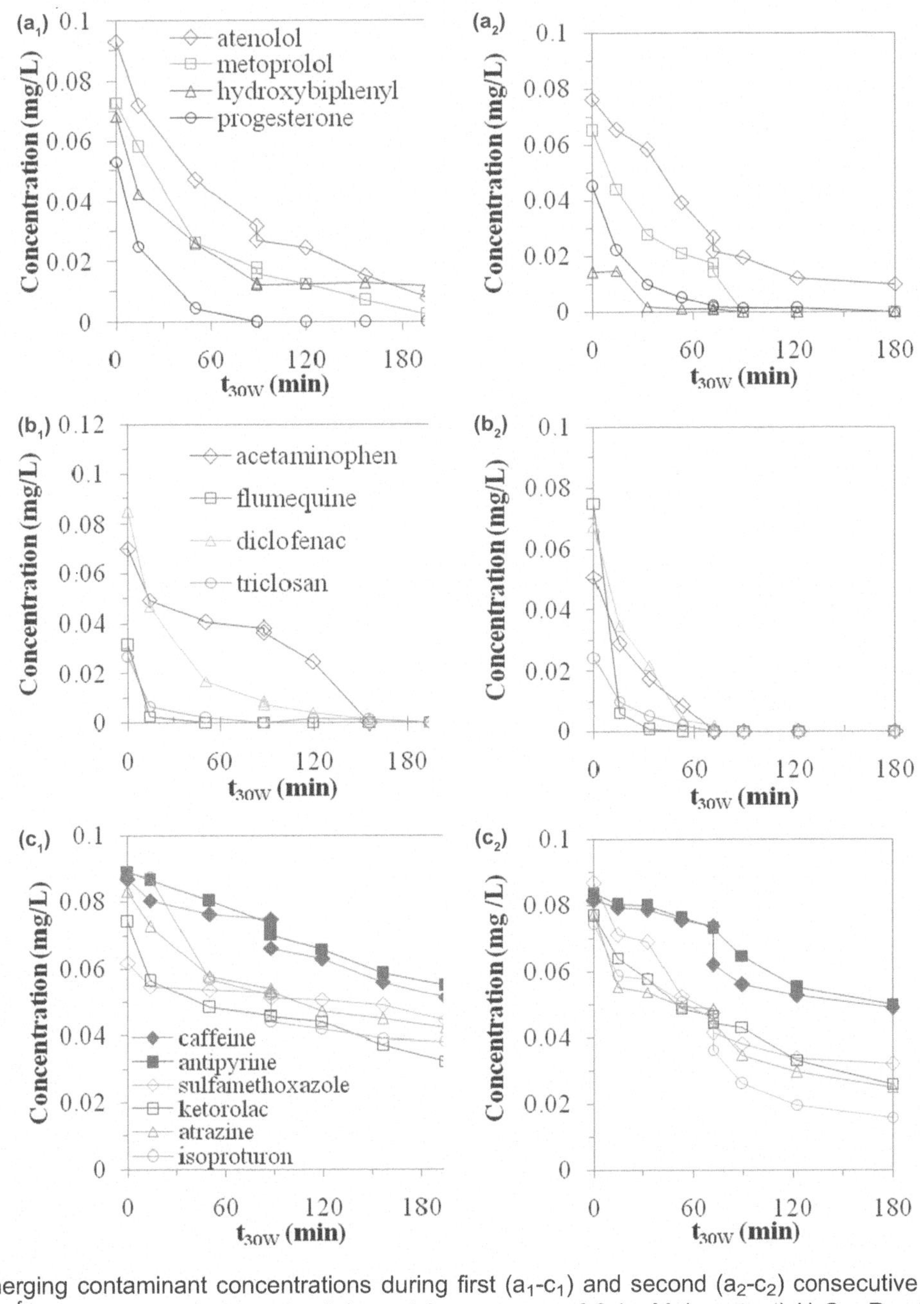

Figure 2.40 Emerging contaminant concentrations during first (a_1-c_1) and second (a_2-c_2) consecutive degradation tests mediated by PVF[f]-TiO_2-Fe oxide under solar light and in presence of 3.1 mM (constant) H_2O_2. Reprinted from Mazille *et al.* (2009a) with permission of Editor.

Figure 2.36 shows the chemical structure of the different compounds: some of the more recalcitrant compounds, namely, caffeine (pKa 10), ketorolac (pKa 10.3), contain tertiary amine which are protonated and thus positively charged under the pH of experiments (<7). The presence of this positive charge and/or the absence of negative or complexing group (such as carboxylic acid or chelating agent), possibly induced repulsion or weak attraction between the substance and the catalyst surface leading to a slow degradation. In contrast, atenolol, metoprolol (although positively charged under experimental condition, pKa 9.6, 9.7, respectively) and triclosan contain a chelating group, flumequin and diclofenac contain a carboxylic acid function and are halogenated (weak bonds). Thus the results suggest that reactivity is controlled by the presence of Fe(III) complexant moieties (carboxylic acid, chelate) rather by the electrostatic attraction/repulsion mechanism. However, a systematic study under more controlled conditions is required to draw a definitive conclusion on the structure reactivity relationship for heterogeneous photo-Fenton systems.

2.3.5 Conclusions

Implement a functionalization of polymer surface can be an efficient strategy to improve the attraction between iron oxide and polymer film supports, leading to more efficient photocatalysts. Besides functionalization process such as Ti-PC can concomitantly lead to the deposition of additional active specie (TiO_2 particles). Among the different functionalization processes studied here, RF-P and more markedly Ti-PC treatments were useful to improve photocatalytic activity. PVF, PE or PET supports led to catalysts showing similar photoactivity ruling out the possibility that the polymer nature influence significantly the catalyst efficiency. The preparation conditions were optimised leading to efficient photocatalysts. The optimised photocatalysts (PVF^f-TiO_2-Fe oxide) were highly stable toward repetitive NA degradations. This outstanding long-term stability was assigned to the discovering of TiO_2 during usage which increased its exposure to light and consequently its synergistic contribution to the overall photocatalytic process. However, the amounts of dissolved iron species leached from catalyst substrate lead to a significant homogeneous photo-Fenton contribution.

This work demonstrated that PVF^f-TiO_2-Fe oxide can be adapted to a pilot-scale solar photoreactor resulting in high performance even with controlled neutral pH for nalidixic acid. However, pesticides were more resistant, with 50% degradation in 1.5 hours. In contrast with homogeneous photo-Fenton, emerging pollutant degradation mediated by PVF^f-TiO_2-Fe oxide/H_2O_2/light was found to be selective. Degradation rates were higher for the emerging pollutants containing moieties (carboxylic acid, chelating group) able to form complexes with Fe(III).

2.4 MBR/AOP TREATMENT OF PHARMACEUTICAL WASTEWATER

2.4.1 Introduction

Wastewater from the chemical and pharmaceutical industries normally contains, besides residuals of the synthesized products, solvents and intermediates which may be non-biodegradable and/or toxic towards the water bodies and the environment in general. Consequences on the aquatic environment were reported to be related to improper disposal of pharmaceutical wastes and limited effectiveness of conventional wastewater treatments towards these compounds (Hirsch *et al.*, 1999; Zuccato *et al.*, 2000; Cunningham *et al.*, 2006). Common characteristics normally encountered in these streams are i) possible coexistence of non-biodegradable solvents, intermediates, synthesis products together with high fractions of easily biodegradable carbon (solvents) ii) high saline content due to acidification/neutralization steps in the production processes (Lefebvre and Moletta, 2006).

If conventional biological wastewater treatments have limited degradation efficacy, conventional and advanced chemical oxidative treatment technologies such as the Advanced Oxidation Processes (AOPs) may be helpful for their known capacity of oxidising and partially mineralising many organic contaminants (Mascolo *et al.*, 1995; Lopez *et al.*, 1997; Gogate and Pandit, 2004a, 2004b; Parsons, 2004). However, AOPs have high operational costs that often hinders their commercial application, therefore their adoption is limited to those cases when more cost-effective systems are not suitable. It follows that specific treatment trains need to be selected for the removal of pharmaceuticals from industrial wastewaters. So far several studies have tested the opportunity of either enhancing biodegradability through ozonation pre-treatment or including ozonation as a final polishing step (Contreras *et al.*, 2003; Hsu *et al.*, 2004; Aparicio *et al.*, 2007). Similarly, another set-up capable of lowering the operational cost has been recently tested for the treatment of real pharmaceutical wastewater, consisting in the combination of biological and solar-based AOP (Sirtori *et al.*, 2009a, 2009b). However, in both these approaches, biological and physico-chemical processes are taken into account separately in the design of the treatment train, and their effects are "added" one to the other. Moreover, in presence of high concentrations of organic compounds (often highly biodegradable, as in the case of many solvents in pharmaceutical industry effluents), chemical pre-treatments can only be sustainable if highly selective towards the target compound. On the other hand, when an AOP is adopted as a final polishing step, the presence of organic by-products in the final effluent can only be avoided by favouring their complete mineralization. Considering that usually the formation/decomposition kinetics of by-products are not very fast in real effluents, intensive post-treatments may be necessary in order to obtain very low residual concentrations, and this leads to high operational costs.

A possible alternative is the integration of ozonation as part of treatment trains that include biological processes. This option may have a strong potential in terms of minimization of oxidation by-product release, if these can be removed in the biological step. A real integration of biological treatment and ozonation can also be considered a safer "multiple barrier approach", with the biological process involved in both the reduction of the pollutants load to the following chemical oxidation process, and the biological removal of the oxidation by-products.

In this view, the membrane bioreactors (MBR) offer unique opportunities in terms of suspended solid-free effluent and possible selection of specialized biomass. Indeed, the performance of ozonation is well know to be negatively affected by suspended solids that act as scavengers toward hydroxyl radicals, the reactive free radicals formed by ozone decomposition in water (Langlais *et al.*, 1991).

In the present study, the above mentioned integrated approach was investigated by placing an oxidation step in the effluent recirculation stream of a lab-scale MBR treating real pharmaceutical wastewater streams. The adopted pharmaceutical wastewaters were characterized by the contemporary presence of high fractions of easily biodegradable carbon and lower amounts of recalcitrant organics as well as a high saline content. The work was focused on the evaluation of the different contributions to the removal of pollutants, and on the detailed investigation about the by-products formation/degradation during both MBR and oxidation by ozonation.

2.4.2 Preliminary investigations

The investigation on the suitability of the proposed integrated systems for the treatment of wastewater resulting from the industrial production of pharmaceuticals required the selection of wastewater characterized by a large fraction of recalcitrant compounds. Considering that the main issue of the performed investigation was the persistence of non-biodegradable compounds in the environment, the driving criterion for selecting the wastewater was low biodegradability of the main target pharmaceutical, its main parent compounds, and its by-products. Therefore, In the preliminary tests three real pharmaceutical wastewaters resulting from the industrial production of three different drugs, namely acyclovir, nalidixic acid and naproxen were investigated in terms of organic parent compounds and metabolites recalcitrant to biodegradation. The wastewater samples were chosen as representative of several different real situations in terms of minor recalcitrant TOC fractions due to parent drug or other organics, and the biodegradation metabolites. Indeed, discharge of these wastewaters into the centralized sewage treatment plant would cause operational drawbacks to biological treatments. Specific objective of the preliminary study was to evaluate the TOC fraction that is not biodegraded, also including the organic metabolites formation. The evaluation was performed through the standard Zahn-Wellens test. In addition, the wastewater composition before and during the test was investigated by LC/MS in terms of both parent compounds and main metabolites. The biodegradability of the parent compounds was also assessed by performing extra Zahn-Wellens tests on synthetic solutions.

2.4.2.1 Experimental

Standard of acyclovir (9-[(2-hydroxyethoxy)methyl]guanine), nalidixic acid (1-ethyl-1,4-dihydro-7-methyl-4-oxo-1,8-naphthyridine-3-carboxylic acid) and naproxen (2-(6-methoxynaphthalen-2-yl)propanoic acid) were purchased from Fluka. Also, two different acyclovir powders (named as ACG and ACY) derived from the synthesis of the drug were provided by the producing company. ACG is the first product obtained from the synthetic process while ACY is derived from a subsequent purification step.

Dissolved organic carbon (DOC) and total nitrogen (TN) were measured on filtered samples (0.45 μm) by a Shimadzu TOC-V_{CSH} analyzer equipped with a TNM-1 Total Nitrogen Measuring Unit and an ASI-V autosampler. COD and Total Phosphorus (TP) were measured by the Lange Cuvette Test on not filtered and filtered solutions. The determination of residual parent drug concentration as well as the intermediate compounds identification was performed by HPLC/MS and HPLC/MS-MS using an Acquity chromatographic system equipped with a diode array detector (Waters) interfaced to an API 5000 mass spectrometer (AB Sciex) by means of an electrospray interface (ESI). Samples (5 μL) were eluted at 0.3 mL/min through a HSS T3 (trifunctional C_{18}) 150 × 2.1 mm, 1.8 μm (Waters) analytical column adopting a gradient from water/methanol 95/5 with 0.05 mM ammonium acetate to methanol with 0.05 mM ammonium acetate. More details are reported elsewhere (Mascolo *et al.*, 2010a).

2.4.2.1.1 Zahn-Wellens tests

Information on the potential suitability of biological wastewater treatment to specific polluted streams can be obtained through the so-called "inherent (potential) biodegradability tests" (Pitter and Chudoba, 1990). These tests provide information concerning the bio-treatability of a substance, and one of the most commonly applied within this type of determinations is the Zahn-Wellens procedure, as reported in the EC protocol C.09 of the Annex V to Directive 67/548/EEC (EC, 1988). The Zahn Wellens test simulates the "behaviour" of a chemical compound in an activated sludge treatment plant, because the experimental conditions of the test are similar to those normally maintained in this process (Lapertot and Pulgarin, 2006). A relevant feature of this category of tests is that poor results obtained in an inherent biodegradability test are considered sufficiently indicative for poor biodegradability, but positive results are not necessarily predictive of biodegradability under real environmental conditions (OECD, 1993). Inherent biodegradability tests such as the Zahn-Wellens for the prediction of carbon removal of specific chemical substances were reported to be reliable and highly indicative for a similar removal in activated sludge systems, although a certain possibility for a significant overestimation of the removal extent was also reported (Guhl and Steber, 2006). In this respect, particular care should be taken when the target compounds may be "hidden" in large amounts of easily biodegradable matrices. In these cases, the biodegradability test needs to be backed by specific analyses aimed at assessing the extent of biotransformation of the target compounds. Elimination of the tested compounds and formation of aerobic degradation products in Zahn-Wellens tests was sometimes reported, and demonstrated biotransformation but incomplete mineralization of the parent compound tested (Haiss and Kümmerer, 2006).

2.4.2.2 Biodegradability of pharmaceutical wastewaters

Three pharmaceutical wastewater samples were obtained from the industrial production of three drugs (acyclovir, nalidixic acid and naproxen) and their main features are listed in Table 2.4. In addition, biodegradability tests with synthetic solutions made with the pure drugs were also performed.

Table 2.4 Chemico-physical parameters of the investigated pharmaceutical wastewaters

Parameter	Unit	Acyclovir wastewater	Nalidixic acid wastewater	Naproxen wastewater
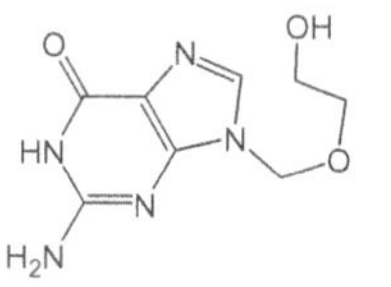 Acyclovir	mg/L	240	–	–
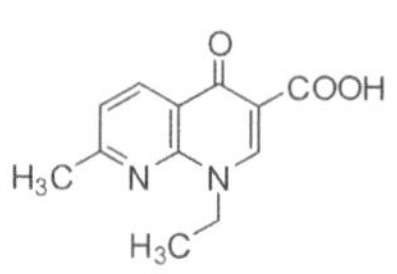 Nalidixic acid	mg/L	–	222	–
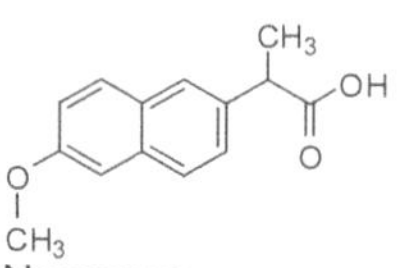 Naproxen	mg/L	–	–	324
acetate	$mg_{CH3COOH}$/L	–	1900	–
TOC	mg/L	9900	3500	35 600
COD	mg/L	26 700	10 900	114 000
TN (Total Nitrogen)	mg/L	1200	160	7500
NH_4^+	gN/L	<0.2	<0.2	<0.2
IC	mg/L	<0.3	<0.5	<0.5
pH		6.7	<1	5.3
Conductivity	mS/cm	41.8	1.6	45

The wastewaters selected for the investigation were characterized by a large range of TC, COD, TN and conductivity (Table 2.4). The TOC profiles of the investigated wastewaters during the duration of the Zahn-Wellens test are shown in Figure 2.41a. For acyclovir wastewater a sharp removal of TOC was obtained in the first 2 days of the test, and in the remaining days the residual TOC fraction was removed at a much slower rate. As for the nalidixic acid and naproxen wastewaters, results showed that they can also be considered biodegradable similarly to what found for acyclovir wastewaters as a TOC fraction greater than 70% was removed (USEPA, 1996). Zahn-Wellens tests were also performed with aqueous solutions made from ACG and ACY powders as well as with the other two drugs. ACG and ACY (Figure 2.41b) showed a similar behaviour that, in turn, is very close to the profile obtained with acyclovir wastewater. Indeed, about 40 and 55% of the initial TOC disappeared in the first three days for ACY and ACG aqueous solution, respectively. Afterwards, the decay rate was much slower and reached a similar final value for both drug solutions. The differences between the TOC removal for ACG and ACY are consistent with the fact that the first still contains some biodegradable organics as impurities leading to higher percentage of TOC removal. However, on the basis of this final value both the acyclovir powders must be considered to be not completely biodegradable. This result suggests that the slow TOC disappearance in the wastewaters at long test times is due to the residual acyclovir or to metabolites. It follows that particular care should be paid to the content of these organics when submitting the pharmaceutical wastewater to a biological treatment. In fact, even though the wastewater (as a whole) showed to be biodegradable, the small fractions of recalcitrant organic compounds should also be carefully evaluated since they could be accumulated into the biological system. The biodegradability profiles of the nalidixic acid and naproxen drugs in solution showed different trends with respect to the corresponding synthetic solutions (Figure 2.41). Indeed, the nalidixic acid resulted to be quite non-biodegradable; its removal being below 40% throughout the whole duration of the test, and this is consistent with the low contribution (4%) of the nalidixic acid to the TOC of the wastewater (Table 2.4), that was found biodegradable. Naproxen, conversely to the other two drugs, showed to be completely biodegraded after the first 10 days of the test.

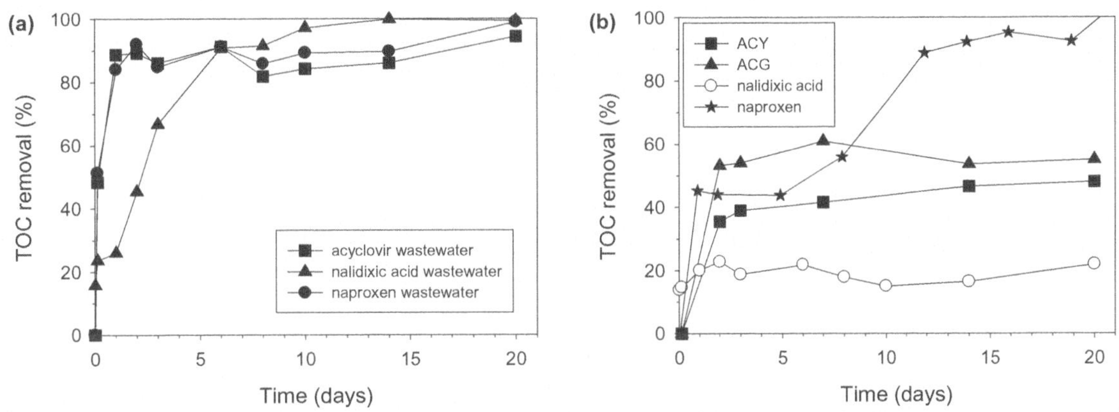

Figure 2.41 TOC removal during Zahn-Wellens tests of pharmaceutical wastewaters deriving from the synthesis of acyclovir, nalidixic acid and naproxen (a) and synthetic solutions made from powders of the same pharmaceuticals (b)

2.4.2.3 Main organics biodegradability in the pharmaceutical wastewaters

The removal of main organic compounds is also an important aspect to be monitored during the Zahn-Wellens tests. it was found that acyclovir was completely removed within just few days (Figure 2.42a). It follows that it can be considered to be very biodegradable. Nalidixic acid showed to be very poorly biologically removed from the wastewater (20%) within the whole duration of the test. As for the naproxen, the results showed that the parent drug was completely removed from the wastewater after an initial acclimatization of few days. As for the results obtained with the aqueous solutions containing only one pharmaceutical compounds, for acyclovir ACG powder showed a very similar decay pathway with acyclovir contained in the wastewater (Figure 2.42b). Instead, ACY showed a slower decay in the first seven days of the test. This unexpected result (i.e. the solution of the compound of highest purity has a slower decay during the first days) can be explained assuming that ACG contains impurities that favour the acyclovir biodegradation, leading to a more prompt biomass response during the first days. However, in both cases the complete removal of acyclovir was obtained before the end of the Zahn-Wellens test. Also, the comparison of ACG and ACY profiles in Figures 2.41b and 2.42b suggest that acyclovir is easily removed biologically but it is not completely mineralized (TOC removal less than 60%). As far as nalidixic acid profile is concerned, Figure 2.42b shows that it was obtained a scarce biodegradability of this compound when present in both wastewater and aqueous solution. However, this result was not evidenced during the Zahn-Wellens test of the wastewater (Figure 2.41a) because nalidixic acid accounts for just 4% of the total carbon. It follows that a relevant limitation of the Zahn-Wellens test arises when small concentrations of target pollutants are "hidden" by large concentrations of biodegradable compounds. In these cases the response in terms of biodegradability has to be confirmed by specific analyses. A lag phase was registered for the synthetic solution made from the powder of naproxen, similarly to what obtained with the correspondent watewater. This suggests that the compound and the other organics contained in the resulting wastewater are suitable for biological treatment. Furthermore, naproxen degradation in the real wastewater occured slower than in the synthetic solution which indeed contains just the drug. It follows that the presence of all the other organics present in the real wastewater negatively affect the naproxen removal.

2.4.2.4 Removal of other organic compounds and metabolites during Zahn-Wellens tests

It is of valuable scientific interest to obtain information about the degradation of the main organic compounds during the Zahn-Wellens tests. In fact, this would allow to check whether some of those are recalcitrant to biodegradation. Beside, it is also interesting to monitor the formation of organic metabolites that can be accumulated during the test. Both these investigations were performed for acyclovir and naproxen wastewaters. Five main compounds were detected during the biodegradability tests, three with the acyclovir wastewater and two with the naproxen one, respectively. their molecular weights are listed in Table 2.5.

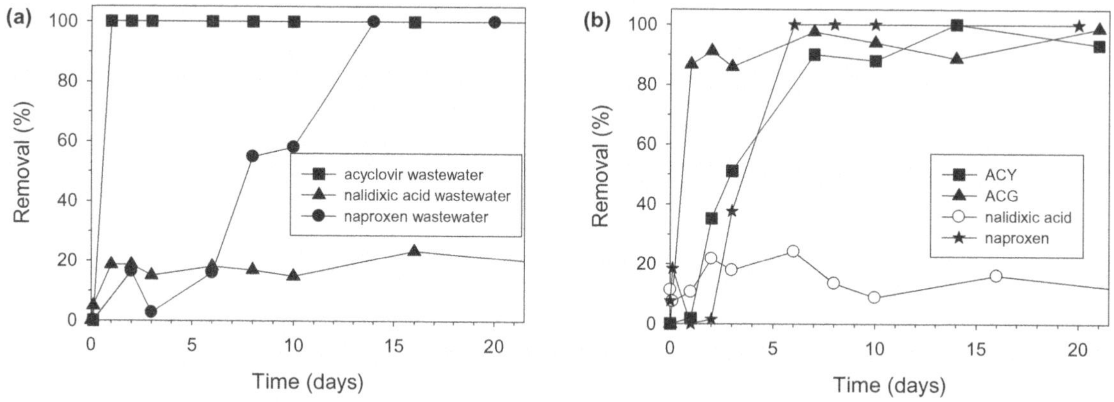

Figure 2.42 Percentage of removal of acyclovir, nalidixic acid and naproxen in the respective wastewaters (a) and with synthetic solutions made from powders of the same drugs (b) during Zahn-Wellens tests

Table 2.5 Metabolites of acyclovir and naproxen wastewaters detected during Zahn-Wellens test

Wastewater	Compound number	Molecular weight
Acyclovir	1	239
Acyclovir	2	221
Acyclovir	3	291
Naproxen	4	216
Naproxen	5	200

As for the acyclovir wastewater, one of the three detected compounds (compound No. 1) was found to be formed during the tests as evidenced from the formation/degradation profiles depicted in Figure 2.43a. Its molecular weight is consistent with the structure of a metabolite of the acyclovir, namely the 9-carboxymethoxymethylguanine that has been already reported to be responsible of the incomplete urinary recovery of acyclovir (Svensson *et al.*, 1997). Profile of compound No. 1 show that it is quite persistent, and tended to accumulate during Zahn-Wellens test. Molecular weight of compounds Nos. 2 and 3 were found to be 221 and 291, respectively. Their decay during the Zahn-Wellens test showed different trends (Figure 2.43a). In particular, compound No. 2 showed to be readily removed in the wastewater while compound No. 3 showed to be quite persistent and remained till the end of the test. The persistence of compound No. 2 is also consistent with the small fraction of TOC that was removed very slowly during the Zahn-Wellens tests.

As for the naproxen wastewater, two main organic compounds were detected and their decays were monitored during Zahn-Wellens tests (Figure 2.43b). Results showed that both the detected organics were completely removed during the biodegradation test. It follows that these compounds do not represent a potential problem for the biological treatment of the naproxen wastewater, confirming what previously found when monitoring the TOC and naproxen concentration during the Zahn-Wellens test.

In conclusion, the proposed approach about the adoption of a standard biodegradability test coupled with LC-MS allowed to assess the real potential of a biological treatment in case of wastewaters where the refractory substances may be masked by major fractions of biodegradable compounds. In fact, such a refractory fraction can accumulate in the biological tank leading to long term toxic effects to the biomass. In this case, it should be considered the adoption of more efficient biological treatments (such as the MBR technology) and/or the integration with an advanced oxidation process, either to enhance the biodegradability or to mineralize the refractory compounds.

Based on the results of the above reported preliminary investigations the wastewaters from the production of acyclovir and nalidixic acid were selected for the subsequent activities and laboratory scale studies as they were characterized by the presence of low biodegradability of main organics and metabolites (acyclovir wastewater) and of low biodegradability of the main target pharmaceutical (nalidixic acid wastewater), respectively.

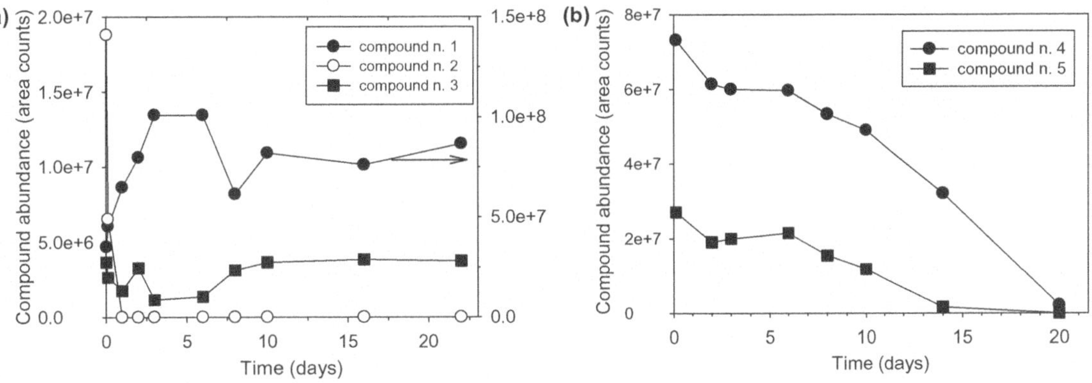

Figure 2.43 Removal/formation profiles of main organic compounds detected/formed during Zahn-Wellens tests performed with acyclovir wastewater (a) and naproxen wastewater (b)

2.4.3 Laboratory scale studies

2.4.3.1 Materials and methods

Several batches of each investigated wastewater (acyclovir and nalidixic acid) deriving from various production cycles were mixed in order to obtain an average sample representative of the two wastewater streams. The chemical-physical characteristics of the obtained wastewaters are reported in Table 2.6. The wastewaters were also characterized in term of organics composition as reported below.

Table 2.6 Chemico-physical characteristics of the pharmaceutical wastewaters employed as influent of the MBR. NH_4^+ and phosphorus (as $(NH_4)_2CO_3$ and $(NH_4)H_2PO_4$) were added in order to provide the necessary nutrient requirement to the biomass

Parameter	Unit	Acyclovir wastewater	Nalidixic acid wastewater
pH		6.3	4
Conductivity	mS cm^{-1}	11.5	7
DOC	mg L^{-1}	4400	775
COD	mg L^{-1}	10 900	2660
TN	mg_N L^{-1}	120	5.4
NH_4^+ (added)	mg_N L^{-1}	347	49
Acetate (as sodium salt)	mg L^{-1}	9300	1900
Acyclovir	mg L^{-1}	154	
Diactetylacyclovir	mg L^{-1}	0.4	
Nalidixic acid	mg L^{-1}		45
Cl^-	mg L^{-1}		2800
PO_4^{3-} (added)	mg L^{-1}	223	19
$SO_4^=$	mg L^{-1}	<0.4	160
Na^+	mg L^{-1}		2000
TSS	mg L^{-1}		496
VSS	mg L^{-1}		264

2.4.3.1.1 MBR, ozonation, and integrated process

A 10 L MBR, equipped with a Zenon ZW1 submerged hollow fiber membrane module having a filtration surface of 0.047 m^2, was operated by using a feed flow rate of 1.6 L d^{-1} and a permeate recirculation flow rate of 4.8 L d^{-1}. The filtration process was operated out-in, and the permeate was extracted from the internal side of the hollow fibers through the application of a gentle negative pressure (never exceeding 0.6 kPa) to one end of the membrane module. To limit the occurrence of clogging and biofouling of the membrane surface, intermittent cycles of suction and relaxation were alternated (suction: 165 min; relaxation: 15 min). In order to obtain similar operating conditions for the MBR alone and for the integrated system, the permeate recirculation was operated since the start-up of the plant. The resulting permeate flow rate was about 0.27 L h^{-1}, with a consequent flux through the membrane of about 5.6 L m^{-2} h^{-1}. Activated sludge from the local wastewater treatment plant and mineral nutrients were used for the inoculum and start-up of the biological reactor. The MBR was operated with a sludge retention time (SRT) of 60 d in the tests with the acyclovir wastewater, whereas in the tests with the nalidixic acid wastewater the SRT was 30 d. The system was maintained under continuous aeration, in order to promote the development of aerobic microbial consortia for the biodegradation of bulk and target compounds. The MBR biomass was acclimated by treating the wastewater stream in the recirculated mode until the steady state was reached. Afterwards, according to the operational scheme reported in Figure 2.44a, the recirculation stream of the MBR was subjected to oxidation (ozonation). Four daily samples (raw wastewater; MBR influent; MBR effluent; ozonation effluent) were withdrawn and analyzed to monitor the system performance.

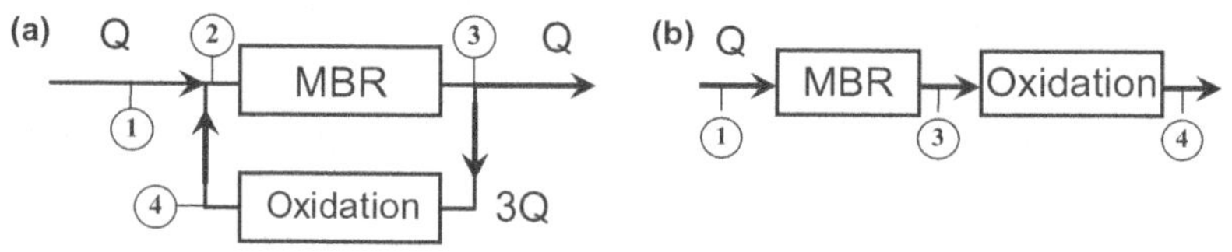

Figure 2.44 Scheme of the investigated integrated wastewater treatment system (a) and the polishing treatment train (b) employed for comparison. Q = 1.6 L/d. Legend: 1 = raw wastewater; 2 = MBR influent; 3 = MBR effluent (oxidation feed); 4 = oxidation effluent

As for the ozonation step, a 7 L reactor was used and fed with ozonated air (ozone concentration, 6.8 and 11 g m^{-3} for acyclovir and nalidixic acid wastewater, respectively; flow, 1 L min^{-1}) supplied from a Fisher 502 ozonator (Germany). As reported above, the daily volume of MBR effluent subjected to ozonation was 4.8 L. This volume was collected daily and ozonated for 15 min which corresponds to an ozone dosage of 1.36 and 2.3 mg L^{-1} min^{-1}. Afterwards, the ozonator was turned off and air was let to bubble for 15 min in order to strip the residual dissolved ozone that could be detrimental for the MBR biomass. Ozone concentration in water and ozone concentration in the inlet and in the off-gas were measured by an amperometric ozone probe (Orbisphere). The same reactor and the same operating conditions were used for carrying out oxidation experiments (Figure 2.44b) with the MBR effluent obtained during the first phase (MBR only).

2.4.3.1.2 Analytical determinations

Dissolved organic carbon (DOC) and total nitrogen (TN) were measured on filtered samples (0.45 μm) by a Shimadzu TOC-VCSH analyzer equipped with a TNM-1 Total Nitrogen Measuring Unit and an ASI-V autosampler. DOC was determined as the difference between Total Carbon and Inorganic Carbon. COD and Total Phosphorus (TP) were measured on non-filtered and filtered samples adopting the Lange Cuvette Tests, with a Lange DR2800 photometer for the quantification. Total kjeldal nitrogen (TKN) and ammonium were determined according to Standard Methods 4500-Norg B and 4500-NH3 C (APHA, 1999).

Nitrite and nitrate were measured by ion chromatography by means of a Dionex AS14A 4 mm × 250 mm column and UV detection (220 nm). Isocratic elution was carried out with a solution of $NaHCO_3$ (10 mM) and Na_2CO_3 (10 mM) at a flow rate of 1.0 mL min^{-1}. Total dissolved solids and total solids were measured according to Standard Methods 2540-D and 2540-B, respectively.

The determination of residual concentrations of the target pharmaceuticals as well as the intermediate compounds identification was performed by HPLC/MS and HPLC/MS-MS using an Acquity chromatographic system (Waters) interfaced to an API 5000 mass spectrometer (AB Sciex) by means of a turboionspray interface. Influent and effluent samples were properly diluted (1/5) and added with an internal standard (benzotriazole at a final concentration of 0.2 mg L^{-1}) before to be injected in the HPLC/MS system. A detailed investigation was performed in order to obtain information about the organics present in the wastewater, as the acyclovir and nalidixic acid represented about 2% and 4% of the DOC, respectively. For this purpose several analytical procedures were employed, based on purification and/or extraction followed by analyses by HPLC/MS. This characterization was mainly addressed to the detection of polar organics as preliminary analyses had revealed that volatile organics represented a minor fraction of the total DOC.

2.4.3.2 Results

In the presence of high concentrations of organic compounds, chemical pre-treatments can only be cost-effective if highly selective towards target compounds. On the other hand, when an oxidation process is used as final polishing step, the presence of organic by-products into the final effluent can only be avoided by favoring their complete mineralization setting a high reaction time and, consequently, with increased operational costs. A possible alternative is the integration of the biological treatment with an oxidation process aimed at improving the biodegradability of refractory compounds.

This option may have a strong potential in terms of minimization of oxidation by-product release if these can be removed in the biological step. In this view, membrane bioreactors (MBR) offer unique opportunities in terms of suspended solids-free effluent, possible selection of specialized biomass, flexibility and compactness. These features make MBR particularly suited to be integrated with O_3 as it is well known that the performance of ozonation is negatively affected by suspended solids that act as scavengers towards hydroxyl radicals (i.e., the reactive free radicals formed by ozone decomposition in water). To test the validity of such hypothesis the integration of an ozonation step with an MBR as either a combined or post treatment configuration was tested.

2.4.3.2.1 MBR and MBR-ozonation performance in treating acyclovir wastewater

The MBR was fed with real wastewater from the batch production of the pharmaceutical compound acyclovir. Due to a purification step of the acyclovir production, employsying acetic acid, most of the organic content of this wastewater is constituted of sodium acetate, therefore nutrients in concentration proportional to the COD were provided as ammonium phosphate and ammonium carbonate to promote biodegradation processes. The system rapidly approached complete nitrification, and despite the high sludge retention time (SRT 60 d), after 56 d it was considered at the steady state in terms of biomass growth. Under steady state conditions about 30–50% of the nitrogen provided to the system escaped as nitrate in the effluent, meaning that the biomass nitrogen requirements for growth had been over-estimated. In other words, the biomass growth rate was lower than expected and this was attributed to the high salinity of the treated wastewater, which may have caused an increase of the maintenance energy, so decreasing the amount of substrate utilized for cellular growth (van Loosdrecht *et al.*, 1999; Lobos *et al.*, 2005). The steady state average effluent conductivity was higher than that of the influent (15.2 ± 0.5 and 11.5 ± 0.2 mS cm^{-1}, respectively), due to the addition of hydrochloric acid to counteract the pH increase related to acetate biodegradation. Under these conditions the observed bacterial yield of the biomass was estimated to be 0.06 ± 0.1 gVSS $gCOD^{-1}$, about half of the value estimated for municipal wastewater in a previous experimental campaign under similar conditions and with the same equipment (Pollice *et al.*, 2008a). The average mixed liquor suspended solids concentration (MLSS) was 7.9 ± 0.9 gSS L^{-1} with a volatile fraction of $80 \pm 4\%$.

The system's COD removal performance was very good, as shown in Figure 2.45. The COD was removed at very high rates with an average removal ratio of $99 \pm 1\%$ and effluent average concentration of 48 ± 12 mg L^{-1}. This result was expected since acetate was the main constituent of the total COD. Just a slight COD decrease was observed in the MBR effluent after the integration with the ozonation process, although the difference between the average effluent concentration of the two systems is below the uncertainty of the two values (51 ± 12 and 44 ± 9 for the MBR-only and for the MBR-ozonation phase, respectively). The average COD of the ozonated effluent was 38 ± 6 mg L^{-1}, and also in this case a minor effect of the ozonation step can be suggested. It is necessary to point out that the benefit deriving from the ozonation step should not be evaluated through the measurement of bulk parameters (COD and/or TOC), because the amount of ozone dosed was too small for mineralization of organics. This was intentionally done, as the pharmaceutical residues are a minor fraction of the total carbon, and the main goal of ozonation was the transformation of these residues into potentially biodegradable by-products.

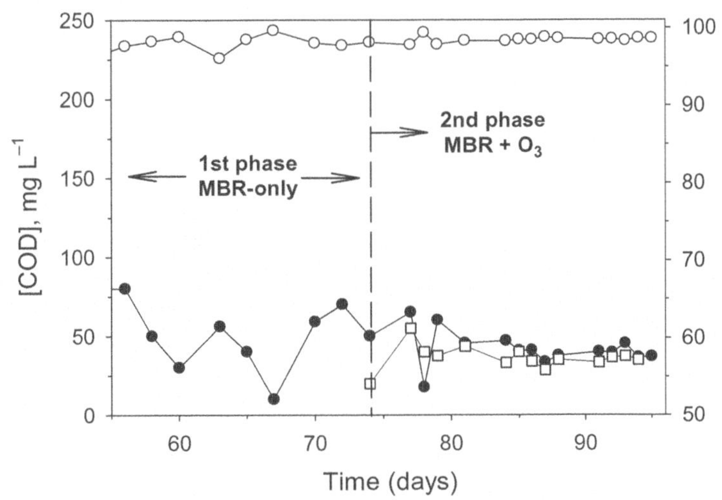

Figure 2.45 Effluent concentration and removal of COD after the treatment with MBR only and with the integrated MBR-ozonation system (after day 74). Concentration in the MBR effluent, •; concentration in the ozonated effluent, □; removal percentage by MBR, ○

In terms of biomass filterability, under steady state conditions this was worse than previously observed in MBR systems treating municipal wastewater with similar sludge concentrations and SRT. The membrane cleaning frequency was around 1.6 cleaning cycles per month and the average capillary suction time was 16 ± 5 s, and both these values are 3–4 times higher than the values previously reported for municipal MBR (Pollice *et al.*, 2008b). This worsened filterability can be reasonably linked to the high salinity of the industrial effluent.

With reference to removal of the target organic compounds, quantitative monitoring of the degradation of acyclovir and acetylacyclovir was performed through the adoption of standards. Figure 2.46a–b shows the concentration profiles of these two compounds in the effluents (MBR-only and MBR-ozonation) and the removal percentages referred to the single treatment steps. Both organics were efficiently removed during the MBR-only phase, with removal percentages very close to 100%, with average acyclovir and acetylacyclovir concentrations in the MBR effluent of 2.5 and 0.05 mg L^{-1}, respectively. As mentioned above, the introduction of ozonation in the MBR recirculation stream caused limited improvement of total COD removal efficiency (Figure 2.45). However, as shown in Figure 2.46a, the ozonation affected acyclovir removal reducing its effluent concentration close to the detection limit (0.5 μg L^{-1}) and resulting in a whole removal efficiency of almost 100% (MBR-ozonation). This is consistent with the fast acyclovir degradation by ozone obtained during separate ozonation experiments (results not shown). As for the acetylacyclovir removal, Figure 2.46b shows that ozonation removed it almost completely, being the removal efficiency in the range 80–100%. During preliminary experiments, acyclovir and acetylacyclovir (dissolved in the MBR effluent) showed similar degradation rates when exposed to ozone. However, when ozone is in large excess (as during semi-batch experiments) the degradation kinetics of organic compounds are known to be "pseudo-first order". It follows that the degradation efficiencies of the two organics during the ozonation of real MBR effluent were different because the concentration of acyclovir was 10-fold higher than acetylacyclovir. Therefore, results about acyclovir and acetylacyclovir degradation suggest that the net result of placing the ozonation step in the recirculation stream of the MBR treatment is that a much lower residual concentration of organics was obtained.

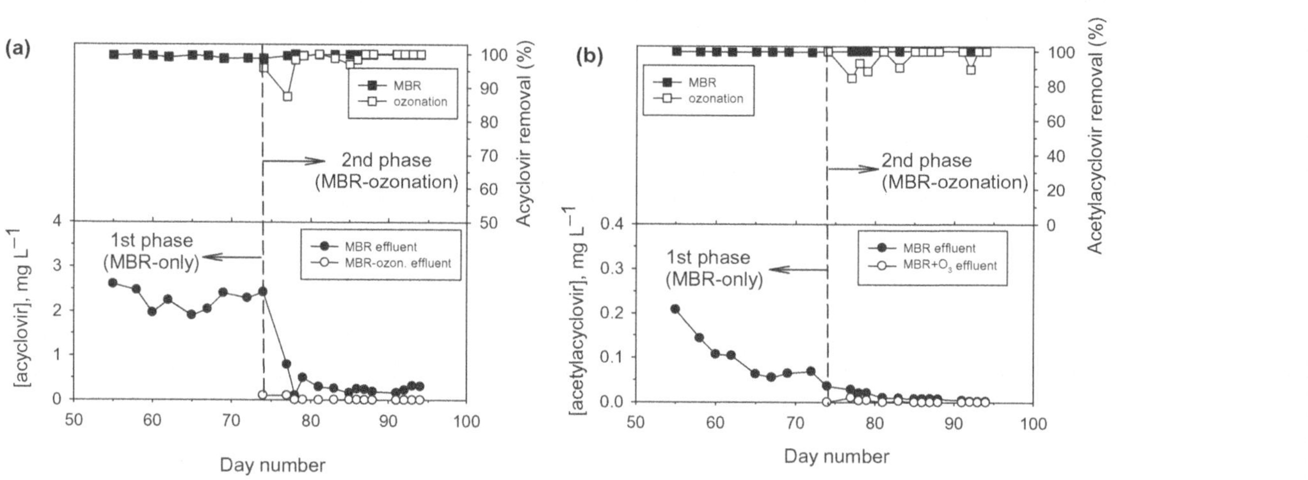

Figure 2.46 Concentration of acyclovir (a) and acetylacyclovir (b) in the effluent and percentage of removal during MBR and MBR-ozonation treatment of the pharmaceutical wastewater

Besides the two main target compounds, the degradation of the other organics listed in Table 2.7 was also monitored during the two investigated phases (MBR-only and MBR-ozonation). However, for these compounds no standards were available, therefore it was only possible to evaluate the relative abundance (normalized to that of the internal standard) as a function of day number. Interestingly, it was found that only 12 compounds out of 27 (namely No. 1–6, 9, 11, 12, 18, 20, 25) showed trends similar to those obtained for acyclovir and acetylacyclovir. Ozonation was also evaluated as a final polishing step (i.e. after the MBR-only treatment), and when 15 min ozonation time of was adopted similar results to those of the integrated system were obtained in terms of removal of acyclovir, acetylacyclovir and other organics. This was expected because all the organics present in the wastewater are easily degraded by ozone (Mascolo *et al.*, 2010b).

Identification of the by-products deriving from both MBR and the ozonation treatments, and investigation of their removal during these two treatments were also performed. This is an important aspect, since by-products may have higher toxicity than their parent compounds. The molecular weights of the organic by-products detected during the two treatments are listed in Table 2.7. All by-products were detected applying the same analytical procedure described for organic compounds of the acyclovir wastewater. All by-products originating during the MBR treatment were detected in negative ion mode only, while the one formed during the ozonation step was detected in positive ion mode. The detection of only one by-product after 15 min ozonation suggests that mainly organic acids (not detectable by HPLC/MS) are formed during this step. It was found that the abundance of all by-products originating from MBR step greatly decreased when MBR was integrated with ozonation. As an example, in Figure 2.47 are depicted the profiles of by-products No. 30 and 33 that showed to be very similar. Their abundance in the MBR effluent dropped significantly (90%) during the first 20 d and reached a constant concentration at longer plant operation time suggesting that the biomass required a long adaptation time. When the ozonation step was also included (second phase) the abundance of this compound in the MBR effluent did not change. Besides, ozonation proved to be very effective as this compound was completely removed (abundance in the ozonation effluent the same of that of raw wastewater).

Table 2.7 Molecular weight (MW) of the organic compounds detected in the raw acyclovir wastewater and after both MBR and MBR-ozonation treatment

Compound number (MW)		
Raw wastewater	**After MBR**	**After MBR-ozonation**
Acyclovir (225)	29 (282)	34 (209)
1, 2 (341)	30 (302)	
3 (255)	31, 32 (345)	
4 (239)	33 (332)	
5, 6 (299)		
7–9 (221)		
10 (307)		
11, 12 (295)		
Acetylacyclovir 13, 14 (267)		
15 (226)		
16 (297)		
17 (266)		
18, 20 (319)		
Diacetylacyclovir 19 (309)		
21–23 (480)		
24 (151)		
25 (366)		
26, 27 (291)		
28 (200)		

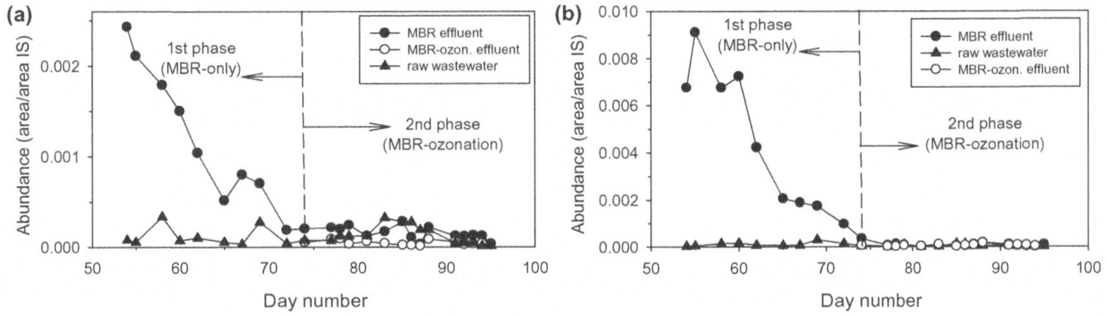

Figure 2.47 Relative abundance of by-products No. 30 (a) and 33 (b) before and after the treatment of the pharmaceutical wastewater with MBR and MBR-ozonation

A different trend was shown by compound No. 34 (Figure 2.48), that was formed in both the MBR and ozonation steps. During the second phase (MBR-ozonation) the abundance of this compound in the MBR effluent was quite low and, more interestingly, even increased in the ozonated effluent. This suggests that this compound was formed during both the MBR and ozonation steps but it was also removed to some extent in the MBR. In fact, the consequence of such an evidence is that the by-product abundance in the MBR effluent did not increase during the second phase as it would be expected considering that the ozonated effluent was recirculated to the MBR. It follows that the increased abundance in the ozonation effluent was compensated by a partial removal in the MBR step. Accordingly, the net results was a rather constant concentration in the MBR effluent during both treatments (MBR-only and MBR-ozonation). In Figure 2.48 the formation/decomposition profile of compound No. 34 during a prolonged ozonation (polishing step) of the MBR-only effluent is also depicted. This test has been done to check whether the by-product removal can be more easily obtained by a single ozonation step after the MBR (i.e. the standard polishing configuration) instead of integrating the two treatments in a more complex set-up. The profile reported in Figure 2.48 shows that similar performances, in terms of minimization of the by-product abundance, are only obtained after applying an ozone dosage higher than 80 mg L^{-1}. It follows that, as the ozone dosage in the integrated MBR-ozonation system was 61 mg L^{-1}, the standard polishing configuration (ozonation after MBR) is associated with a higher operational cost than the integrated MBR-ozonation system. These results demonstrate the beneficial effect of ozonation when integrated with the MBR for organic by-products removal. It is also worth noting that no by-products were detected by HPLC/MS when the integrated MBR-ozonation process reached steady state conditions. This suggests that only low molecular weight organic acids could be present in the effluent, as also no TOC reduction was observed. Such low molecular weight organic acids can be detected by ion chromatography (Mascolo *et al.*, 2005). It follows that no ecotoxicity issue should be associated with the final effluent, as it is known that these organic acids do not present any toxicity concern.

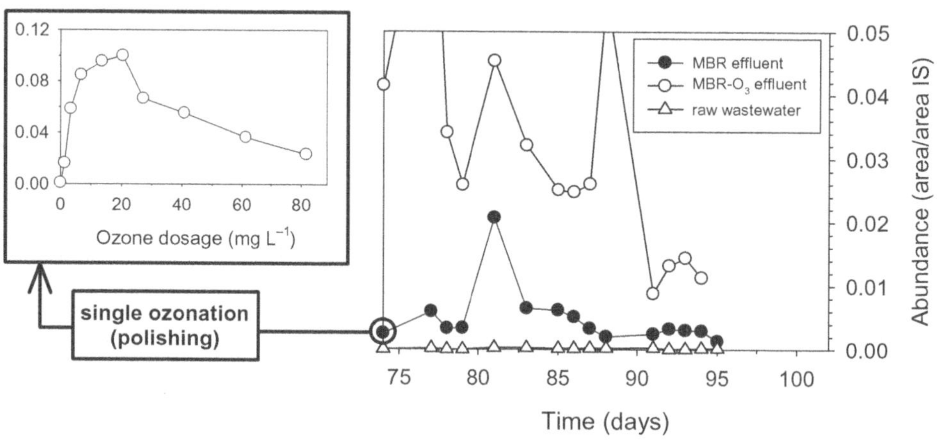

Figure 2.48 Relative abundance of by-product No. 34 in the influent and effluents during MBR and MBR-ozonation treatment of the pharmaceutical wastewater and degradation profile of the same by-product during prolonged ozonation (polishing)

In conclusion, the removal of acyclovir and most organic compounds present in the raw wastewater by the integrated MBR–ozonation was comparable to those obtained by the two treatments in series (i.e. ozonation after MBR), but the integrated system was very effective in removing the ozonation by-product. Filterability of the MBR biomass was somewhat lower than those observed under similar operating conditions in the treatment of municipal wastewater. This was attributed to the high salinity of the treated industrial effluents, and reflected in higher membrane cleaning frequencies.

2.4.3.2.2 Single and integrated process performance in treating nalidixic acid wastewater

The sole MBR was initially operated for acclimating the biomass and investigating the effectiveness of the biological treatment towards the wastewater. Results showed, as expected on the basis of previous Zahn-Wellens test, a COD removal above 90%, but the nalidixic acid concentration remained unvaried until the chemical oxidation step was integrated in the recirculation stream. The mixed liquor filterability was not affected by the introduction of the integrated processes, and the CST remained mostly unvaried across the experimental periods. Saline conditions and the presence of the antibacterial compound resulted in very limited growth of the MBR biomass: despite the high load of readily biodegradable substances (0.88 $g_{COD}\ g_{VSS}^{-1}$, mostly due to acetate and other unknown biodegradable solvents), the observed yield only resulted in about 39 $mg_{VSS}\ g_{CODremoved}^{-1}$.

In order to sustain the biomass growth, nutrients were added to the feed as ammonium and ortho-phosphate. Accurate nitrogen mass balances suggested the unexpected occurrence of denitrification in the completely aerated MBR. In particular, before introducing the ozonation step, about 52% of the influent nitrogen was removed with the wasted biomass, while the effluent accounted for about 13%. Therefore, denitrification accounted for about 35% of the nitrogen removal. Therefore the amount of nitrogen provided to the system to sustain the biomass growth was largely in excess.

This can be explained considering that, in the initial design of the experiment, it was difficult to estimate the expected biomass growth because the extent of inhibition due to the presence of nalidixic acid and oxidation by-products was unknown. Thus a higher biomass growth has been considered to avoid any possible nutrients limitation. Then the influent nitrogen amount was not adjusted during the test in order to avoid changes of the experimental conditions with consequences on the interpretation of the results. Microbiological characterizations confirmed the presence of Thauera sp., that include some strains previously reported as aerobic denitrifiers (Sholten *et al.*, 1999; Schühle *et al.*, 2003). To investigate the degradation of nalidixic acid and organics present in the MBR effluent and by-product formation, preliminary experiments were also performed with ozone separately. These preliminary experiments were aimed at defining the optimal ozonation conditions for nalidixic acid degradation.

The activation of a chemical oxidation process in the recirculation stream of the MBR did not have apparent effects on most of the commonly measured gross parameters. The COD removal remained unvaried during the integrated process and the good performance in terms of nalidixic acid removal with ozonation was mostly hidden by the limited contribution of this compound to the total COD fed to the system (Figure 2.49).

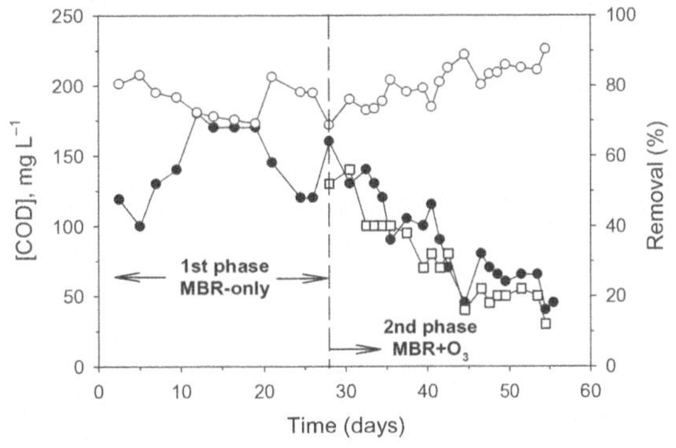

Figure 2.49 Concentration of COD and percentage of removal during MBR + ozone of the nalidixic acid pharmaceutical wastewater. Concentration in the MBR effluent, ●; concentration in the ozonated effluent, □; removal percentage by MBR, ○

The DOC profiles (effluent concentration and removal percentages) during MBR and MBR-chemical oxidation phases are reported in Figure 2.50. The MBR effectively removed the DOC, and also the integration with ozonation caused an improvement of the DOC removal efficiency of the MBR step, that raised up to 80% at the end of the integration phase. Besides, the removal percentage of DOC by the single ozonation step was practically zero. This is consistent with the low amount of ozone dosed in order to induce a transformation of the chemical structure of the organics with the objective to improve the biodegradability of the resulting by-products.

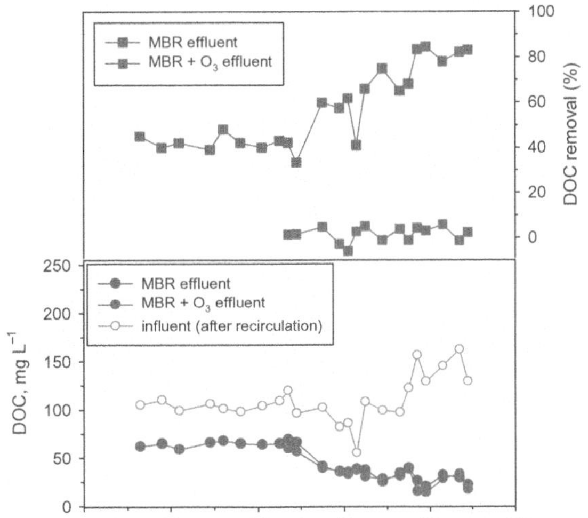

Figure 2.50 Concentration of DOC and percentage of removal during MBR + ozone of the nalidixic acid pharmaceutical wastewater

In terms of the different nitrogen compounds, the nitrogen mass balance was consistent with a denitrification of about 40% of the influent nitrogen. Moreover, similar disturbing effects on aerobic denitrification process were observed after the introduction of the chemical oxidation. In particular, as shown in Figure 2.51, effluent nitrate concentrations tended to increase when the integrated process was operated, possibly due to increased competition for substrates from the dominant heterotrophic strains caused by the depletion of the antibiotic. This was more evident when ozonation was adopted, that also resulted in partial inhibition of nitrification (see effluent NH_4-N concentration).

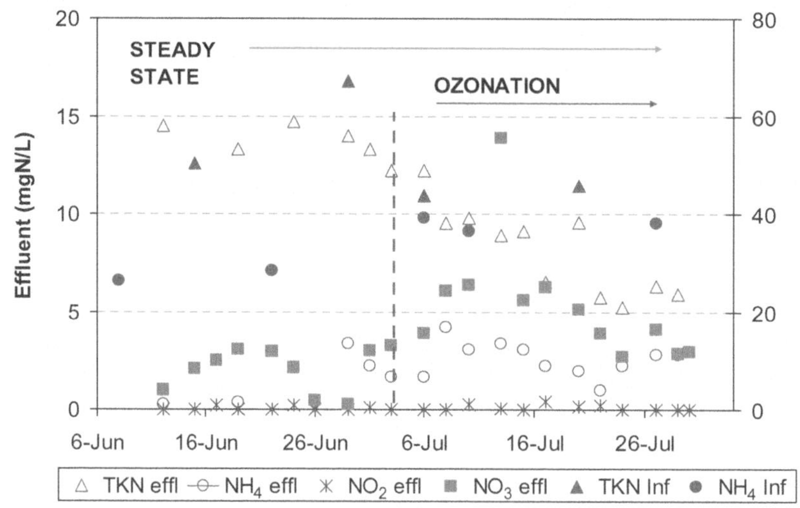

Figure 2.51 Nitrogen compounds before and after MBR integration with ozonation

The trend of nalidixic acid concentration as well as its removal percentage were obtained during both the MBR-only and the MBR-chemical oxidation phases. Results showed that during the first phase (MBR only) the removal efficiency of nalidixic acid is practically zero as the MBR influent concentration equals that of the effluent even though the COD removal percentage of the MBR plant is very good (Figure 2.49). This is consistent with the fact that nalidixic acid (antibacterial compound) is recalcitrant for the biomass. As soon as the ozonation step was placed in the recirculation stream of the MBR (second phase) the nalidixic acid in the final effluent (MBR-ozonation) disappeared completely. This was clearly due to the ozonation step that allowed to remove nalidixic acid efficiently, being the average removal percentage practically 100%. This is also consistent with the preliminary batch experiments performed with the MBR effluent that showed the effectiveness of ozone in removing nalidixic acid.

The ozone consumed from the MBR effluent operated with the real wastewater was also monitored during each ozonation cycle. It was found that the amount of ozone consumed by the wastewater decreased over the cycles due to the fact that a single ozonation was carried out on an effluent whose part of it was already ozonated (see scheme of Figure 2.44a) causing the MBR effluent to have a lower ozone demand.

A detailed investigation was performed in order to obtain information about the organics present in the wastewater, as nalidixic acid itself represented about 4% of the DOC. For this purpose several analytical procedures were employed, based on purification and/or extraction followed by HPLC/MS and HPLC/MS-MS analyses. This characterization was mainly addressed to the detection of polar organics as preliminary analyses revealed that volatile organics represented a minor fraction of the total DOC. In Table 2.8 are presented the main organic compounds detected in the raw wastewater. Furthermore, another detailed analytical investigation was carried out aimed at identifying the degradation by-products during ozonation of the MBR effluent. In Table 2.8 the detected by-products are also listed. By-products were first identified using HPLC/MS using a full scan method. Once the mass spectra were obtained, in order to define an analytical method as specific and sensitive as possible for monitoring the organics during the MBR and MBR + O_3 operation, a HPLC/MS-MS procedure including multiple reaction monitoring was optimized.

The formation/degradation trend of all identified organics reported in Table 2.8 was monitored during the phases in which MBR was integrated with ozonation. As mentioned above, during the MBR-ozonation integration an ozone dosage of 103 mg L^{-1} was employed. The degradation profile of each organic compound has to be compared with that obtained by running a single ozonation step for a prolonged time after the MBR process (see scheme in Figure 2.44b). This in order to check whether the organics removal is easier to be achieved by a single chemical oxidation step after the MBR (i.e. the standard polishing configuration) than by integrating the two treatments in a more complex set-up. Because of the high number of organics identified in the MBR effluent, in the following discussion only results for a limited number of them is reported. The behaviour of the identified organics, in fact, can be grouped in two main categories, namely the compounds that are removed by a similar extent by both the integrated system and the standard polishing configuration (chemical oxidation after MBR) and those for which integrated system led to worse results than the polishing configuration. Therefore, the results reported in the following discussion are representative of the trends found for all identified organics.

Table 2.8 Organic compounds identified in the real pharmaceutical wastewater and organic by-products identified during treatment of MBR effluent of nalidixic acid real wastewater by ozonation

Compound number	MW	By-product number	MW
Nalidixic acid	232	1	116
1, 4	152	2, 5	180
2	108	3, 16, 37	194
3, 6	206	4	200
5, 17	208	6, 8	254
7, 12	136	7, 19, 29	184
8, 10, 39	190	9, 17	210
9	166	10, 31, 50	208
11	157	11, 21, 33	224
13	224	12	196
14	180	13	268
15, 42	188	14, 22, 27, 34	238
16, 18	164	15	188
19, 35	222	18	212
20	295	20	166
21, 26	230	24	216
22, 27, 34	204	39, 52	222
23	210	26	130
24, 31, 41, 43	218	28, 54	178
25	220	30, 36	266
28, 48	234	32	236
29	266	35	316
30	186	38	252
32, 38, 44	248	40	289
33, 37	264	41	250
36, 45	246	42, 45	206
40	276	43, 48	264
46	260	44	189
47	216	46	220
		47	262
		49	192
		51, 53	278
		55	234

The results obtained for the organic compound No. 34 are displayed in Figure 2.52. It is possible to note that the MBR step was not able to remove this compound being its intensity in the MBR effluent the same of that measured in the influent after recirculation. The same situation was also obtained when the intensity of this compound was lowered, i.e. when MBR was integrated with ozonation. It follows that compound No. 34 can be considered not biodegradable under the employed MBR conditions no matter its abundance in the influent. On the other side, ozonation was very effective in removing this organic compound. A comparison of the integrated MBR-ozonation system with respect to the standard polishing configuration (single ozonation step after the MBR) can be made taking into consideration the residual compound abundance obtained with a polishing performed with an ozone dosage of 103 mg L^{-1} with respect to that registered in the effluent of the integrated system at steady state. Results of Figure 2.52 show that the conventional configuration gave slightly better minimization of the organic than the integrated system.

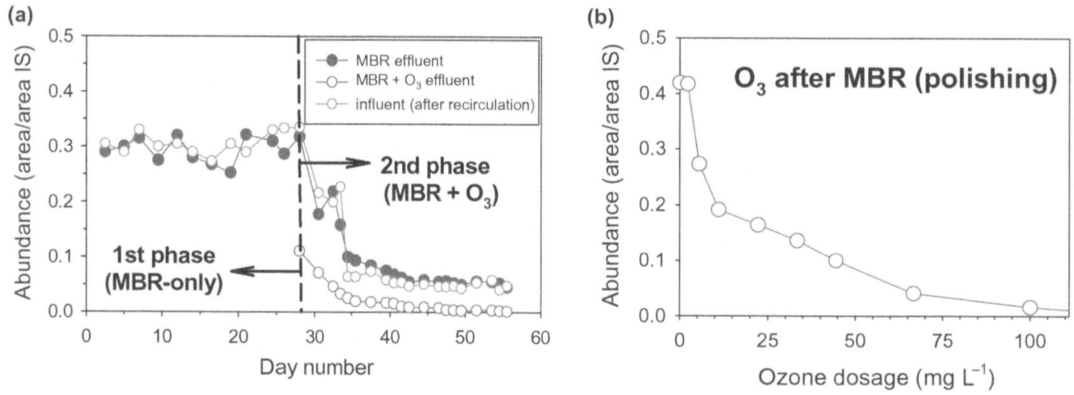

Figure 2.52 Relative abundance of organic compound No. 34 in the effluent and influents during treatment of real wastewater with MBR integrated with ozonation (a) and decay profiles of the same compound during prolonged polishing (b)

The results obtained for the organic compound No. 16 are depicted in Figure 2.53 and are representative of those organics that are more efficiently removed with the standard polishing configuration with respect to the integrated MBR ozonation system. Also for this compound the MBR step was not able to obtain any removal as its abundance in the MBR effluent was the same of that measured in the influent after recirculation. The same situation was also obtained when the intensity of this compound lowered as a consequence of the MBR integration with ozonation treatment. It follows that also the compound No. 16 can be considered not biodegradable under the employed MBR conditions no matter its abundance in the influent. As for the performance of the single oxidation step during integration with MBR, ozonation was very effective in removing this organic compound since a complete removal of the compound was measured as soon as the integrated system reached steady state conditions.

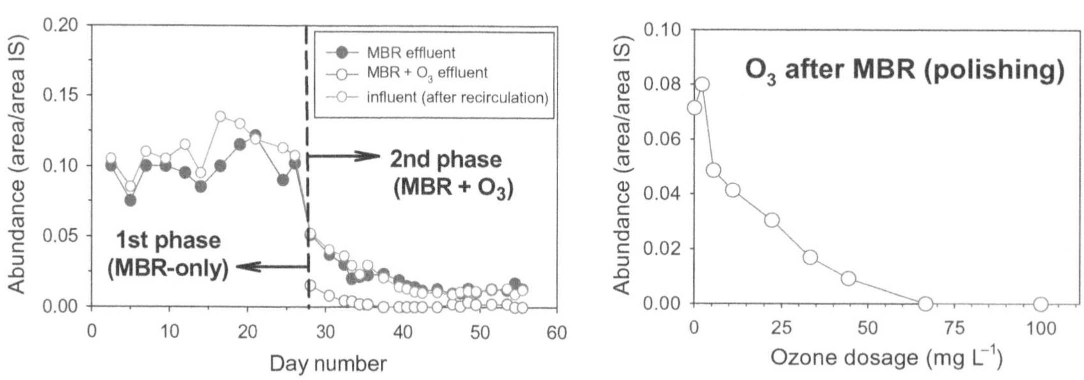

Figure 2.53 Relative abundance of organic compound No. 16 in the effluent and influents during treatment of real wastewater with MBR integrated with ozonation and decay profiles of the same compound during prolonged polishing

During treatment with ozone an exponential decay was obtained leading to a complete removal of the compound after an ozone dosage of 65 mg L^{-1} was applied. At the same ozone dosage employed during integration (103 mg L^{-1}) the polishing after MBR led to an almost complete disappearance of the compound. This is a consequence of the fact that the compound is not removed at all during the biological step and just the oxidation treatment was able to induce a relevant degradation of it. It follows that as the final effluent of the integrated system is that of the MBR (Figure 2.44a) it will always be a fraction of the compound present in such an effluent.

The formation/degradation trend of all identified by-products was monitored during the period in which the integrated MBR-ozonation process was operated. For each by-product the formation/degradation profile was compared to that obtained during an ozonation carried out at the same ozone dosage. This is necessary in order to check whether the by-product removal is easier to be achieved by a single ozonation step after the MBR (i.e. the standard polishing configuration) than by integrating the two treatments in a more complex set-up. Results showed that in terms of by-products minimization for the majority of identified compounds the integrated system performed better than the standard polishing configuration (i.e. chemical oxidation after MBR) and just for a couple of by-products the reverse situation was observed. Formation/degradation profiles of by-product No. 36 (Figure 2.54) are representative of such a finding. In fact, its formation/degradation profile during the prolonged ozonation carried out on the MBR effluent (polishing step) shows that it is quickly formed reaching a maximum after an an ozone dosage of 70 mg L^{-1}. At higher dosages it is degraded at comparable rates to those of its formation. During the MBR-ozonation integration, the performance of the single MBR step in removing this by-product was very good.

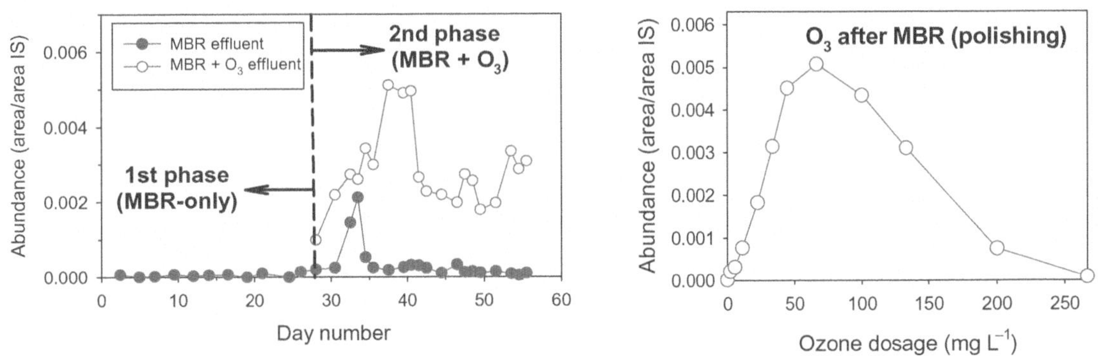

Figure 2.54 Relative abundance of by-product No. 36 in the effluents during treatment of real wastewater with MBR integrated with ozonation and decay profiles of the same compound during prolonged polishing

MBR step when integrated with ozonation lead to a practically complete removal of the compound as soon as the integrated system reached steady state conditions. As for the comparison of the integrated MBR-ozonation system with respect to the correspondent polishing configuration (single ozonation step after the MBR), for this by-product it is evident that the integrated system leads to much better results than the conventional set-up. In fact, the residual compound intensity obtained with a polishing step using an ozone dosage of 103 mg L^{-1} led to higher by-product abundance with respect to those measured during MBR-ozonation integration. Therefore, the obtained results showed that for the minimization of by-product No. 36 the employment of the integrated system gave much better results than those obtained using ozonation as a polishing step after the MBR.

As above reported, for most of identified by-products a similar situation of by-product No. 36 was observed where integration of MBR with ozonation led to a better minimization of by-products. A different situation was observed for a limited number of compounds and by-product No. 7, whose formation/degradation profiles are displayed in Figure 2.55, is well representative of such a situation.

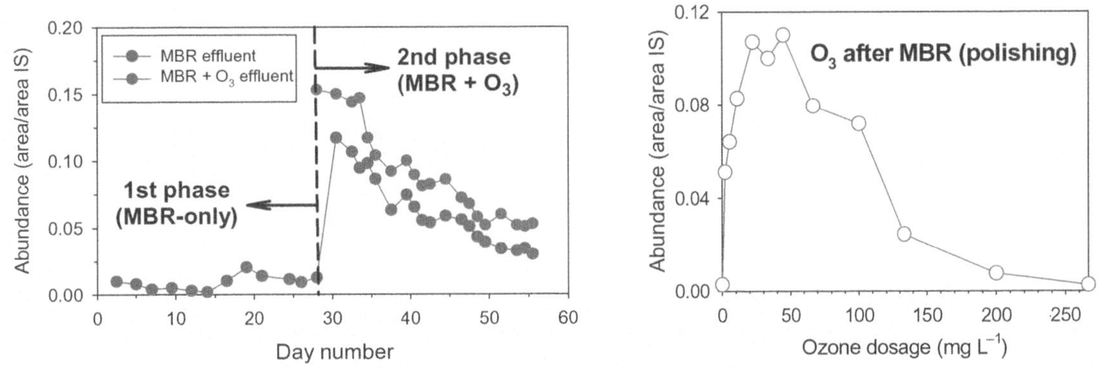

Figure 2.55 Relative abundance of by-product No. 7 in the effluents during treatment of real wastewater with MBR integrated with ozonation and decay profiles of the same compound during prolonged polishing

For such a compound MBR-ozonation allowed a better minimization of the by-product in the final effluent with respect to the conventional polishing set-up. In fact, the by-product formation/degradation profile during the prolonged ozonation carried out on the MBR effluent (polishing step) shows that it is quickly formed reaching a maximum after 40 mg L^{-1} ozone dosage. At higher dosages it is degraded at a slower rate than that observed for its formation. During the MBR-ozonation integration, the single MBR step allowed to remove less than 50% of this by-product. As for the comparison of the integrated system with respect to the polishing configuration (single ozonation step after the MBR), for this by-product it is evident that just the MBR-ozonation integration lead to better results than the conventional set-up. In fact, comparing the residual compound intensity obtained during polishing using 103 mg L^{-1} of ozone dosage with respect to those measured in MBR effluent of the integrated systems, it is evident that ozonation as a polishing step led to a residual by-product abundance much higher than that of the integrated MBR-ozonation system.

2.4.4 Conclusions

Experimental application of integrated MBR-ozonation process to pharmaceutical wastewater streams was successfully performed during different long term tests aimed at investigating their feasibility and effectiveness. The performance of MBR-ozonation was assessed by operating dedicated pilot plants for treating real pharmaceutical wastewaters. Besides the removal of main pharmaceutical compounds and of the main organics, the fate of by-products, the influence of the integrated processes on biomass characteristics, and all the relevant process feature were deeply investigated. Results showed the substantial effectiveness of MBR-ozonation in terms of organic by-products minimization, and showed the synergistic effect of the chemical oxidation step placed in the recirculation stream of the biological system. As a matter of fact, the advantages of the integrated configuration with respect to the use of chemical oxidation as a post-treatment were confirmed. They are mainly the possibility of limiting the dosage of chemical oxidants to the amounts required for improving the biodegradability of the recalcitrant compounds, allowing then the biological system to remove the resulting by-products. In the case of nalidixic acid this was successfully achieved with both the integrated system configurations tested. In particular, the experimental activity showed that:

- Although the MBR-alone removed most of the influent COD (mainly sodium acetate), nalidixic acid remained completely undegraded. Moreover, due to the presence of excess nitrogen with respect to metabolic needs, nitrification activity and aerobic denitrification were observed and confirmed by nitrogen balances and microbiological determinations.
- The presence of high salinity and the antibacterial compound resulted in very limited biomass production (about 40 $mg_{VSS}\,g^{-1}_{COD\ removed}$), although the organic load was quite high (about 0.9 $g_{COD}\,g^{-1}_{VSS}$). These extreme operating conditions led to the selection of a highly active consortium mainly composed of filamentous bacteria, small flocs and diffuse biomass. Even under these conditions, the membrane filtration performed well and the MBR technology demonstrated the feasibility of a biological process based on bacterial strains that are usually difficult to retain in conventional systems.
- In terms of by-products formation/degradation, the integration of MBR with ozonation lead to better results than those obtained using the conventional treatment configuration, i.e. polishing after biological oxidation by MBR, for the compounds that were slowly formed during the ozonation step. Instead, for a limited number of by-products that were promptly formed and then completely removed with an ozone dosage of 60 mg L^{-1} the conventional treatment scheme was as effective as the integrated MBR-ozonation system or, for just few by-products, even more effective.
- Most organics present in the wastewater were found not to be biodegradable, therefore the integrated system showed a lower performance than the conventional set-up for their minimization. This is a consequence of the fact that the oxidation treatment was able to induce a relevant extent of degradation of those compounds that were not removed at all during the biological step. It follows that (Figure 2.44a) the final MBR effluent of the integrated system will always contain a fraction of undegraded compounds. For those compounds for which MBR was able to obtain a certain grade of removal the integrated MBR-ozonation system lead to better results than the conventional set-up.

In conclusion, from a technological standpoint the integration of membrane bioreactor and chemical oxidations with ozonation was observed to be extremely beneficial for the complete removal of pharmaceutical organic pollutants and most of their oxidation by-products. The integrated process can be proposed for industrial exploitation in the treatment of highly saline real wastewater produced by pharmaceutical industries and characterized by the presence of unbiodegradable pharmaceuticals in presence of large amounts of biodegradable compounds. Membrane separation was confirmed as a powerful tool to obtain biomass selection and acclimatization of specific bacterial groups to extreme environmental conditions as those tested. As a matter of fact, thanks to the efficient biomass separation, membrane filtration also demonstrated a relevant role in influencing the population dynamics and offering the unique opportunity to develop biological removal process mainly based on the activity of not-floc-forming biomass (that is not possible with conventional systems). Integrating chemical oxidation with biological treatment was observed to be an effective strategy for enhanced removal of recalcitrant compounds and represents an improvement in the current state of the art.

2.5 BIOZO TREATMENT OF LANDFILL LEACHATE

2.5.1 Introduction

Leachate composition is determined by factors such as the type and age of waste, the moisture content of the waste, amount of water intrusion, environmental temperature and landfill management (Kylefors, 2002). Landfills typically undergo a three-step development that includes an aerobic/anoxic phase (months), an anaerobic acidogenic phase (years) and an anaerobic methanogenic phase (decades). Partly stabilized leachate can be characterized with high ammonium concentrations, moderately high COD concentration and with a high alkalinity buffer (e.g. Kylefors, 2002). As for the possible xenobiotic organic chemicals (XOCs), Beun *et al.* (2004) present a comprehensive assessment, showing BTEX, C3-benzenes, bicyclo compounds, napthalenes, chlorinated aliphatics, phenols (chloro-, methyl-, dimethyl, nonyl-), pesticides and phthalates in landfill leachate derived from ten different landfill sites in Denmark. Compared to biological processes implemented to remove nitrogen, phosphorus and COD in municipal and industrial sewage treatment systems, less well established is the practice of optimising secondary and tertiary biological and chemical treatment processes to remove XOCs, e.g. human antibiotics (Plósz *et al.*, 2010a). Biotransformation of most XOCs of environmental and toxicological significance is carried out through the process of cometabolism, i.e. transformation of a non-growth XOC substrate by growing cells in the presence or absence of a growth substrate (Horváth, 1972; Criddle, 1993). Xu *et al.* (2005) report on the occurrence and biological removal of 17 organochlorine pesticides, 16 PAHs and technical 4-nonylphenol in landfill leachate, and show acenaphthylene in the highest concentration (595 ng/L) amongst the 16 PAHs detected. Anaerobic membrane bioreactor technology is shown to have high XOC removal efficiency both in terms of total organic constituents and trace organic contaminants (Xu *et al.*, 2005).

Biological treatment of partially stabilized leachate, used in this study, can be limited by the low amount of biodegradable COD, bioavailable phosphorous and by compounds, inhibiting biochemical processes, such as nitrification. The high alkalinity buffer content can additionally deteriorate biological nitrogen removal by increased pH in the anoxic and aerobic reactors. One way to overcome such operational problems is to assess the potential of non-biological treatment processes to improve leachate treatment. Methods collectively referred to in the literature as Advanced Oxidation Processes (AOPs) are shown to have the potential to treat pollutants at low and high concentrations, and much research and development work has been undertaken on this topic since the 1970s (Parsons, 2004).

Ozonation in combination with UV, hydrogen-peroxide, etc., proved to be efficient in a number of industrial applications (Kurniawan *et al.*, 2006). High energy costs associated with the complete mineralization of XOCs using, for instance, ozonation suggest the possible benefits derived from advanced oxidation coupled with biological treatment. Wang *et al.* (2004) conducted studies with ozone and O_3/H_2O_2 using aged landfill leachate, and found that hydrogen-peroxide dosing had little effect on COD reduction. Ratio of the BOD_5 to COD is shown to increase to maximum 120% of the initial COD concentration, and the potential of chemical nitrification was assessed to be approx. 24 mg/L nitrate-nitrite-nitrogen per g O_3 dosed. Geenens *et al.* (2000) and Haapea *et al.* (2002) assessed the impacts of ozonation on leachate treatment and its potential to improve biological treatment of landfill leachate. The above researchers found that ozonation can be effective in decreasing toxicity, for instance, on autotrophic nitrifying bacteria, and in decreasing inert soluble COD as well as increasing the amount of biodegradable organic matters in leachate.

Compared to biological processes used to remove nitrogen, phosphorus and COD in municipal and industrial sewage treatment systems, less well established is our scientific knowledge and engineering capabilities to design biological treatment processes to remove xenobiotic organic compounds, XOCs, from sewage (Grady *et al.*, 1999). Bacteria residing in biofilm employed in sewage treatment systems can exhibit increased resistance to chemical toxins, e.g. XOCs, in comparison with free bacterial cells in suspension, i.e. activated sludge (Henriques and Love, 2007). The mechanism responsible for the failure of the toxin to penetrate the biofilm is mostly due to adsorption onto and diffusion limitation by extracellular polymeric substances. Biofilm reactor technologies can be distinguished as non-submerged systems, i.e. trickling filters, rotating biological contactors; and as submerged, fixed-bed biofilm reactors and fluidised biofilm reactors (Henze *et al.*, 2008). The latter group comprises, for instance, aerobic granular sludge (de Kreuk and van Loosdrecht, 2006), and the concept of moving-bed biofilm reactor, MBBR (Ødegaard *et al.*, 1994).

Within the biofilm, microbial competition and selection are impacted by the substrate availability that is a function of substrate mass transport (diffusion) and the location of bacteria. Furthermore, under plug-flow conditions, in different zones of a fixed-bed reactor, biofilm with different biotic structure and different metabolic potential can develop (Flemming *et al.*, 2000). Wobus and Röske (2000) speculate that the exposure of microbial communities to different substrates under controlled conditions can potentially enhance the metabolic activity of bacteria, and can improve their capacity to cope with different growth conditions.

However, in biofilm systems, such longitudinal microbial heterogeneity is considered as a disadvantage, because, under shock-loading conditions, substrates can bleed into section with low biomass concentration that may not be well adapted to the given substrate, such as the high ammonia concentration shock (Boller and Gujer, 1986). Kaballo *et al.* (1995) compares fixed-bed biofilm reactors operated in continuous- (plug-flow conditions) and in sequencing batch mode (homogenous conditions), and used *p*-chlorophenole, as xenobiotic compound. In the continuous-flow system, very sharp biomass stratification was obtained, and biomass growth could only be observed up to only half of the depth of the packed tower, which possibly resulted in a significant mass transfer limitation near the influent point.

Despite such inefficiencies, the continuous-flow fixed bed system is shown to have identical *p*-chlorophenole removal capacity to the one operated in batch mode under shock-loading conditions, characterised with influent concentrations altered from 10 to 30 mg/L, i.e. 3-fold increase. One way to achieve homogeneously distributed biofilm in the entire reactor volume is to use MBBR technology, instead of fixed-bed systems, whereby bacteria attached to carrier elements, density slightly lower than water, can be effectively fluidised using an optimum filling ratio.

For single reactions, described and followed using one rate expression coupled with the stoichiometric and equilibrium expressions, mixed flow always requires larger volume than plug-flow does for any given conversion (Levenspiel, 1972). To achieve plug-flow conditions in full-scale systems, completely stirred tank reactor (CSTR) cascades can be used (Grady *et al.*, 1999). Plósz (2007) presents a kinetic optimization method to evaluate optimal denitrification bioreactor stages in activated sludge systems. It has been shown that as a function of the biodegradable substrate composition, the stage system design with three optimized reactor compartments can effectively increase reaction rates in activated sludge denitrification zones. Plósz (2007) additionally demonstrates the flexibility of the optimized three-stage pre-denitrification configuration over two- or single stage arrangements, in terms of nitrogen removal capacity, when applied to greatly varying, 4-fold increase in the influent readily biodegradable substrate conc., wastewater quality. Joss *et al.* demonstrates that the implementation of reactor cascades can significantly improve the removal of non-sorbing ($K_d \leq 0.1\ L\,g^{-1}$ suspended solids) and biologically degradable compounds compared to a single mixed reactor. Moreover, for mitigating strong influent fluctuations, the role of microbial biomass used as a sorbent and its impact on the elimination of, for instance, chlorophenols, has been shown by Tsezos and Bell (1989). An advantage of biofilm systems over activated sludge is that the treatment efficiency is less influenced by the secondary solid-liquid separation process. We note, however, that the settling behaviour of biofilm detached from MBBR is not well known from literature yet, unlike for other biofilm systems, e.g. aerobic granules (Nor Anuar *et al.*, 2007).

2.5.1.2 Objectives

We present the BIOZO concept, developed at the Norwegian Institute for Water Research (NIVA), that combines controlled ozonation, used as advanced oxidation process (AOP), with biological landfill leachate (partly stabilised) treatment, in the secondary treatment step. Results obtained in a screening study (GC-ToF-MS) were used to compile a priority list of xenobiotic organic chemicals (XOCs) in leachate based on relative occurrence, showing poly-aromatic hydrocarbons (PAHs) as the predominant fraction. We assess the efficiency of the combined system, in terms of nitrogen, COD and PAHs removal efficiency. A novel, staged moving-bed biofilm reactor (STRAKINES) design was implemented in the laboratory- and pilot-scale experiments. This biological treatment technology developed offers an opportunity to enhance the removal of nitrogen and xenobiotic organic pollutants, e.g. PAHs, from wastewater in biofilm systems by means of controlling bacteria heterogeneity through reactor stages. A methodology to assess the size of reactor compartments using reaction kinetics data is developed. We used the moving-bed biofilm technology, and a continuous-flow biofilm system was developed, in which, plug-flow conditions are approximated in the pre-denitrification zone by employing three continuous stirred tank reactor compartments. The sizes of the pre-anoxic biofilm reactor stages are assessed using data obtained in batch nitrate-nitrite uptake tests, and its efficiency is contrasted against data obtained with single stage arrangement in a laboratory-scale experiment. We demonstrate that organic substrate and nitrogen removal efficiencies can be improved using reactor compartments, as a result of higher reaction rates and because bacteria can become more effectively acclimatised to utilise a broader range of organic substrates. Experimental results show further benefits of the staged biofilm reactor design with regard to the solids settling behaviour. Finally, main biofilm microbial components were quantified by Fluorescence *In-Situ* Hybridization (FISH). Biological treatment shows to be an effective means to remove PAHs detected in the aqueous phase. PAH removal takes in most part place in the pre-anoxic zone, thereby decreasing toxicity exhibited by PAH on autotrophic nitrifier bacteria in the aerobic zone. Additionally, we show an assessment of factors that can affect nitrification and denitrification in the moving bed biofilm system. Results obtained in the laboratory-scale, continuous-flow biofilm system suggest that, in the anoxic reactors, nitrogen removal can be limited by the low influent biodegradable substrate concentrations, whereas ammonia oxidation, via nitrite and nitrate, is shown to be adversely impacted by the high alkalinity buffer and high ammonia concentrations, in the aerobic unit. The high inert soluble COD content of the leachate can additionally jeopardise meeting effluent criteria for biological leachate treatment systems. In the presence of high concentration of compounds, exhibiting scavenging effects on the oxidative processes, coupling controlled ozonation with biological treatment is shown to be an effective means using an optimum ozone dosage of approx. 0.5 mg/L O_3. We demonstrate that this dose can decrease the effluent soluble COD concentration and pH profile in the aerobic reactors, improve the biodegradable COD production from inert compounds, and induce chemical nitrification. The specific cost of ozonation evaluated is 0.36–0.73 euro/m^3 with 911 mg/L average effluent soluble COD measured in the biofilm system. Controlled ozonation was tested using the AOP installed in the nitrate recirculation line (Position 1) and, subsequently, between the pre-anoxic and aerobic zones (Position 2), in the pilot-scale system. Ozonation installed in Position 2 is shown to be superior over Position I in terms of COD, PAH and nitrogen removal efficiencies. We additionally demonstrate the potential of intermittent sludge ozonation as a means to decrease PAH concentrations in sludge wasted and to improve nitrogen removal in the BIOZO system.

2.5.2 Materials and methods

2.5.2.1 Laboratory- and pilot-scale experiments

A novel continuous-flow, anoxic–aerobic moving-bed biofilm system (Ødegaard *et al.*, 1994) employing K1-carriers (KrügerKaldnes, Norway) has been developed in laboratory-scale (Plósz *et al.*, 2010b) (Figure 2.56), and implemented in pilot-scale as shown in Figure 2.56 and Figure 2.57, respectively.

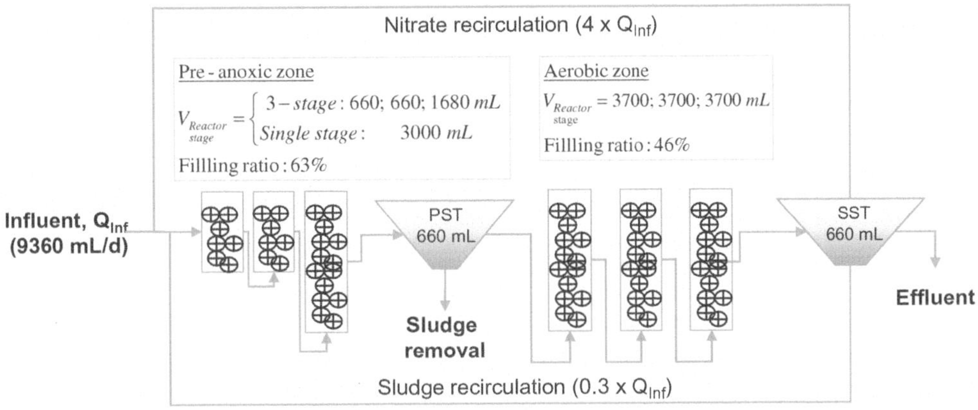

Figure 2.56 Layout of the laboratory-scale biofilm system developed

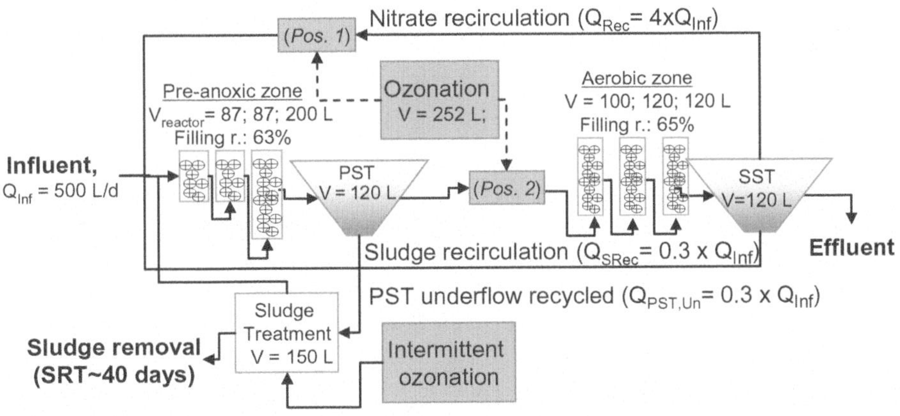

Figure 2.57 Flow scheme of the BIOZO system, including the ozonation either in Position 1 or Position 2 in the secondary treatment line, secondary settling tank (SST) and sludge treatment unit with intermittent sludge ozonation, which the underflow of the primary settling tank (PST) is recycled through.

In the BIOZO system, kinetically optimised three-stage pre-anoxic MBBR cascade is employed (Plósz, 2007; Plósz *et al.*, 2010c) upstream to the aerobic zone. Biofilm detached is separated in the primary settling tank (PST) installed between the pre-anoxic and the aerobic zones. The sludge collected is transferred to the sludge treatment unit (STU), from which the liquid phase is redirected to the inlet of the anoxic reactors. The underflow derived from the secondary settling tank (SST) conveys biofilm detached in the aerobic zone, first, into the influent stream (enhancing the sorption of possibly inhibitory compounds onto the solid phase), and then to the pre-anoxic zone, thereby optimising the utilisation of biodegradable organic compounds in the system. The system was operated around room temperature (20 ± 4°C) using a heating unit equipped with a thermostat installed in the influent stream in pilot-scale. The laboratory- and pilot-scale systems were operated around room temperature (20 ± 4°C). The continuous-flow, laboratory-scale experiment was carried out for the period of 119 days, out of which, in the first 91 days, the system was operated using an anoxic zone, including three unequal reactor stages. The pre-anoxic zone was then replaced with a single-stage configuration, having equal total working volume to the staged arrangement. The biofilm contained on the K1 carriers used in the staged setup was utilised in the single-stage system with an identical filling ratio, 63%. In the pilot-scale system, we only used a kinetically optimised three-stage pre-anoxic reactor configuration, and the system was running for 230 days.

2.5.2.2 Batch experiments using biofilm

Nitrate utilisation rate (NUR) batch tests were run using 600 mL system effluent spiked with 70 mL raw and membrane filtered (0.1 μm) leachate. The experiments were carried out at a temperature of 20°C in 1-L bioreactors sealed from the atmosphere (Plósz *et al.*, 2003), and to each batch reactor, we added approximately 215–220 mL of K1 carriers, derived from the continuous-flow system. Characteristics of the biofilm used in the batch experiments are described in Table 2.9.

Table 2.9 Experimental setups and the characterization of biofilm derived from the different reactor stages and used in the batch experiments

1st experiment **Biofilm carriers:** **single reactor stage**	**2nd experiment** **Biofilm carriers:** **Reactor stage 1**	**3rd experiment** **Biofilm carriers:** **Reactor stage 2**	**4th experiment** **Biofilm carriers:** **Reactor stage 3**
Biofilm characteristics • 21.3 ± 0.96 mgCOD/carrier • 5.1 gCOD/L biomass concentration	Biofilm characteristics • 17.7 ± 2.56 mgCOD/carrier • 4.8 gCOD/L biomass	Biofilm characteristics • 16.1 ± 1.45 mgCOD/carrier • 4.4 gCOD/L biomass concentration	Biofilm characteristics • 14.6 ± 0.9 mgCOD/carrier • 3.9 gCOD/L biomass concentration
Liquid composition: 700 mL effluent; 80 mL leachate spike and 80 mL spike containing 368 mg Na-acetate	Liquid composition: 610 mL effluent; 80 mL leachate spikes	Liquid composition: 610 mL effluent; 60 mL leachate spikes	Liquid composition: 610 mL effluent; 80 mL leachate spikes

2.5.2.3 Batch ozonation experiments

In the laboratory-scale experiments, a Modular 4 HC ozone generator (Wedeco AG, Germany) with an internal potentiometer and pure oxygen feed was used to produce ozone during the tests (Fig. 2.58). The potentiometer was set to 3 and 10 during the testing, while the feed gas flow rate was kept constant at 15 l/h, resulting in an ozone concentration in the feed gas of 83 mg/l and 210 mg/l, respectively. The ozone contact reactor (radius: 0.028 m; height: 0.75 m) was filled with 1.5 litre effluent sample taken from the biological treatment system that was rapidly and continuously circulated using a pump.

Ozone was supplied at the bottom of the contactor through a stone diffuser. Samples were collected at contact times in the range 10–120 min. To remove any residual oxidants, samples were immediately sparged with nitrogen gas for approx. 10 min. In the tests, when samples for biodegradability evaluation were collected, the entire water volume was sampled. All off-gas volume produced between each sampling was collected in sealed aluminium bags, and were analysed for residual ozone.

For pilot-scale experiments, advanced oxidation in the secondary treatment line was carried out using 0.4–0.6 g O_3 per L^{-1} as influent leachate produced by a GSA 40-06 WEDECO ozone generator. We assessed the potentials of sequential ozonation of the sludge wasted (0.05 g O_3/g TSS per day) to optimise the biodegradability of organic compounds in the overflow recycled into the pre-anoxic zone. The dimensions and operating conditions used in the laboratory-scale system are presented by Plósz *et al.* (2010d).

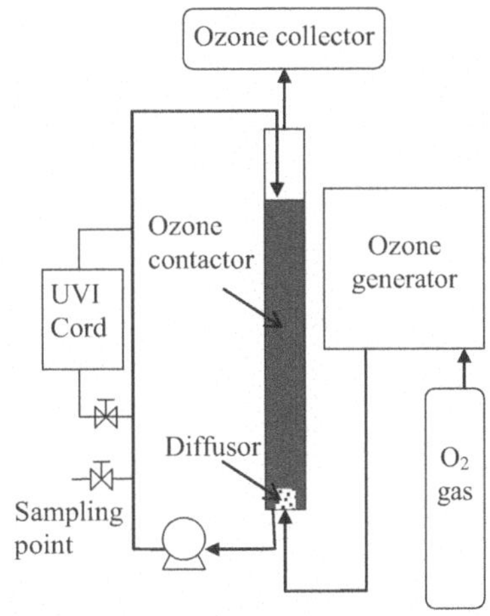

Figure 2.58 Experimental layout of the laboratory-scale ozone contactor

2.5.3 Analytical methods

2.5.3.1 Conventional water quality parameters

COD, ammonium, nitrate and nitrite were analysed according to Standard Methods (APHA, 1995) using HACH-Lange test kits and a HACH-Lange DR2800 spectrophotometer for quantification. The ozone concentration in the off-gas was analysed using Dräger test tubes (Stephan Dräger GmbH, Germany).

2.5.3.2 Screening for xenobiotics

1 L samples of influent and effluent and 5 g of sludge were double solvent extracted with dichloromethane. Extracts were evaporated under nitrogen to approximately 1 ml before clean up through a 5% deactivated alumina column. Analysis was by gas chromatography-time of flight-mass spectrometry (GC-ToF-MS) (GCT, Micromass USA). The GC injector was 250°C, and the oven temperature initially at 60°C, ramped to 280°C at 5°C/min for separation using a ZB-5ms column, (30 m × 0.25 mm id × 0.25 μm film thickness, (Phenomenex UK). The source and electron energy were 180°C and 70 eV respectively and the instrument was operated with a resolution of 8400.

2.5.3.3 Sample analyses for PAHs

Samples analyses were performed at the laboratory of NIVA accredited by the Norwegian Accreditation, according to the requirements of NS-EN ISO/IEC 17025 for the determination of PAHs and PCBs (analytical methods are presented by Vogelsang *et al.*, 2006). The sample extracts were analysed by GC/MS, and the MS detector was operated in selected ion monitoring mode (SIM). The GC was equipped with a 30 m capillary column with a stationary phase of 5% phenyl polysiloxane, and the inlet was operated in the splitless mode. The initial column temperature of 60°C was raised to 310°C in steps. Quantification of individual components was performed by using the internal standard method. The sixteen PAHs determined were acenaphthene, acenaphthylene, anthracene, benz[a]anthracene, benzo[a]pyrene, benzo[b]fluoranthene, benzo[ghi]perylene, benzo[j+k]fluroanthene, chrysene, dibenz[a,c+a,h]anthracene, fluoranthene, fluorene, indeno[1,2,3-cd]pyrene, naphthalene, phenanthrene and pyrene.

2.5.3.4 Toxicity measurements

The acute toxicity is measured using the ToxTrak^{TM} toxicity test kit of HACH Lange. The method is based on the reduction of resazurin, a redox-active dye, by bacterial respiration. When it is reduced, resazurin changes colour from blue to pink. Toxic substances can inhibit the rate of resazurin reduction. A chemical accelerant is added to shorten the reaction time (45–75 min). *E. coli* was used as inoculum in the tests. For selected samples primarily taken in the pilot-scale system, the growth inhibition test using *Pseudokirchneriella subcapitata* is used (OECD 201, ISO 8692).

2.5.3.5 Characterisation of the landfill leachate used for system development

The partially stabilised landfill leachate used in the laboratory and pilot-scale experiments derived from the final effluent of the Lindum AS site situated South-West of Oslo, Norway. In Table 2.10, we show typical pH and pollutant concentrations in the partially stabilised landfill leachate used in a continuous-flow lab-scale experiment to assess biological treatment. The composition of the partially stabilised landfill leachate used in the lab-scale experiment can be characterised with total influent COD concentration between 1700 and 3000 mg/L, out of which, approx. 71% was measured as soluble.

Table 2.10 Composition of the stabilised leachate used in the continuous-flow experiment

	Average	Min.	Max.
pH (–)	7.6	6.88	8.54
NH_4-N (mg/L)	198	55	288
NO_3-N (mg/L)	4.36	0.41	12.4
NO_2-N (mg/L)	0.75	0.04	4.88
COD_{Total} (mg/L)	2408	1768	2968
$COD_{Soluble}$ (mg/L)	1649	788	2544

2.5.4 Results

2.5.4.1 Characterisation of the landfill leachate used for system development

2.5.4.1.1 Conventional water quality parameters

The partially stabilised landfill leachate used in the laboratory and pilot-scale experiments derived from the final effluent of the Lindum AS site situated South-West of Oslo, Norway. In Table 2.10, we show typical pH and pollutant concentrations in the partially stabilised landfill leachate used in a continuous-flow lab-scale experiment to assess biological treatment. The composition of the partially stabilised landfill leachate used in the lab-scale experiment can be characterised with total influent COD concentration between 1700 and 3000 mg/L, out of which, approx. 71% was measured as soluble.

2.5.4.1.2 Xenobiotics

XOCs occurring in the leachate can effectively inhibit biological treatment processes, such as nitrification and denitrification (Geenens *et al.*, 2000; Plósz *et al.*, 2010d). Furthermore, various degradation products of xenobiotic organic chemicals can occur in leachate, which are important indicators of on-going degradation (Baun *et al.*, 2004). Samples were taken from the raw influent, lab-scale system's liquid effluent and waste sludge and analysed for organic micro-pollutants by GC-ToF-MS. We note that the GC-ToF-MS method is a qualitative and not quantitative estimation method. All of the samples showed evidence of unresolved complex mixtures (UCM) as demonstrated by the humps in the chromatograms (data not shown). The UCM highlights the complex nature of the samples containing large numbers of unresolved peaks. Some compound peaks, however, were identified that are shown in Table 2.11.

Results obtained in the screening study do not show any evidence of the occurrence of degradation products reported by Baun *et al.* (2004) – benzylsuccinic acid, nonylphenol-ethoxycarboxylates and monoesters of o-phthalic acid. Most of the XOC detected in the samples taken from the influent, effluent and sludge are poly-aromatic-hydrocarbons (PAHs).

2.5.4.1.3 PAHs occurrence

In Figure 2.59a, values of the influent, total PAHs concentration are 7375 ng/L, on average, in the 30-days sampling period examined, including twelve measurement days. In Figure 2.59b, influent concentration values are shown for the fifteen PAHs selected to assessing the potential of the BIOZO system to remove priority pollutants from landfill leachate. Unfortunately, analytical results obtained on naphthalene (possible carcinogen) are only available in the solids samples shown later. Based on the PAH concentrations encountered in the samples, two main groups can be distinguished. Group A, characterised with influent concentrations of 254–1710 ng/L, includes acenaphthene, fluorene, phenanthrene, anthracene, fluoranthene and pyrene.

These compounds have been found to be toxic to a range of aquatic organisms with effects being attributed to loss of membrane integrity and narcosis (Munoz and Tarazona, 1993; Schirmer *et al.*, 1998; Loibner *et al.*, 2004). Several of the compounds detected in concentrations up to approx. 93 ng/L, Group B, are "reasonably anticipated to be carcinogens" – benzo[a]anthracene, benzo[a]pyrene, Benzo[b]fluoranthene, Benzo[k]fluoranthene, Dibenz[a,h]anthracene and Indeno [1,2,3-cd]pyrene – NTP (2005). Furthermore, the IARC (2009) have determined that benzo[a]pyrene is a confirmed carcinogen and that crysene is a possible carcinogen. This is a striking evidence of the necessity to develop and implement robust treatment technologies in landfill leachate management.

Table 2.11 Prioritised XOCs in landfill leachate, CAS numbers, compound characteristics and occurrence

Compound	CAS no.	Characteristics	Reference	Influent	Effluent	Sludge
Acenapthene	83-32-9	Irritant	(Faust, 1994)	X	X	X
Anthracenedione	84-65-1	Irritant	(ITII, 1988)	X	X	
Anthracene	120-12-7	Photosensitising	(ILO, 1983)	X	X	X
Phenanthracene	85-01-8	Phototoxic, narcotic	(Bellas *et al.*, 2008)	X	X	X
Benzophenone	119-61-9	Hepatotoxic, nephrotoxic	(Chhabra, 2000)	X	X	X
Bromoimidazole	2302-25-2	24h LD50 in rats: 1.7–3.4 millimoles/kg**	(Verschoyle *et al.*, 1987)	X	X	X
Cyclopenta-phenanthrene	80455-52-3	Mutagenic	(Morrochi *et al.*, 1996)	X	X	X
Di-isopropyl-naphthalene	24157-81-1	Hepatotoxic	(Hasegawa *et al.*, 1982)	X	X	X
Fluorene	86-73-7	Low toxicity	(Bellas *et al.*, 2008)	X	X	X
Methyl-octahydro-anthracene	N/A	N/A	N/A	X	X	X
Methylthio benzothiazole	21564-17-0*	7 day NOEC in crustacean C. dubia: 1.21 mg/L**	(Nawrocki *et al.*, 2005)	X	X	X
Phthalates	EDF-150	Estrogenic	(Nakai *et al.*, 1999)	X	X	X
Phthalic-acids	88-99-3	Estrogenic	(Pavan *et al.*, 2001; Ema *et al.*, 1997)	X	X	X
Tetramethyl-butylphenol	98-54-4	Narcotic	(Russom *et al.*, 1997)	X	X	X

**This is the code for TCMTB, the parent compound from which this one is a metabolite. The CAS number of MTBT is unknown for the authors*
***This is the lethal dose for 50% of test group or the no observable effect concentration for these chemicals and not the type of toxicity*

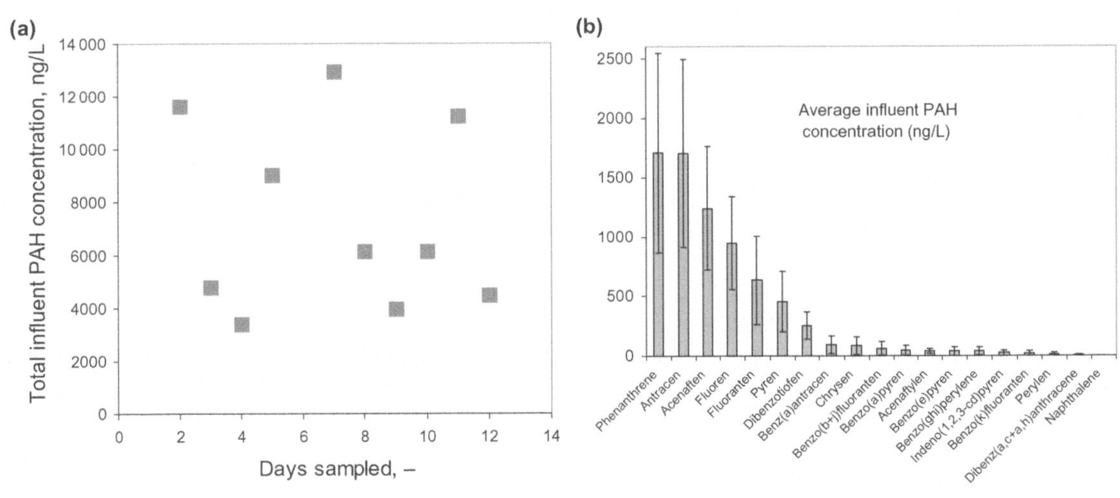

Figure 2.59 Values of the total PAH concentration (a) and the average individual PAH concentrations with standard deviation (b) in raw partly stabilised leachate

2.5.4.2 *Biological landfill leachate treatment*

2.5.4.2.1 Biodegradation kinetics

The occurrence of PAH compounds in landfill leachate can potentially inhibit biological nitrogen and COD removal. Biotransformation of most XOCs of environmental and toxicological significance is carried out through the process of cometabolism, i.e. transformation of a non-growth XOC substrate by growing cells in the presence or absence of a growth substrate (Horváth, 1972; Criddle, 1993), illustrated in Figure 2.60.

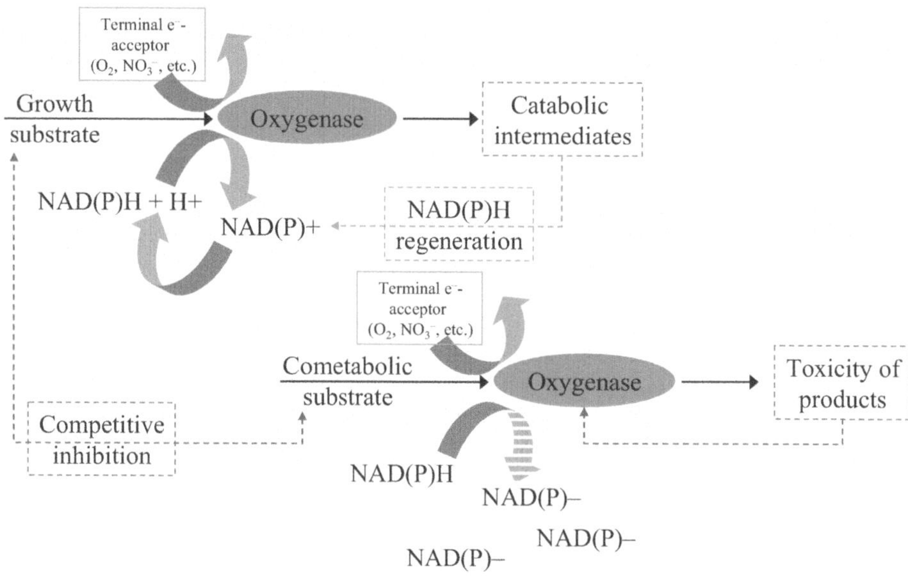

Figure 2.60 Schematic description of microbial cometabolism of xenobiotics compounds in the presence of growth substrates

Growth substrates can be oxidised at the expense of terminal electron acceptors, such as oxygen and nitrate also using the reductive energy sources derived from terminal electron donors. These can be regenerated by the catabolic intermediates produced through the oxidation process that is not the case for cometabolic substrates, and cells can effectively exhaust their reductive energy, in the absence of growth substrates. Furthermore, direct toxicity of the cometabolic substrate, toxicity of the oxidation process's product, and competitive inhibition between the growth and the cometabolic substrates to the non-specific enzyme active site can effectively hinder the transformation of XOCs.

From an engineering viewpoint, the question arises whether XOCs encountered in landfill leachate can have a significant impact on biological treatment processes, e.g. nitrogen removal, and if they can; how to take them into account in the system design. From the prospective of environmental risk, another question is whether biofilm systems could be optimised to remove a wider range of organic chemicals from wastewaters, and degrade more effectively XOCs.

In order to assess factors, influencing biological nitrogen and COD removal within the presence of xenobiotic compounds, we evaluated the anoxic biodegradation kinetics in a series of batch tests using raw and 0.1 μm filtered landfill leachate samples as shown in Table 2.4. In Figure 2.61a, the nitrate-nitrite nitrogen elimination rate values are plotted as a function of the time elapsed. Data obtained suggest two consecutive degradation steps using organic substrates contained in the filtered and the raw leachate spikes. In Figure 2.61a, results obtained show that an initial relatively fast nitrate elimination rates prevail until approximately 75 minutes elapsed since the reactor start-up, and rate values subsequently drop to an endogenous respiration level. Following the plateau in the rate profile using filtered leachate, caused by possible toxic impacts lasting for approx. 50 min, a second relatively fast nitrate elimination period, accompanied with some nitrite accumulation (data not shown), can be observed. Using raw leachate, only a short lag period is revealed between the two rate peaks. External substrate depletion leads to a final endogenous respiration phase close to the end of the experiments.

In the first period, the organic substrates oxidised at high denitrification rates, possibly soluble-colloidal readily biodegradable substrates, are denoted as S_1. In the second period, the relatively slowly biodegradable substrates are shown as S_2. Based on the results shown in Figure 2.61, a possible explanation of the kinetic behaviour observed is hypothesised, suggesting the inhibition of denitrification processes by one or more toxic cometabolic biodegradation product(s) formed. Degradation products XOC can be detected in landfill leachate, e.g. benzylsuccinic acid related to toluene degradation, nonylphenol-ethoxycarboxylates generated through nonylphenol-ethoxylates degradation, and monoesters of o-phthalic acid derived from the degradation of diesters of o-phathalic acid (Baun *et al.*, 2003; Baun *et al.*, 2004). Their possible occurrence and impact on nitrogen removal should be assessed in further research. Degradation products can be more polar and thus more water soluble than the parent compounds – a factor that can enhance inhibitory impacts on heterotrophic anoxic metabolism. Sorption onto particulate matter can decrease the aqueous concentration of XOC, thereby also allowing denitrification to occur at higher rates using raw leachate (see rate values between the two peaks in Figure 2.59a). Inspection of Figure 2.61 also suggests that, in the presence of XOCs, biodegradation kinetics can effectively hinder denitrification in a continuous stirred tank reactor under anoxic conditions (ACSTR), an effect that can be mitigated by reactor compartmentalisation.

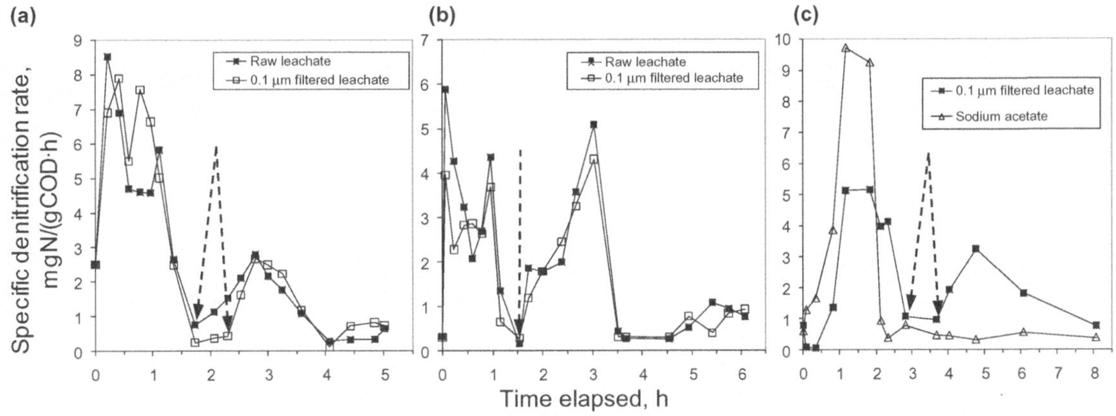

Figure 2.61 Values of the specific denitrification rate obtained using raw and filtered landfill leachate in batch experiments carried out with biofilm carriers derived from the stage 1 (a), from the stage 3 (b) and from the single-stage pre-anoxic system (c) plotted as a function of the time elapsed. Batch reactors were spiked with raw leachate and with 0.1 µm membrane-filtered leachate (A, B, C) and leachate filtered and sodium-acetate (D) as carbon sources

2.5.4.2.2 The conceptual approach

A conceptual approach for optimising denitrification in a continuous biofilm system is formulated. According to Grady *et al.* (1999), let us first describe steady-state mass balance around an ACSTR, containing biofilm of thickness, L_f, and biomass at concentration $X_{B,Hf}$ as

$$\frac{1}{\tau_{AX}} = \left(\frac{F}{V_{AX}}\right) = \eta_{eO} \cdot X_{B,Hf} \cdot L_f\left(\frac{A_S}{V_{AX}}\right)\left[\frac{\hat{q}_H \cdot S_{N,0}}{(K_{S,N} + S_{N,0}) \cdot (S_{N,0} - S_{N,e})}\right] + \frac{X_{B,Hb} \cdot \hat{q}_H \cdot S_{N,0}}{K_{S,N} + S_{N,0}}, \quad (1)$$

where the effectiveness factor is η_{eO}, A_S denotes the biofilm surface area in the reactor, $\hat{q}_H$ is the maximum substrate consumption rate, V_{AX} is the anoxic bioreactor volume, F is the flow rate of influent and effluent, τ_{AX} is the hydraulic retention time in the anoxic bioreactor, $S_{N,0}$ and $S_{N,e}$ are the influent and effluent bulk liquid nitrate-nitrite-nitrogen concentrations, respectively. In Eq. (1) $X_{B,Hb}$ denotes the biomass dispersed, $K_{S,N}$ is the Monod half-saturation coefficient for nitrate-nitrite concentration as shown by the ASM1 (Henze *et al.*, 1985). For a desirable conversion, ($S_{N,0}$–$S_{N,e}$), the kinetically optimum sizes of reactors in-series can be assessed by minimising Eq. (2) as a function of the denitrification rate, r_D (Plósz *et al.*, 2003).

$$\tau_{AX} = \left(\frac{\sum_{i=1}^{e} V_{AX,i}}{F}\right) = (S_{N,0} - S_{N,1}) \cdot \left(\frac{1}{-r_{D,1}}\right) + (S_{N,1} - S_{N,2}) \cdot \left(\frac{1}{-r_{D,2}}\right) + \cdots + (S_{N,(e-1)} - S_{N,e}) \cdot \left(\frac{1}{-r_{D,e}}\right), \quad (2)$$

where the number of the anoxic stages is denoted with e. In a series of biofilm ACSTRs, $X_{B,Hf}$, L_f, A_S and the maximum denitrification rate can alter from stage to stage. For design purpose, we propose to use $\tau_{AX,i}$ design values assessed from $-1/r_D$–S_N curves obtained using biofilm derived from a reactor, continuously receiving raw wastewater. Compared to the design $-1/r_{D,i}$ values obtained for stages $i > 1$, the $-1/r_{D,i}$ values subsequently obtained in operation shall be decreased, as a result of improved microbial adaptation to organic substrates available in each stage, thereby also introducing a design safety factor.

This effect is demonstrated in Figure 2.61b by the high denitrification rates obtained relative to values in Figure 2.61a shown at the second peaks. A generic assessment, carried out by Scuras *et al.* (2004), reveal three stages as an optimum number for kinetically optimised reactors. This is in good agreement with the results obtained by Plósz (2007) for anoxic activated sludge reactors. In Figure 2.62a, reciprocal denitrification rate values plotted as a function of the nitrate-nitrite concentrations removed are shown. Based on Eq. (2), the shaded rectangular areas represent the anoxic single stage or the three-stage reactors in terms of hydraulic retention time, τ. In order to determine the optimum bioreactor configuration, the optimum sizes of the ACSTRs in-series, ($e-1$), can be assessed applying the method of maximisation of rectangles, touching arbitrary curves (Plósz, 2007). For a total of three reactor stages, to achieve a given conversion rate, the last ACSTR stage, i.e. $e = 3$, can additionally be assigned.

In Figure 2.62b and Figure 2.62c, the reciprocal denitrification rate values obtained and plotted as a function of the nitrate–nitrite concentration removed in the batch experiment (Figure 2.61) are shown. For the three reactor stages, $AX_{1,2,3}$, the τ-distribution includes 1.7, 1.7 and 4.3 hours, respectively (Figure 2.62b). The effectiveness of the staged system is contrasted against the nitrate-nitrite-nitrogen removal efficiency of single stage reactor arrangement, shown by the rectangular area in Figure 2.62c. In the continuous-flow system (Figure 2.56), in addition to the staged reactor design, denitrification efficiency can be further improved by recirculation streams. In Figure 2.62c, at approx. 40 mg/L N removed, data obtained with membrane filtered leachate suggest a significant increment in the reciprocal rate values, i.e. higher inhibition on denitrification processes. This translates to higher τ_{AX} values required to achieve equal nitrogen removal efficiency obtained with raw leachate. This outcome suggests omitting the primary solids-liquid separation process in biological leachate treatment systems.

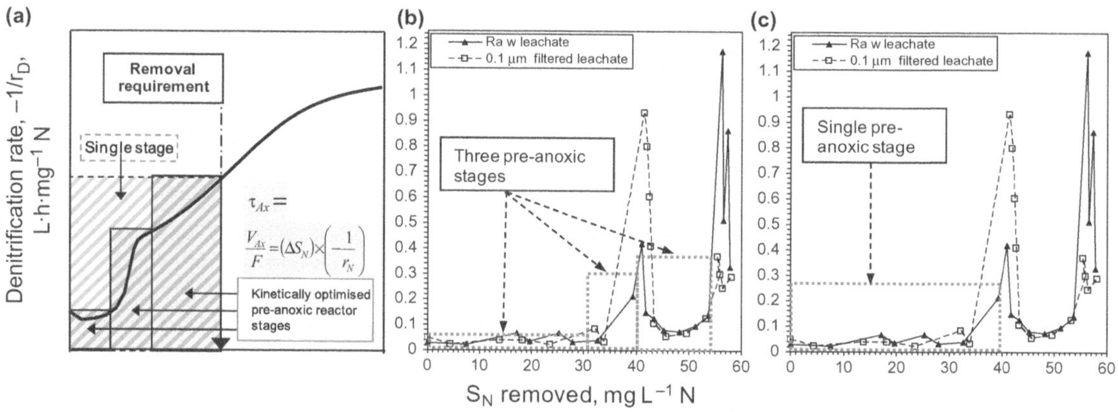

Figure 2.62 Illustration of the pre-anoxic reactor stages, based on Plósz, 2007 (a) and reciprocal denitrification rates, calculated using values shown in Figure 2.59b, and plotted as a function of the nitrate, nitrite concentrations removed with dashed lines showing the increased nitrogen removal achieved using the three possible biofilm reactor stages (b) as compared to the single stage arrangement (c)

To assess the feasibility of the kinetic optimisation of reactor configuration for biofilm systems, the continuous-flow, anoxic-aerobic, two-stage moving-bed biofilm reactor system developed (Figure 2.56) has been operated, first, using a three-stage reactor configuration. Biofilm carriers were thereafter placed into a single reactor vessel, and the experiment was continued, having the total working volume of the pre-anoxic zone, identical to that of the three stages.

2.5.4.2.3 Substrate characterisation and microbial acclimatisation

In line with the continuous-flow experiments, we assessed the biodegradable organic fractions in the leachate and evaluated the acclimatisation state of the bacteria residing in the biofilm using batch experimental data (Table 2.4). Compared to the results obtained using the biofilm carriers derived from the stage 1, AX_1 (Figure 2.61a), we obtained very similar results using biofilm carriers taken for the reactor stage 2, data not shown. Characteristically different kinetics is suggested by data obtained using biofilm derived from the third relatively large compartment, AX_3 (Figure 2.61b). The plot of the specific denitrification rates against the time elapsed shows denitrification rates, approx. half of that observed with AX_1 and AX_2 biofilm using the first group of substrates, S_1. For the biofilm derived from AX_3, a very short inhibited lag phase and the second peak characterised with relatively high denitrification rates, approx. 5–6 mgN/gCOD·h rate, can be observed (Figure 2.61b). These results suggest an improved acclimatisation state of AX_3 biofilm to the prevailing toxicity. We draw attention to the comparable biomass content of carriers derived from AX_3 relative to $AX_{1,2}$ and the single stage (see in Table 2.4), a crucial factor to prevent possible break-through events under shock-loading conditions. Results presented in Figure 2.61c were obtained in a set of batch tests using biofilm carriers taken from the single-stage anoxic reactor, AX_{SS}, at the end of the continuous-flow experiment. The two batch reactors were spiked with filtered landfill leachate and with known amount of sodium-acetate as readily biodegradable substrate, respectively. Utilising acetate as terminal electron acceptor (Figure 2.61c), the maximum denitrification rate value obtained with biofilm derived from the single-stage pre-anoxic reactor, is between 9.0 and 9.5 mgN/gCOD·h. This value is approx. twice as high as the rate obtained utilising S_1 (Figure 2.61c). Biofilm derived from AX_1 can exhibit denitrification rates, 6.5–8 mgN/gCOD·h using S_1 that are comparable to that obtained with acetate (Figure 2.61a).

The observed heterotrophic yield coefficient, $Y_{H,Obs}$, mgCOD biomass produced/mg COD substrate, accounts for decay in both the biofilm and the dispersed bacteria. Values of $Y_{H,Obs}$ was calculated using the estimated amounts of nitrate and nitrite nitrogen reduced (ΔS_N) and the soluble biodegradable carbon sources, $\Delta COD_{Soluble,biodegradable}$ in the term

$$Y_{H,Obs} = 1 - \frac{2.86 \cdot \Delta S_N}{\Delta COD_{Soluble,biodegradable}} \quad (3)$$

For the relatively small AX_1 and AX_2, as well as for the AX_{SS}, $Y_{H,obs}$ values obtained are uniformly 0.41, and for the AX_3 reactor stage, $Y_{H,obs}$ is 0.52. In Figure 2.63a, values of the biodegradable organic fraction suggest an improved ability of the AX_3 biofilm to distinguish a larger fraction of organic matter available as biodegradable, as compared to $AX_{1,2}$ and AX_{SS}.

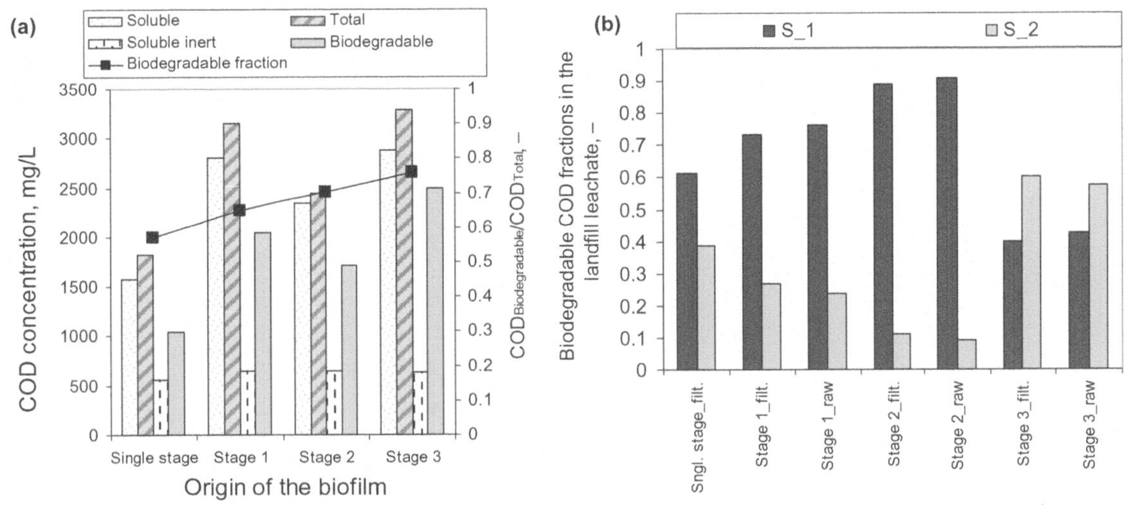

Figure 2.63 Values of the total, soluble, biodegradable and soluble inert COD concentrations in the leachate spiked, as well as of ratios between the soluble inert COD and biodegradable COD to the total COD concentrations (a); and values of the biodegradable COD fractions, S1 and S2, calculated using the nitrate respiration rate–time elapsed plots (b) obtained in the batch experiments

Based on the observations made by Kaballo *et al.* (1995) and Lemming *et al.* (2000), the improved acclimatisation state can be associated with higher XOC degrading potential of the biofilm. We note that, compared to $AX_{1,2}$ and AX_{SS}, only the biofilm derived from AX_3 could remove, i.e. hydrolyse and oxidise as terminal electron acceptor, an approx. 8–10% higher amount of organic substrates, contained in raw leachate, than from filtered leachate (data not shown). In other cases, the difference was less than 4% between the biodegradable fraction obtained using raw and filtered leachate spikes. In Figure 2.63b, the relative biodegradable COD fractions are shown. For biofilm derived from AX_1, AX_2 and AX_{SS}, the relative fraction of S_1 organic substrates is higher than S_2. In contrary, biofilm derived from the AX_3 is shown to oxidise a large relative portion of S_2 than S_1. The S_1 than S_2 fractions obtained using raw and filtered samples are very similar (Figure 2.63b). In Figure 2.64a, three different experimental phases, A, B and C, are distinguished. In Phase A, detached biofilm was removed from the system separately via the PST and SST. In Phase B and Phase C, sludge is wasted solely via the PST. In Phase C, the system was operated using a single stage pre-anoxic reactor arrangement. In Figure 2.64a, values of the effluent soluble COD to the influent, total COD ($COD_{Inf,Tot}$) concentration ratio ($COD_{Eff,sol}/COD_{Inf,tot}$) obtained in continuous-flow experiments are plotted as a function of the time elapsed (Figure 2.64a).

The average inert soluble COD fraction detected in the system's effluent is 30 and 35% using the three-stage pre-anoxic zone in the laboratory- and pilot-scale experiments, respectively. After switching to the single pre-anoxic stage, in laboratory-scale, an increasing tendency in the $COD_{Eff,sol}/COD_{Inf,tot}$ value can be observed that reaches an approx. 50% in about 18 days of operation. An important thing about Figure 2.64a, is that in only 28 days, after switching operation from using $AX_{1,2,3}$ to AX_{SS}, denitrifiers can effectively lose their ability to metabolise a wider range of substrates, and become more sensitive to the possible product toxicity, as shown in Figure 2.63a. Note that, in Figure 2.61b, irrespective to the impact of using filtered or raw leachate, the denitrifiers in AX_3 biofilm can readily recover from the inhibited period, which is not the case for AX_{SS} (Figure 2.61c) or AX_1 (Figure 2.61a).

Values of the nitrification and denitrification efficiencies, in terms of nitrate-nitrite-nitrogen concentration produced and removed, respectively, are plotted in Figure 2.64b. The difference between the nitrified and denitrified nitrogen concentrations, i.e. the residual effluent nitrate concentration, suggests a considerable deterioration in the nitrogen removal capacity in Phase C. Based on the results shown in Figure 2.61 and Figure 2.63, nitrogen removal capacity in the three stage system, shown in Figure 2.64b, is superior to that observed in the single stage reactor as a result of higher average denitrification rates and of the improved microbial acclimatisation state. According to (Ødegaard *et al.*, 2007) nitrification rates of approx. 1.8 g NH_4-$N/m^2 \cdot d$ can be expected for moving bed biofilm reactors at 20°C and a DO concentration of 6 mg/L when treating domestic wastewater. In our laboratory-scale system, the obtained nitrification rate was 0.90 ± 0.32 gNH_4-$N/m^2 \cdot d$.

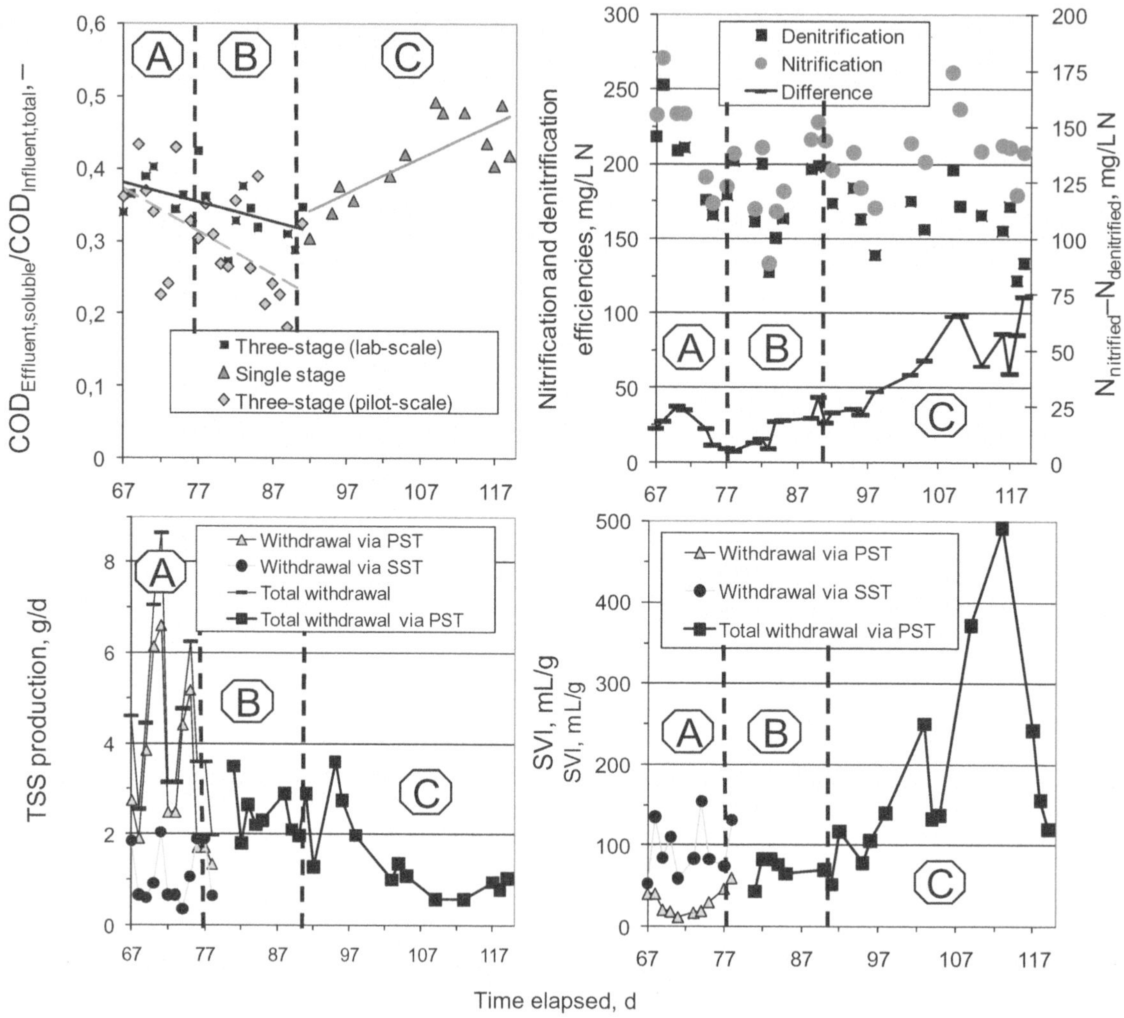

Figure 2.64 Values of the soluble inert to the influent total COD concentration ratio in the laboratory- and pilot-scale experiments (a); nitrification and denitrification efficiencies obtained (b); values of the solids production obtained in the system (c) and sludge volume index (SVI) measured (d) plotted as a function of the time elapsed. Solids removal from the system separately via the PST and SST (Phase A) and from solely via the PST (Phase B and Phase C); alteration in the pre-anoxic reactor arrangement from three-stage to a single stage (Phase C)

2.5.4.2.4 Microbial selection in biofilm by pre-anoxic reactor staging

Biofilm samples taken from the three anoxic reactor stages were analysed by FISH. Despite the presence of toxic pollutants in the feeding, the percentage of *Bacteria* respect to the total cells determined by DAPI staining (%EUB338mix/DAPI) is always higher than 75% in all screened anoxic compartments (Figure 2.65).

This means that a large fraction of the biomass is actively involved in the biological reactions prevailing in the system. The detectability of bacteria by FISH is indeed often utilised as a simple "gross parameter" to assess the overall physiological state of bacterial communities. It was found to be in agreement with reactor performances and with conventional respirometric tests applied for microbial activities measurements in previous experiences (Kurisu *et al.* 2002; Witzig *et al.*, 2002; Eschenhagen *et al.*, 2003; Li *et al.*, 2006). A lower percentage value for *Bacteria* is obtained in the first compartment, compared to the second relatively small reactor stage that suggests somewhat increased inhibitory effect in AX1.

The FISH analysis was additionally performed to characterize the overall phylogenetic composition and diversity of the bacterial populations in the biofilm community. As shown in Figure 2.63, most of *Bacteria* were additionally identified by group-specific FISH probes. The analysis revealed a selected biomass mainly composed by *Proteobacteria* (mostly belonging to *Betaproteobacteria* and *Gammaproteobacteria*) with only a marginal presence of *Alphaproteobacteria* and of members of *Flavobacteria/Bacteroidetes/Flexibacter* group. The relative abundance of single bacterial groups differed, particularly, in regard to *Betaproteobacteria* that is 35% BET42A/DAPI in AX3, compared to 13% and 30% in AX1 and AX2, respectively (Figure 2.65). Further evidence of microbial stratification can be observed in the relatively high presence, 35% GAM42A/DAPI, of *Gammaproteobacteria* in AX1 and AX2, characterised with relatively high readily biodegradable COD concentrations. In AX3, the value of GAM42A/DAPI obtained is only 22%. Still very little is known about the identity of abundant denitrifiers in wastewater treatment systems. The main reason for this is that most studies have been carried out after isolation and cultivation of denitrifying species, and these isolates do not necessarily represent the denitrifiers actually dominant in such systems. Bacteria affiliating in the *Betaproteobacteria*, were recently shown to be potential denitrifiers in the municipal WWTP with N- and P-removal by applying a combination of microautoradiography (MAR) and FISH. (Thomsen *et al.*, 2007).

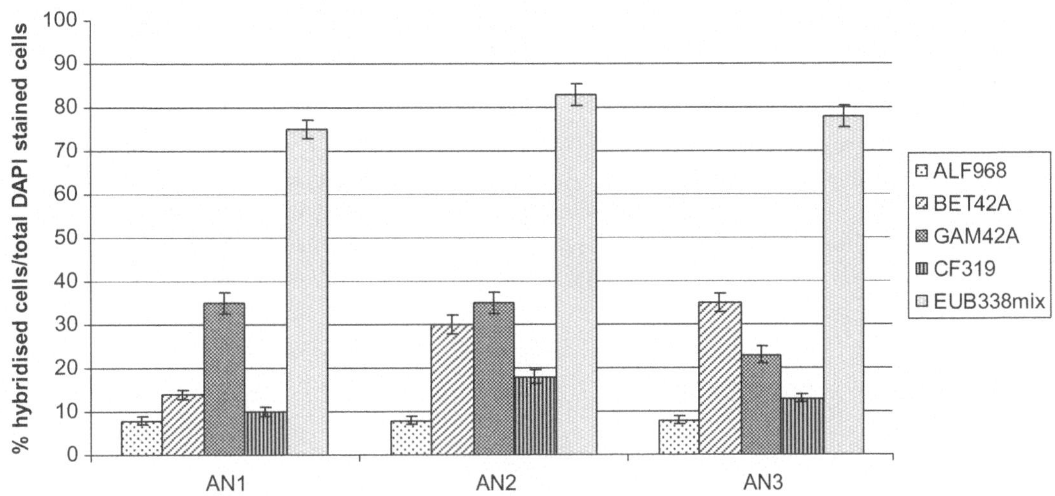

Figure 2.65 Microbial population compostion of biofilm samples from the three anoxic compartments (AN1, AN2 and AN3). The amount of Bacteria and of cells binding group- specific probes are estimated by FISH and reported as percentage of total DAPI stained cells

In particular, mainly members of *Thauera* spp. and *Azoarcus* spp. were found as abundant denitrifiers by using culture-independent methods and reported to be the denitrifying bacteria in the anoxic zones of wastewater treatment plant (Morgan-Sagastume *et al.*, 2008). The molecular analysis of biofilms samples was therefore refined by means of FISH probes with higher specificity within *Betaproteobacteria.* THAU646 and AZO644 probes, specific for *Thauera* spp. and *Azoarcus* spp. respectively, were applied on biofilm sample taken from the three anoxic reactor compartments. No cells were highlighted, which outcome suggests that other denitrifying species should prevail in our system. Moreover, previous evidences showed that in activated sludge, denitrification was potentially conducted by a diverse, versatile population including not only *Betaproteobacteria* (i.e. *Thauera*, and *Azoarcus*) but also some *Alphaproteobacteria* and *Gammaproteobacteria*, as indicated by the assimilation of $^{14}CO_2$ by these probe-defined groups with a complex substrate mixture as an electron donor and nitrite as an electron acceptor in MAR-FISH tests (Morgan-Sagastume *et al.*, 2008). Therefore the possibility of having multiple denitryfing populations with different adaptation capabilities in our system cannot be excluded. This will require ***further investigation by adopting a more complex approach to study the* in situ *ecophysiology*** of this heterogeneous functional group.

2.5.4.2.5 Excess solids removal and solids settling behaviour

In Figure 2.64c, we show the amount of biofilm detached from the aerobic and anoxic zones. Solids separated in the PST contain anoxic biofilm detached and inert particulates derived from the influent. Values of the anoxic sludge production varied between 2–6.5 g/d, and values specific to the daily COD load in the system were between 0.06–0.32 $gTSS/gCOD_{Inf}$. Sludge production of the aerobic zone obtained was between 0.5 and 2 g/d that translates to 0.2–0.95 $gTSS/gNH_4N_{Inf}$, values specific to the influent ammonia load. Total solids removed from the biological system varied between 1.5 and 6.5 g/d.

In order to optimise the anoxic usage of biodegradable organic substrates contained in the aerobic biofilm detached, operation was altered by recirculating the underflow of the aerobic SST into the anoxic reactor, thereby removing solids only from the PST. According to Figure 2.64c, as a result of hydrolysing additional amounts of particulate organics in the anoxic zone, sludge production was reduced, thereby decreasing the overall solids wastage rate to 2–3.5 g/d. As a result of installing the single-stage pre-anoxic zone (period C), the solids production was decreased to 0.5–1 g/d, and the amount of biofilm, residing on the carriers increased (see Table 2.4). The settling behaviour of withdrawn sludge was characterised by daily measurement of the sludge volume index (SVI). Figure 2.64d demonstrates that dewaterability of biomass detached from the three-stage anoxic zone was very good (SVI = 20–60 mL/g). For aerobic biofilm, SVI values varied from 50 to 150 mL/g, in period A. In Phase B, average SVI values obtained with sludge removed only via the PST was 75 mL/g. In Phase C, the dewaterability of the excess sludge severely deteriorated, resulting in SVI values obtained from 150 to 500 mL/g. Solids-liquid separation was further compromised by increased nitrate-nitrite concentrations in the PST, causing floating sludge and higher COD load on the aerobic zone (data not shown). Providing good biofilm settleability, the 3-stage pre-anoxic reactor configuration can also effectively decrease the particulate organic substrate load of the nitrification zone that can further improve nitrification and thus nitrogen removal in the system.

2.5.4.3 Assessment of controlled ozonation as a means to optimise biological landfill leachate treatment

2.5.4.3.1 COD and nitrogen removal

Based on the results shown in Figure 2.66, approximately 71 per cent of the $COD_{inf,t}$ was soluble COD out of which approx. 50% was detected as biodegradable soluble COD. Values of soluble COD concentration in the system's effluent thus were high, and varied between 400 and 1162 mg/L. We note that, in the case of discharging treated leachate into sewer systems, such organic matter cannot be biodegraded in subsequent municipal wastewater treatment plants either, but can potentially decrease treatment efficiencies by having inhibitory impacts on the bacteria (Jönsson *et al.*, 2000).

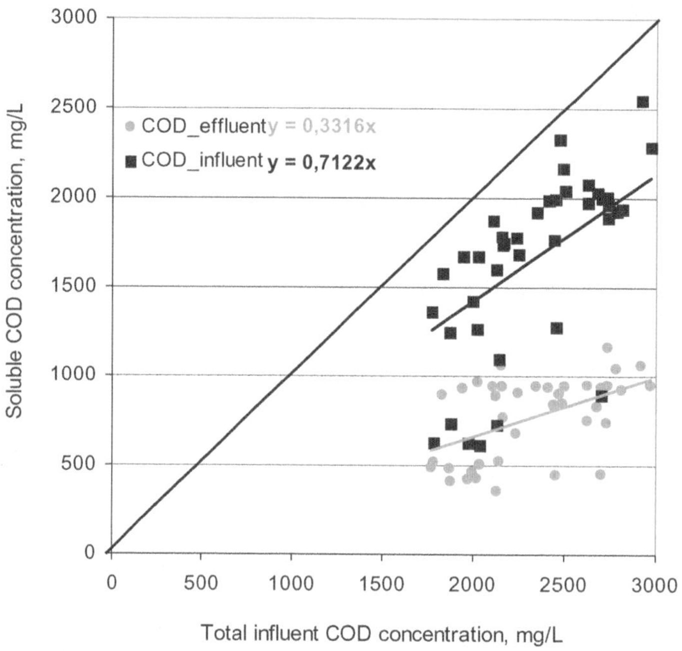

Figure 2.66 Values of soluble COD concentration in the influent and effluent to the biological treatment system plotted as a function of the total COD concentration in the influent

Consequently, reduction of the effluent soluble COD is identified as an optimisation objective for coupling biological treatment with AOPs at the landfill site. Biodegradable carbon sources serve as terminal electron donors for denitrification, and, were removed in the pre-anoxic zone of the system shown in Figure 2.56. In Figure 2.67, the nitrate-nitrite-N concentration is plotted as a function of the soluble biodegradable COD concentration, and its value is calculated as NO_3-N + 0.6 * NO_2-N.

This is because the reduction of 1 g NO_2-N to 1 g N_2-gas needs the same amount of electrons (biodegradable COD) as the reduction of 0.6 g NO_3-N to 0.6 g N_2 gas. In Figure 2.67, below a value of approx. 300 mg/L soluble COD removed in the anoxic zone, the average nitrate-nitrite-N concentration in the effluent of the pre-anoxic zone can sharply increase. Denitrification capacity can thus be limited by the biodegradable COD available in the pre-anoxic zone, and the increase of biodegradable COD is selected as an optimisation objective for ozonation. Inhibitory impacts on biodegradation is another factor, limiting denitrification capacity using carbon sources contained in landfill leachate (Plósz *et al.*, 2010d), a factor that can be mitigated by implementing kinetically optimised staged bioreactors. Values of pH in the influent raw leachate varied between 6.88 and 8.54. In the effluent of the pre-anoxic zone and in the system's effluent pH was measured between 8.1–9.0 and 6.45–8.6, respectively. In Figure 2.68, values of the produced nitrate-nitrogen concentration specific to the highest nitrification rate measured in the aerobic zone are plotted as a function of the pH measured in the effluent of the aerobic reactors. Sensitivity of the nitrifying autotrophic biomass to pH beyond normal physiological range, characterised with values between 7 and 8 (Grady *et al.*, 1999), is shown to be high in the biological treatment system.

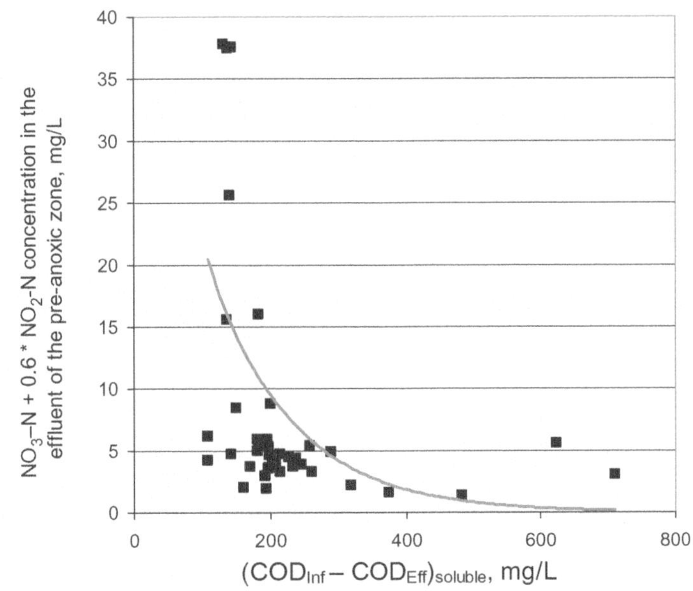

Figure 2.67 Values of the NO_3-N + 0.6 * NO_2-N concentration plotted against the soluble biodegradable COD concentration removed in the pre-denitrification reactor

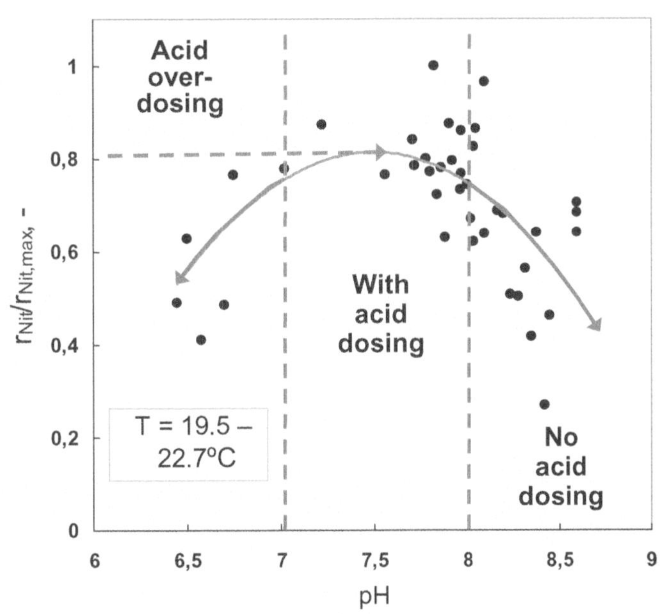

Figure 2.68 Effect of pH, measured in the effluent of the aerated reactors, on the maximal nitrification activity in the aerobic zone of the continuous-flow laboratory-scale system

At a level of 80% of the maximum nitrification potential detected in the system, conditions can be characterised with minimum pH inhibition. To control pH in the aerobic zone, we dosed on average, approx. 260 ml 0.1 M HCl per day, when treating 9.4 litres of leachate, into the influent of the nitrifying reactors. Feedback pH-control was used by means of pH values measured in the effluent of the first aerobic reactor stage. Without acid dosing, pH increases up to 8.7 in the aerobic effluent, and nitrification potential deteriorates to 20–60% of the maximum activity.

Decreasing the pH in the aerobic zone therefore is chosen as an optimisation objective for the assessment of ozonation. When acidic solution was overdosed due to problems with the pH-probe, which may happen during normal operation, ammonium oxidation potential was reduced to 40% at a pH as close to neutral as 6.5. This may be explained by a combination of the low NH_3 concentration, the actual substrate for the nitrifiers, at this pH and by the possibly even lower pH within the biofilm caused by the diffusion-limited transfer of H^+ produced by the nitrifiers (Siegrist and Gujer, 1987). Effluent NH_4-N concentration values, plotted against the specific nitrification rate (Figure 2.69), increase as nitrification deteriorates that can mostly be attributed to the high pH.

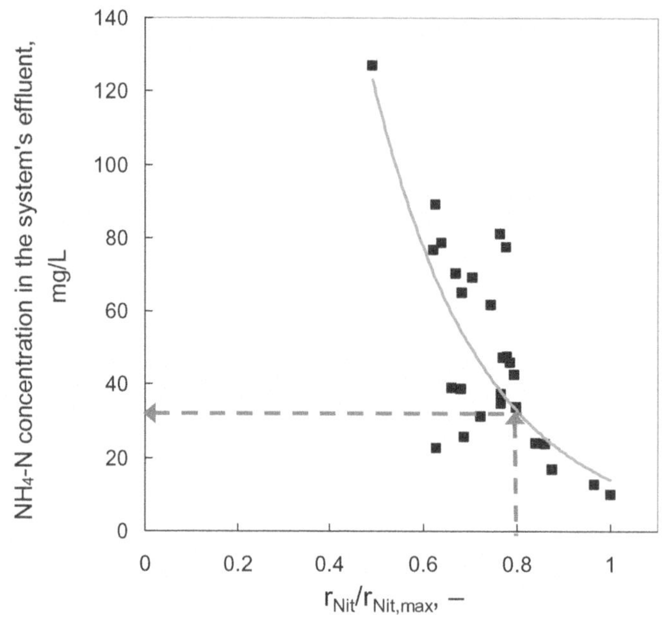

Figure 2.69 Values of the effluent ammonium-nitrogen concentration plotted as a function of the biological nitrification activity

Nevertheless, under optimum pH conditions, effluent NH_4-N concentration is detected up to 30 mg/L N that warrant further optimisation of the nitrification capacity in the system. In Figure 2.70, we summarise shortages of the biological leachate treatment and the potential benefits that can be achieved by coupling biological landfill leachate treatment with ozonation.

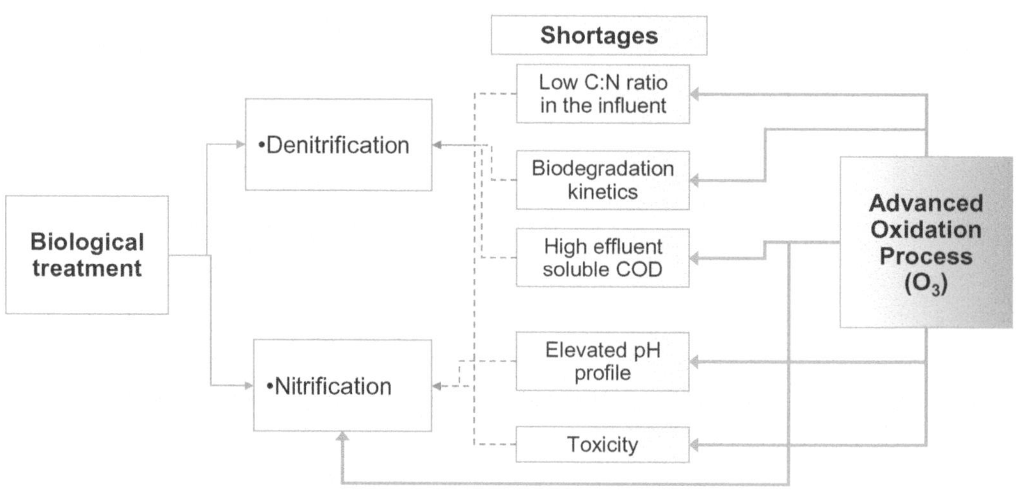

Figure 2.70 Optimisation objectives of coupling biological landfill leachate treatment with ozonation

2.5.4.3.2 Ozone transfer and consumption in batch experiments

Biologically treated clarified effluent samples were used in the experiments with ozone that was supplied directly in the contactor through a stone diffuser (see Figure 2.58). Results shown in Figure 2.64 were obtained using two different ozone concentrations (83 mg/l and 210 mg/l) in the influent gas to the contactor.

The ozone concentration in the off-gas collected at the top of the contactor increased exponentially during the initial stage of each test (see Figure 2.71a), and gradually ceased to increase towards the end of the test, and reached a level somewhat below the influent ozone concentration. Hence, more ozone was added than absorbed by the system (Figure 2.71b). Ozone adsorption occurred with maximum 0.5 g O_3 added to a litre of sample, and its value saturated at approx. 1.45 g/L O_3. A large fraction of the non-biodegradable organic matter in leachate is humic and fulvic acids (Kurniawan *et al.*, 2006). The chromophoric or aromatic nature of these compounds contributes considerably to the brownish colour of the leachate, a factor that can deteriorate the efficiency of UV treatment in combination with ozonation. As shown in Figure 2.72, ozonation lead to a decolourisation of the leachate and decreased UV absorbance that is caused, in part, by destruction of chromophoric structures within the humic matter (Kurniawan *et al.*, 2006).

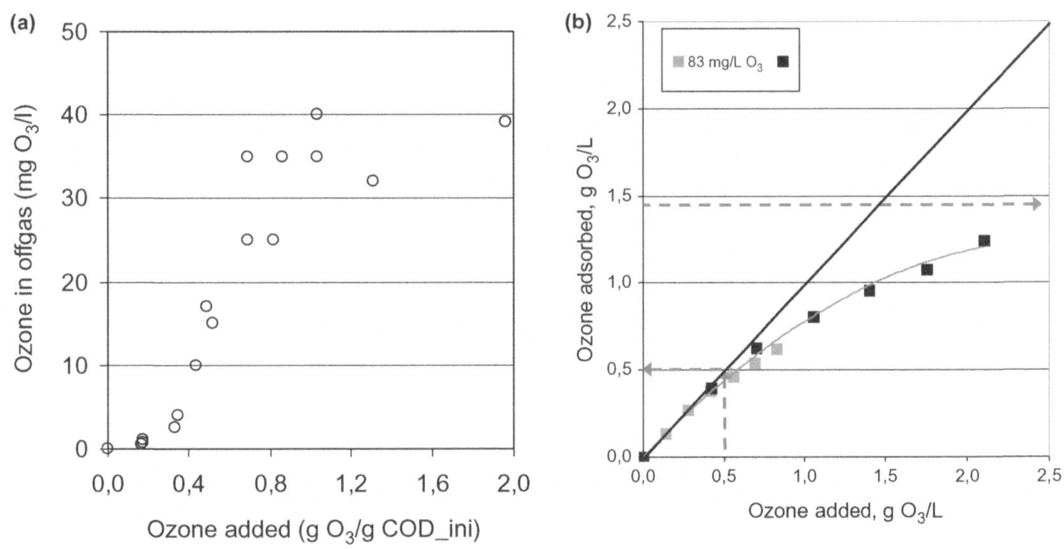

Figure 2.71 Values of ozone concentration in the off-gas from the ozone contactor plotted as a function of the ozone added specific to the initial COD concentration (a); and values of the amount of ozone absorbed plotted as a function of the ozone added to a litre of sample (b)

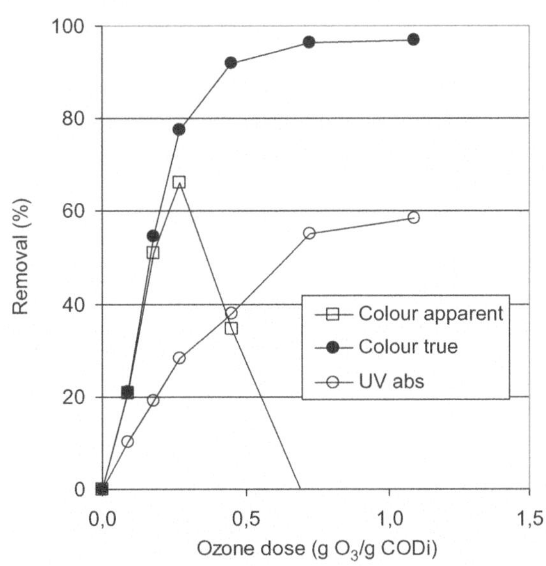

Figure 2.72 Apparent and true colour removal and removal of UV absorbance during ozonation with 83 mg/l ozone in the influent gas flow

In accordance with visual observations, the sudden decline in the removal of apparent colour (non-filtered samples) with increasing ozonation, shown in Figure 2.72, indicates the development of precipitates during the ozonation. This outcome suggests that, in full-scale systems, solids-liquid phase separation, installed downstream to ozone contact zones should be considered as an optimisation means, particularly, to optimise ozonation combined with UV treatment.

Ozonation of organic matter, such as humic substances, leads to the formation of low molecular weight acids as acetic acid or oxalic acid, and most of these acids are easily biodegradable (Karrer *et al.*, 1997; Derco *et al.*, 2002). In Figure 2.73, total and soluble COD and organic acid concentration (OAc) and pH profiles are plotted as a function of time elapsed. As shown in Figure 2.73a, following a short initial increase, soluble COD and OAc concentration values were decreased as a function of time elapsed. In the second experiment using higher ozone dosing rate, accumulation of initial soluble COD and OAc were not detected, and convergence of the total and soluble COD concentrations are shown to occur faster, thereby reducing the concentration of particulate organics to a minimum value at the end of the experiment. If pH is higher than 9, ozone reacts with bicarbonate ions that are thus converted to carbonate, and it slows down the kinetic rate of the oxidation reaction. Thus bicarbonate is also considered as a scavenger molecule to •OH radical formation.

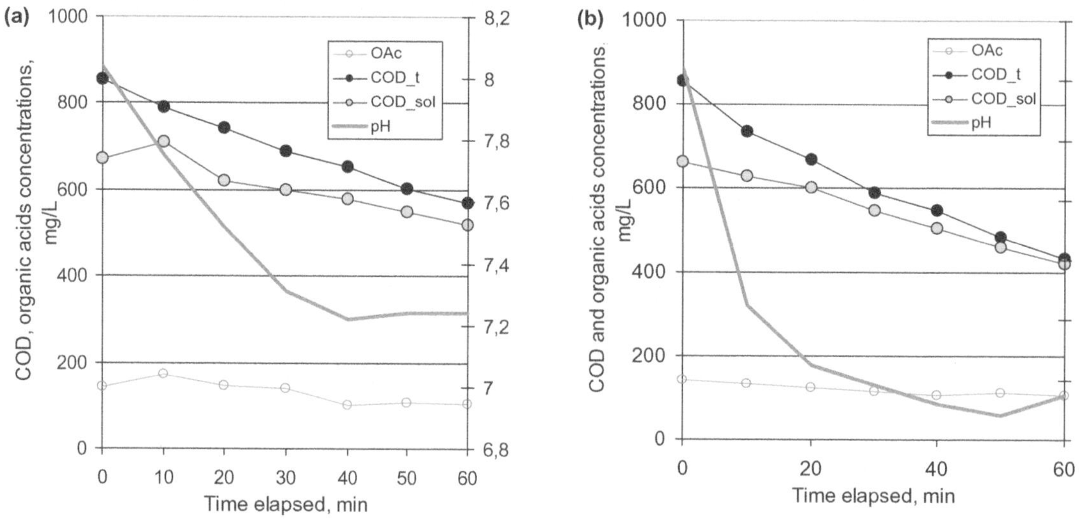

Figure 2.73 Values of the total and soluble COD and acidic-acid concentrations and pH measured using 83 mg/L (a) and 210 mg/L (b) O_3 concentrations in the gas flow produced by the ozone generator

When exposed to pH ranging from 8 to 9, in the presence of OH− ions, ozone rapidly decomposes into more reactive •OH radicals. In basic conditions, both OH− and •OH act as the catalysts for the ozone decomposition process into intermediate compounds that are also highly reactive. In acidic pH, ozone undergoes selective electrophilic attack on the specific part of the organic compounds that have C=C bonds and/or aromatic rings and decomposes them into carboxylic acid and aldehydes as the end products (Kurniawan *et al.*, 2006). Depending upon ozonation intensity, values of pH are shown to initially decrease from 8.1 to 6.9–7.2, and subsequently remained approximately constant until the end of the two experiments. Higher ozonation intensity correlates well with faster depression of pH, and causes lower pH value at the end of the experiment shown in Figure 2.74.

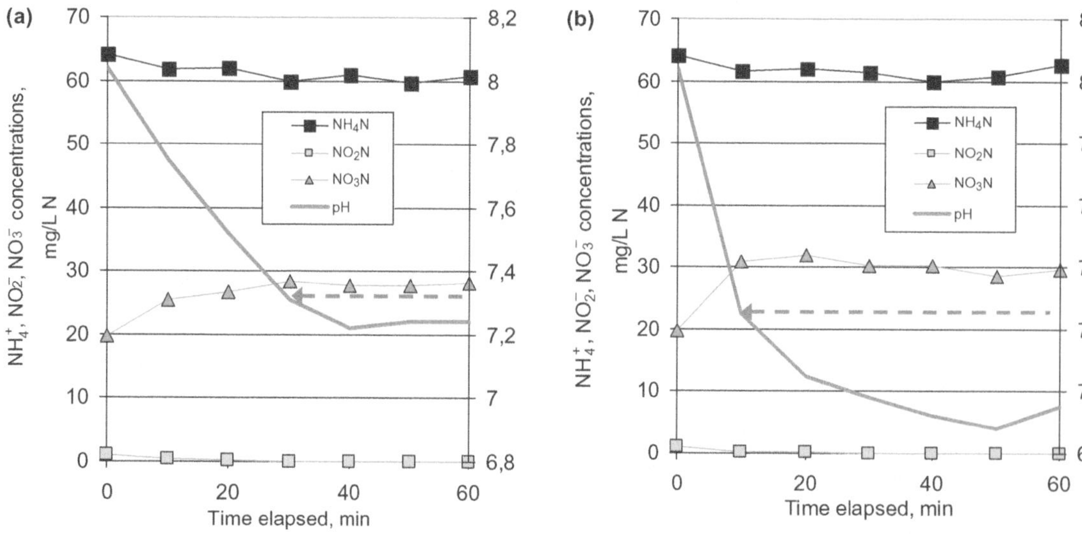

Figure 2.74 Values of the ammonia, nitrate-nitrite-N concentrations and pH values measured using 83 mg/L (a) and 210 mg/L (b) O_3 concentrations in the gas flow produced by the ozone generator

Values of the ammonium, nitrate, nitrite and pH are plotted as a function of the time elapsed in Figure 2.74, and results show increasing nitrate and slightly decreasing ammonium concentration values. Ozonation can potentially catalyse chemical nitrification (Haag *et al.*, 1984), and the reaction for chemical nitrification is

$$NH_3 + 4O_3 \longrightarrow NO_3^- + 4O_2 + H_2O + H^+, \tag{4}$$

where the equilibrium concentration of ammonia can be expressed as

$$NH_3 + H^+ \longrightarrow NH_4^+, \tag{5}$$

and the reaction can be characterised with a pK_a value of 9.23 in distilled water at 20°C. Nitrite is oxidised by ozone according to

$$NO_2^- + O_3 \longrightarrow NO_3^- + O_2, \tag{6}$$

and, as shown in Figure 2.74, it does not accumulate in the liquid phase. At pH levels below the pKa = 9.23, the ammonia oxidation rate can be considered relatively slow (Langlais *et al.*, 1991), which is the case for biologically treated effluents used in this study. With regard to the equilibrium ammonia concentration in the leachate sample used in the experiments, an approx. 3.7 mg/L N value is calculated. With 83 and 210 mg/L O_3 in the gas phase, pH drops to 7.2 and 6.9 in the liquid phase, and ammonia concentration is calculated to be 0.4 and 0.2 mg/L at the end of the experiments, respectively.

The insignificant alterations in the ammonium concentration values, plotted in Figure 2.74, suggest ammonium production through the oxidation of organic matter catalysed by ozonation. With the two ozonation intensities, nitrate concentration values are shown to increase with about 8–10 mg/L N while pH decreases, and chemical nitrification rate considerably deteriorates below pH 7.3 (equilibrium NH_3 concentration around 0.48 mg/L N), beyond which pH further decreases and plateaus in the neutral range (6.9–7.2). Beyond pH = 7.3, ammonium concentration values does not seem to change considerably, that can be explained by the fact that, the further the pH shifts toward acidic conditions, ozone selectively attacks C=C double bonds over forming radicals that can potentially deteriorate ammonification from organic matter.

Values of the NO_3^-:NH_4^+ molar ratio and pH plotted against the experimental time elapsed show that, independently from the two different pH levels measured, beyond pH = 7.3, values of the molar ratio converge to approx. 0.47. The above findings suggest that the chemical nitrification process can be optimised by means of controlling ozonation via pH monitoring. From an operation point of view, further benefits can be gained by neutralising the pH of the ozonated solution that can minimise the acid dosing demand in the system.

Effects of ozonation on the biodegradability of leachate were assessed by measuring the biological oxygen demand during the course of seven days (BOD_7). Results obtained in BOD tests are shown in Figure 2.75.

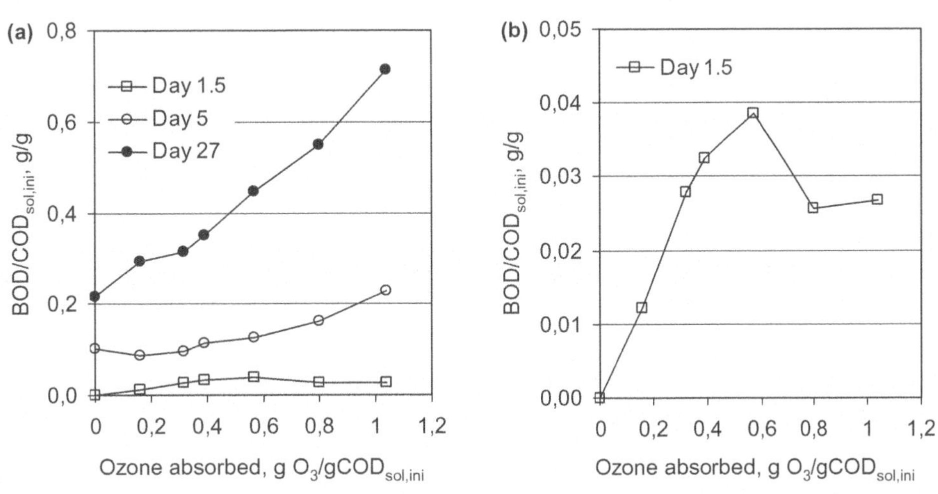

Figure 2.75 Values of the BOD/COD$_{ini}$ ratio plotted against the amount of ozone absorbed in the liquid phase after different incubation times (a) and an excerpt of Figure 2.75a shown in (b)

Values of the BOD to the initial soluble COD content in each sample are plotted as a function of the ozone dosages at three different incubation times of the BOD measurements (day 1.5, day 5 and day 27). Ozone dosing was calculated as the amount of ozone absorbed per g COD in the initial leachate solution.

Results obtained after 1.5 day incubation illustrate the effect of ozonation on the ratio of readily biodegradable carbon production, and indicate a minor but apparent optimum ozonation dosage at ca. 0.6 g absorbed ozone per g COD_{ini}. According to data obtained after 5 days, this amount of ozone can produce biodegradable COD up to 10–13% of the initial concentration.

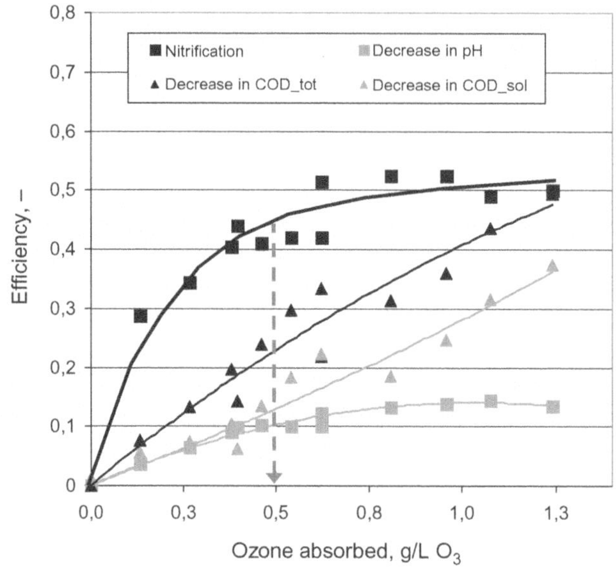

Figure 2.76 Values of efficiency plotted as a function of the ozone absorbed. Dash line indicates optimum value assessed in this study

From the standpoint of biological treatment, Figure 2.76 suggest an optimum ozone dosage of approx. 0.5 mg/L O_3 that can effectively decrease soluble COD concentration and pH in the liquid, and increase biodegradable COD yield and chemical nitrification rate. With regard to ozone adsorbed versus the ozone added to our systems (see Figure 2.71), higher ozone dosages appear to be less efficient, and thus 0.5 mg/L O_3 is an optimum from the perspective of operational costs.

2.5.4.4 Cost assessment

In order to assess additional costs related to ozonation, we used specific values of 1 and 2 euros per kg O_3 produced (Ried, 2008). Effluent soluble COD concentration values were measured in the laboratory-scale continuous-flow biological system that was fed with leachate taken from a landfill during the course of six months. In Figure 2.77, values of the specific cost are plotted as a function of the soluble COD concentrations measured in the effluent of the biological system. Operation costs are obtained between 0.34–1.16 euros per m^3 leachate treated, and average cost values with 1 and 2 euros per kg O_3 were obtained to be 0.36 and 0.73 €/m^3, respectively using an average soluble effluent COD value of 911 mg/L. Operation costs obtained for ozonation should be contrasted against costs related to external substrate dosing (~0.4 €/kg COD as methanol), acid dosing (>0.03 €/m^3 leachate treated) and to increased aeration capacity for nitrification.

2.5.5 Combining biological wastewater treatment with controlled ozonation – the BIOZO system

In the BIOZO concept, we propose integrating the controlled ozonation process in the secondary treatment step. The benefits of using this technology over ozonation installed in the tertiary step can be sought in the recirculation streams in the secondary step (nitrate and sludge recirculation) – more effective oxidation process – and in the effective biodegradation of residual XOCs and their oxidation products derived from the ozonation step. For integration of AOP in the secondary treatment step, two possible positions were identified. Position 1 was selected as a possible location for AOP (Figure 2.57) because of (i) the relatively low alkalinity and ammonium/ammonia concentration, i.e. lower scavenging effect; (ii) the need to increase the biodegradation substrate concentration and thus the C:N ratio in the anoxic zone. We note that this system configuration is the elevated dissolved oxygen (DO) concentration in the influent of the anoxic zone (on average 4–5 mg/L) that can effectively cause metabolic inhibition of denitrication processes in

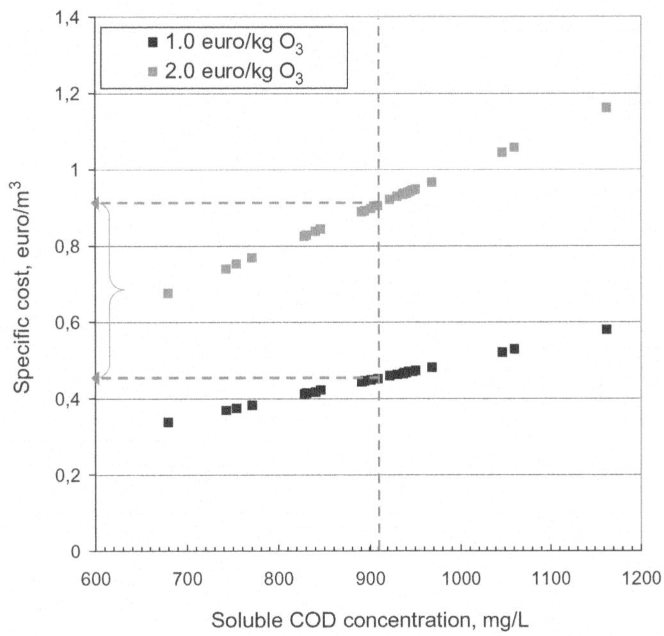

Figure 2.77 Values of ozonation cost specific to the volume of leachate treated plotted against the soluble COD concentration in the biological system's effluent stream (Dashed lines indicate treatment costs with an average soluble effluent COD of 911 mg/L)

the staged pre-anoxic zone, i.e. less COD for nitrogen removal. However, that the DO influx cannot cause significant kinetic inhibition of denitrification as demonstrated by Plósz (2007). Controlled ozonation was additionally assessed in Position 2, because of (i) the more effective XOC removal in the AOP unit situated in the main line and not in the recirculation stream – five times recirculation in the AOP instead of four; (ii) its potential to reduce toxicity in the influent of the aerobic zone, and thus the inhibitory impacts on the autotrophic nitrifiers, thereby optimising nitrogen removal in the system; (iii) the potential of nitrifying bacteria to degrade XOCs; (iv) the need to decrease the pH and alkalinity in the aerobic zone; (v) the high oxygen concentration in the effluent of the AOP.

2.5.5.1 COD removal

The biodegradable fraction of the soluble COD was approx. 50%, resulting in an effluent soluble COD concentration around 600–1000 mg/L. Landfill leachate used in the lab-scale experiment as influent was collected from the same pumping station, where the pilot-scale system was deployed. Influent and effluent solids COD concentration values measured in the lab- and pilot-scale systems, employing only biological treatment; show good agreement. Data plotted in Figure 2.78 indicate significant benefits that can be derived from ozonation combined with biological treatment to decreased non-biodegradable soluble COD in the treated effluent. The experimental assessment of combined operation suggest that Position 2, i.e. ozonation implemented upstream the aerobic zone (average ozone dose: 0.51 $mgO_3/mgCOD_{Soluble}$), can significantly increase the treatment efficiency as compared to operation using the AOP in Position 1 (average ozone dose: 0.48 $mgO_3/mgCOD_{Soluble}$). Using controlled ozonation in Pos. 2 can ensure effluent soluble COD concentrations between 500 and 800 mg/L even with total influent COD concentrations above 4000 mg/L.

2.5.5.2 PAHs and nitrogen removal

In Figure 2.79a, efficiency values of the total PAH removal from the aqueous phase plotted as a function of the influent PAH mass-load suggest an almost complete micro-pollutant elimination in the BIOZO system (~98%). Mass-balance calculations done around the three pre-anoxic reactor stages reveal approx. 83% of the influent PAH removed via biodegradation and possible abiotic reactions and through sorption onto the biofilm and subsequent solids removal in the PST. Aerobic PAH removal capacity obtained is approx. 13% that was assessed employing ozonation installed in Pos. 1. According to Figure 2.79a, data obtained using ozonation upstream to the aerobic zone (Pos. 2) show an approx. 12% PAH removal capacity, i.e. eliminating the bulk of the PAHs from the influent of the aerobic zone, thereby further decreasing the toxicity on autotrophic nitrifying bacteria.

In Figure 2.79b, specific nitrogen removal efficiency values plotted as a function of the total influent nitrogen concentration suggest that combined operation can be superior over sole biological treatment. Data obtained with ozonation installed in Pos. 2 (Figure 2.79b) suggest some additional benefits, possibly, as a result of decreased toxicity on nitrifiers, and due to the reduced oxygen input into the pre-anoxic zone. Average effluent PAH concentration, presented in Figure 2.80, suggest values <40 ng/L for individual compounds in the system's effluent.

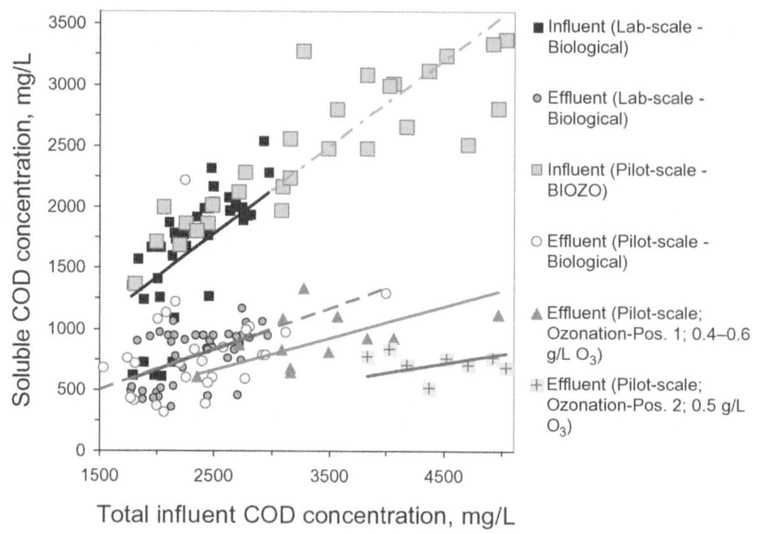

Figure 2.78 Values of the soluble COD concentration plotted as a function of the total influent COD obtained in the laboratory- and pilot-scale experiments

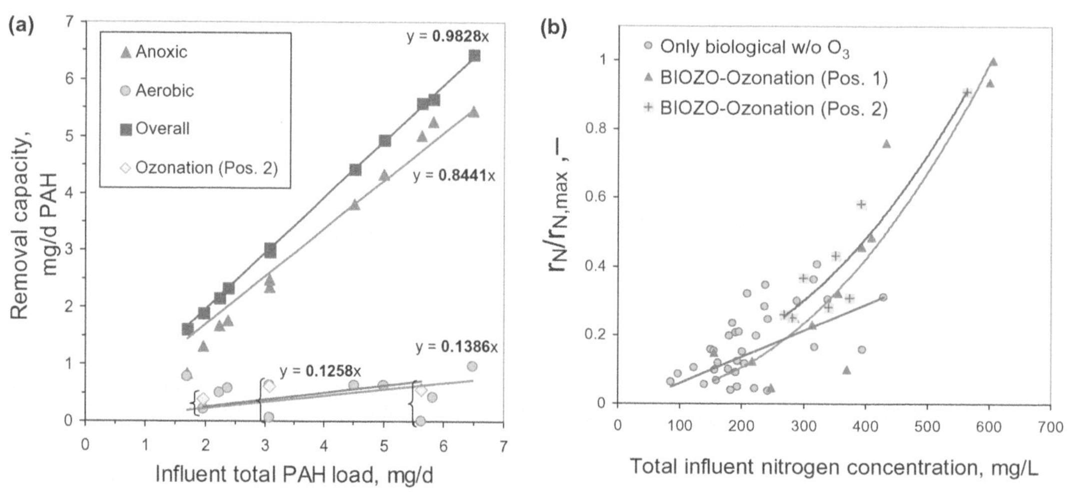

Figure 2.79 Values of the PAH removal capacity plotted as a function of the influent total daily PAH mass-load with left braces showing the effects of ozonation in Pos. 2 (a); values of the specific nitrogen removal efficiency (rN/rN,max) in the BIOZO system plotted as a function of the influent total nitrogen concentration (b)

In Figure 2.80, results obtained in the effluent stream of the BIOZO system show PAHs in significantly lower concentrations than the conservative effluent criteria suggested by the Norwegian Pollution Control Authority for PAH discharge for municipal treatment works in Norway. We note that effluent criteria values are assessed by assuming that a given discharge shall not deteriorate the water quality, prevailing in the recipient water body. The strictest criteria for each PAHs can be achieved in most cases via combined biological-advanced oxidation treatment. However, in the cases of chrysene, indeno[1,2,3-cd]pyrene, dibenz[a,c+a,h]anthracene and benzo[ghi]perylene the discharge limit can only be kept using ozonation installed in Pos. 2.

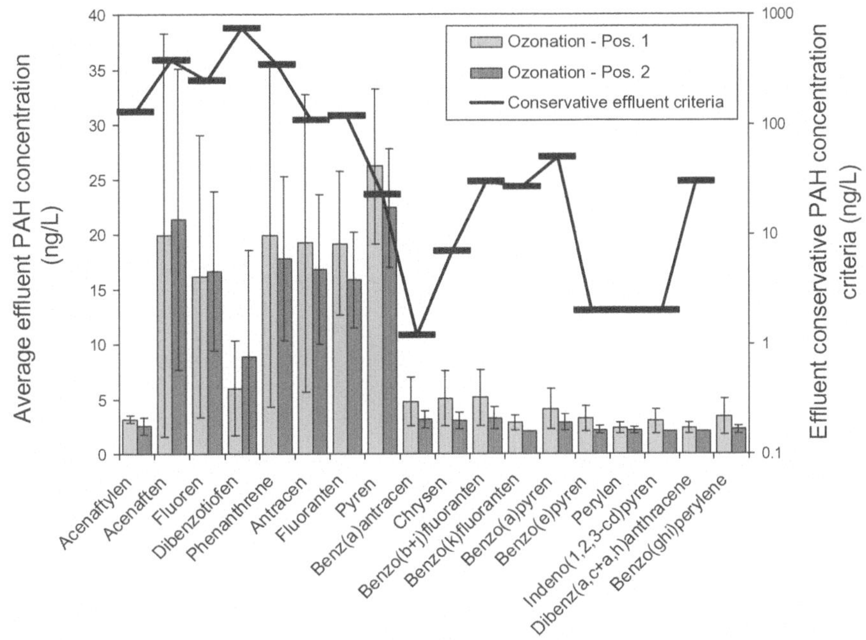

Figure 2.80 Values of the average effluent PAH concentrations with standard deviation and suggested effluent criteria and plotted as a function of the PAH compounds detected. Values of the conservative effluent criteria are assessed assuming that a given discharge shall not deteriorate the water quality, prevailing in the recipient water body

2.5.5.3 *Toxicity in the raw and treated landfill leachate*

Chitosan has been shown to have strong adsorbtion properties for a number of heavy metals. The removal ratios of a number of selected heavy metals using a 50 mg/l dosage of chitosan at pH 4.5, pH 5.0 and pH 6.8 (no pH adjustment) is shown in Figure 2.81a. The Toxtrak (HACH) measurements, shown in Figure 2.82b, indicate that the leachate itself was somewhat, but not to a very high extent, toxic, and the toxic response of the bacteria used in the test (E. coli) was an increased activity (negative inhibition). At a 50 mg/l chitosan dosage the treated leachate still produced a toxic response in the test bacteria.

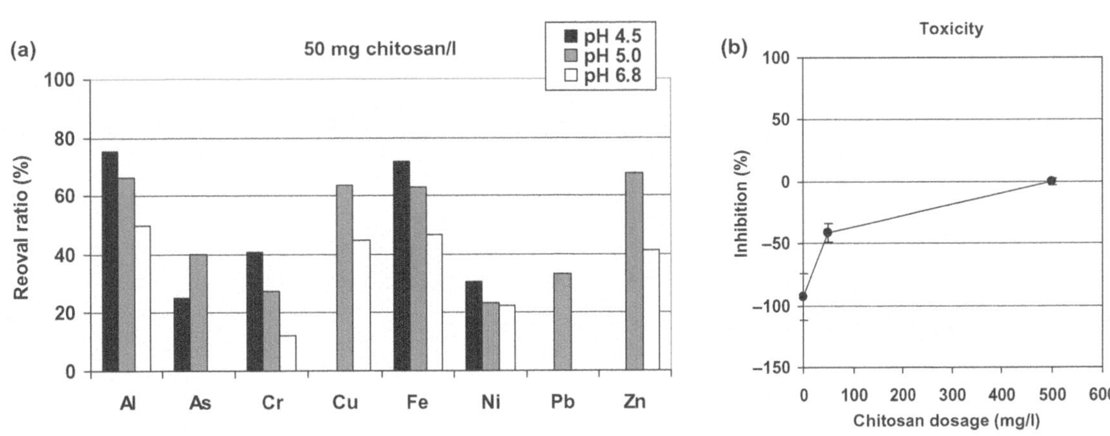

Figure 2.81 Removal ratios of heavy metals (a) with chitosan and toxicity (b)

In the system's influent, the particulate organics fraction was identified as a source of biodegradable organic carbon, and the pre-treatment of landfill leachate using chitosan-flocculation was omitted. In the pilot-scale experiment, samples were taken from the raw influent and from the effluent of the BIOZO system, and the toxicity reduction was assessed using the growth inhibition test ISO8692 (Figure 2.82). Results obtained suggest a moderately toxic influent, i.e. 50% decrease in the microbial growth is obtained at a dilution rate of approximately 14% (EC50 = 14%) that is in close agreement with the results obtained with the ToxTrak tests. An approximately three-fold decrease in the toxicity (EC50 = 35%) can be achieved using the BIOZO system, and the discharge of the treated effluent into sewer networks shall thus not pose significant risk on the biological treatment processes, in the central sewage treatment works.

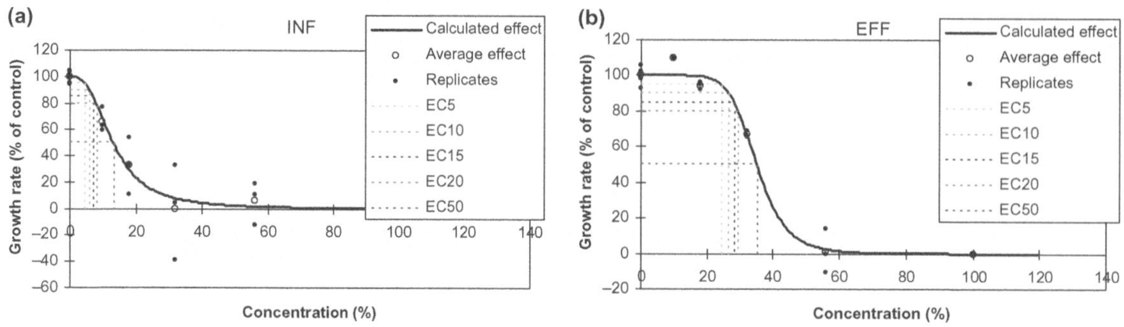

Figure 2.82 Toxicity assessment using the reduction of growth rate in *Pseudokirchneriella subcapitata* (ISO 8692)

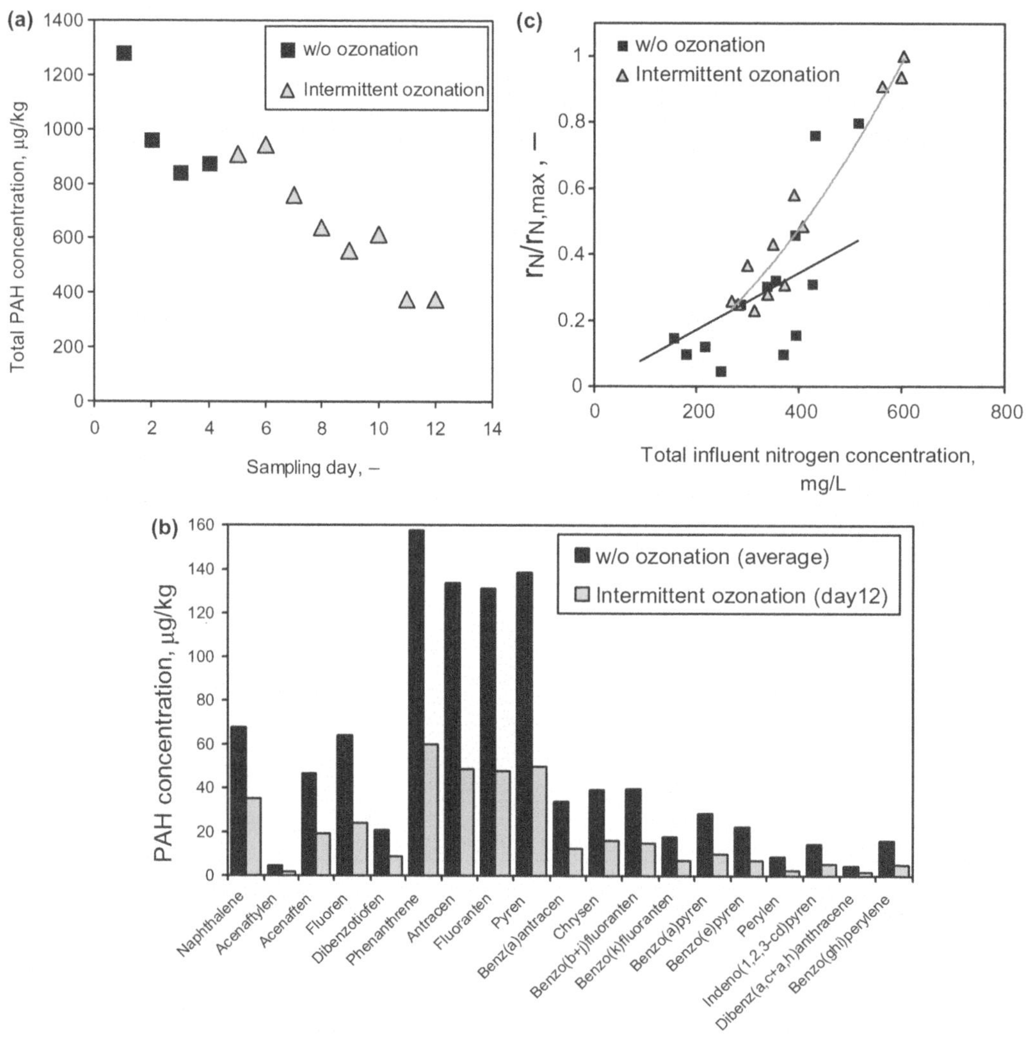

Figure 2.83 Values of the total PAH solids concentration (a) and the individual PAH concentration obtained with and without intermittent sludge ozonation (b); values of the specific nitrogen removal efficiency (rN/rN,max) in the BIOZO system plotted as a function of the influent total nitrogen concentration (c)

2.5.5.4 PAHs removal from the solid phase and nitrogen removal

Data plotted in Figure 2.59a show high variation in the influent total PAH concentration without any decreasing tendency. The PAH content of the sludge wasted through the underflow of the PST is a factor that can influence the overall removal in of the sludge treatment unit (STU), a concentration not measured in the campaign. We hypothesise that the total PAH content of solids in the influent of the STU is a relatively unvarying parameter. The PAHs sorbed and bound on the biofilm have a relatively long residence time in the pre-anoxic zone before being removed via biofilm detachment and solid-liquid separation in the PST. In the STU, see Figure 2.57, fermentation of and methanogenesis from solids continuously conveyed from the PST is possibly inhibited by the relatively high influent nitrate concentration (average: 15 mg/L N). As a results of biological sludge treatment, values of the total PAH concentration measured in the solids wasted from the STU are shown in Figure 2.83a. Additionally, we assess the potential of intermittent sludge ozonation (ISO) as a means to decrease the PAH content of sludge wasted using 0.05 g O_3/g TSS per day (duration: approx. 20 min/d). Data shown in Figure 2.83a, suggest that ISO can effectively decrease the total PAH solids concentration to an average 38% of that obtained in untreated sludge, within nine operation days.

In Figure 2.83b, data obtained on the individual PAH solids concentrations suggest >50% increase in the removal using intermittent sludge ozonation (data obtained on sampling day 12) as compared to only biological sludge treatment (average value calculated using data obtained on sampling days 1–4). Total PAH solids concentration is decreased predominantly as a result of effective transformation of Group A compounds. We note that, since the intermittent ozonation is not detrimental to biological activities in the sludge treatment unit, the biotransformation processes not explicitly studied, can further improve the quality of sludge wasted (SRT = 40 days; specific solid wastage: ~0.25–0.3 g TSS/L leachate treated). To increase the biodegradable COD available for denitrification using the products of the sludge treatment processes, the PST underflow is recirculated to the pre-anoxic zone via the sludge treatment unit. The impact of biodegradable substrates conveyed from the sludge treatment unit to the pre-anoxic zone on the overall specific nitrogen removal is illustrated in Figure 2.83c.

2.5.6 Conclusions

The results summarised in this report are concluded in the following remarks:

- Biological landfill leachate treatment can be an effective, but insufficient means to remove the bulk of COD and nitrogen from landfill leachate.
- The STRAKINES biofilm technology developed can offer the opportunity to enhance the XOC removal from both municipal and industrial wastewater through the optimisation of microbial acclimatisation and increased reaction rates, achieved in the staged system.
- Controlled ozonation, used as advanced oxidation process, can be an effective means to optimise biological sewage treatment, in terms of COD, N and PAHs removal.
- To optimise the XOCs removal, the STRAKINES biological sewage treatment process can effectively decrease the chemical dosing requirements of the chemical treatment process integrated into the secondary treatment step.
- In the combined biofilm-ozonation system (BIOZO) developed (Figure 2.84), ozonation installed between the pre-anoxic and aerobic zones is a solution that is superior over installing the AOP in the nitrate-recirculation line, in terms of COD, PAH and nitrogen removal efficiency.
- The BIOZO system can provide with an effective biological barrier (downstream nitrification zone and recirculation via the nitrate and sludge streams to the pre-anoxic zone) for the removal of oxidation by-products derived from the AOP unit.
- Intermittent sludge ozonation is demonstrated as a robust means to decrease the PAH solids concentration in the sludge produced and to improve nitrogen removal in the system.

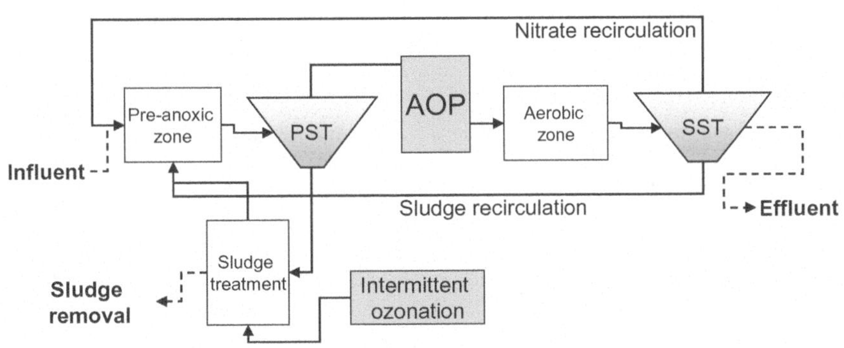

Figure 2.84 The BIOZO system – Biological treatment combined with AOP between the pre-anoxic and aerobic zones

REFERENCES

Ajona J. A. and Vidal A. (2000). The use of CPC collectors for detoxification of contaminated water: design, construction and preliminary results. *Solar Energy*, **68**, 109–120.

Amann R. I., Ludwig W. and Schleifer K. H. (1995). Phylogenetic identification and in situ detection of individual microbial cells without cultivation. *Microbiol Rev.*, **59**,143–169.

Amat A. M., Arques A., García-Ripoll A., Santos-Juanes L., Vicente R., Oller I., Maldonado M. I. and Malato S. (2009). A reliable monitoring of the biocompatibility of an effluent along an oxidative pre-treatment by sequential bioassays and chemical analyses. *Wat Res.*, **43**, 784–792.

Andersen H. and Ledin A. (2009). Treatment of estrogenic chemicals in biologically treated sewage by chlorine dioxide oxidation. International Conference on Xenobiotics in the Urban Water Cycle–XENOWAC, 11–13 March, 2009, Paphos, Cyprus.

Andreozzi R., Raffele M. and Nicklas P. (2003). Pharmaceuticals in STP effluents and solar photodegradation in aquatic environment. *Chemosphere*, **50**, 1319–1330.

Aparicio M. A., Eiroa M., Kennes C. and Veiga M. C. (2007). Combined post-ozonation and biological treatment of recalcitrant wastewater from a resin-producing factory. *Journal of Hazardous Materials*, **143**(1–2), 285–290.

APHA (1995). 19th edn, Standard Methods for the Examination of Water and Wastewater, Water Environment Federation, American Public Health Association, Washington DC, USA.

APHA (1999). 20th edn, Standard Methods for the Examination of Water and Wastewater, American Public Health Association (APHA), American Water Works Association (AWWA) and Water Environment Federation (WEF), Washington, DC.

Augugliaro V., Litter M., Palmisano L. and Soria J. (2006). The combination of heterogeneous photocatalysis with chemical and physical operations: A tool for improving the photoprocess performance. *J. Photochem. Photobiol.*, **7**, 127–144.

Bacardit J., Stötzner J., Chamarro E. and Esplugas S. (2007). Effect of Salinity on the Photo-Fenton Process. *Ind. Eng. Chem. Res.*, **46**, 7615–7619.

Baun A., Ledin A., Reitzel L. A., Bjerg P. L. and Christensen T. H. (2004). Xenobiotic organic compounds in leachates from ten Danish MSW landfills - chemical analysis and toxicity tests. *Water Res.*, **38**(18), 3845–3858.

Baun A., Reitzel L. A., Ledin A., Bjerg P. L. and Christensen T. H. (2003). Natural attenuation of xenobiotic organic compounds in a landfill leachate plume (Vejen, Denmark). *J Contam Hydrol.*, **65**(3–4), 269–291.

Bellas J., Saco-Álvarez L., Nieto Ó. and Beiras R. (2008). Ecotoxicological evaluation of polycyclic aromatic hydrocarbons using marine invertebrate embryo-larval bioassays. *Marine Poll. Bull.*, **57**, 439–502.

Blanco J. and Malato S. (2003). Solar Detoxification. UNESCO Publishing, France.

Blanco J., Malato S., Fernández P., Vidal A., Morales A., Trincado P., de Oliveira J. C., Minero C., Musci M., Casalle C., Brunotte M., Tratzky S., Dischinger N., Funken K.-H., Sattler C., Vincent M., Collares-Pereira M., Mendes J. F. and Rangel C. M. (2000). Compound parabolic concentrator technology development to commercial solar detoxification applications. *Sol. Energy*, **67**, 317–330.

Blanco-Galvez J., Fernández-Ibáñez P. and Malato-Rodríguez S. (2007). Solar photocatalytic detoxification and disinfection of water: recent overview. *J. Solar Energy Eng.*, **129**, 4–15.

Boller M. and Gujer W. (1986). Nitrification in tertiary trickling filters followed by deep-bed filters. *Water Res.*, **20**(11), 1363–1373.

Borer P., Sulzberger B., Hug S. J., Kraemer S. M. and Kretzschmar R. (2009). Photoreductive dissolution of iron(III) (Hydr)oxides in the absence and presence of organic ligands: Experimental studies and kinetic modeling. *Environmental Science and Technology*, **43**, 1864–1870.

Bozzi A., Yuranova T., Mielczarski J., Lopez A. and Kiwi J. (2004). Abatement of oxalates catalyzed by Fe-silica structured surfaces via cyclic carboxylate intermediates in photo-Fenton reactions. *Chemical Communications*, **19**, 2202–2203.

Chan C., Ko -M. T. -M and Hiraoka H. (1996). Polymer surface modification by plasmas and photons. *Surface Science Reports*, **24**, 1–54.

Chen F., He J., Zhao J. and Yu J. C. (2002). Photo-Fenton degradation of malachite green catalyzed by aromatic compounds under visible light irradiation. *New Journal of Chemistry*, **26**, 336–341.

Chen R. and Pignatello J. J. (1997). Role of quinone intermediates as electron shuttles in Fenton and photoassisted Fenton oxidations of aromatic compounds. *Environ. Sci. Technol.*, **31**, 2399–2406.

Chhabra R. S. (2000). NTP technical report on the toxicity studies of benzophenone CAS. No. 119–61–9. Administered in feed to F344/N rats and B6C3F mice. Toxicity Report Series **61**, 1–53, A1–13.

Colina-Márquez J., Machuca-Martínez F. and Li Puma G. (2009). Photocatalytic mineralization of commercial herbicides in a pilot-scale solar CPC reactor: Photoreactor modeling and reaction kinetics constants independent of radiation field. *Environmental Science and Technology*, **43**, 8953–8960.

Comninellis C., Kapalka A., Malato S., Parsons S. A., Poulios I. and Mantzavinos D. (2008). Perspective advanced oxidation processes for water treatment: advances and trends for R&D. *J. Chem. Technol. Biotechnol.*, **83**, 769–776.

Contreras S., Rodriguez M., Momani F. A. F., Al Sans C. and Esplugas S. (2003). Contribution of the ozonation pre-treatment to the biodegradation of aqueous solutions of 2,4–dichlorophenol. *Water Research*, **37**(13), 3164–3171.

Criddle C. S. (1993). The kinetics of cometabolism. *Biotechnol Bioeng.*, **41**(11), 1048–1056.

Cunningham V., Buzby M., Hutchinson T., Mastrocco F., Parke N. and Roden N. (2006). Effects of human pharmaceuticals on aquatic life: nest steps. How do human pharmaceuticals get into the environmental, and what are their effects? *Environ. Sci. Technol.*, **40**, 3457–3462.

De Kreuk M. K. and van Loosdrecht M. C. M. (2006). Formation of aerobic granules with domestic sewage. *J Environ Eng-ASCE*, **132** (6), 694–697.

De Laat J., Truong Le G. and Legube B. (2004). A comparative study of the effects of chloride, sulfate and nitrate ions on the rates of decomposition of H2O2 and organic compounds by Fe(II)/H2O2 and Fe(III)/H2O2. *Chemosphere*, **55**, 715–723.

Deliyanni E. A., Nalbandian L. and Matis K. A. (2006). Adsorptive removal of arsenites by a nanocrystalline hybrid surfactant-akaganeite sorbent. *Journal of Colloid and Interface Science*, **302**, 458–466.

Derco J., Gulyásová A. and Horňák M. (2002). Influence of ozonation on biodegradability of refractory organics in a landfill leachate. *Chem. Pap.*, **56**(1), 41–44.

Dhananjeyan M. R., Kiwi J., Albers P. and Enea O. (2001). Photo-assisted immobilized Fenton degradation up to pH 8 of azo dye orange II mediated by Fe3+/Nafion/glass fibers. *Helvetica Chimica Acta*, **84**, 3433–3445.

EC (1988). Dir 88/303/EEC. OJ L 133 1988. http://ecb.jrc.ec.europa.eu/documents/Testing-Methods/ANNEXV/C09web1988.pdf.

Ema M., Miyawaki E., Harazono A. and Kawashima K. (1997). Developmental toxicity evaluation of phthalic acid, one of the metabolites of phthalic acid esters, in rats. *Toxicol. Lett.*, **93**, 109–115.

Eschenhagen M., Schupplerb M. and Roskea I. (2003). Molecular characterization of the microbial community structure in two activated sludge systems for the advanced treatment of domestic effluents. *Water Res.*, **37**, 3224–3232.

Everett M. L. and Hoflund G. B. (2006). Chemical alteration of poly(vinyl fluoride) Tedlar® induced by exposure to vacuum ultraviolet radiation. *Applied Surface Science*, **252**, 3789–3798.

Faust B. C. and Hoigne J. (1990). Photolysis of Fe(III)-hydroxy complexes as sources of OH radicals in clouds, fog and rain. *Atmos. Environ.*, **24A**, 79–89.

Faust R. A. (1994). Toxicity summary for acenaphthene. Oak Ridge Laboratory. http://risk.lsd.ornl.gov/tox/rap_toxp.shtml,

Feng J., Hu X. and Yue P. L. (2003). A novel laponite clay-based Fe nanocomposite and its photo-catalytic activity in photo-assisted degradation of orange II. *Chemical Engineering Science*, **58**, 679–685.

Fenton H. J. H. (1894). Oxidation of tartaric acid in presence of iron. *J. Chem Soc.*, **65**, 899–910.

Fernández E. and Cárdenas A. M. (1990). The Mechanism of Photohaemolysis by Photoproducts of NXA. *J. Photochemistry and Photobiology B: Biology*, **4**, 329–333.

Flemming H. C., Szewzyk U. and Griebe T. (2000). Biofilms: Investigative Methods and Applications. Technomic publishing, Lancaster, Pennsylvania, USA.

García M. T., Ribosa I., Guindulain T., Sánchez-Leal J. and Vives-Rego J. (2001). Fate and effect of monoalkyl quaternary ammonium surfactants in the aquatic environment. *Environ. Pollut.*, **111**, 169–175.

Gaya U. I. and Abdullah A. H. (2008). Heterogeneous photocatalytic degradation of organic contaminants over titanium dioxide: a review of fundamentals, progress and problems. *J. Photochem. Photobiol. C*, **9**, 1–12.

Geenens D., Bixio B. and Thoeye C. (2000). Combined ozone-activated sludge treatment of landfill leachate. *Wat. Sci. Technol.*, **44**(2–3), 359–365.

Gernjak W., Fuerhacker M., Fernández-Ibáñez P., Blanco J. and Malato S. (2006). Solar photo-Fenton treatment—Process parameters and process control. *Applied Catalysis B: Environmental*, **64**, 121–130.

Gibbs M. M. (1979). A simple method for the rapid determination of iron in natural waters. *Water Research*, **13**, 295–297.

Gogate P. R. and Pandit A. B. (2004a). A review of imperative technologies for wastewater treatment I: oxidation technologies at ambient conditions. *Adv. Environ. Res.*, **8**(3–4), 501–551.

Gogate P. R. and Pandit A. B. (2004b). A review of imperative technologies for wastewater treatment II: hybrid methods. *Adv. Environ. Res.*, **8**(3–4), 553–597.

Golub M. A., Wydeven T. and Cormia R. D. (1989). ESCA study of several fluorocarbon polymers exposed to atomic oxygen in low Earth orbit or within or downstream from a radio-frequency oxygen plasma. *Polymer*, **30**, 1571–1575.

Goswami D. Y. (1995). Engineering of solar photocatalytic detoxification and disinfection processes. In: Advances in Solar Energy, K. W. Böer (ed.), 165–209, Am. Solar En. Soc.

Grady C. P. L. Jr., Daigger G. T. and Lim H. C. (1999). Biological Wastewater Treatment. 2nd edn, Marcel Dekker, New York, USA.

Guhl W. and Steber J. (2006). The value of biodegradation screening test results for predicting the elimination of chemicals' organic carbon in waste water treatment plants. *Chemosphere*, **63**, 9–16.

Guiberteau A., Rodríguez M. I., Martínez M. A., Ortiz J. M. and Galeano T. (2007). Square wave adsorptive stripping voltametric determination of the mixture of NXA and its main metabolite (7–hydroxymethyl nalidixic acid) by multivariate methods and artificial neural network. *Talanta*, **72**, 932–940.

Gültekin I. and Ince N. H. (2007). Synthetic endocrine disruptors in the environment and water remediation oxidation processes. *J. Env. Manag.*, **85**, 816–831.

Haag W. R., Hoigné J. and Bader H. (1984). Improved ammonia oxidation by ozone in the presence of bromide ion during water treatment. *Wat. Res.*, **18**(9), 1125–1128.

Haapea P., Korhonen S. and Tuhkanen T. (2002). Treatment of industrial landfill leachates by chemical and biological methods: ozonation, ozonation+hydrogen peroxide, hydrogen peroxide and biological posttreatment for ozonated water. *Ozone: Sci. Eng.*, **24**(5), 369–378.

Haiss A. and Kümmerer K. (2006). Biodegradability of the X-ray contrast compound diatrizoic acid, identification of aerobic degradation products and effects against sewage sludge micro-organisms. *Chemosphere*, **62**, 294–302.

Hasegawa H., Sato M. and Tsuruta H. (1982). Toxicological studies on di-isopropylnaphthalenes KMC and 1-phenyl-1-xylyl-ethanes SAS. II. Biochemical examination of rats administered KMC and SAS. *Ind. Health*, **20**(4), 283–296.

Heberer T. (2002). Occurrence, fate and removal of pharmaceutical residues in the aquatic environment: a review of recent research data. *Toxicol Lett.*, **131**, 5–17.

Henriques I. D. S. and Love N. G. (2007). The role of extracellular polymeric substances in the toxicity response of activated sludge bacteria to chemical toxins. *Water Res.*, **41**(18), 4177–4185.

Henze M., Grady C. P. L. Jr, Gujer W., Marais G. V. R. and Matsuo T. (1987). A general model for single-sludge wastewater treatment systems. *Water Res.*, **21**(5), 505–515.

Henze M., van Loosdrecht M. C. M., Ekama G. A. and Brdjanovic D. (2008). 1st edn, Biological Wastewater Treatment, IWA Publishing, London, UK.

Hernando M. D., Fernández-Alba A. R., Tauler R. and Barceló D. (2005). Toxicity assays applied to wastewater treatment. *Talanta*, **65**, 358–366.

Hildebrant A., Lacorte S. and Barceló D. (2006). Sampling of water, soil and sediment to trace organic pollutants at a river-basin scale. *Anal. Bioanal. Chem.*, **386**, 1075–1088.

Hirsch R., Ternes T., Haberer K. and Kratz K. L. (1999). Occurrence of antibiotics in aquatic environment. *Sci Total Environ.*, **225**, 109–118.

Hollender J., Zimmermann S. G., Koepke S., Krauss M., McArdell C. S., Ort C., Singer H., von Gunten U. and Siegrist H. (2009). Elimination of organic micropollutants in a municipal wastewater treatment plant upgraded with a full-scale post-ozonation followed by sand filtration. *Environ. Sci Technol.*, **43**(20), 7862–7869.

Horváth R. S. (1972). Cometabolism of the herbicide 2,3,6–trichlorobenzoate by natural microbial populations. *Bull. Environ. Contam Toxicol.*, **7**, 273–276.

Hsu Y.-C., Yang H.-C. and Chen J.-H. (2004). The enhancement of the biodegradability of phenolic solution using preozonation based on high ozone utilization. Chemosphere, **56**(2), 149–158.

Huang C.-P., Chen C.-R., Huang Y.-F., Lu Y.-W. and Huang Y.-H. (2009). Reductive dissolution and oxidative catalysis of an immobilized iron oxide in the presence of catechol and phenol. *Journal of Molecular Catalysis A: Chemical*, **304**, 121–127.

IARC (2009). Agents reviewed by the IARC monographs. Volumes p1–100A.

Ikehata K. and Gamal El-Din M. (2006). Aqueous pesticide degradation by hydrogen peroxide/ultraviolet irradiation and Fenton-type advanced oxidation processes: a review. *J. Environ. Eng. Sci.*, **5**, 81–135.

ILO (1983). International Labour Office. Encyclopedia of Occupational Health and Safety. *Vols. I&II.* International Labour Office, Geneva, Switzerland.

ITII (1988). Toxic and hazardous industrial chemicals safety manual. The International Technical Information Institute, Tokyo, Japan.

Jönsson K., Grunditz C., Dalhammar G. and La Cour Jansen J. (2000). Occurrence of nitrification inhibition in Swedish municipal wastewaters. *Wat. Res.*, **34**(9), 2455–2462.

Joss A., Siegrist H. and Ternes T. A. (2008). Are we about to upgrade wastewater treatment for removing organic micropollutants? *Water Sci Technol.*, **57**(2), 251–255.

Joss A., Zabczynski S., Göbel A., Hoffmann B., Löffler D., McArdell C. S., Ternes T. A., Thomsen A. and Siegrist H. (2006). Biological degradation of pharmaceuticals in municipal wastewater treatment: Proposing a classification scheme. *Water Res.*, **40**(8), 1686–1696.

Kaballo H. P., Zhao Y. and Wilderer P. A. (1995). Elimination of p-chlorophenol in biofilm reactors – a comparative study of continuous flow and sequenced batch operation. *Water Sci Technol.*, **31**(1), 51–60.

Karrer N. J., Ryhiner G. and Heinzle E. (1997). Applicability test of combined biological-chemical treatment of wastewater containing biorefractory compounds. *Wat. Res.*, **31**(5), 1013–1020.

Kavitha V. and Palanivelu K. (2004). The role of ferrous ion in Fenton and photo-Fenton processes for the degradation of phenol. *Chemosphere*, **55**, 1235–1243.

Kim G. G., Kang J. A., Kim J. H., Kim S. J., Lee N. H. and Kim S. J. (2006). Metallization of polymer through a novel surface modification applying a photocatalytic reaction. *Surface and Coating Technology*, **201**, 3761–3766.

Klamerth N., Miranda N., Malato S., Agüera A., Fernández-Alba A. R., Maldonado M. I. and Coronado J. M. (2009). Degradation of emerging contaminants at low concentration in MWTPs effluents with mild solar photo-Fenton and TiO2. *Catalysis Today*, **144**, 124–130.

Klavarioti M., Mantzavinos D. and Kassinos D. (2009). Removal of residual pharmaceuticals from aqueous systems by advanced oxidation processes. *Environ. Int.*, **35**, 402–417.

Kosjek T. and Heath E. (2008). Applications of mass spectrometry to indentifying pharmaceuticals transformation products in water treatment. *TRACs*, **27**, 807–820.

Kuhn R. (2010). Metaproteomic approach to functional MBR biomass characterization. PhD thesis, Università degli Studi di Bari.

Kurisu F., Satoh H., Mino T. and Matsuo T. (2002). Microbial community analysis of thermophilic contact oxidation process by using ribosomal RNA approaches and the quinone profile method. *Water Res.*, **36**, 429–438.

Kurniawan T. A., Lo W. H. and Chan G. Y. S. (2006). Radicals-catalyzed oxidation reactions for degradation of recalcitrant compounds from landfill leachate. *Chem. Eng. J.*, **125**(1), 35–57.

Kylefors K. (2002). Predictions of leaching from municipal solid waste MSW and measures to improve leachate management at landfills. PhD thesis, Luleå University of Technology, Luleå, Sweden.

Langlais B., Reckhow D. A. and Brink D. R. (1991). Ozone In Water Treatment, Application, and Engineering. Lewis Publishers, Chelsea, MI, USA.

Lapertot M. and Pulgarin C. (2006). Biodegradability assessment of several priority hazardous substances: Choice, application and relevance regarding toxicity and bacterial activity. *Chemosphere*, **65**, 682–690.

Lapertot M., Pulgarín C., Fernández-Ibañez P., Maldonado M. I., Pérez-Estrada L., Oller I., Gernjak W. and Malato S. (2006). Enhancing biodegradability of priority substances (pesticides) by solar photo-Fenton. *Water Res.*, **40**, 1086–1094.

Larrea L., Albizuri J., Abad A., Larrea A. and Zalakain G. (2007). Optimizing and modelling nitrogen removal in a new configuration of the moving-bed biofilm reactor process. *Wat. Sci. Tech.*, **55**(8–9), 317–327.

Lefebvre O. and Moletta R. (2006). Treatment of organic pollution in industrial saline wastewater: A literature review. *Water Res.*, **40**, 3671–3682.

Legrini O., Oliveros E. and Braun A. M. (1993). Photochemical Processes for Water. *Treatment. Chem. Rev.*, **93**, 671–698.

Levenspiel O. (1972). Chemical Reaction Engineering. Wiley, New York, USA.

Li H., Yang M., Zhang Y., Yu T. and Kamagata Y. (2006). Nitrification performance and microbial community dynamics in a submerged membrane bioreactor with complete sludge retention. *J Biotechnol.*, **123**, 60–70.

Lobos J., Wisniewski C., Heran M. and Gramsick A. (2005). Effects of starvation conditions on biomass behaviour for minimization of sludge production in membrane bioreactors. *Water Sci. Technol.*, **51**(6–7), 35–44.

Loibner A. P., Szolar O. H. J., Braun R. and Hirmann D. (2004). Toxicity testing of 16 priority polycyclic aromatic hydrocarbons using lumistox®. *Environ. Toxicol. Chem.*, **23**(3), 557–564.

Lopez A., Mascolo G., Tiravanti G. and Passino R. (1997). Degradation of herbicides (ametryn and isoproturon) during water disinfection by means of two oxidants (hypochlorine and chlorine dioxide). *Water Sci. Technol.*, **35**(4), 129–136.

Loy A., Maixner F., Wagner M. and Horn M. (2007). ProbeBase - an online resource for rRNA-targeted oligonucleotide probes: new features. *Nucleic Acids Res.*, **35**, D800–D804.

Lv X., Xu Y., Lv K. and Zhang G. (2005). Photo-assisted degradation of anionic and cationic dyes over iron(III)-loaded resin in the presence of hydrogen peroxide. *Journal of Photochemistry and Photobiology A: Chemistry*, **173**, 121–127.

Malato S., Blanco J., Vidal A., Alarcón D., Maldonado M. I., Cáceres J. and Gernjak W. (2003). Applied studies in solar photocatalytic detoxification: an overview. *Solar Energy*, **75**, 329.

Malato S., Fernández-Ibañez P., Maldonado M. I., Blanco J. and Gernjak W. (2009). Decontamination and disinfection of water by solar photocatalysis: Recent overview and trends. *Catal Today*, **147**, 1–59.

Mantzavinos D. and Psillakis E. (2004). Enhancement of biodegradability of industrial wastewaters by chemical oxidation pre-treatment. *J. Chem. Technol. Biotechnol.*, **79**, 431–454.

Marttinen S. K., Kettunen R. H., Sormunen K. M., Soimasuo R. M. and Rintala J. A. (2002). Screening of physical–chemical methods for removal of organic material, nitrogen and toxicity from low strength landfill leachates. *Chemosphere*, **46**, 851–858.

Mascolo G., Balest L., Cassano D., Laera G., Lopez A., Pollice A. and Salerno C. (2010a). Biodegradability of pharmaceutical industrial wastewater and formation of recalcitrant organic compounds during aerobic biological treatment. *Bioresource Technology*, **101**, 2585–2591.

Mascolo G., Laera G., Pollice A., Cassano D., Pinto A., Salerno C. and Lopez A. (2010b). Effective organics degradation from pharmaceutical wastewater by an integrated process including membrane bioreactor and ozonation. *Chemosphere*, **78**, 1100–1109.

Mascolo G., Lopez A., Bozzi A., Sonnante G. and Tiravanti G. (2002). By-products formation during the ozonation of the reactive dye Uniblu-A. *Ozone Science & Engineering*, **24**(6), 439–446.

Mascolo G., Lopez A., James H. and Fielding M. (2001). By-products formation during disinfection of herbicide (isoproturon) contaminated water. I: ozonation. *Wat. Res.*, **35**(7), 1695–1704.

Mascolo G., Lopez A., Foldenyi R., Passino R. and Tiravanti G. (1995). Prometryne oxidation by sodium hypochlorite in aqueous solution: kinetics and mechanism. *Environ. Sci. Technol.*, **29**, 2987–2991.

Mascolo G., Balest L., Guzzella L., Locaputo V. and Lopez A. (2007). Degradation of carbofuran in groundwater by Uv and UV/H2O2: comparison of by-products formation and their toxicity. Micropol & Ecohazard 2007, 5th IWA Specialized Conference on Assessment and Control of Micropollutants/Hazardous Substances in Water. *Frankfurt (Germany), 17–20 June*, **2007**, 454–455.

Mascolo G., Lopez A., Detomaso A. and Lovecchio G. (2005). Ion chromatography-electrospray mass spectrometry for the identification of low-molecular-weight organic acids during the 2,4–dichlorophenol degradation. *J. Chromatogr. A*, **1067**, 191–196.

Mazille F., Schoettl T. and Pulgarin C. (2009a). Synergistic effect of TiO2 and iron oxide supported on fluorocarbon films. Part 1: Effect of preparation parameters on photocatalytic degradation of organic pollutant at neutral pH. *Applied Catalysis B: Environmental*, **89**, 635–644.

Mazille F., Lopez A. and Pulgarin C. (2009b). Synergistic effect of TiO2 and iron oxide supported on fluorocarbon films. Part 2: Long term stability and influence of reaction parameters on photoactivated degradation of pollutants. *Applied Catalysis B: Environmental*, **90**, 321–329.

Meyer T. and Wania F. (2007). What environmental fate processes have the strongest influence on a completely persistent organic chemical's accumulation in the Arctic? *Atmos. Environ.*, **41**, 2757–2767.

Mezcua M., Malato O., García-Reyes J. F., Molina-Díaz A. and Fernández-Alba A. R. (2009). Accurate-mass databases for comprehensive screening of pesticide residues in food by fast liquid chromatography time-of-flight mass spectrometry. *Anal. Chem.*, **81**, 913–929.

Mittal K. L. (1996). Polymer Surface Modification: Relevance to Adhesion. VSP Utrecht, The Netherland.

Morgan-Sagastume F., Nielsen J. L. and Nielsen P. H. (2008). Substrate-dependent denitrification of abundant probe-defined denitrifying bacteria in activated sludge. *FEMS Microbiol Ecol.*, **66**, 447–461.

Morrochi A., Minuti L., Morozzi G., Pampanella L. and Taticchi A. (1996). Synthesis and mutagenicity of some cyclopenta[c]-phenanthrenes and indeno[c]phenanthrenes. *Carcinogenesis*, **17**(9), 2009–2012.

Muñoz I., Malato S., Rodríguez A. and Doménech X. (2008). Integration of environmental and economic performance of processes. Case study on advanced oxidation processes for wastewater treatment. *J. Adv. Oxid. Technol.*, **11**, 270–275.

Munoz M. J. and Tarazona J. V. (1993). Synergistic effect of two- and four-component combinations of polycyclic aromatic hydrocarbons: phenanthrene, anthracene, naphthaline and acenaphthene on Daphnia magna. *Bull. Environ. Contamin. Toxicol.*, **50**, 363–368.

Muñoz R. and Guieysee B. (2006). Algal-bacterial processes for the treatment of hazardous contaminants: A review. *Water Res.*, **40**, 2799–2815.

Music S., Krehula S., Popović S. and Skoko Ž (2003). Some factors influencing forced hydrolysis of FeCl3 solutions. *Materials Letters*, **57**, 1096–1102.

Nakai N., Tabira Y., Asai D., Yakabe Y., Shimyozu T., Noguchi M., Takatsuki M. and Shimohigashi Y. (1999). Binding characteristics of dialkyl phthalates for the estrogen receptor. *Biochem. Biophys. Res. Comm.*, **254**, 311–314.

Nawrocki S. T., Drake K. D., Watson C. F., Foster G. D. and Maier K. J. (2005). Comparative aquatic toxicology evaluation of 2-thiocyanomethylthiobenzothiazole and selected degradation products using Ceriodaphnia dubia. *Arch. Environ. Contam. Toxicol.*, **48**, 344–350.

Neyens E. and Baeyens J. (2003). A review of classic Fenton's peroxidation as an advanced oxidation technique. *J. Haz. Mater. B*, **98**, 33–50.

Nogueira R. F. P., Oliviera M. C. and Paterlini W. C. (2005). Simple and fast spectrophotometric determination of H2O2 in photo-Fenton reactions using metavanadate. *Talanta*, **66**, 86.

Nor Anuar A., Ujang Z., van Loosdrecht M. C. M. and de Kreuk M. K. (2007). Settling behaviour of aerobic granular sludge. *Water Sci Technol.*, **56**(7), 55–63.

NTP (2005). 11th edn, National Toxicology Program, Report on Carcinogens. Public Health Service, US Department of Health and Human Services, Washington, DC.

Ødegaard H., Rusten B. and Wessman F. (2007). Optimization of nitrogen removal by the use of combined pre- and post-denitrification. Proceedings from the 2007 Nordic Wastewater Conference in Hamar, Norway, 125–134.

Ødegaard H., Rusten B. and Westrum T. (1994). A new moving-bed biofilm reactor–applications and results. *Water Sci Technol.*, **29**, (10–11), 157–165.

OECD (1993). OECD Guidelines for testing of chemicals. Organisation for Economic Cooperation and Development (OECD), **vol. 1**. Paris.

Parra S., Sarria V., Pulgarín C., Malato S. and Peringer P. (2000). Photochemical vs. coupled photochemical-biological flow system for the treatment of two biorecalcitrant herbicides: metobromuron and isoproturon. *Appl. Catal. B: Environ.*, **27**, 153–168.

Parsons S. (2004). Advanced oxidation processes for water and wastewater treatment. IWA publishing, London.

Pavan B., Biondi C., Ferretti M. E., Lunghi L. and Paganetto G. (2001). Phthalic acid mimics 17β-estradiol actions in WISH cells. *Toxicol. Lett.*, **115**, 157–164.

Pignatello J. J., Oliveros E. and MacKay A. (2006). Advanced oxidation processes for organic contaminant destruction based on the fenton reaction and related chemistry. *Critical Reviews in Environmental Science and Technology*, **36**, 1–84.

Pignatello J. J. (1992). Dark and Photoassisted Fe3+-Catalyzed Degradation of Chlorophenoxy Herbicides by Hydrogen Peroxide. *Environ. Sci. Technol.*, **26**, 944–951.

Pitter P. and Chudoba J. (1990). Biodegradability of organic substances in the aquatic environment. CRC Press, Boca Raton, FL, USA, 306.

Plósz B. G. (2007). Optimization of the activated sludge anoxic reactor configuration as a means to control nutrient removal kinetically. *Wat. Res.*, **41**(8), 1763–1773.

Plósz B. G., Langford K. H., Vogelsang C. and Rossetti S. (2010c). Impact of xenobiotics organic micro-pollutants on biological nitrogen removal in a kinetically optimised biofilm system, treating landfill leachate. Manuscript submitted to Chemosphere.

Plósz B. G., Leknes H. and Thomas K. V. (2010a). Impacts of competitive inhibition, parent compound formation and partitioning behaviour on antibiotic micro-pollutants removal in activated sludge. *Env. Sci. Technol.*, **44**(2), 734–742.

Plósz B. G., Langford K. H., Heiaas H. H., Macrae K., Liltved H., Lopez A. and Vogelsang C. (2010b). Occurrence of xenobiotic organic micro-pollutants in landfill leachate and PAHs removal from the liquid and sludge phases in a biofilm system combined with ozonation. *Water Sci Technol.*, **61**(12), 3188–3197.

Plósz B. G., Vogelsang C., Liltved H., Ried A. and Lopez A. (2010d). Ozonation as a means to optimise biological nitrogen removal from landfill leachate. *Ozone Sci. Eng., Accepted – OSE No.* 09–18.

Plósz, B. G., Jobbágy A. and Grady C. P. L. Jr. (2003). Factors influencing deterioration of denitrification by oxygen entering an anoxic reactor through the surface. *Water Res.*, **37**(4), 853–863.

Pollice A., Laera G. and Blonda M. (2004). Biomass growth and activity in a membrane bioreactor with complete sludge retention. *Water Research*, **38**, 1799–1808.

Pollice A., Laera G., Saturno D., Giordano C. and Sandulli R. (2008a). Optimal sludge retention time for a bench scale MBR treating municipal sewage. *Water Sci. Technol.*, **57**(3), 319–322.

Pollice A., Laera G., Saturno D. and Giordano C. (2008b). Effects of sludge retention time on the performance of a membrane bioreactor treating municipal sewage. *J. Membrane Sci.*, **317**, 65–70.

Pulgarín C. and Kiwi J. (1996). Overview on photacatalytic and electrocatalytic pretreatment of industrial non-biodegradable pollutants and pesticides. *Chimia*, **50**, 50–55.

Ried A. (2008). Personal communication with the director of application development at ITT WEDECO GmbH, Herford, Germany.

Russom C. R., Bradbury S. P., Broderius S. J., Hammermeister D. E. and Drummond R. A. (1997). Predicting the toxic action from chemical structure: acute toxicity in the fathead minnow pimephales promelas. *Environ. Toxicol. Chem.*, **16**(5), 948–967.

Sarria V., Parra S., Adler N., Péringer P., Benitez N. and Pulgarin C. (2002). Recent developments in the coupling of photoassisted and aerobic biological processes for the treatment of biorecalcitrant compounds. *Catalysis Today*, **76**, 301–315.

Sarria V., Parra S., Invernizzi M., Péringer P. and Pulgarin C. (2001). Photochemical-biological treatment of a real industrial biorecalcitrant wastewater containing 5-amino-6-methyl-2-benzimidazolone. *Water Science and Technology*, **44**, 93–101.

Schirmer K., Dixon D. G., Greenberg B. M. and Bols N. C. (1998). Ability of 16 priority PAHs to be directly cytotoxic to a cell line from rainbow trout gill. *Toxicology*, **127**, 129–141.

Schühle K., Gescher J., Feil U., Paul M., Jahn M., Schagger H. and Fuchs G. (2003). Benzoate-Coenzyme A Ligase from Thauera aromatica: an Enzyme Acting in Anaerobic and Aerobic Pathways. *Journal of Bacteriology*, **185**(16), 4920–4929.

Scott J. P. and Ollis D. F. (1995). Integration of chemical and biological oxidation processes for water treatment: review and recommendations. *Environ. Prog.*, **14**, 88–103.

Scott L. L. (1999). Prevention of perinatal herpes: prophylactic antiviral therapy. *Clin. Obstet. Gynecol.*, **42**, 134–148.

Scuras S. E., Jobbágy A. and Grady C. P. L. Jr. (2001). Optimization of activated sludge reactor configuration: kinetic considerations. *Water Res.*, **35**(18), 4277–4284.

Shannon M. A., Bohn P. W., Elimelech M., Georgiadis J. G., Mariñas B. J. and Mayes A. M. (2008). Science and technology for water purification in the coming decades. *Nature*, **452**, 301–310.

Sholten E., Lukow T., Auling G., Kroppenstedt R. M., Rainey F. A. and Diekmann H. (1999). Thauera mechernichensis sp. nov., an aerobic denitrifier from a leachate treatment plant. *Int. J. Syst. Bacteriol.*, **49**, 1045–1051.

Siegrist H. and Gujer W. (1985). Demonstration of mass transfer and pH effects in a nitrifying biofilm. *Wat. Res.*, **21**(12), 1481–1487.

Sirtori C., Zapata A., Oller I., Gernjak W., Agüera A. and Malato S. (2009a). Decontamination of industrial pharmaceutical wastewater by combining solar photo-Fenton and biological treatment. *Water Res.*, **43**, 661–668.

Sirtori C., Zapata A., Oller I., Gernjak W., Agüera A. and Malato S. (2009b). Solar photo-Fenton as finishing step for biological treatment of a pharmaceutical wastewater. *Environ. Sci. Technol.*, **43**, 1185–1191.

Standard Methods for the Examination of Water and Wastewater, 20th edn, (1995). American Public Health Association (APHA), American Water Works Association (AWWA) and Water Environment Federation (WEF), Washington, DC, USA.

Suty H., De Traversay C. and Cost M. (2004). Applications of advanced oxidation processes: present and future. *Wat. Sci. Technol.*, **49**, 227–233.

Svensson J. O., Barkholt L. and Säwe J. (1997). Determination of acyclovir and its metabolite 9–carboxymethoxymethylguanine in serum and urine using solid-phase extraction and high-performance liquid chromatography. *J. Chromatogr. B*, **690**, 363–366.

Ternes T. A., Stuber J., Herrmann N., McDowell D., Ried A., Kampmann M. and Teiser B. (2003). Ozonation: a tool for removal of pharmaceuticals, contrast media and musk fragrances from wastewater? *Water Res.*, **37**(8), 1976–1982.

Thomsen T. R., Kong Y. and Nielsen P. H. (2007). Ecophysiology of abundant denitrifying bacteria in activated sludge. *FEMS Microbiol Ecol.*, **60**, 370–382.

Tsezos M. and Bell J. P. (1989). Comparison of the biosorpion and desorption of hazardous organic pollutants by live and dead biomass. *Water Res.*, **23**(5), 561–568.

United States Environmental Protection Agency, (1996). Prevention, Pesticides and Toxic 503 Substances (7101). Fates, Transport and Transformation Test Guidelines OPPTS 504 835.3200 Zahn-Wellens/EMPA Test. EPA 712–C-96–084.

Van Loosdrecht M. and Henze M. (1999). Maintenance, endogenous respiration, lysis, decay and predation. *Water Sci. Technol.*, **39**(1), 107–117.

Verschoyle R. D., Brown A. W., Ray D. E. and Bird I. (1987). The relationship between uncoupling of oxidative phosphorylation and neuronal necrosis within the CNS in rats dosed with trihalogenated imidazoles. *Toxicol. App. Pharmacol.*, **89**, 179–182.

Vogelsang C., Grung M., Jantsch T. G., Tollefsen K. E. and Liltved H. (2006). Occurrence and removal of selected organic micropollutants at mechanical, chemical and advanced wastewater treatment plants in Norway. *Water Res.*, **40**(19), 3559–3570.

Wang F., Gamal El-Din M. and Smith D. W. (2004). Oxidation of Aged Raw Landfill Leachate with O3 Only and O3/H2O2: Treatment Efficiency and Molecular Size Distribution Analysis. *Ozone Sci. Eng.*, **26**(3), 287–298.

Witzig R., Manz W., Rosenberger S., Krüger U., Kraume M. and Szewzyk U. (2002). Microbiological aspects of a bioreactor with submerged membranes for aerobic treatment of municipal wastewater. *Water Res.*, **36**, 394–402.

Wobus A. and Röske I. (2000). Reactors with membrane-grown biofilms: their capacity to cope with fluctuating inflow and with shock loads of xenobiotics. *Water Res.*, **34**(1), 279–287.

Xu Y. P., Zhou Y. Q., Wang D. H, Chen S. H., Liu J. X. and Wang Z. J. (2008). Occurrence and removal of organic micropollutants in the treatment of landfill leachate by combined anaerobic-membrane bioreactor technology. *J Environ Sci-China*, **20**(11), 1281–1287.

Xue X., Hanna K., Despas C., Wu F. and Deng N. (2009). Effect of chelating agent on the oxidation rate of PCP in the magnetite/H2O2 system at neutral pH. *Journal of Molecular Catalysis A, Chemical*, **311**, 29–35.

Yamashita T. and Hayes P. (2008). Analysis of XPS spectra of Fe2+ and Fe3+ ions in oxide materials. *Applied Surface Science*, **254**, 2441–2449.

Yuranova T., Rincon A. G., Bozzi A., Parra S., Pulgarin C., Albers P. and Kiwi J. (2003). Antibacterial textiles prepared by RF-plasma and vacuum-UV mediated deposition of silver. *Journal of Photochemistry and Photobiology A: Chemical*, **161**, 27–34.

Zapata A., Velegraki T., Sanchez-Perez J. A., Mantzavinos D., Maldonado M. I. and Malato S. (2009). Solar photo-Fenton treatment of pesticides in water: Effect of iron concentration on degradation and assessment of ecotoxicity and biodegradbility. *Applied Catalysis B Environmental*, **88**, 448–454.

Zhang Y. and Zhou J. L. (2008). Occurrence and removal of endocrine disrupting chemicals in wastewater. *Chemosphere*, **73**, 848–853.

Zuccato E., Calamari D., Natangelo M. and Fanelli R. (2000). Presence of therapeutic drugs in the environment. *Lancet*. **355**, 1789–1790.

Chapter 3

Membrane-based processes

Bruce Jefferson, Marc Pidou, Olivier Autin, Jitka MacAdam, Simon A. Parsons, Christoph Bayer, Martin Follmann, Thomas Melin, Ingrid M. Wienk and Folkert Petrus Cuperus

3.1 BACKGROUND

Membranes first entered our consideration in the early twentieth century although it is not until the development of asymmetric cellulose acetate membranes in the 1960s that acceptance of their value as a treatment technology was truly realised. Today, membrane technology is commonly used in industrial wastewater treatment especially where difficult to treat wastewaters are encountered and/or water recycling and resource recovery are being considered. For example, reverse osmosis and ultrafiltration are both widely used within the pharmaceutical industry to remove pyrogens. The application of membranes to industrial wastewater treatment has been constrained principally by the costs involved. As a consequence, the majority of membranes sales are for highly specific tasks with high value products. For example, almost half of all membrane sales are accounted for by hemodialysis used for kidney machines in hospitals. However, advances in membrane technology are greatly increasing the competitiveness of utilising the technology for large scale duties. Indeed, in many cases industrial wastewater treatment is seen as an early adopter of new technology and indeed there is a track record of novel membrane processes first gaining acceptance in this sector.

Perhaps the most significant example of this has been the uptake of membrane bioreactors which integrate biological treatment with a membrane. The technology was developed by Dorr-Oliver in the late 1960s and consisted of a side stream UF membrane linked to a traditional bioreactor for treatment of ship board wastewater (Judd, 2010). Application is now widespread for industrial applications across all sectors and whilst recent growth in uptake of the technology has been most striking for municipal applications, industrial adoption of the technology continues to grow as a healthy rate. Overall, global growth rates of MBRs is between 9.5 and 12% with the market value for MBRs estimated to be become around \$0.5 billion by 2013.

Adoption of membrane based processes for industrial application is driven by a number of features including: the need to produce consistent treated water quality even when the influent is highly variable, especially for reuse applications; the application of intensive technology is common place and discharge costs of the untreated wastewater are increasing. Overall this leads many new schemes to utilise membrane processes. In the case of industrial wastewater reuse the adoption of membrane based processes is considered largely successful (Judd and Jefferson, 2003). Examples include the clean up of wash water in a car paint facilities, treatment of wastewater from bank note production, processing of sewage effluent for boiler feed water, food processing wastewater and dye house effluent treatment. In the majority of reported case studies, the application of membranes has been shown to be cost effective with pay back times between 9 months and 5 years (Judd and Jefferson, 2003). The economic success is related both to reduction in discharge costs in addition to water recycling and resource recovery. This has been shown to be in the form of raw material recovery and/or heat which can represent as much as 50% of the economic benefit observed.

Whilst the success of membrane bioreactors for industrial wastewater treatment has beenestablished the efficacy of the technology is constrained by the biodegradability and ultimately, the toxicity of the wastewater to be treated: the technology is unable to treat toxic or recalcitrant organic matter which limits the appropriateness to some industrial wastewater. To a degree this can be managed by the use of pre treatment, commonly chemical oxidation processes such as ozone which alter the nature of the toxic chemicals (Parsons, 2004). However, there is a limiting scope for such approaches in some cases as the costs of the chemical additions become prohibitive. This leads to the direct application of chemical processes for industrial wastewater treatment where the lack of biological component enables different operating conditions to be employed greatly enhancing efficiency and reducing costs. Consequently, there is an opportunity to utilise membrane in combination with chemical reactions to achieve equivalent benefits to those observed for biological systems in terms of intensification, footprint, robustness and treatment quality. Looking forward there are a number of potential new technologies based around membranes which offer promise for industrial treatment (ref). The majority can be described as either membrane contactors (MC) where the membrane is combined with chemical scrubbing to enhance mass transfer or membrane chemical reactors (MCR) where membranes are used to directly enhance chemical oxidization technologies (Figure 3.1):

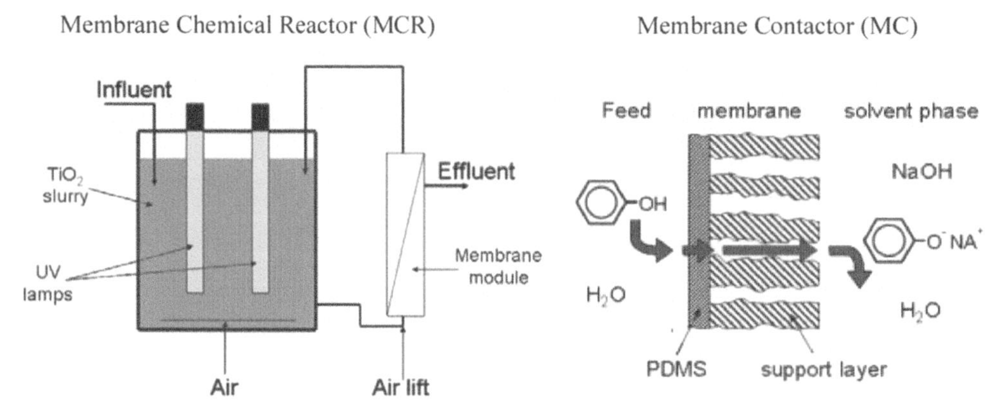

Figure 3.1 Schematic of MCR and MC

Similar technologies have been used for other applications, typically very dilute systems and whilst these provide some insights direct translation is not possible as they represent a 5–6 orders of magnitude decrease in organic concentration to be treated compared to the many industrial applications. As with all novel technology it is important to understand the true potential they offer and as such the work outlined in this chapters develops a clear design approach suitable for industrial wastewater treatment (and recovery) enabling there potential to be properly considered.

3.2 RECOVERY OF PHENOLIC COMPOUNDS WITH MEMBRANE CONTACTORS

3.2.1 Introduction

The removal of phenol from industrial wastewater has hitherto not been solved satisfactorily. The usual approaches for removing dilute organic compounds from water all suffer from serious drawbacks in this case:

- Liquid–Liquid Extraction: Due to the comparatively polar nature, phenol is poorly soluble in non-polar solvents. However, the choice of a mildly polar solvent leads to increased contamination of the aqueous phase with solvent. This leads to a new separation problem.
- Stripping: The high boiling-point and reasonable solubility in water renders phenol poorly strippable.
- Nanofiltration: Due to the low molecular weight, nanofiltration is not suitable for the separation of phenol from water.
- Biological Treatment: This class of methods is only applicable to wastewater with low phenol content (toxicity against activated sludge microorganisms: EC50 = 270 mg/L; from: Neumegen *et al.*, 2005).

Other process options include reverse osmosis, activated carbon adsorption and wet oxidation. They share prohibitively high costs. Further, a recycle of separated compounds is not feasible. As a solution to this problem we examined the reactive extraction of phenol in a membrane contactor. The solvent is an aqueous solution of sodium hydroxide. This process' operating principles are described below.

Porous membranes that are commonly used as membrane contactors are either wetted by aqueous (polar) or organic fluids (non-polar). A preference for one of these molecules depends on the hydrophobicity or hydrophilicity of the membrane material. This behaviour leads to an immobilization of the phase boundary at the membrane surface if two immiscible phases are brought into contact via the membrane. Are however, both substances of similar type, no phase boundary is formed and the two streams mix with each other. For such a case, mixing can be prevented by applying a dense coating onto the porous membrane. Since phenol is going to be separated, rubbery coatings, for instance of silicone, are preferred (Wesselingh and Krishna, 2006).

Phenol is a weak acid and reacts rapidly with caustic soda to sodium phenolate (Figure 3.2), which is made available in the porous layer. Phenolates are practically insoluble in the dense coating, which prevents back-diffusion. Additionally, phenolates do not contribute to the driving force for phenol, which improves the rate of transport for phenol across the dense layer. This effect is termed *mass transfer enhancement* due to species conversion. At a very high pH value at the stripping side, far more OH ions are present (at the interface between dense coating and porous support) than needed to convert phenol to phenolate. Hence, mass transfer enhancement becomes maximal. In contrast, at a lower stripping pH value, for instance at neutral pH for the class of phenolic compounds, chemical equilibrium favors neutral phenols. Since phenol is not converted under such conditions, mass transfer enhancement is absent.

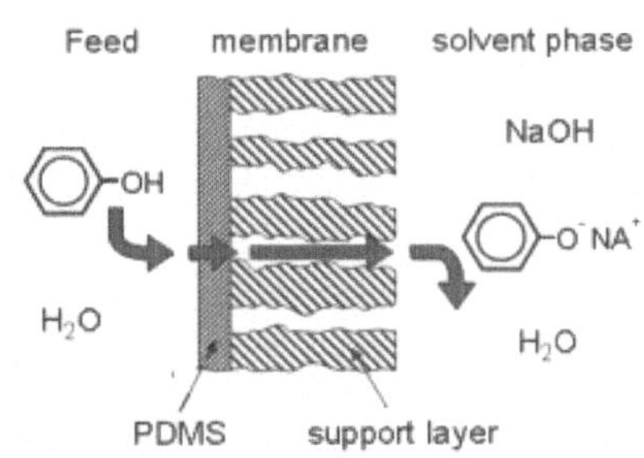

Figure 3.2 Functional principle of a coated membrane contactor

At intermediate pH values (typically between pH 11 and pH 14), a change in stripping pH triggers a change in mass transfer performance. This change in mass transfer corresponds to a change in location of the reaction zone inside the porous support where the conversion to phenolate takes place. With rising pH the reaction zone shifts towards the dense layer, thus the length of travel for (neutral) phenol is shortened and as a consequence the rate of mass transfer increased.

Preliminary work was performed by Brookes *et al.* (1995) with (symmetric) tubular silicone rubber membranes for various weak organics acids (e.g. phenol) and bases (e.g. aniline). Gruber and Melin (1999) made use of asymmetric membranes applied for the removal of phenol as model compound. Later, the group of Prof. Livingston performed work exploring some of the basic principles of this process (e.g. Ferreira *et al.*, 2005). As part of their work, a first step was taken towards up-scale and full-scale implementation (Ferreira *et al.*, 2002). The first implementations were based on very long and rather thick silicone tubes, which were immersed into the stripping solvent. Since neither the membrane properties nor the flow in the vessels and the membranes were optimized, the overall target of this sub-project was to design, test and model coated membrane contactor systems to recover phenolics from well-defined wastewaters so that industrial end-users can evaluate technical and economic features of such a process.

3.2.2 Membrane performance

In the course of the INNOWATECH project, the dutch company Solsep BV developed composite membranes, which suit particularly well the target application (type 030605.0713). Tests based on the selective transport of gas proved that these membranes were stable in a caustic regime (pH 13) for more than a year. Since it displayed superior performance in preliminary tests, this membrane's transport properties were investigated in greater detail. To this end, an experimental set-up was devised that enabled steady-state measurements of the transport rates of phenol and water under varying conditions.

A concentration-controlled, iso-thermal and iso-volumetric test cell with an effective membrane diameter of 90 mm was used. Both sides are continuously stirred to ensure proper mixing of the respective compartments. Mass fluxes into and out of the feed and stripping side compartments were measured gravimetrically with Acculab balances of type VIC-5101. Phenol concentrations on both membrane sides were determined with a UV-Photospectrometer of type Varian Cary 50 equipped with Hellma QS-174 cuvettes. Alkalinity was recorded with a pH meter Knick 913 pH hooked up to pH probe of type SE-101. All devices were operated with LabView. Variations in stirring rate (Wilson, 1915) did not yield conclusive results. Hence, the effect of the boundary layer is insignificant compared to the membrane resistance.

Following the assembly, the set-up was tested for proper sealing of both the feed and the permeate side. Before first use, the membranes are immersed into CO_2 for 30 minutes to replace are present inside the porous support. Afterwards, they are placed into a sodium hydroxide solution to replace the CO_2 with the stripping solution. Aqueous solutions of phenol were prepared gravimetrically (Acculab-210.4 and Acculab VIC-5101) from dried material (99% purity). The photometer was calibrated by pumping solutions with known concentrations through the cuvettes. The pH-meter was calibrated at 3 points for the pH-range of interest.

A constant flux of water was added to the feed compartment to improve concentration control. However, a theoretical sensitivity study illustrated that this (artificial) flux of water into to the feed compartment enlarges the sensitivity towards measurement error at the same time. Nonetheless, if the in- and outgoing fluxes in the respective quantity (e.g. phenol concentration) show a significant spread, sensitivities remain moderate, i.e. smaller than 2. This means that a 1% error in the flux measurement may result in a maximum error in the measurement mass transfer coefficient of 2%.

As shown in Figure 3.3, the mass transfer coefficient easily increases ten-fold by increasing the pH of the stripping side bulk solution. The maximum mass transfer coefficient was obtained at approximately pH 13.6, which equals to a hydroxide ion concentration of 0.4 mol/L. Under the conditions employed, no flux of water across the membrane was observed.

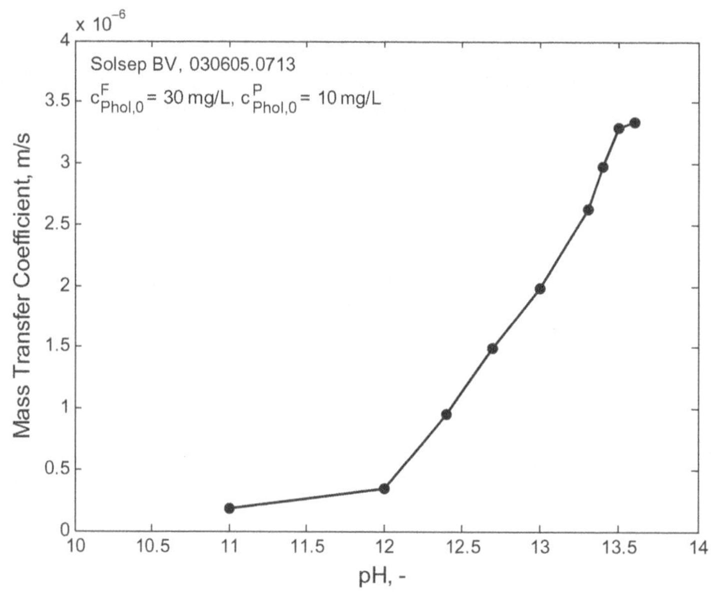

Figure 3.3 Permeability for phenol

Clearly, mass transfer enhancement may become costly in terms of stripping-side OH ion concentration. The increase in mass transfer from pH 13.5 to pH 13.6 appears to be rather small, but equates to an increase in OH ion concentration from 0.25 mol/L to 0.4 mol/L. Further, one shall keep in mind that an increase in phenol concentration in the feed leads to a flux increase and thus requires a higher OH ion concentration in the stripping solution. In short, operation at significantly higher phenol fluxes (e.g. due to an increase in concentration on the feed side or due to an improved dense layer) are mostly likely limited by the strength of the stripping solution if the transport resistance on the stripping side is not reduced accordingly.

3.2.3 Membrane contactor modules

Up to now, hollow fiber membrane modules are generally employed in contacting applications. As the example "Reverse Osmosis" illustrates, spiral wound membrane modules have virtually fully substituted all other types of membrane modules in the field of seawater desalination because they can be cheaply built, easily deployed with large membrane areas and sustain high trans-membrane pressures. In contrast, spiral wound modules are not used to date for 4-end applications, e.g. as membrane contactors. They are particularly ill-suited for conventional membrane contactors due to insufficient pressure control inside the module. Since pressure drop (and as a consequence the break-through to the other membrane side as for conventional contactors) is not of relevance for coated membrane contactors, it was one of the objectives of this sub-project to explore the limits of 4-end spiral wound modules (Figure 3.4).

Spiral wound modules consist of a central permeate pipe with membrane pockets wound around it. A membrane pocket consists of two flat sheets of membrane material glued to each other with a spacer in between to keep the membrane sheets apart at a predefined distance. A so called feed spacer is placed (but not glued) in between two wound membrane pockets to ensure also a well defined flow channel for the wastewater that contacts the membranes' dense layer. The stripping solution (NaOH in this case) enters the 4-end module and the membrane pockets through the central tube, is directed by additional glue lines and leaves the pocket and the module again through the central tube.

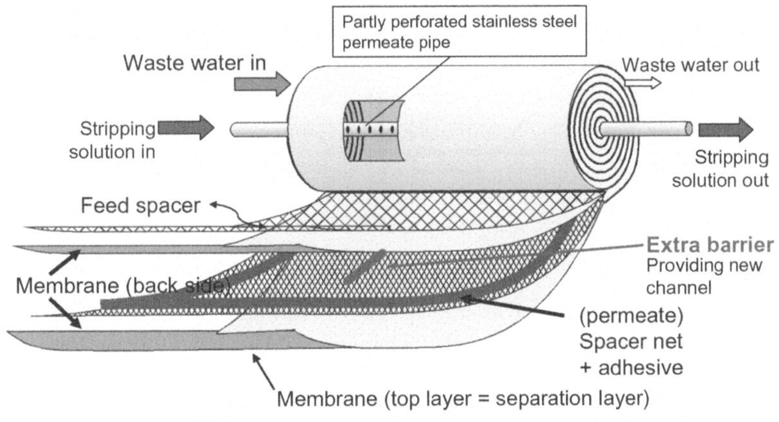

Figure 3.4 4-end spiral wound module

The left-hand image in Figure 3.5 replicates what is known in heat exchanger design a co- or counter-current flow pattern (relative to the feed side). In contrast, the right-hand image depicts a cross-flow pattern. These two designs are geometrically similar if the channel to the left of the glue line (F) is twice as wide (width: 2g) as each distribution channel (width: g) to the right of the glue line (Figure 3.5a). If the area E is decreased in length and the distribution channels scaled proportionally, one ends up with the flow pattern as shown on the right hand-side in Figure 3.5.

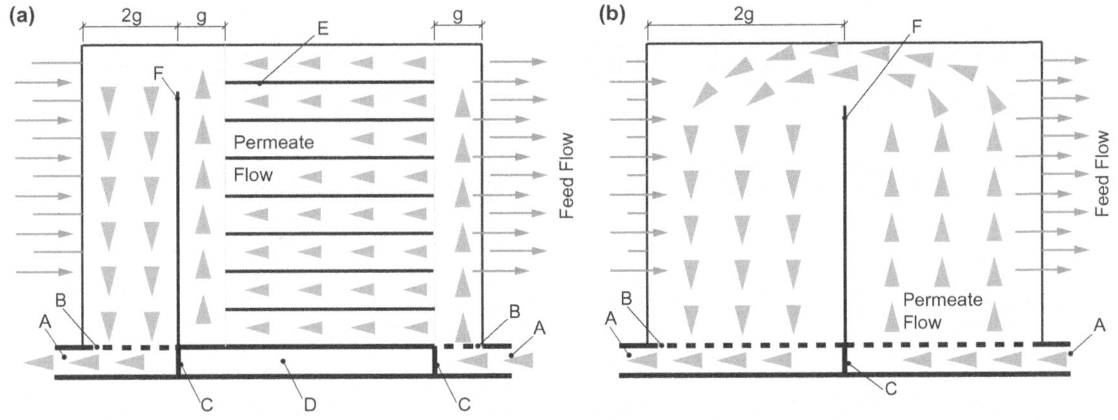

Figure 3.5 Stripping side flow pattern: co-/counter-current flow (a) and cross flow (b); A, permeate inlet/outlet; B, perforation of the permeate tube; C, plug to inhibit flow; D, closed section; E, glue lines to ensure counter-current permeate flow; F, dividing glue line to ensure flow perpendicular to the central tube; g/2g, width of distribution channels oriented perpendicular to the tube

It was expected that the optimal module design for this very application is a blend of these two extremes, because the cross-flow design pattern should be prone to maldistribution effects and should thereby exhibit lower mass transfer performance. The co-/counter-current design pattern on the other hand should offer a more robust behaviour towards maldistribution because of the better flow control due to additional glue lines. Although, it is more prone to pressure drop due to the rather small distribution channels of width g and 2g.

The prototype membrane modules under investigation are 20 inch long, 2.5 inch in diameter and have approximately 0.4 m^2 of membrane area with spacers as distance pieces between two layers of membranes. One of the preliminary modules built by SolSep BV is shown in Figure 3.6. These modules can be operated at a pressure of up to 5 bar, whereas the pressure difference between the permeate inlet and outlet must not exceed 0.5 bar and the permeate pressure level must always be lower than the feed pressure level. Otherwise the glue lines may be fractured and consequently, the permeate side may leak slightly to the feed side which was found not to be very significant in every-day operation, but corrupts mass transfer measurements.

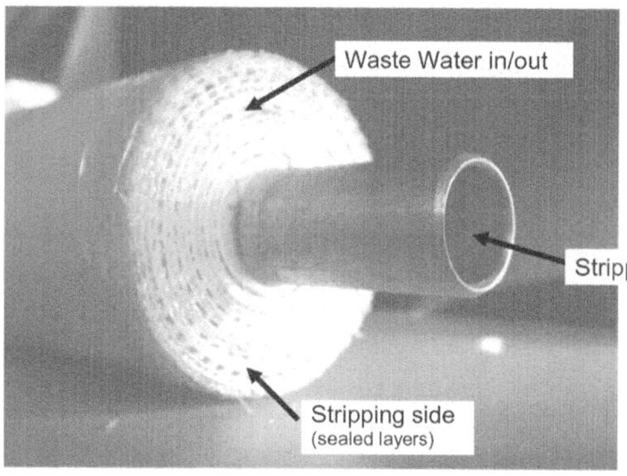

Figure 3.6 Spiral wound module used for preliminary testing

Whereas mass transfer and pressure drop is well characterized for feed-side flow conditions (Schock and Miquel, 1987), a study targeting flow and mass transfer in spacer-filled channels at low Reynolds numbers as they are relevant for extraction was not previously done.

3.2.4 Flow in spacer-filled channels

Materials and Methods

Spacer nets were characterized with respect to pressure drop, mass transfer and mixing/dispersion. Available for testing were one woven and two extruded spacer nets (Figure 3.7). Key parameters are summarized in Table 3.1.

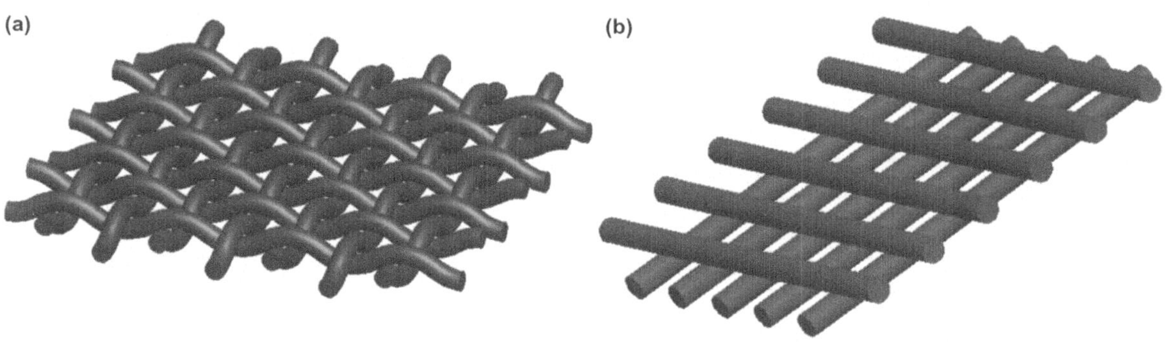

Figure 3.7 Schematic drawings of woven (a) and extruded spacers (b)

Table 3.1 Key parameters of the investigated spacer nets

Spacer	Height, mm	Mesh size, mm	Mesh angle	Porosity	Type
Conwed X04410	1.0	2 × 2	90°	0.75	Extruded
Naltex NO1329_90pa	0.5	1.4 × 1.4	90°	0.8	Extruded
Filmoor PBT	0.4	0.2 × 0.2	90°	0.35	Woven

Pressure drop is measured with the difference in hydrostatic level across the spacer mesh. Mass transfer was determined with both the limiting current technique and the copper dissolution technique. Mixing/dispersion was monitored with the help of an acidic and an alkaline stream, both charged with a pH indicator which changes color around pH 7. Details are provided below.

Start-up Procedure

To avoid entrapment of air, all set-ups were subsequently flushed with CO_2, aqueous caustic soda, deionized water and finally with the respective test solutions. Further, all test solutions were at all times N_2 stripped to avoid the formation of bubbles in the test cell.

Pressure drop measurements of spacers

The pressure drop Δp across a spacer mesh is determined via the difference in hydrostatic level Δh:

$$\Delta p = \rho g \Delta h$$

where ρ is the density of water and g the gravitational acceleration. A test cell similar to the one in Figure 3.9 was used.

Dispersion measurements

Observation of lateral mixing was based on research performed by Hartung (1975). Streams of hydrochloric acidic and sodium hydroxide flow first in parallel separated by an impermeable barrier (Figure 3.8). Both streams were marked with a pH indicator, which shifts its colour at pH 7 (30 mg/L of each neutral red and methylene blue). As the barrier ends, the streams start to mix. The points in space of neutral pH were observed as colour-change boundary. Different acid to alkaline concentration ratios shift the location of the colour-change boundaries.

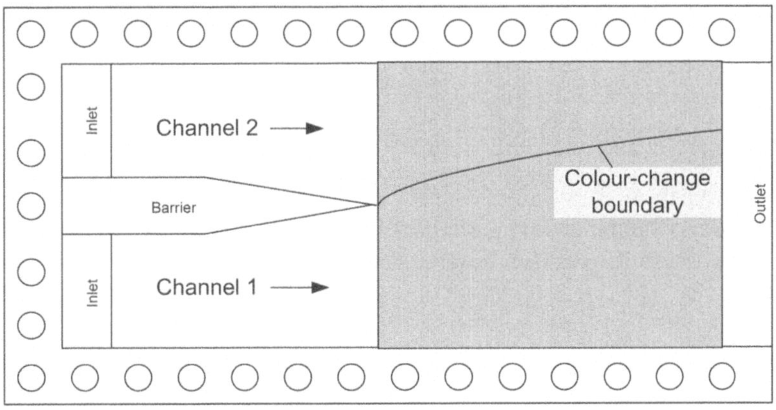

Figure 3.8 Schematic drawing of the dispersion test cell

The experimental set-up consists of a measuring cell for spacer samples either of size 250 mm × 80 mm or of 250 mm × 150 mm, 1 L feed containers continuously stirred, temperature controlled and stripped with N_2 as well as a calibrated tubing pump (Ismatec MCP Process equipped with 12 roller/2 stopper pump head). The flow rates were randomly chosen to avoid experimental bias. Images were taken with conventional digital cameras placed inside a housing with indirect lighting.

Copper dissolution measurements

The rate of mass transfer between the channel wall and fluid bulk were deduced with the copper dissolution technique proposed by Gregory and Riddiford (1960). Sulphuric acid (1 mol/L) containing potassium dichromate contacts a copper surface. Dichromate ions oxidize the copper according to the following equation:

$$3Cu + K_2Cr_2O_7 + 7H_2SO_4 \longrightarrow 3CuSO_4 + K_2SO_4 + Cr_2(SO_4)_3 + 7H_2O$$

where Cr^{4+} is reduced to Cr^{3+}. The decline in Cr^{4+} concentration yields the mean mass transfer coefficient k:

$$k = \frac{\dot{V}}{A} \ln \frac{C_\alpha}{C_\omega}$$

where C_α and C_ω stand for the inlet and outlet concentrations of Cr^{4+}. A represents the copper surface area and $\dot{V}$ is the volumetric flow rate. A thorough description of the method can be found in the work by Gruber and Melin (2003). These authors noted convection due to density gradients. They improved the method by decreasing the dichromate concentration by a factor of 100 to 0.3 mmol/L versus Gregory and Riddiford (1960). Apparently, for systems which are unaffected by natural convection, the concentration level can be freely chosen up to 30 mmol/L.

The experimental set-up consisted of the measuring cell for spacer samples of size 200 mm × 80 mm (Figure 3.9), a 1 L feed container continuously stirred, temperature controlled and stripped with N_2 as well as a calibrated tubing pump (Ismatec MCP Process equipped with 12 roller/2 stopper pump head). The copper plate is 170 mm × 80 mm in size. The flow rates were randomly chosen to avoid experimental bias. To ensure steady state, time series of samples were analyzed. Cr^{4+} concentrations were measured at 350 nm with a UV photospectrometer (Varian Cary 50 equipped with a Hellma 115-QS cuvette).

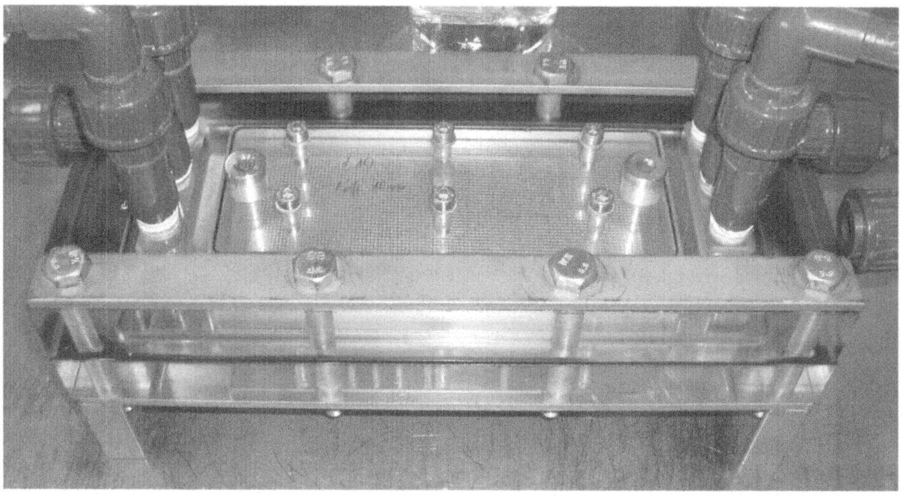

Figure 3.9 Assembled copper dissolution test cell for spacer characterisation

Limiting current measurements

Mass transfer measurements from the bulk towards channel walls can also be performed with the limiting current technique. This method is based on a redox reaction induced by a voltage difference between two electrodes. With increasing potential difference, the rate of reaction increases until reaching a limit. The rate limiting step is diffusive mass transfer of the reactants between the electrodes. The boundary layer mass transfer coefficient k can be calculated from the limiting current I_{lim} according to:

$$k = \frac{I_{\text{lim}}}{nFC_bA}$$

where n is the number of electrons participating in the reaction on the measuring electrode, F is Faraday's constant, C_b is the concentration of $[Fe(CN_6)]^{3-}$ in the fluid bulk and A is the area of the measurement electrode. A thorough description of the method can be found in the article by Li *et al.* (2004).

The experimental set-up consists of the measuring cell for spacer samples of size 200 × 80 mm (Figure 3.10), a 1 L feed container continuously stirred, temperature controlled and stripped with N_2 as well as a calibrated tubing pump (Ismatec MCP Process equipped with 12 roller/2 stopper pump head). The platinum electrodes are connected to the DC power source Agilent E3640A. The current is tracked with the ampere meter Agilent 34401A. The cathode as measurement electrode is smaller than the anode and 70 mm × 40 mm in size.

The flow rates were randomly chosen to avoid experimental bias. For a given flow rate the order of applied voltages was randomly chosen as well. Steady state was assumed when the electrical current was stable within 0.1 mA. The set-up was shielded from sunlight to eliminate UV-light triggered degradation of cyanide-ions.

Results

Pressure Drop

Pressure drop, which needs to be overcome for the transport of a given amount of fluid through a device or system, is a measure for energetic efficiency.

For the set of spacers investigated, the Conwed spacer provides the largest cross section for flow through. Consequently, its pressure drop is lowest (Figure 3.11). The extruded spacers in 45° arrangement to the direction of flow (meant: filament orientation of the spacer, cp. Figures 3.11 and 3.12) show generally higher pressure drop, which cannot be observed for the woven spacer (Filmoor). The pressure drops of the extruded spacers are much lower than for the woven spacer, most likely due to higher porosities.

Dispersion

Maldistribution of liquid in membrane modules may decrease system performance. Maldistribution stands for the uneven spatial distribution of liquid, so that the residence time distribution is significantly broadened. Broad residence time distributions lead to uneven and thus inefficient use of the stripping solvent and might even reduce the effective membrane area available for mass transport. To identify the optimal spacer arrangement, dispersion experiments were performed.

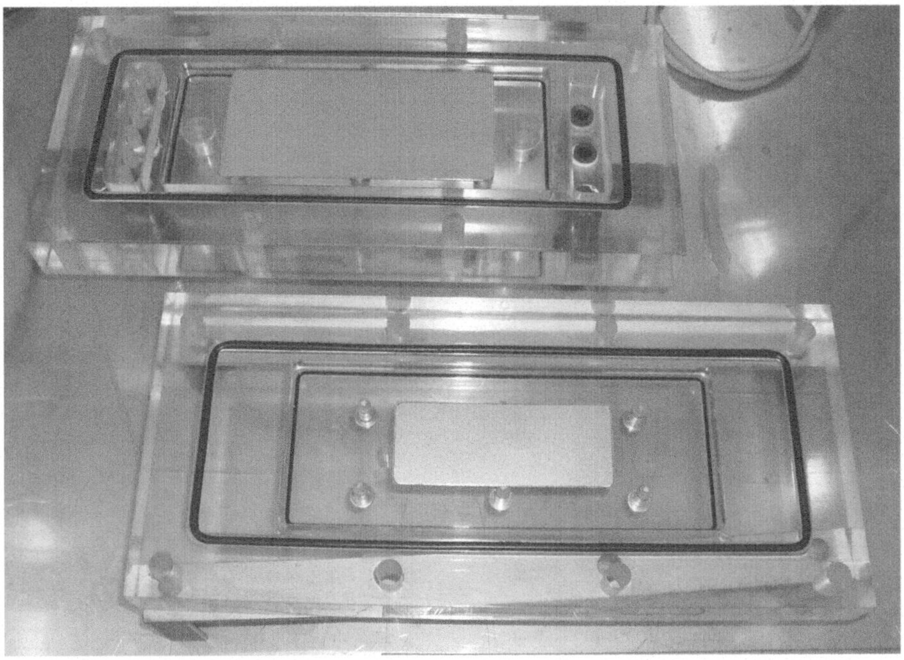

Figure 3.10 Disassembled limiting current test cell

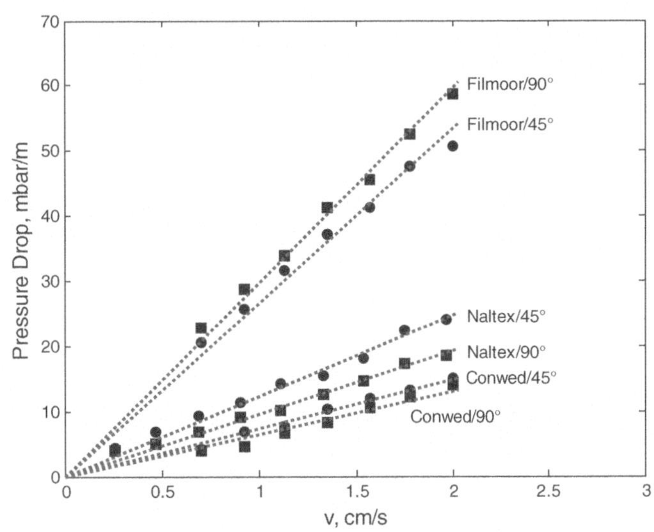

Figure 3.11 Pressure drop for the spacers investigated

Figure 3.12 shows images of the fluid distribution for the 3 spacer nets evaluated. The upper left hand image depicts fluid flow perpendicular and parallel (90° attack angle) to the filaments of the Conwed spacer. The flow clearly follows the filament channels (dots follow one filament). The image on the right hand side shows the Conwed spacer arranged at 45° attack angle. Spreading of the tracer can be clearly observed. The lower right-hand image shows the same experiment for the Naltex spacer. Spreading can be observed, but it is not as pronounced as for the Conwed spacer. If the Naltex spacer is arranged at 90°, spreading cannot be monitored. Independent of the attack angle, the Filmoor spacer does not exhibit any tracer spreading. The findings for the spacers arranged at 90° and for the Filmoor spacer are independent of flow velocity. In case of the extruded spacers arranged at 45°, the color-change boundary shifts its location with a change in flow velocity. The higher the velocity becomes, the more pronounced the dispersion/mixing effect is.

Mass transfer with the limiting current technique

Since concentration boundary layers contribute to the overall mass transfer resistance, their contribution was quantified with the limiting current method. As can be deduced from Figure 3.13, the Conwed spacer performs twice as good for a given flow velocity in comparison to the Filmoor and the Naltex spacer. In general, mass transfer coefficients range from 5E-6 up to 3E-5 m/s.

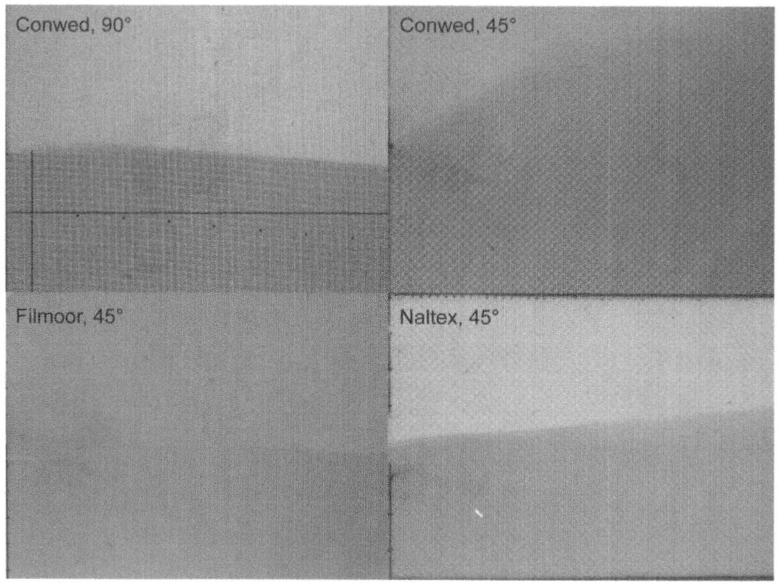

Figure 3.12 Dispersion in spacer filled channels (v = 2 cm/s)

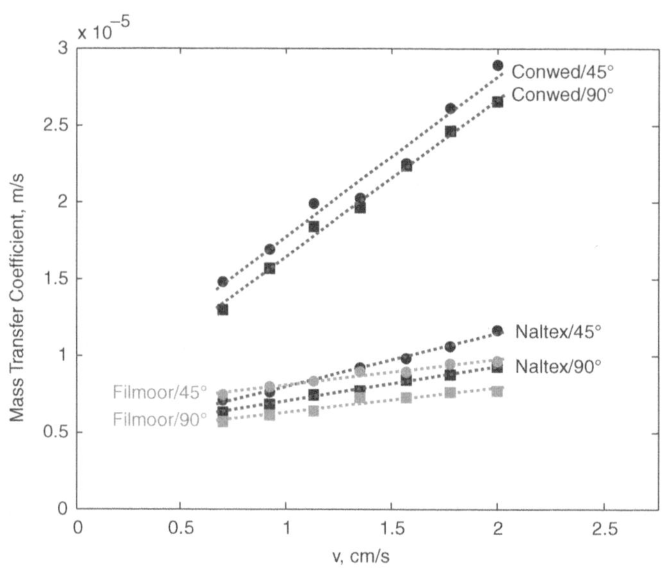

Figure 3.13 Concentration boundary layer mass transfer coefficients with the limiting current method

Mass transfer with the copper dissolution technique

Mass transfer coefficients were also obtained by the copper dissolution method. The primary objective was to check the validity of the method against the limiting current technique for mass transfer measurements in membrane modules equipped with copper foils. Besides that, the copper dissolution method depends only on one surface and not on the transfer between two surfaces. Hence, it was expected to see differences especially for the Conwed spacer due to the channelling effect described before.

The measured mass transfer coefficients (Figure 3.13) are slightly below the values obtained with the limiting current technique at attack angles of 90° (orientation of the spacer filaments in contact with the copper surface: (A) parallel/(B) perpendicular to the direction of flow). However, the data obtained at 45° attack angle are significantly lower. This may be explained by the channelling of the fluid. At first the fluid is in contact with the copper surface until it reaches the channel boundary (cp. Figure 3.14). There, it is flipped upwards into the filament channels that do not have contact. As a consequence, the fluid is on average in contact with only half the copper surface, which explains the low mass transfer coefficient. Hence, channelling can be observed also with the copper dissolution mass transfer measurements.

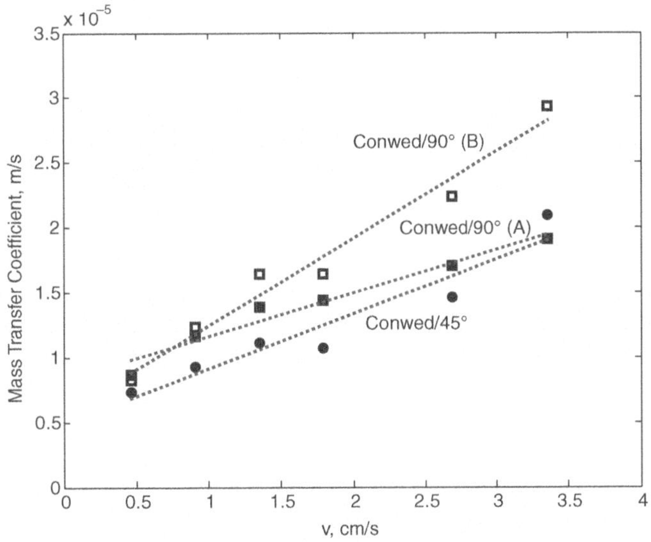

Figure 3.14 Mass transfer coefficients obtained with the copper dissolution technique: Conwed spacer; (A) filaments in contact with the copper surface are parallel or (B) perpendicular to the direction of flow

Shown in Figure 3.15 are the copper dissolution measurements for the Naltex spacer. For identical flow rates, the mass transfer coefficients tend to be higher than with the limiting current method. The Naltex spacer arranged at 45° shows the same behaviour as with the limiting current method. Hence, channelling is not the predominant effect as for the Conwed spacer, which supports the dispersion experiments. It should be noted that the flow range covered with the copper dissolution technique is larger than with the limiting current technique. The upper flow limit arises from the decreasing change in Cr^{4+} concentration with increasing flow rate, which renders results of even higher flow rates uncertain. The lower limit stems from the time it takes to obtain a stable signal.

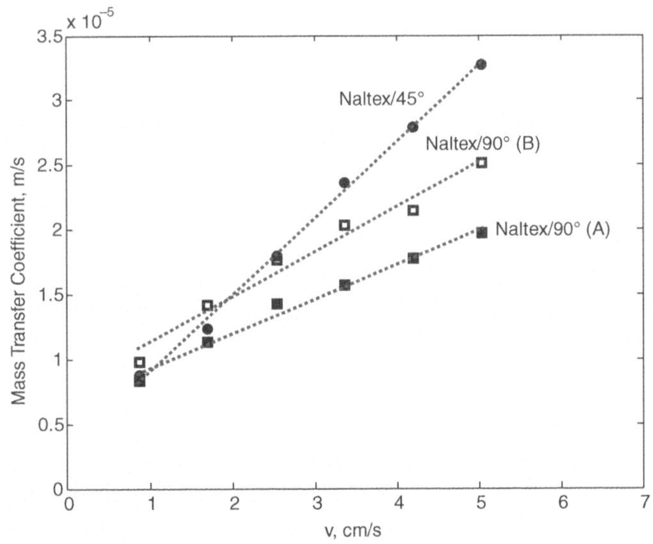

Figure 3.15 Mass transfer coefficients obtained with the copper dissolution technique: Naltex spacer; (A) filaments in contact with the copper surface are parallel or (B) perpendicular to the direction of flow

3.2.5 Mass transfer investigations inside spiral wound modules

The previous studies characterized spacers if deployed under well-defined conditions (e.g. geometry well known). Besides quantifying performance parameters, these tests showed that channelling between spacer filaments is a distinct characteristic of extruded spacers. This feature may be exploited to avoid maldistribution in spiral wound modules. Figure 3.16 illustrates the spacer arrangements and their effect on fluid distribution. At 90° attack angle, the fluid follows the filaments into and out of the corners of a membrane module. In contrast, at 45° attack angle, the fluid tends to short-cut the corner. The fluid in these corners remains stagnant; its residence time becomes very long. Therefore, extruded spacers at 90° attack angle are the preferred option.

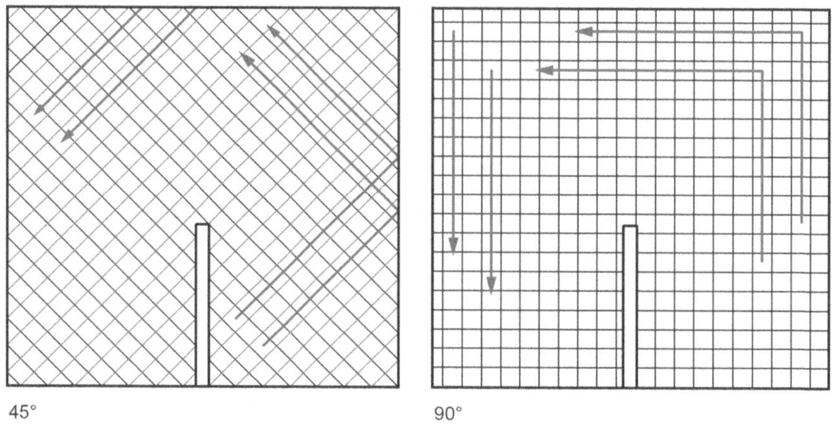

Figure 3.16 Flow distribution for different spacer arrangements (arrows show fluid flow)

Not considered so far is the effect of flow patterns (i.e. glue lines) in spiral wound modules as schematically shown in Figure 3.5. The upcoming section utilizes the copper dissolution method to investigate the effect of maldistribution on boundary layer mass transfer in spiral wound modules.

Materials and Methods

The rate of mass transfer between the walls of the module and fluid bulk may be deduced with the copper dissolution technique proposed by Gregory and Riddiford (1960). Sulphuric acid (1 mol/L) containing potassium dichromate contacts a copper surface. Dichromate ions oxidize the copper according to the following equation:

$$3Cu + K_2Cr_2O_7 + 7H_2SO_4 \longrightarrow 3CuSO_4 + K_2SO_4 + Cr_2(SO_4)_3 + 7H_2O$$

where Cr^{4+} is reduced to Cr^{3+}. The decline in Cr^{4+} concentration yields the mean mass transfer coefficient k:

$$k = \frac{\dot{V}}{A} \ln \frac{C_\alpha}{C_\omega}$$

where C_α and C_ω stand for the inlet and outlet concentrations of Cr^{4+}. A represents the copper surface area and $\dot{V}$ is the volumetric flow rate. A thorough description of the method can be found by Gruber and Melin (2003). Gruber and Melin (2003) noted convection due to density gradients. They improved the method by decreasing the dichromate concentration by a factor of 100 to 0.3 mmol/L versus Gregory and Riddiford (1960). Apparently, for systems which are unaffected by natural convection, the concentration level can be freely chosen up to 30 mmol/L. The inlet concentration in these experiments was chosen as to achieve a good signal-to-noise ratios.

The experimental set-up consisted of the copper modules (Figure 3.17), a 20 L feed container continuously stirred, temperature controlled and stripped with N_2 as well as tubing pumps of type Ismatec Ecoline VC-380, which were continuously controlled based on the flow rate measured with Endress + Hauser Promag 53H-02 flow meters. The copper plate is 170 mm × 80 mm in size. The flow rates were randomly chosen to avoid experimental bias. To ensure steady state, time series of samples were analyzed. Cr^{4+} concentrations were measured at 350 nm with a UV photospectrometer (Varian Cary 50 equipped with a Hellma 115-QS cuvette).

Results

The results obtained are documented in Figure 3.18. Surprisingly, the (simpler) cross-flow design shows superior performance over the co-/counter-current design. Depending on the point of operation, it shows 25% to 40% better mass transfer. This means that the cheaper design performs better, which is a very important finding with respect to technical implementation, since membrane area is the main contributor to cost.

Figure 3.17 Modules equipped with copper foils for investigation with the copper dissolution technique

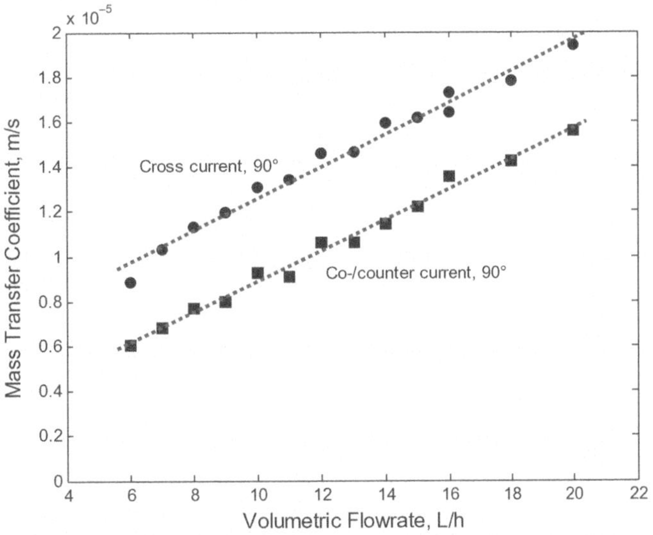

Figure 3.18 Mean mass transfer coefficients of copper spiral wound modules equipped with Naltex spacers

3.2.6 Mass transfer investigations inside spiral wound modules

Based on the results presented above, a prototype of an optimized membrane module was built (Figure 3.19). The spacer experiments show that woven spacers do not induce cross-mixing. Thus, the flow in channels defined by Filmoor spacers is highly segregated. Maldistribution due to differences in relative pressure drop or due to dead spots are not effectively eliminated. The Conwed spacer shows strong channelling behaviour. As a consequence it spreads the fluid perpendicular to the main direction of flow if arranged at 45°. However, the fluid layers flow above each other, essentially bringing two fluids into contact only. Thus, the mixing of individual layers of fluid appears to be irrelevant. The behaviour of the Naltex spacer is intermediate to the Filmoor and the Conwed spacer. Channeling of the Naltex spacer is less distinct, but the fluid layers seem to be better mixed than by the Conwed spacer. Mass transfer coefficients for all spacers are sufficiently high and pressure drop sufficiently low for use in spiral wound membrane modules.

The Naltex spacer is thinner than the Conwed spacer. Thinner spacers lead to lower module hold-up, which eases solute enrichment in the permeate. Most suitable for extraction applications appears to be the Naltex spacer, as such representing thin extruded spacers. Therefore, the optimised prototype module was finally equipped with Naltex spacers.

The copper module tests did show a significant performance difference between the two flow patterns. The most simple design is superior: the cross current flow pattern. However, this design is prone to maldistribution at the end of the membrane pockets (at the end of the glue line baffle, cp. Figure 3.5) where the fluid reverses its flow direction to flow back towards the permeate tube outlet. Flowing along the filament channels is apparently most efficient and enables flow "into" the corners of the module pocket and thus avoids short-cuts (Figure 3.16), if the 90° spacer arrangement is chosen. Thus, maldistribution is as far as possible eliminated.

Figure 3.19 Prototype membrane module

The optimized membrane module (Figure 3.19) was further used for measurements with the standard test system phenol. The results are shown in Figure 3.21.

Materials and Methods

The module was tested in a plant for mass transfer and fluid dynamics measurements. It consisted mainly of two reactor vessels – one for the feed and the other for the permeate side – equipped with water jackets for temperature control (see Figure 3.20, left-hand side). In these vessels aqueous solutions for the feed and the stripping side of the membrane contactor were stored and conditioned to keep the concentration levels of the phenols constant over time (if required). The concentration of the phenols and its salt, the phenolates, is measured online with UV spectrometers and flow through cuvettes (Varian Cary50 with Hellma QS-174.400). For this purpose the overlapping UV-absorption peaks (at 270 nm and at 287 nm) require calibration of cross-dependencies. The feed and the stripping solutions were fed to the membrane module on the hydraulics side of the plant (right-hand side in Figure 2.12) by two peristaltic pumps (Ismatec Ecoline VC-380) for each the feed and the stripping side. The volumetric flow rates were continuously recorded with flow meters (Endress + Hauser Promag 53H-04 or 53H-02) and adjusted by software-based feedback control. In order to quantify the decrease in pressure across the membrane module the pressure levels at the four in- and outlets were recorded. Due to the higher volumetric flow rates on the feed side, 12 mm-tubing was used on this side, whereas on the permeate side tubes with a diameter of 6 mm are chosen. This was to avoid significant changes in pressure up- and downstream of the membrane module. In order to protect the membrane module from damage caused by pressure pulses (e.g. malfunctioning pumps) on the permeate side, a safety valve connects the permeate with the feed side.

Results

According to Figure 3.21, the results of the feed and the permeate mass balances are in good agreement. Variations of the volumetric flow rate in the permeate (the stripping side) do not yield significant changes in the performance parameter: the mass transfer coefficient. The changes are within the margin of error of measurement. However, low flow rates in the feed (5 L/h) result in significantly lower mass transfer coefficients. This is probably due to a by-pass and consequently of low flow rates through the module itself.

3.2.7 Conclusions

For the membrane contactor system, several membranes were developed and the one with the best performance regarding stability and mass transfer was chosen. Spacer nets were tested with respect to pressure drop, flow distribution and mass transfer. To this end, several experimental set-ups were specifically developed. Thin extruded spacers were found to be most suitable for extraction applications. The copper dissolution technique, which is based on a fast surface reaction limited by the transport of Cr^{4+} to the copper surface, was further employed to investigate the effect of different flow patterns. The differences between the different flow patterns are insignificant, so that the most simple pattern was chosen. To minimize the effect of fluid maldistribution in spacer filled channels, the spacer netting is preferably oriented in parallel to the permeate tube.

Figure 3.20 Pilot plant

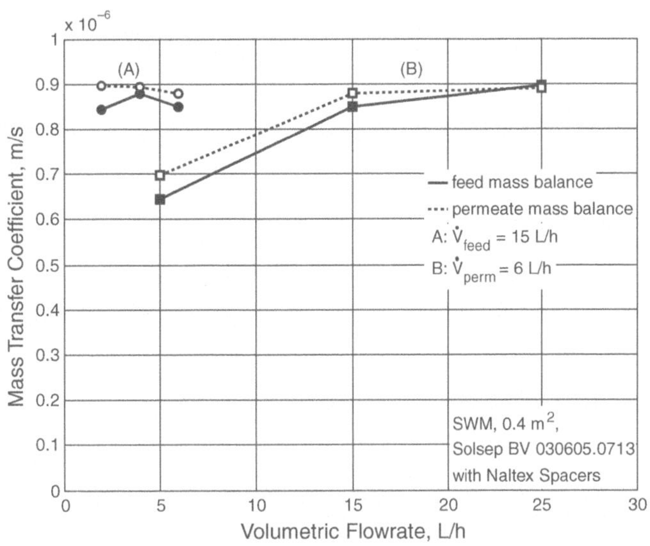

Figure 3.21 Performance data of the optimized membrane module

According to these findings, a prototype module was built. Even though the final performance tests show very good mass transfer performance, one shall keep in mind that the economic viability of such systems depends strongly on separation targets and the size of the volumetric streams to be treated.

3.3 MEMBRANE CHEMICAL REACTOR DESIGN

3.3.1 Introduction

The current state of the art in advanced wastewater treatment is represented by membrane bioreactors (MBRs) for intensive biotreatment (Judd, 2010), and advanced oxidation processes (AOPs) for intensive chemical treatment (Parsons *et al.*, 2005). The rapid market penetration of submerged MBR technology since the first commercial installation some 15 years ago is striking and demonstrates the potential for new technology uptake if appropriate drivers are met. Municipal MBR operations are dominated by submerged systems with polymeric membranes reflected in part by the increasing scale of operation now being consider; the largest operating MBR in 2010 was 110 MLD at Shen Ding River, Beijing (Judd, 2010). In contrast, industrial wastewater treatment remains at a relatively small scale and uptake in the market continues to be evenly spread across a wide range of design flows. Continued uptake of MBRs into this market is expected to grow year on year as increasing costs of water supply and wastewater disposal drive desire for enhanced treatment at source (Judd, 2010). The nature of the wastewaters and the scale of operation in the industrial context means that a broader range of configurations are more commonly used including side stream operation and the use of ceramic membrane materials. However, the efficacy of the technology can be constrained by limitations of utilising a biological reactor to drive pollutant removal: the technology is unable to treat toxic or recalcitrant organic matter. It is against this background that the merits of linking membranes to advanced oxidation processes should be assessed.

AOPs are broadly defined by the use of hydroxyl radicals (OH·) which act as a very powerful oxidising agent. Key AOPs include heterogeneous and homogeneous photocatalysis based on near ultraviolet (UV) or solar irradiation, electrolysis, ozonation, Fenton's reagent, ultrasound (US) and wet air oxidation (WAO) (Comninellis *et al.*, 2008). Whilst the potential to use the above processes has been demonstrated concerns remain around the use of chemicals and/or energy associated with them. In part this has led implementation to focus on treating dilute and difficult pollutants such as pesticides (Malato *et al.*, 2000) and other recalcitrant compounds (Crittenden *et al.*, 1997).

In relation to the above, the coupling of membrane processes and AOPs is most strongly linked to heterogeneous photo-catalysis where semi conductor materials such as titanium dioxide (TiO_2) are used as the catalyst. The process is initiated by adsorption of photons onto the catalyst surface with an energy greater than the band gap of the catalyst resulting in molecular excitation between the filled valance band and the empty conduction band. The result is the generation of mobile electrons in the higher band (E_{cb}) and electrons holes in the lower valance band (E_{vb}) leading to the potential for both oxidative and reductive reaction pathways:

$$\mathrm{TiO_2} \xrightarrow{hv} e^- + h^+$$

The reaction is reversible leading to the potential for recombination of the electron and hole pair. This leads to a major inefficiency within the process and indicates that only a small proportion of adsorbed photons will leads to pollutant removal. The overall reaction pathway involves a number of steps which ultimately leads to the key reaction:

$$\mathrm{OH}^{\cdot}_{\mathrm{ad}} + RX_{\mathrm{ad}} \longrightarrow intermediats$$

Where RX_{ad} refers to the adsorbed pollutant onto the surface of the catalyst. The $OH^{\cdot}$ is very reactive and thus can completely mineralise organics to carbon dioxide and water. However, the complexity of organic structures being considered means that multiple $OH^{\cdot}$ are required with the possibility of desorption of the intermediates from the catalyst before complete mineralisation occurs. As such partial oxidation is likely with the intermediates being more reactive than the parent compounds. Direct experimental evidence for this has recently been reported during the treatment of natural organic matter where application of AOPs has increase disinfection by product formation potential over that of the raw water (Bond *et al.*, 2009; Phillipes *et al.*, 2010).

Modelling the process in terms of the rate of photoconversion leads to an expression of the form (de Lasa *et al.*, 2005):

$$r_1 = \frac{V}{W_{\mathrm{br}}}\frac{dC_i}{dr} = \sum_k v_{1,k}R_k$$

Where V is the total reactor volume (L), C_i is the concentration of chemical species i ($\mathrm{g \cdot L^{-1}}$), t is the time (s), $v_{i,k}$ is a dimensionless stochiometric coefficient for i species involved in the reaction step k, R_k is the rate of photoconversion of step k based on the weight of irradiated catalyst, W_{irr}. A key parameter within this equation is the weight of irradiated catalyst reflecting that only a proportion of the reactor volume is actively involved in pollutant removal and as such any expression defining a given reactor must reflect this through either weight of catalyst or reactor volume that is active rather than the total available. Overall a series of expression can be defined relating the rate to a given reactor geometry:

$$r_1''' = r_1' \frac{V_{irr}}{V} = r_1' \frac{A_{irr}}{V} = r_1 \frac{W_{irr}}{V}$$

Work to dates has shown that three factors influence the rate of photoconversion:

$$r_{i,in}''' = f_1(C_{i,in}) f_2(C_c) f_3(P_a)$$

Where $C_{i,in}$ is the initial concentration of pollutant species i, C_c is the catalyst concentration and P_a is the rate of adsorbed photons. Importantly, the conversion rate is expected to increase with catalyst loading up to a limiting value beyond which no further improvement will be seen. This reflects the point beyond which adding further catalyst does not increase the W_{irr} but just adds catalyst to the unreactive component (W_d). Evidence suggests that back scattering is complete at relatively low concentrations and so the maximum catalyst load to achieve an optimum reaction rate is likely to be relatively low. In addition a minimum concentration is required as initially catalyst will coat the surfaces of the technology such that below a certain starting concentration insufficient free catalyst will be available to drive a reaction. The exact concentrations concerned depend on the reactor geometry and wastewater being treated but to illustrate, in the case of phenol degradation in a 6.6 cm diameter glass tube, the limiting TiO_2 concentrations were 0.014–0.14 $g \cdot L^{-1}$ (Salaices *et al.*, 2001). The upper value equates to a 0.005% radiation transmission demonstrating the significance of catalyst concentration on photon penetration into the reactor and hence the active reactor volume. Converting such expressions to the rate of removal requires inclusion of an adsorption terms to reflect the dual nature of the process:

$$r_r'' = r_1''(1 + K')$$

Where R_T''' is the overall rate of pollutant removal and K' is a dimensionless adsorption term:

$$K' - \frac{K^* W q_{1,m}}{V}$$

Where K^* is equilibrium constant, $q_{1,m}$ is the maximum amount of adsorbed pollutant and W is the total weight of adsorbent.

3.3.2 Description of the technology

The technology that delivers a coupling of an advanced oxidation process to a membrane separation device is similar to a MBR but utilises a chemical reaction pathway to remove the pollutant rather than a biological pathway. To reflect this connection the technology will be referred to as a membrane chemical reactor (MCR).

The MCR combines three different aspects: initially the pollutant adsorbs onto the surface of the catalyst (stage 1: adsorption) after which the catalyst may become activated by a photon whereby it either oxides or reduces the adsorbed contaminant (stage 2: Photo-catalysis). Subsequently, the reacted catalyst needs to be separated out of the fluid and recycled back to the start of the process (stage 3: filtration) (Figure 3.22). To maximise stages 1 and 2, nano-particle (NP) sized catalysts are preferred (5–50 nm) as they deliver specific surface areas in excess of 250 $m^2 \cdot g^{-1}$. However, as soon as the catalyst particles are added into a water solution, the nano-particles aggregate into cluster sizes between 1000–10 000 nm. Consequently, microfiltration/ultrafiltration membrane are only required to retain the NPs and prevent loss with the treated effluent.

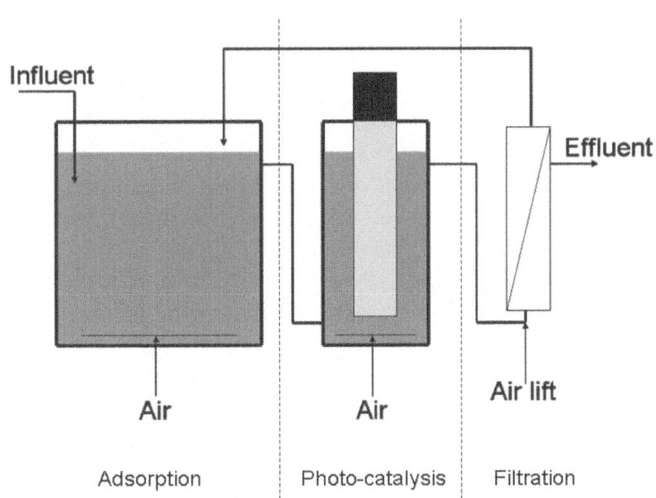

Figure 3.22 Process steps involved in MCR technology

In defining any embodiment of the MCR concept, the key features that must be specified include: (1) reactor geometry, (2) irradiation source, (3) catalyst type, loading and residence time and (4) membrane selection and operating parameters. Perhaps the most defining characteristic is whether slurry or immobilised catalyst systems are used. In the case of industrial wastewater treatment, the high pollutant concentration and scale means that slurry reactors systems are by far the most effective. In fact the majority of reactors used for photcatalysis in water treatment are of a well mixed slurry configuration (de Lasa *et al.*, 2005) with explanation provided in the case of solar systems which indicate a 3–5 times greater rate constant in slurry as opposed to immobilised systems (Wyness *et al.*, 1994). A number of slurry reactor systems have been reported including *Taylor Vortex Reactor* (Karpel *et al.*, 1997), *Turbulent Slurry Reactor* (Butters and Powell, 1995) and *Slurry Annular Reactor* (Andreozzi *et al.*, 2000) to name but a few. All the different designs involve methods of trying to efficiently expose the largest proportion of catalyst to the irradiation source in the minimum residence time. For instance in the case of annular reactors, the reactor contains two concentric tubes with the inner tube being transparent to radiation with the TiO_2 slurry flowing through the channel created between the two tubes (Figure 3.23a). The reactor design has advantages in that the irradiation is symmetrical and the flow patterns are relatively controlled. As such annular reactors are by far the most common and are considered most appropriate for industrial wastewater treatment (de Lasa *et al.*, 2005). However, operating flow rates are limited by the narrowness of the gap and the reactor system is modular making larger flow rate systems very expensive. As an alternative open tank systems have been developed where either the lamp and/or the membrane are submerged into a mixed tank (LeClech *et al.*, 2006; Fu *et al.*, 2006; Rivero *et al.*, 2006). These have the advantage of enabling much greater flows and provide a scale up basis that is non linear although mixed reactors suffer from poor mixing environments with corresponding large inactive reactor volume proportions (Figure 3.23b). The irradiation source can be either artificial (a UV lamp) or solar as 4–5% of solar irradiation exists at wavelength suitable to excite the catalyst. Whilst examples of solar AOPs exist for industrial pesticide removal (Malato *et al.*, 2000) in the majority of industrial wastewater cases the high feed concentration restricts operation to artificial irradiation.

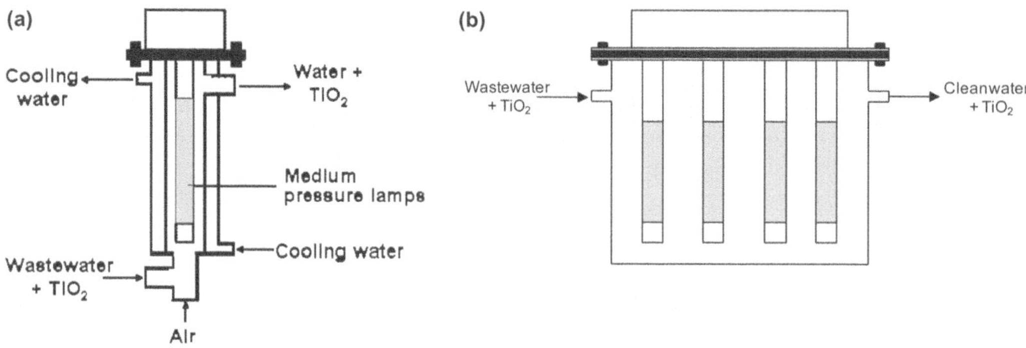

Figure 3.23 Reactor configurations: Annular and Completely mixed

Appropriate design of the membrane is perhaps best realised when it becomes passive within the system. Otherwise, fouling will control the overall process rate such that the reactor is being underutilised. Comparison across the literature reveals a wide variety of membrane systems have been considered including all major geometries (Flat sheet, tubular, hollow fibre), all main membrane materials and both submerged and side stream configurations. The key consideration relates to the link between the membrane and the reactor: clean TiO_2 has a very low fouling propensity such that sustainable fluxes in excess of 100 LMH are possible (Rivero *et al.*, 2006). However, partially regenerated TiO_2 (still coated with organics) generates much greater fouling such that overall membrane operation is greatly enhanced by effective photo-catalytic reactor design (Pidou *et al.*, 2009).

The majority of work to date reporting on design and operation of MCR technologies has focussed on dilute pollutants reflecting where the technology is likely to be implemented first. Such systems are concerned with removing trace levels of recalcitrant compounds and as such low catalyst concentrations are typically used (50–200 $mg \cdot L^{-1}$), maximising active reactor volume fractions. However, treatment of industrial wastewaters typically concerns much higher strength solutions and as such higher catalyst concentrations are required to ensure sufficient contact area to drive adsorption. Consequently, current understanding of design and operation issues for treating industrial wastes is much less understood. The following section details finding related to use of a MCR for treatment of industrial wastewater in order to understand the key issues faced when developing MCR technology for this application. The findings are based on two MCR pilot plants: an annular reactor with four different gap sizes and a mixed tank with submerged UV lamps (Figure 3.24). A series of synthetic (acid orange 7 dye, bisphenol A, nalidixic acid) and real (metal working fluid [MWF], landfill leachate, textile wastewater, pharmaceutical wastewater and laundry wastewater) were tested in both reactors with the impacts of residence time, lamp type and catalyst (type and concentration) considered in each case. Characteristics of the five main wastewaters used throughout the trial reveal large variations in conductivity and variation in pH (Table 3.2). This is expected to influence the nature of the relationship between the adsorbed organics and the TiO_2. A factor not explored here but that requires consideration is the inherent variability in industrial wastewaters. For instance, metal working fluid treatment facilities often take waste from a variety of sources leading to large changes in chemical composition from batch to batch. Operation of MCR technology must be capable of mitigating such variations and so an understanding of the impact of changing conductivity, pH and hydrophobicity are likely to be crucial.

Figure 3.24 Pilot plants used during the investigations (a) annular and (b) completely mixed

Table 3.2 Details of some of the wastewaters used to demonstrate MCR design factors

	COD (mg/L)	DOC (mg/L)	pH	Zeta potential (mV)	Conductivity (mS/cm)
MWF	1181	372	8.5	−20	3
Bisphenol A	361	128	6.5	–	0.2
Laundry Wastewater	2612	908	7.9	−16	1.5
Nalidixic acid	2147	650	7.8	−22	9.5
Textile Waste	342	90	9.3	−18	2.9

3.3.3 Factors influencing design: Results

Wastewater type

Comparative assessment of the annular MCR configuration at a fixed gap width and residence time demonstrates that the effectiveness of the MCR is highly source water specific (Figure 3.25). Overall three groups can be identified: (1) *easy to treat* identified by removal efficiencies greater than 80%, (2) *difficult to treat* with removal efficiencies less than 40% and (3) *moderate to treat* with removal efficiencies between 40 and 80%. For a wastewater to be defined as easy to treat the organic molecules within the wastewater must have both a high adsorption potential onto TiO_2 as well as a high photocatalytic reaction rate constant (Table 3.3). Whereas if treatment is limited by adsorption or photocatalytic reaction rate the wastewater is defined as moderately treatable. Finally for wastewaters to be defined as difficult to treat either one or both of the two factors must be limiting.

Unfortunately, in real industrial wastewaters it is not possible to directly know the individual compounds held within and as such empirical testing is likely to be required. The data presented in Figure 3.25 is based on a single set of testing conditions that were found to be effective for grey water to act as a bench mark as these condition have been shown to make consideration of MCR technology economically viable. As such it should be recognised that improved removal in all the trialled wastewaters is possible by optimising the conditions for each specific wastewater type. This can mean increased catalyst load when adsorption is the limiting condition or increasing photocatalytic reaction time if photocatalysis is rate controlling feature. To illustrate, in the case of the MWF, the removal during the comparative trial was 50% whereas during later optimisation work a maximum removal of 85% was observed (Figure 3.32).

Three industrial wastewaters (Landfill leachate, pharmaceutical wastewater and laundry wastewater) and two municipal wastewaters (aerobic and anaerobic) were seen to have low inherent treatability through the MCR. Comparative trials with UV/H_2O_2 revealed much higher levels of treatment suggesting the process was not limited by $OH^{\bullet}$ oxidation. To illustrate in the case of the pharmaceutical wastewater, the comparative trial delivered a removal of 40% whereas application of an equivalent UV dose and 1 $g \cdot L^{-1}$ of H_2O_2 yields 80% removal. Thus adsorption onto the TiO_2 (or readsorption of the intermediates) appears to be a critical issue on overall treatment. This is supported by previous characterisation studies which have shown all the identified wastewaters to be relatively hydrophilic. A similar picture is seen with the treatment of natural organic matter where hydrophilic NOM water sources (lowland waters) are more poorly treated than hydrophobic NOM water sources (moorland sources). However, a number of the tested wastewaters showed a high potential for treatment in the MCR including the textile wastewater and the bisphenol-A solution. Ultimately, it is clear that the effectiveness of MCRs is industrial wastewater specific. Wastewaters graded as easy or moderate appear treatable with the later group requiring optimisation to assess the minimum reaction conditions to drive sufficient removal.

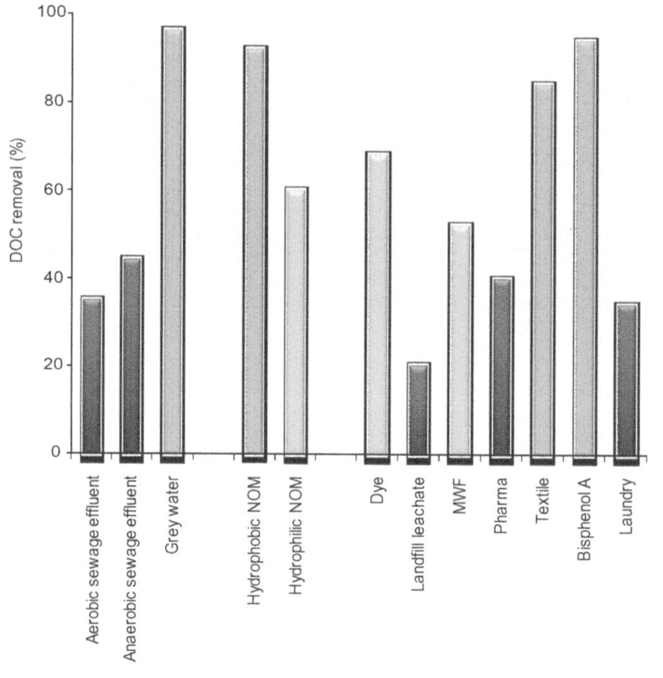

Figure 3.25 Comparative treatment efficacy for different wastewater types

Table 3.3 Ranking of wastewater properties in relation to treatability in the MCR

Grade	Hydrophobicity	Reaction rate
Easy	Hydrophobic	High
Moderate	Hydrophobic	Medium-low
Moderate	Hydrophilic	High- medium
Difficult	Hydrophilic	Low

Choice of catalyst and UV lamps

The performance of three commercially available media in terms of adsorption, photo-catalytic reaction and membrane fouling demonstrates significant difference based on catalyst selection (Figure 3.26). The media tested were Hombikat UV100, Degussa P25 and Adsorbia GTO (Table 3.4) with each media providing different properties in terms of size and composition. The composition is important as anatase and rutile have different photo catalytic properties primarily seen in a differences in there band gap energy (3 eV for rutile and 3.2 eV for anatase) resulting in excitation wavelengths of 413 and 388 nm respectively. The wider band gap energy for anatase results in superior activity and reduced electron hole recombination making it the preferred catalyst in most cases.

The impact of catalyst selection is demonstrated for the case of treating a metal working fluid at a starting COD concentration of 1000 $mg \cdot L^{-1}$ in a completely mixed MCR configuration. In terms of adsorption (dark light) the Hombikat UV1000 exhibited far greater adsorptive capacity across the dose range tested at an equilibrium time of 24 hours (Figure 3.26) consistent with it much smaller particle size (Table 3.4). However, the Adsorbia GTO provides more removal than the Degussa P25 even though its particle size was much greater. To illustrate, COD removal of

16.4, 4.3 and 2.9% was observed for a dose of 5 $g \cdot L^{-1}$ for the Hombikat UV1000, Adsorbia GTO and Degussa P25 respectively. The switch in order reflects the impact of porosity which has been designed into the Adsorbia GTO media akin to an activated carbon structure. In such cases the process is dominated by intra particle diffusion whereas for the other two media, surface adsorption dominates. The Irregularities in the curves reflect changes in the aggregate properties as the catalyst dose is increased which changes the available adsorption contact area.

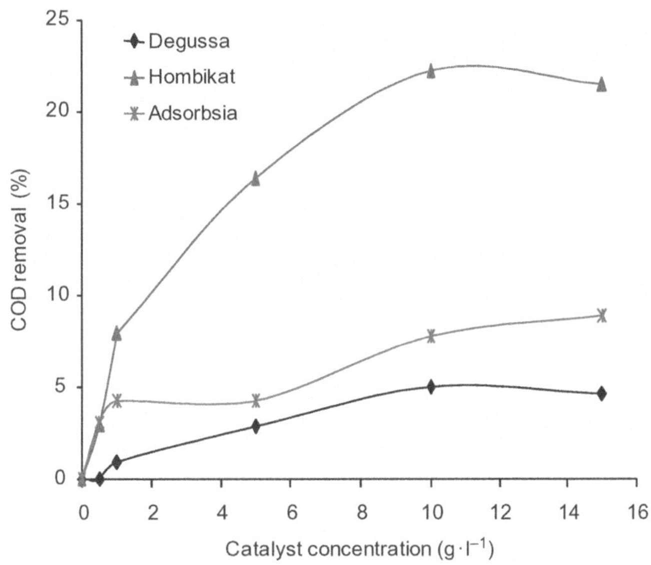

Figure 3.26 Comparison of adsorption for different TiO_2 nanoparticles demonstrated for the treatment of a metal working fluid

Table 3.4 Characteristic of three commercial TiO_2 media

	Hombikat UV100	Degussa P25	Adsorbia GTO
Size	5 nm	25 nm	5% > 2 mm; 10% < 0.25 mm
Composition	75% anatase 25% rutile	100% anatase	

Comparison between media in terms of photo-catalytic reaction effectiveness demonstrates a different pattern where there is no real difference between the media at low irradiation doses (<5 $J \cdot cm^{-2}$). Above that the Degussa P25 is significantly more effective delivering a near linear increase in removal as a function of dose (Figure 3.27). In contrast the other two media have either plateaued or increased only slightly indicting that Degussa P25 is likely to be the most appropriate for industrial wastewater treatment where higher irradiation doses are expected to be needed. To illustrate the difference between the media, at an irradiation dose of 13.8 $J \cdot cm^{-2}$ COD removals of 30, 20 and 8% were recorded for the Degussa P25, Homikat UV100 and the Adsorbia GTO respectively. No difference was observed in terms of fouling propensity when tested with a 1 m long, tubular membrane consisting of 10 PES fibers of 5 mm internal diameter and 0.05 μm pore size (Figure 3.28). Overall, catalyst choice is seen to influence initial adsorption and photo-catalytic activity but have no real influence over membrane operation enabling an appropriate media to be selected based on the balance of needs. In the case of industrial wastewater treatment, the need for large reaction rates drives preference for the Degussa P25. Recent advances have seen a new range of TiO_2 based media being developed where the base catalyst is doped with metals and metals oxides such as platinum (Vorontsov *et al.*, 2000) altering the ease of separation of the charges enhancing activity and enabling solar light to be more effectively utilised. To date there has been only limited reports on their effectiveness at pilot scale but looking forward doped catalyst are likely to be crucial in development of successful MCR technologies and the relative merits against the three components outlined here will need to be determined.

The radiation source used delivers near ultraviolet radiation in the 200–400 nm wavelength range with sub division into UVA (315–400 nm), UVB (280–315 nm) and UVC (200–280 nm). Artifical UV sources are classified according to Table 3.5 with the majority of application to date utilising low or medium pressure lamps. A key consideration of using MCR technology is the energy required to run the technology as both UV and membrane processes are perceived to be high energy using components. Energy Efficiency is usually quoted in terms of the Electrical Energy per Order (EEO) which is a semi empirical parameter which defines the kWh required to reduce the concentration of a compound in

1000 gallons of water by one order of magnitude (De Lasa *et al.*, 2005):

$$EEO - \frac{Pt3785}{V\log\left(\frac{C_{in}}{C}\right)}$$

Where P is the lamp emitted power (kW), t is the irradiation time (s), V is the reactor volume, C_{in} is the initial concentration and C is the concentration after time t. Accordingly, small EEO describe a more efficient treatment. Care

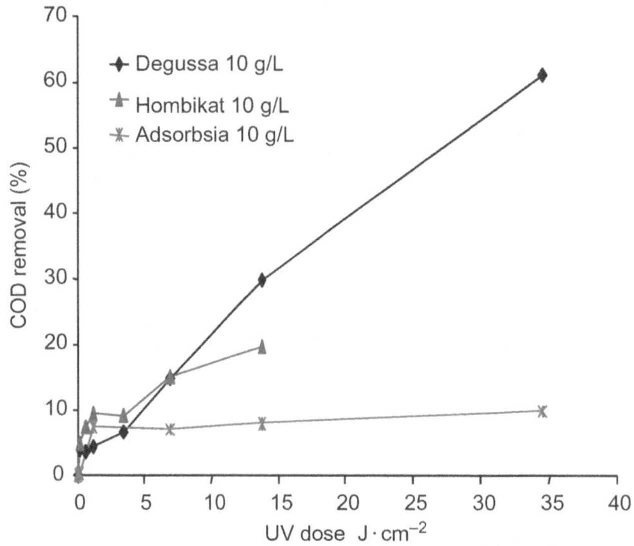

Figure 3.27 Comparison of photo-catalytic reaction effectiveness for different TiO_2 nanoparticles demonstrated for the treatment of a metal working fluid

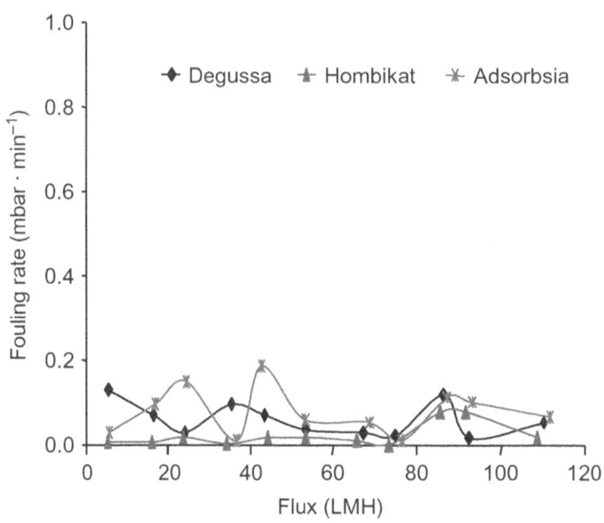

Figure 3.28 Comparison of fouling propensity for different TiO_2 nanoparticles demonstrated for the treatment of a metal working fluid

Table 3.5 UV light sources (adapted from Bolton *et al.*, 1995)

	Low pressure	Medium pressure	High pressure
Life time (d)	>5000	>2000	>3000
Output range	80% in narrow range around 254 nm	Broad but not much below 250 nm	Strong below 250 nm
Energy Density	1 W/cm	125 W/cm	250 W/cm
Electrical energy to photon energy	30%	15%	30%

should be taken as the above equation assumes first order kinetics and is sensitive to the method of concentration measurement (specific compound or bulk organic parameter). Importantly when based on specific chemicals, the energy associated with treatment of intermediate products is ignored and so reduces the EEO compared to measures based on COD or TOC. Whilst such an approach appears generally appropriate, the assumption of first order kinetics for the photocatalytic treatment of industrial wastewaters is less certain potentially limiting its use as a comparative guide.

Comparison of the different wavelength lights is shown in the case of low pressure lamps fitted into the completely mixed MCR configuration and tested for the treatment of both MWF and an artificial dye solution made up of acid orange 7 dye [AO7] (Figure 3.29). For both wastes the UV-B lamps are seen to be most effective with the lowest EEO of between 400–9000 $kWh \cdot m^{-3}$ compared values of 700 and 1400 $kWh \cdot m^{-3}$ in the case of UVA for treatment of MWF and AO7 respectively. However, during the trials it was clear that the lamp were more powerful than required as the temperature of the water increased significantly. Consequently, much lower power lamps will potentially deliver a similar level of treatment, vastly reducing the reported EEOs.

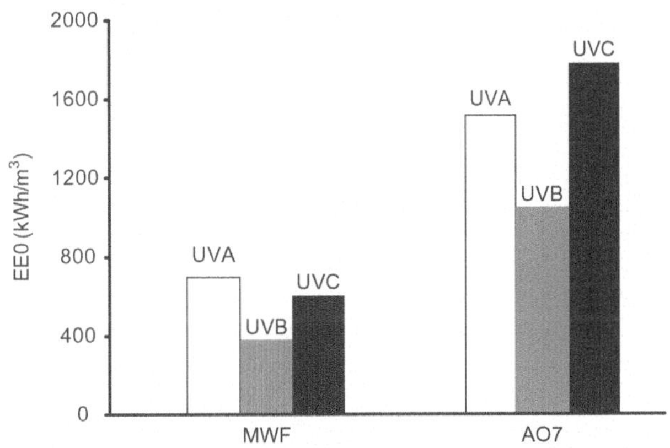

Figure 3.29 Comparison of UV source type on the treatment of two industrial wastewaters

Reactor loading and catalyst dose

The applied catalyst dose influences two aspects of the process: adsorption and photocatalytic reaction. The latter is known to be maximised at relatively low catalyst doses with the photocatalytic reaction rate remaining constant once a maximum dose is exceeded. This reflects that only a proportion of the applied catalyst is active in the photocatalytic reaction linked to exposure from the UV light source. Consequently, a significant proportion of the catalysis audit is inactive. The exact proportion that is active relates to the working volume near the lamp surface and decreases with dose due to the light blocking and reflecting properties of TiO_2. For instance, at a dose of 0.35 $g \cdot L^{-1}$ radiation transmission is at around 0.005% of that in a clean liquid (De Lasa *et al.*, 2005). Whilst photocatalytic reaction is not improved with increasing dose, adsorption is as the capacity and rate of adsorption are directly influenced by the available contact surface area (Figure 3.26). In the case of industrial wastewater treatment of metal working fluid, an optimum catalyst dose of 10 $g \cdot L^{-1}$ has been identified which is significantly higher than used in most other applications (50–200 $mg \cdot L^{-1}$). At such doses it is postulated that photocatalytic activation occurs only a few nano-particle diameters into the reactor and as such reactor designs must be able to ensure sufficient mixing to enable the entire catalyst audit to enter the activation zone during the residence time in the reactor.

Illustration of the impact of catalyst dose on removal is provided in the case of AO7 treatment in an annular MCR configuration with a 2.5 mm annular gap between the lamp and the outer wall of the reactor (Figure 3.30). AO7 offers an additional issue in that the dye adsorbs UV light and so reduces the amount available to drive photocatalysis. This can be seen in the case of the concentrated trial where no removal was observed below catalyst doses of 50 $g \cdot L^{-1}$. Above 60 $g \cdot L^{-1}$ removal was found to vary between 36 and 46% indicating that a certain ratio of catalyst to available organics is required to ensure effective photocatalysis. A similar picture is observed at lower initial organic concentrations where removal increases from 43% at a dose of 5 $g \cdot L^{-1}$ up to a maximum of 95% removal at a dose of 10 $g \cdot L^{-1}$ before dropping off slightly at higher doses and ranging between 75–90% thereafter up to the maximum dose trialled of 130 $g \cdot L^{-1}$. Optimum removal in the case of AO7 occurred at dose ratios of 0.014 and 0.01 $gTiO_2 \cdot mg_{COD}^{-1}$ for the high and low feed strength systems respectively. A similar trend was observed in the case of treatment of a metal working fluid in the same MCR configuration with the exception that even at the higher COD loading concentration some removal occurred (28%), reflecting the impact of direct adsorption of UV irradiation by AO7. In the case of the MWF fluids, optimum removal occurred at a dose ratio of 0.014 and 0.01 $gTiO_2 \cdot mg_{COD}^{-1}$ for the high and low feed strength systems respectively yielding removals of 46 and 81%. The shape of the observed curves indicates that it is possible to under or over dose the catalyst due to limited adsorption and UV shielding respectively. The latter factor is also related to mixing as the optimum dose reflects the best catalyst concentration to ensure as much of the TiO_2 enters the activation zone as possible. In addition there appears to be a limitation on the strength of wastewater that this configuration of MCR can effectively treat.

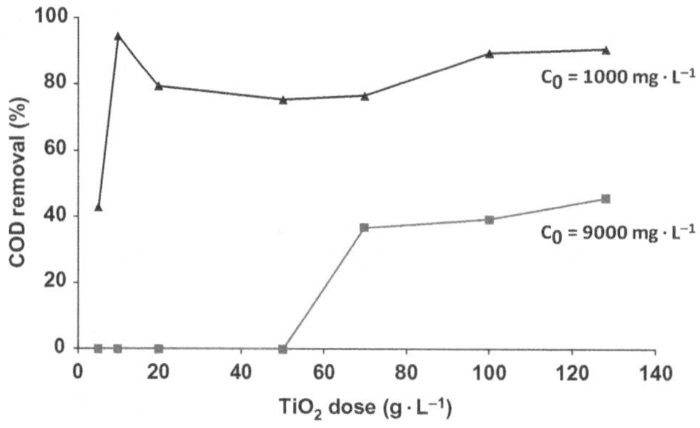

Figure 3.30 Impact of catalyst dose on removal of AO7 in an annular MCR configuration

In practice it is likely that a fixed dose of catalyst will be used and then the process must operate over the range of feed strength is encountered. This makes understanding the variations in the feed important in designing the system. The impact of changing feed strength is illustrated in the case of an annular MCR configuration with a 2.5 mm annular gap operated at a fixed residence time of 20 minutes, irradiated with a 600 W medium pressure lamp and a catalyst dose of $10\ g \cdot L^{-1}$ (Figure 3.31). Similar trends are observed for both wastewater types presented in that a maximum removal is seen at a specific feed concentration with removal decreasing at either lower or higher feed strengths. For instance, in the case of the metal working fluid, removal increased from 56% at very low doses up to a maximum of 81% at $1000\ mg \cdot L^{-1}$ feed and then gradually dropped down to 28% at a feed strength of $8835\ mg \cdot L^{-1}$. Similarly, in the case of AO7 treatment, removal increased from 68% at very low doses up to a maximum of 95% at a feed strength of $1000\ mg \cdot L^{-1}$ but then dropped off significantly such that no removal was effectively seen beyond a feed strength of $5000\ mg \cdot L^{-1}$ and only 18% at a feed strength of $2800\ mg \cdot L^{-1}$. Comparing the two industrial wastewaters provides some interesting key features:

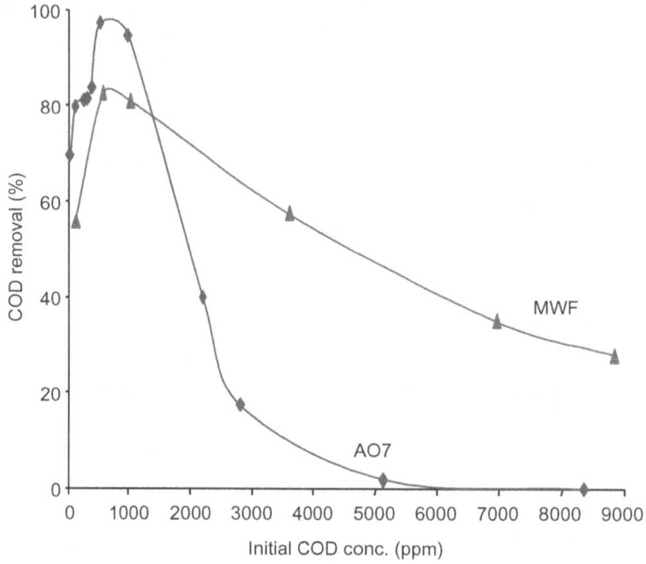

Figure 3.31 Impact of feed concentration on treatment efficacy at a fixed catalyst dose

Removal is sub optimal at catalyst doses commonly utilised (50–200 $mg \cdot L^{-1}$) in other systems where dilute pollutant concentrations are encountered (Chrion *et al.*, 2000; Konstantinous *et al.*, 2001) such that an additional 20% removal can be obtained by operating at the optimum catalyst dose. In the two cases here, an optimum dose is observed around 0.01 $gTiO_2 \cdot mg_{COD}^{-1}$ which seems to be an effective guide to define the catalyst dose required for such applications. There appears to be a relatively narrow optimum range away from which removal drops off. In the current case this appears to be related to a change in feed strength of around 700 $mg \cdot L^{-1}$ such that the optimum catalyst dosing range is 0.006–0.0125 $gTiO_2 \cdot mg_{COD}^{-1}$. The nature of the drop off is very different for the two industrial wastewater with removal in the dye based systems dropping off far more rapidly as a function of catalyst dose than for the metal working fluid. This indicates the waste specific nature of the process with the exact relation determined by the relative hydrophobicity, natural absorbance of UV and the impact of the nano-particle aggregates that form. In particular, significant under dosing of catalyst in the AO7 case results in no treatment and would make photocatalysis look unable to operate for such wastewaters.

Photocatalysis occurs when the catalyst adsorbs photons and as such only catalyst exposed to the light sources will be activated. This represents, perhaps, the most critical limitation of photocatalysis with either artificial or solar irradiation sources as reactors designs require large contact irradiation zone areas to enable sufficient flow to be treated. There are two basic technological solutions to the problem: (1) use a narrow gap such as in an annular reactor (2) use larger mixed tanks with sufficient mixing. The impact of configuration choice is illustrated for treatment of a metal working fluid wastewater with both systems operating at a catalyst dose of 10 $g \cdot L^{-1}$. The annular reactor contained a 5 mm gap representing working volume of 0.077 L compared to the mixed tank that had a working volume of 9 L. If residence time is adjusted so that they treat the same flow, 4.5 $L \cdot h^{-1}$, the mixed system delivers slightly higher treatment efficiency at 22% compared to the annular system which delivered 16%, although both are very low (Figure 3.32). The issue is available contact time which can be improved by reducing the residence time. For instance, for the annular reactor, increasing the average residence time to 10 minutes results in a significant improvement in removal to 46% (Figure 3.32) whereas a residence time in excess of 20 hours (batch operation) was required to achieve an equivalent removal in the mixed system.

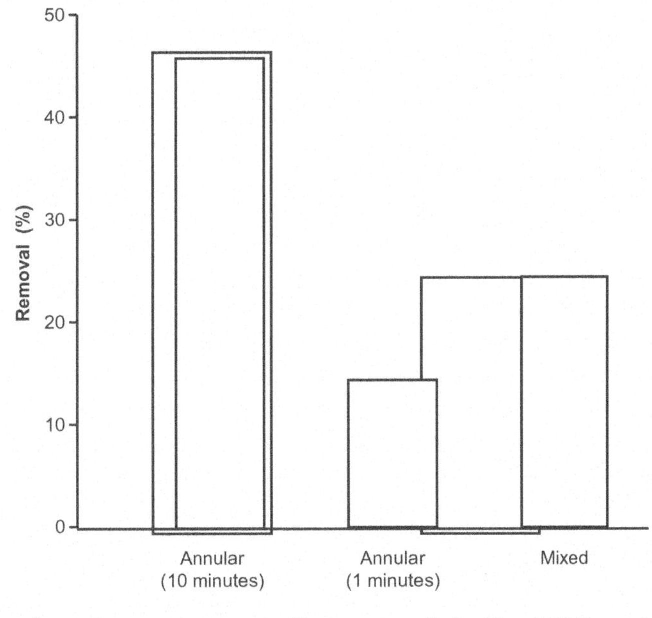

Figure 3.32 Impact of MCR configuration on treatment efficacy when fed with a MWF wastewater

Ultimately a balance exist between using better mixing/smaller annular gaps (to improve treatment efficiency) and residence time to ensure appropriate treatment at a suitable flow rate per unit. The balance is illustrated for a metal working fluid treated in an annular MCR configuration where the annular gap size and the residence time have both been varied (Figure 3.33). The MCR is treating a metal working fluid at an initial concentration of 1000 $mg \cdot L^{-1}$ and contains a catalyst concentration of 10 $g \cdot L^{-1}$ placing the dose ratio within the optimum range. As expected increasing residence time improved removal efficiency for all annular gap sizes. For example, with the largest gap of 15 mm, COD removals of 13, 20, 31 and 40% were observed for residence times of 1, 5, 10 and 20 minutes respectively. Improvement was also seen by decreasing annular gap size with removal levels of 31, 45, 55 and 66% observed for gap sizes of 15, 5, 2.5 and 1 mm respectively at a residence time of 10 minutes. Interestingly, the impact of annulus gap size appears less critical than residence time over the ranges examined here. A similar picture was observed for equivalent experiments but with AO7. Key differences were that not such a marked improvement was seen as a function of gap size and/or residence time. In this case a residence time of 20 minutes was required before high removal levels were observed irrespective of annular gap size.

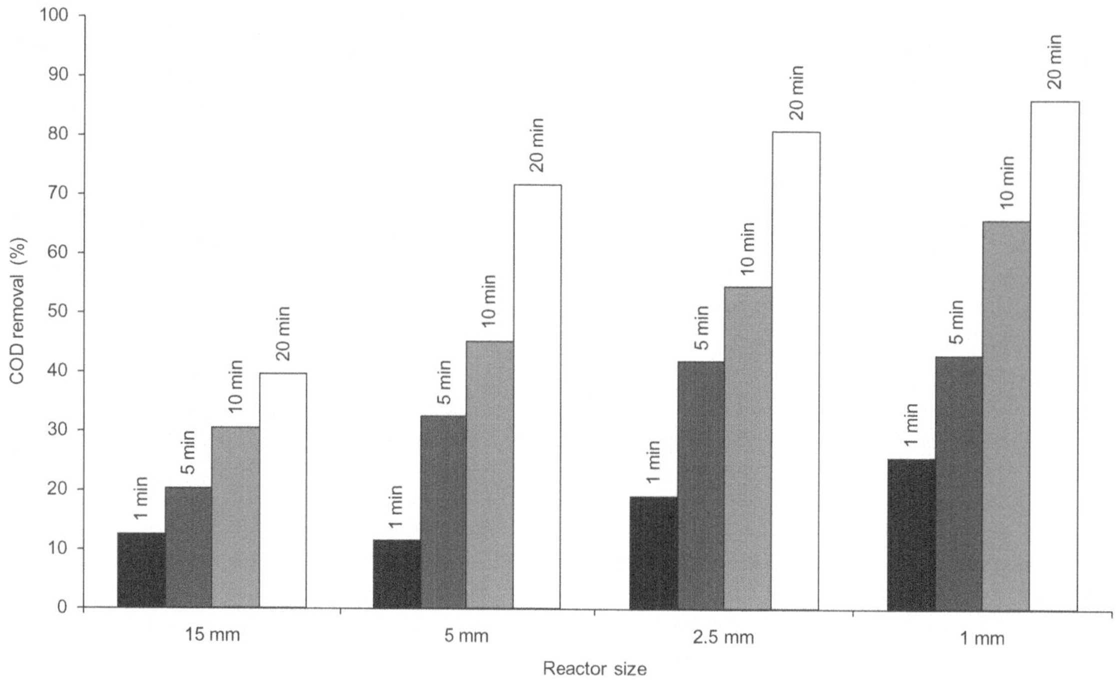

Figure 3.33 Impact of annular width and residence time on treatment efficacy when fed with a MWF wastewater

A number of general features can be observed: Removal is poor for residence times of less than 5 minutes irrespective of annular gap size. Further, there is a big difference between 15 mm and 5 mm gap sizes which is less obvious as the annulus gets progressively smaller. High removal efficiencies can be obtained for industrial wastewaters if smallish gaps and long residence times are used. This is especially significant as the two industrial wastewaters detailed in this section where previously identified as demonstrating moderate treatability in the comparative trials (Figure 3.25) such that more inherently treatable industrial wastewaters are likely to require shorter residence times to achieve the same total level of removal.

The observed difference reflect the relative effectively of slurry transport to the photocatalytic activation zone which is effectively only a few nanoparticle diameters away from the lamp surface (<200 nm). Ultimately this becomes a mixing problem and involves ensuring all of the catalyst is exposed for a sufficient length of time within the activation zone during passage through the reactor. Unfortunately, to date, no adequate mixing model exists to adequately describe slurry mixing to a fixed point within a reactor (the lamp surface) restricting theoretical developments especially in industrial wastewater applications where high catalyst audits are required. A number of techniques have been empirically tried to improve treatment efficacy by imparting turbulence either through the use of a slurry distribution systems up front of the reactor zone to ensure intense mixing (Salacies *et al.*, 2001) or gas bubbling (Phillipes *et al.*, 2010). The later is a further adaptation of MBR operation where coarse gas bubbling is utilised in tubular and flat sheet configuration to generate slug flow bubbling which generates high levels of shear at the membrane surface (Judd, 2010). In fact slug flow is extremely effective as it generates very high levels of variation in the shear pattern as the front and back ends of the slug pass over a given point on the surfaces constraining the bubble shape. It is the variation in shear rate that has been attributed to the effectiveness of slug flow operation in MBRs (Judd, 2010) such that aeration is now commonly operated in an intermittent manner rather than continuously. For instance, in further experiments with the annular MCR configuration treating a metal working fluid, the impact of no aeration and medium ($10\ L \cdot min^{-1}$) to high ($20\ L \cdot min^{-1}$) level aeration was examined. The difference between no (pumped flow) and aeration (gas lift operation) was significant with an almost doubling of the removal efficiency observed at high residence times. However, small differences were seen between the two mixing intensities although this reflects the initially high rate used which was set by the minimum gas lift rate required for effective flow through the reactor. A combined pump and air lift system appears the most favourable as much lower gassing rates will be required. This is akin to side stream tubular membrane operation where combined pump/gas lift systems have been effectively utilised.

Membrane Operation

A key requirement for all slurry MCR technology is the need to separate the nano-particles from the treated water prior to discharge. A number of systems have been trialled including centrifugation, prolonged batch settling and coagulation (De Lasa *et al.*, 2005; Chong *et al.*, 2010) although the most common method is to utilise a membrane separation device

(Molinari *et al.*, 2002; LeClech *et al.*, 2006; Rivero *et al.*, 2006). A wide range of membranes have been utilised including side stream and submerged configurations, hollow fibre, tubular, flat plate module geometries and both polymeric and ceramic materials. The primary particle size of the catalyst nano-particles can be as low as 5 nm (Table 3.4) leading some researchers to focus on nanofiltration (NF) membranes. However, in water environments the nano-particles naturally aggregate such that effective sieve sizes akin to porous ultrafiltration membrane and microfiltration membranes have been shown to be suitable. This has a great advantage as they provide much higher clean water permeabilities providing the potential for lower energy, higher flow operation than using NF systems as long as the membrane can be operated without excessive fouling.

The majority of practical studies into membrane fouling of porous membranes involve the measurement of the fouling rate (or permeability decline) as a function of different fixed flux rates (Le Clech *et al.*, 2003). Analysis of such an approach defines a critical flux [Jc] (strictly the weak form critical flux) corresponding to the flux rate beyond which fouling become rapid and represents the upper practical limit that the membrane can be operated at. Beyond that, longer term fouling studies, below Jc, define a sustainable flux policy which provides a picture of how long the membrane can be operated at between cleaning cycles under any given operating condition (Pollice *et al.*, 2005; Brooks *et al.*, 2006). Understanding the limiting operation conditions for any membrane then involves conducting such fouling trials to establish the impact of backwash frequency, chemical cleaning and foulant propensity of the system. This is a well establish practice in the field of MBRs (Judd, 2010) but is much less understood concerning MCR systems. Early work has established that TiO_2 in water has a low inherent fouling propensity such that operating fluxes in excess of 100 LMH ($L \cdot m^{-2} \cdot h^{-1}$) can be operated at without significantly fouling in a tubular side stream gas lift system (Rivero *et al.*, 2006). This contrasts, MBR systems where operating fluxes of around 15–30 LMH are common and suggests that the membrane component of the system should not have as significant an influence on the overall process suitability as in MBRs. Fouling is determined by the size of the NP aggregates and the inherent fouling propensity they exhibit. As such different industrial wastewaters are likely to exert different inherent fouling propensities. This is illustrated in the case of four industrial wastewaters treated in an annular MCR configuration connected to a PES polymeric tubular membrane with a 50 nm pore size (Figure 3.34). Relatively limited fouling is observed with the landfill leachate and the AO7 resulting in critical fluxes of 51 and 67 LMH. In contrast, much greater fouling was observed for the pharmaceutical wastewater and especially the metal working fluid with critical fluxes of 35 and 15 LMH respectively. Importantly, these are all substantially lower than with clean TiO_2 and in the case of the MWF below normal operating fluxes for MBRs. Equivalent tests conducted on the mixed tank configuration MCR on the pharmaceutical and landfill leachate wastewaters identified critical fluxes of 46 and 40 LMH respectively which is different than in annular reactor configuration, however the two systems provided different levels of treatment so direct comparison is complicated. Additional trials to establish sustainable flux plots for the two industrial wastewaters indicated very different fouling profiles even though the critical fluxes were similar (Figure 3.35). The profiles presents a classical sustainable flux curve where fouling is low and increases slowly up to a critical time where after fouling become rapid. This is explained in terms of localised flux increasing beyond the critical flux or a critical deposited mass (Judd, 2010). Critical times for the pharmaceutical wastewater trials were 0, 1900 and 3000 minutes compared to 750, 1700 and >4000 minutes for the landfill leachate at 10, 20 and 30 LMH respectively. For similar fluxes in MBRs, critical times range between 4320–33120 minutes suggesting a more demanding operating cycle is likely to be required in the MCR. The combination of high critical fluxes but lower critical times suggest that fouling in MCRs behaves differently to that in MBRs restricting our ability to utilised existing knowledge on membrane operation.

No universal fouling model exists for MBRs reflecting the complex nature of the fouling process and the fact that a wide range of foulants can cause the observed responses. However, a relationship exists between the operation of the bioreactor and membrane fouling. As such the technology requires effective integration of the two in order to achieve optimum operation and the lowest possible energy use. An equivalent picture can be seen in the MCR where fouling is strongly linked to feed water characteristics and reactor operation. As with MBRs, no agreed fouling model exists for MCR operation and currently fouling is reported empirically for each study. A direct link between the effectiveness of the photocatalytic reactor and the membrane can however be observed; in fact there is a more clearly defined link than in MBRs where fully reacted TiO_2 presents a fouling propensity similar to that of clean unused TiO_2, whereas poorly treated TiO_2 presents a much greater fouling propensity (Figure 3.36). Indeed, this relationship has been detailed in terms of photocatalytic residence time and source water type in the case of grey water treatment in a mixed tank configuration MCR (Pidou *et al.*, 2009). Fouling rate curves were seen to move progressively towards the clean TiO_2 line as the photocatalytic reaction time was increased. Thus an operational balance exists enabling a reduction in the energy demand associated with membrane operation to offset increased energy in running at longer residence times in the photocatalytic reactor. The above also indicates that the reactor configuration is likely to influence membrane operation through differences in treatment efficiency. Comparison of fouling in annular and mixed tanks MCR configurations confirms this observation in that the critical fluxes are usually higher in the annular configuration. This reflects that more of the TiO_2 audit is likely to contact the lamps when operating annular reactors whereas a proportion of the coated TiO_2 is likely to pass through the mixed tank reactor before being exposed to UV irradiation.

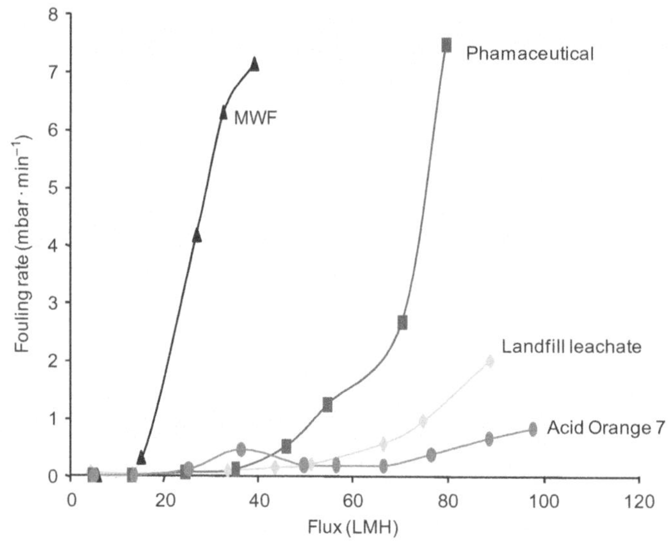

Figure 3.34 Impact of industrial wastewater type on the fouling propensity of TiO_2

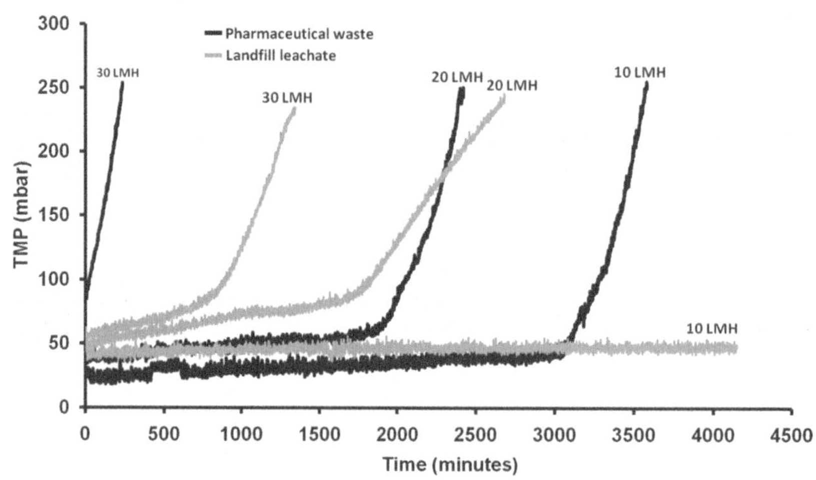

Figure 3.35 Sustainable flux profiles for two industrial wastewaters in the MCR

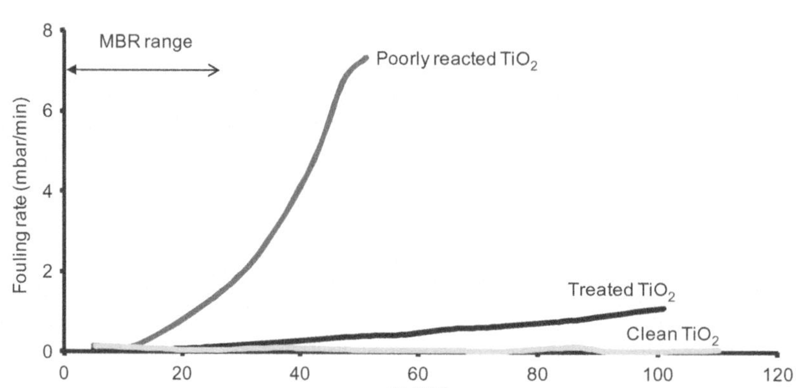

Figure 3.36 Sustainable flux profiles for two industrial wastewaters in the MCR

The other major design decision is in the types of membrane which can have a dramatic influence on the overall ease of membrane operation. Conventional practice in MBRs is to use polymeric submerged membranes as they provide the least expensive option. In the case of MCR operation a relatively big difference between fouling profiles indicates a different strategy may be beneficial. To illustrative the relative impact of membrane selection, data from fouling trials of three membranes are presented based on the use of an annular MCR configuration treating a metal working fluid wastewater (Figure 3.37). The membranes trialled were a hollow fibre polymeric membrane for use in a submerged configuration, and two side stream tubular membranes, one polymeric and one ceramic (Table 3.6). Maximum fouling was observed in the case of the submerged polymeric hollow fibre membrane were fluxes above 18 LMH resulted in very rapid fouling. In contrast, the two tubular membranes were able to operate up to 30 LMH before either started to demonstrate significantly fouling. Beyond 30 LMH a clear difference was observed in that the ceramic membrane started to foul whereas the polymeric membrane continued to operate with low fouling rates. Differences between the fouling profiles of the two tubular membranes can be linked to the surface properties of the materials with the polymer materials providing a much smoother surface with an average roughness around 5 nm which is considerably smaller than the primary catalyst particle size used at 25 nm (Table 3.4). Whereas, average roughness for the ceramic membrane was around 100 nm which is substantially larger than the primary particle size and thus likely to exert a large influence over the fouling behaviour. Overall, the tubular membranes operated more effectively than the hollow fibre which reflects the more controlled hydrodynamics that can be utilised due to the confined nature of the channels they contain. In contrast submerged hollow fibre systems are non confined and thus operate with very different hydrodynamics. In particular, the principle fouling control mechanism is thought to be related to fibre vibration as opposed to classical back lift with the later appearing to be more effective in the case of treatment of the metal working fluid.

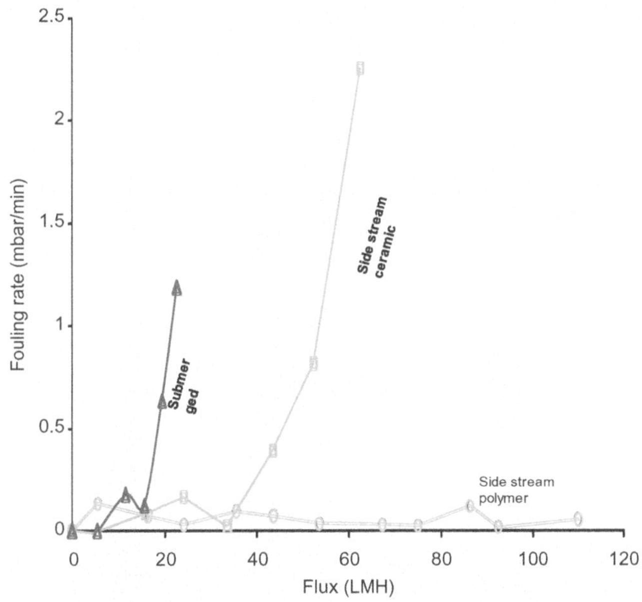

Figure 3.37 Impact of membrane selection on the fouling propensity of TiO_2

Table 3.6 Details of membrane modules tested

	Tubular polymeric	Tubular ceramic	Hollow Fibre polymeric
Channel	10–5–circle	20–5–Square	Fibre
Surface area (m^2)	0.16	0.35	0.18
Pore size (nm)	50	100	40
Ra (nm)	4.9 nm	104 nm	–
Increased surface area due to roughness	5%	25%	–

The choice of membrane configuration also requires consideration of the wastewater type being treated (Figure 3.37). Equivalent trials with other industrial wastewaters supports the broad statements above that tubular polymeric membranes appear to enable the highest sustainable flux to be used. However, the differences observed between membrane configurations in the case of the metal working fluid do not appear when considering other industrial wastewaters. For instance, Jc when treating a textile wastewater were all below 15 LMH with the submerged hollow fibre providing the highest value. However, comparison of the entire fouling curve for each membrane configuration and industrial wastewater indicates that the polymeric tubular membrane is likely to provide the lowest fouling rates in all cases. To illustrate, in the case of the textile wastewater fouling rates for the hollow fibre and polymeric tubular membrane at a flux of 38 LMH were 0.7 mbar · min^{-1} and 0.2 mbar · min^{-1} respectively.

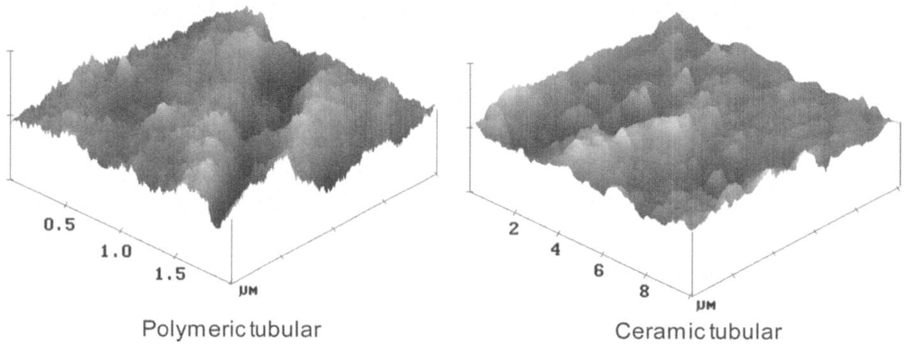

Figure 3.38 AFM images of the different surfaces

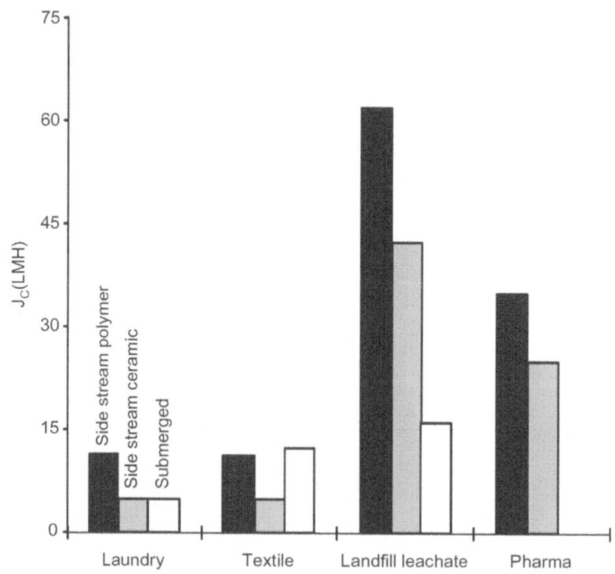

Figure 3.39 Impact of membrane selection on the fouling propensity of TiO_2

Impact of nanoparticle aggregation

The differences observed in terms of treatment and fouling when comparing the different industrial wastewaters under identical MCR operating conditions indicates the importance of the characteristics of the nanoparticle aggregates that form in the different wastewaters. There are three importance aggregate characteristics that influence performance of photocatalytic reactors: size, shape and strength of the aggregates. These properties control treatment through influence on the available contact area for adsorption, the conditions of light scattering in the slurry and the proportion of the aggregate that is activated when in the irradiation zone. In addition how the aggregates responds to changes in shear pattern will influence the amount of fines in the water which is known to strongly influence membrane fouling and as such detailed knowledge of these factors is important in designing effective MCR technologies. Methods for measuring these properties are well defined (Jarvis *et al.*, 2005a; Jarvis *et al.*, 2005b) but care is required with some specific issues. TiO_2 is a relatively heavy particle and thus larger aggregates tend to settle out in the tubing of the sizing instruments. In addition, most particle sizing instruments have a maximum concentration beyond which they do not operate. The majority of studies utilise some form of laser diffraction such that in the case of dilute photocatalytic processes, the TiO_2 concentration is around the upper limit and so direct measurement of the aggregate is possible. Whereas in the case of industrial wastewater treatment systems, the more concentrated slurries used require dilution before measurement. These two factors generate a degree of uncertainty in the measured aggregate sizes which is often overlooked. The situation if further complicated as the size of the aggregate change in relation to the turbulence level they encountered and this too can be different between different particle sizing instruments and laboratory standard practices and as such the shear rate used during operation, sampling and especially measurement should always be quoted.

The catalyst nano-particles will aggregate in water to around a medium aggregate size between 1000–10 000 nm at a shear rate of 70 s^{-1} which means the aggregate diameters can be up to 40–200 times the primary particles size. Application to industrial wastewater treatments leads to much larger aggregates which can exceed 300 000 nm at 70 s^{-1} (Figure 3.40). Comparison of the optical and SEM images indicates that the aggregates form different shapes and density structures in relation to the wastewater being treated (Figures 3.41–3.42). The shape will have a significant influence on treatment as only the proportion of the aggregate that enters the irradiation zone will undergo photocatalysis. The sizes of aggregates seen and there shapes exceed the likely size of the irradiation zone and thus for complete regeneration of the TiO_2 during photocatalysis, each floc will require multiple cycles within the irradiation zone. Exposure of the aggregates to elevated shear rates generates classical floc strength curves (Figure 3.40) where a log-log plot of median aggregate size against shear rate generates a regression line that can be used to define the aggregate. The gradient of the line reflects the strength of the aggregate (strength factor) with steeper gradients indicating weaker aggregates. Comparison of the floc strength curves of the different aggregates indicates that the TiO_2 in water was the strongest systems with a floc strength factor of 0.41 and the textile wastewater was the weakest aggregate with a strength factor of 0.63. Analysis of the floc strength curves in relation to the different wastewater characteristics of each source water does not indicate any trend between wastewater character and aggregate properties with the exception of an increase in floc strength as pH increased across the wastewaters. This lack of a clear relationship between water characteristics and floc properties is consistent with research findings about aggregates in other fields (Jarvis *et al.*, 2005a). However, two important observations can be made: (1) increasing turbulence will decrease the aggregate size and so should enhance MCR performance and (2) there is a limited benefit in doing that as wastewater type has a controlling influence on overall aggregate property and will prevent shear rate enhancement from resolving treatment for all industrial wastewater types.

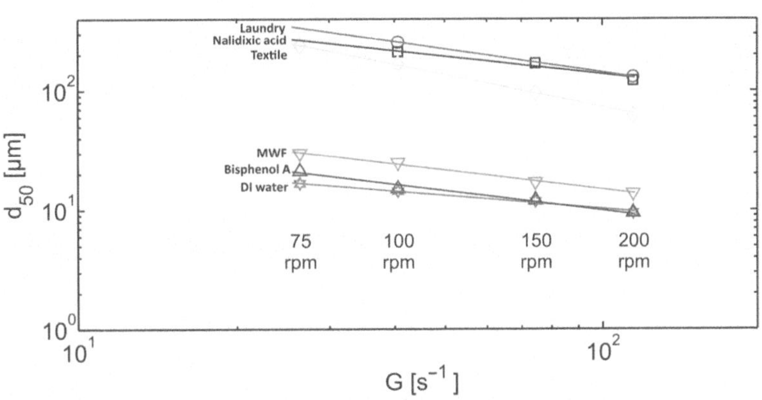

Figure 3.40 Floc strength curves for TiO_2 with different industrial wastewaters

3.3.4 Conclusions

The work outlined has demonstrated the ability to treat a wide range of wastewaters with a MCR based technology. The complex interactions that control the process result in wide variations in performance for different wastewaters based on there chemical properties. Ultimtely this leads to the need for site specific optimisation to ensure effective treatment. Whilst treatment efficacy has been shown, economic viability remains a challenge. The current required operating conditions mean that application will be economcially feasible only for wastewaters with high disposal costs. A typical example of this is for metal working fluids where current discharge costs are £40–80 $\cdot$ m^{-3}. Adoption of MCR technology can offer sufficient treatment to enable pay back times that approach the desire range in such cases. This mirrors development of other novel membrane technology in the past and provides cautious optimism that the technology has a viable future. Looking forward, the challenge is to reduce the required energy to drive photcatalysis. In part this will be achived through imporved reactor design and the development of more effective catalysts. In

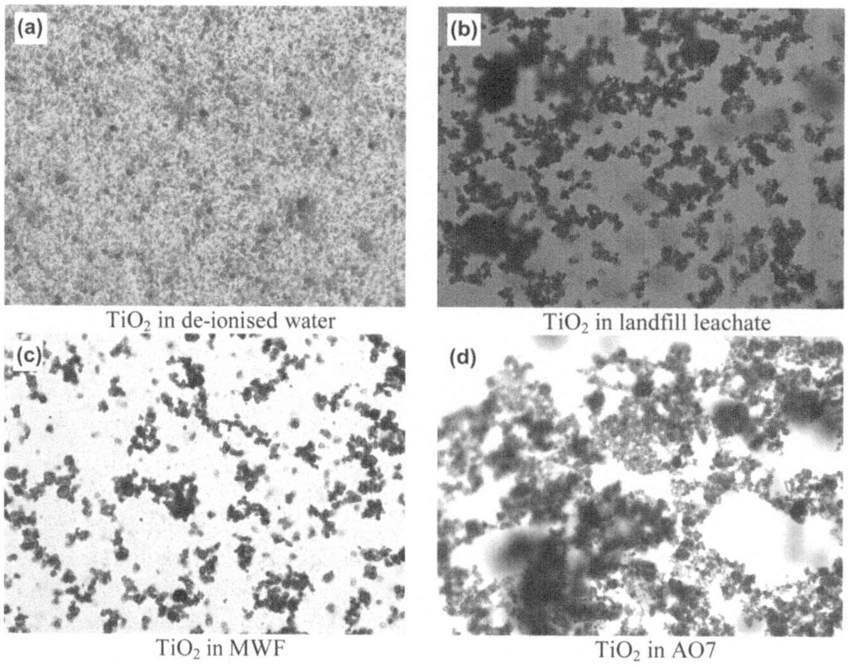

Figure 3.41 Optical microscope images of TiO_2 with different industrial wastewaters

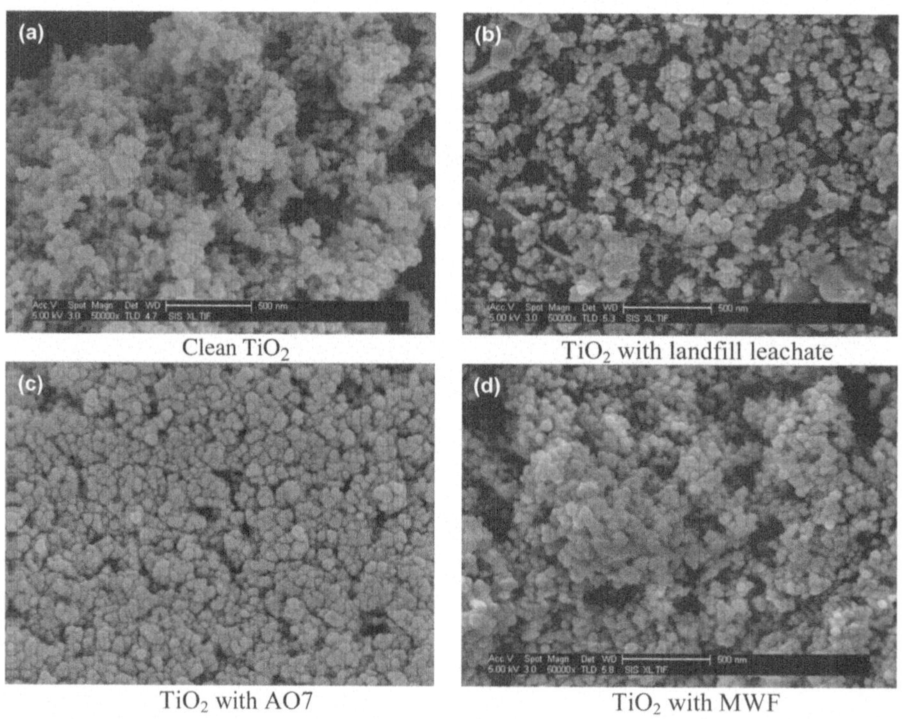

Figure 3.42 SEM images of TiO_2 with different industrial wastewaters

particular the doping of catalysts to enable solar irridation to work effectively offers great potential. The high strength and hence high doses requried in industrial applications are likley to limit there overall suitability and as such it is likely that solar systems will be most effective for low strength applications. Recent advances that offer greatest promise relate to UV LED irridiation sources which provide the potential to reduce energy demand by orders of magnitude. The technology is currently in the early stages with only a few initial publications yet development of MCR technology based on UV LED appears teh most likley to provide widespread suitability (Figure 3.43).

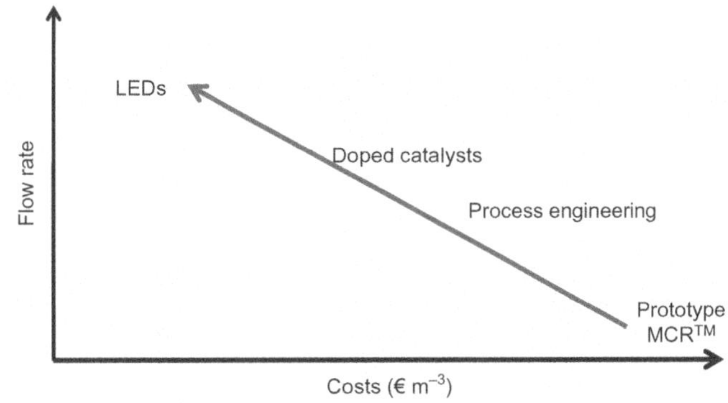

Figure 3.43 Proposed road map for development of the MCR for industrial wastewater treatment

REFERENCES

Andreozzi R., Caprio V. and Insola A. (2000). Photocatlytic oxidation of 4-nitrophenol in aqueous TiO_2 slurries: an experimental validation of literature kinetic models. *J. Chem. Tech. Biotech.*, **75**, 131–136.

Bolton J. R., Safarzadeh-Amiri A. and Cater S. R. (1995). The detoxification of waste water streams using solar and artificial UV light sources. In: Alterative Fuels and the Environment, F. S. Sterret (ed.), Lewish Publishers, Boca Raton, Fl.

Bond T., Goslan E. H., Jefferson B., Roddick F., Fan L. and Parsons S. A. (2009). Chemical and biological oxidation of NOM surrogates and effect on HAA formation. *Water Res.*, **43**, 2615–2622.

Brookes P. R. and Livingston A. G. (1995). Aqueous-aqueous extraction of organic pollutants through tubular silicone rubber membranes. *J. Mem. Sci.*, **104**, 119–137.

Brookes A., Jefferson B., Guglielmi G. and Judd S. J. (2006). Sustainable flux fouling in a membrane bioreactor: impact of flux and MLSS. *Separ. Sci. Technol.*, **41**, 1279–1291.

Butters B. E. and Powell A. L. (1995). Method and system for photocatyltic decontamination, US patent 5462674.

Chiron S., Fernandez-albam A., Rodriguez A. and Garcia-Calvo E. (2000). Pesticide chemical oxidiation. start of the art. *Water Res.*, **34**, 366–377.

Chong M. N., Jin B., Chow C. W. K. and Saint C. (2010). Recent developments in photocatalytic water treatment technology: a review. *Water Res.*, **44**, 2997–3027.

Comninellis C., Kapalka A., Malato S., Parsons S. A., Poulios I. and Mantzavinos D. (2008). Advanced oxidation processes for water treatment: advances and trends for R&D. *J. Chem. Technol. Biotechnol.*, **83**, 769–776.

Crittenden J. C., Liu J., Hand D. and Perram D. L. (1997). Photocatlytic oxidation of chlorinated hydrocarbons in water. *Water Res.*, **31**, 429–438.

De Lasa H., Serrano B. and Salaices M. (2005). Photocatalytic Reaction Engineering. Springer, New York.

Ferreira F. C., Han S., Boam A. and Livingston A. G. (2002). Membrane aromatic recovery system (MARS): lab bench to industrial pilot scale. *Desal.* **148**, 267–273.

Ferreira F. C., Peeva L. G. and Livingston A. G. (2005). Mass transfer enhancement in the membrane aromatic recovery system (MARS): experimental results and comparison with theory. *Chem. Eng. Sci.* **60**, 1029–1042.

Fu J. M., Wang Z., Jin L. and An D. (2006). A new submerged membrane photocatalysis reactor (SMPR) for fulvic acid removal using a nano-structured photocatalyst. *J. Haz. Mats.*, **131**, 238–242.

Gregory D. P. and Riddiford A. C. (1960). Dissolution of copper in sulphuric acid solutions. *J. Electrochem. Soc.* **107**, 950–956.

Gruber R. and Melin T. (1999). Flüssigextraktion von Phenol in einem beschichteten Membrankontaktor, *7. Aachener Membran Kolloquium - Preprints*, 9.-11.3.1999, 389–392.

Gruber R. and Melin T. (2003). Radial mass-transfer enhancement in bubble-train flow. *Int. J. Heat Mass Trans.*, **46**(2003), 2799 – 2808.

Hartung K.-H. (1975). Untersuchung Technischer Mischvorgänge Mittels pH-Indikatoren (Investigation of Technical Mixing Processes with pH Indicators). Ph.D. thesis, RWTH Aachen University.

Jarvis P., Jefferson B. and Parsons S. A. (2005a). Measuring floc structural characteristic. *Reviews in Environmental Science and Biotechnology*, **4**(1–2), 1–18.

Jarvis P., Jefferson B., Gregory J. and Parsons S. A. (2005b). A review of floc strength and breakage. *Water Res.*, **39**, 3121–3137.

Judd S. J. (2010). The MBR Book: Principles and Applications of Membrane Bioreactors for Water and Wastewater Treatment. Elsevier, London.

Karpel N., Leitner V., DeBras E., Foucault E. and Bousgarbies J. L. (1997). A new photochemical reactor design for the treatment of absorbing solutions. *Wat. Sci. Technol.*, **35**, 1753–1755.

Konstantinous I. K., Sakellarides T. M., Sakkas V. A. and Albanis T. A. (2001). Photocatalytic degradation of selected s-Triazine herbicides and organopohosphorus insecticides over aqueous TiO2 suspensions. *Environ. Sci. Technol.*, **35**, 698–405.

Le Clech P., Jefferson B., Chang I. S. and Judd S. J. (2003). Critical flux determination by flux-step method in a submerged membrane bioreactor. *J Membrane Sci.*, **227**, 81–93.

Le Clech P., Lee E.-K. and Chen V. (2006). Hybrid photocatalysis/membrane treatment for surface waters containing low concentrations of natural organic matter. *Water Res.*, **40**, 232–330.

Li F., Meindersma W., de Haan A. B. and Reith T. (2004). Experimental validation of CFD mass transfer simulations in flat channels with non-woven net spacers. *J. Mem. Sci.*, **232**, 19–30.

Malato S., Blanco J., Fernández-Alba A. R. and Agűera A. (2000). Solar photocatalytic mineralisation of commercial pesticides: Acrinathrin. *Chemo.*, **40**, 403–409.

Malato S., Fernández-Ibáñez P., Maldonado M. I., Blanco J. and Gernjak W. (2009). Decontamination and disinfection of water by solar photocatalysis: recent overview and trends. *Catalysis Today*, **147**, 1–59.

Molinari R., Palmisano L., Driloi E. and Schiavello M. (2002). Studies on various reactor configurations for coupling photocatalysis and membrane processes in water purification. *J. Membrane Science*, **206**, 399–415.

Mozia S. (2010). Photocatlytic membrane reactors (PMRs) in water and wastewater treatment, a review. *Separation and Purification Technology*, **73**, 71–91.

Neumegen R., Fernandez-Alba A. R. and Chisti Y. (2005). Toxicities of triclosan, phenol, and copper sulphate in activated sludge. *Environ. Toxicol.*, **20**, 160–164.

Parsons S. A. (2004). Advanced Oxidation Processes for Water and Wastewater Treatment. IWA publishing, London.

Phillipes K. K., Hans C., MacAdam J., Jefferson B., Hart J. and Parsons S. A. (2010). Photocatalytic oxidation of natural organic matter surrogates and the impact in trihalomethane formation potential. *Chemosphere*, **81**, 1509–1516.

Pidou M., Parsons S. A., Raymond G., Jeffrey P., Stephenson T. and Jefferson B. (2009). Fouling control of a membrane coupled photocatalytic process treating greywater. *Water Res.*, **43**, 3932–3939.

Pollice A., Brookes A., Jefferson B. and Judd S. (2005). Sub-critical flux fouling in membrane bioreactors – A review of recent literature. *Desalination*, **174**, 221–230.

Pramauro E., Vincent M., Augugliaro V. and Palmisano L. (1993). Photocatlystic degradation of monuron in TiO2 aqueous dispersions. *Environmental Science and Technology*, **27**, 1790.

Rivero M. J., Parsons S. A., Jeffrey P., Pidou M. and Jefferson B. (2006). Membrane chemical reactor (MCR) combining photocatalysis and microfiltration for grey water treatment. *Water Sci. Technol.* **53**(3), 173–180.

Salaices M., Serrano B. and de Lasa H. (2001). Photocatalytic conversion of organic pollutants. Extinction coefficients and quantum efficiencies. *Ind. Eng. Chem. Res.*, **40**, 5455–5464.

Schock G. and Miquel A. (1987). Mass transfer and pressure loss in spiral wound modules. *Desal.* **64**, 339–352.

Vorontsov A. V., Stoyanova I. V. and Kozlov D. (2000). Kinetics of the photocatalytic oxidation of gfaseous acetone over platinised titanium dioxide. *J. Catal.*, **189**, 360–369.

Wesselingh J. A. and Krishna R. (2006). Mass Transport in Multicomponent Mixtures chap. 16, VSSD, Delft.

Wilson E. E. (1915). A Basis for rational design of heat-transfer apparatus. *Trans. ASME* **37**, 47–70.

Wyness P., Klausner J. F. and Goswami D. Y. (1994). Performance of non concentrating solar photocatalytic oxidation reactors. Shallow pond configuration. *J. Solar Energy Eng.*, **116**, 8–13.

Chapter 4

Tailoring and assessment of wastewater treatment processes

Mats Almemark, Mats Ek, Uwe Fortkamp, Christian Junestedt, Klara Westling and Maria Ekenberg

4.1 BACKGROUND

The overall objective of tailoring and assessing wastewater treatment processes is to find the ecologically most sustainable and economically most feasible treatment systems for given wastewaters. This means that entire systems shall be assessed and optimised, not just individual process steps. This in turn requires two things:

1. Different combinations of technologies should be studied, taking into account their interaction with each other in treatment trains, and their possible interaction with the process generating the wastewater.
2. A holistic assessment is necessary, one which takes into account upstream processes (supply of commodities) and downstream processes (disposal of wastes and impact of the discharged wastewater).

Every industrial wastewater is unique. Therefore, in order to get the optimal treatment with regard to effluent quality as well as to environmental impact and cost, each treatment process for an industrial wastewater has to be individually designed and optimised – tailor-made. In the INNOWATECH project advanced treatment processes for specific industrial wastewaters have been studied in laboratory and pilot-plant scale. In general, the number of process designs which may be studied and optimised experimentally, are for obvious reasons limited, as are the number of wastewaters which may be studied in that way and the possibilities to modify the upstream industrial processes for testing purposes. Therefore, the method to approach item #1 above has been to build assessment models of the studied processes and of other processes not studied in the project. Such models enable scenario calculations assessing various combinations of unit operations. The assessments may also be extended to wastewaters not studied experimentally.

The treatment of a wastewater is an industrial process just like other industrial activities. It consumes resources and generates wastes. It may and should thus be assessed according to the IPP concept (Integrated Product Policy), i.e. the consequences of upstream and downstream processes should be included. Upstream processes are the acquisition of the resources, downstream processes the treatment of the wastes and the recovery of valuable materials for open-loop or closed-loop recycling schemes. The total net efficiency of a wastewater treatment should be calculated as the technical treatment efficiency of the plant (benefits, i.e. avoided impacts in the receiving waters) with deduction of the environmental impacts caused on site and elsewhere by the processes of operating and supplying the plant and by the processes of disposing of the wastes, such as sludge (costs, i.e. induced impacts, mostly at other locations than the plant itself).

Mathematically such an assessment can be translated into minimising I in equation (1) below.

$$I = t + w \tag{1}$$

I = the total environmental (or economic) impact of water use for a given application, disregarding the supply of the water.

t = the environmental (or economic) impact of treating the wastewater from the application.

w = the environmental impact of discharging the wastewater from the application to a receiving water.

The total net efficiency of a wastewater treatment should be calculated as the sum of the parameters $w + t$ in equation 1 compared to w_0, which is the impact of the wastewater in the receiving water, if it is discharged untreated.

In all probability the impact t of a given treatment will rise as the impact w of the wastewater in the receiving water is decreased. As w approaches zero, t may rise steeply and at some point Δt will probably exceed Δw. At that point t may even exceed w_0, i.e. from a global point of view the treatment does more harm than good. It just moves damage to the environment from one location to another.

Obviously the parameter t is a function of the technology used and of the operating conditions selected for that technology. The minimum of $(t + w)/w_0$ will also depend on the supply chain of any energy and consumables and construction materials used.

The factor deciding the eco-efficiency of a highly advanced treatment process applied to an industrial wastewater may well be ecotoxicity, namely the potential ecotoxiciy and human toxicity of specific compounds removed by the treatment and the potential toxicity of heavy metals and other toxic compounds discharged from power plants and chemical industries supplying energy and chemicals to the treatment.

In the literature there are several published holistic assessments of wastewater treatment processes, for municipal wastewaters among others Emmerson *et al.* (1995), Fahner *et al.* (1998), Hospido *et al.* (2004), Lundin *et al.* (2000), Ortiz *et al.* (2007), Roeleveld *et al.* (1997), Tangsubkul *et al.* (2005), Tarantini and Ferri (2001), Strömberg and Paulsen (2002), Vidal *et al.* (2002) and Ødegaard (1995), for industrial wastewaters García-Montaño *et al.* (2006), Jørgensen *et al.* (2004) and Ortiz (2006). These authors except Emmerson *et al.* and Fahner *et al.* used the life-cycle assessment methodology as described for instance in Guinée (ed.) (2002) to assess the treatment plants. Results have been used in a traditional way to compare different treatment technologies for the same application and to pin-point the most important environmental aspects of the treatments. Some of the authors, most explicitly Fahner *et al.* but also Emmerson *et al.* Vidal *et al.* and Roeleveld *et al.* addressed the issue whether or not the environmental benefits of the treatments were greater or more important than the environmental burdens of building and operating the treatment plants. The environmental benefit of a wastewater treatment is usually removal of eutrophication and toxicity. In order to compare the value of this benefit to the burden of other types of emissions caused by the treatment operation normalization was used. The quantified environmental impacts from the treatments and from the impact of the wastewater on the receiving water were divided by reference values, such as the average annual emissions in the region.

Data from a modelled treatment technology was used as basis for assessment by Vidal *et al.* They used the ASM1 model to obtain data for two hypothetical nitrogen removal technologies (anoxic tank + aeration tank and a so called oxidation ditch). The assessments were then used to decide whether or not nitrogen removal would be beneficial from an environmental point of view. García-Montaño *et al.* and Ortiz used laboratory data and data from pilot-plants to calculate design data for real but still non-existing plants. These designs were then used as bases for assessments.

4.2 METHODOLOGY

4.2.1 Specific objectives

In the INNOWATECH project the specific objectives of tailoring and assessment of wastewater treatment processes are the following:

- Verify that the advanced treatment processes investigated in the project are environmentally beneficial, i.e. that the avoided impacts are more harmful than the induced impacts.
- Verify that the advanced treatment processes developed in the project (new technologies) are environmentally and economically more efficient than treatment processes which were available at the start of the project and could have been used for the studied applications (existing technologies).
- Improvement analysis. Pin-point the weak spots of the studied processes and suggest improvements where possible.
- Compile a set of treatment technology models and use them to assess treatment cases not studied in the project.

4.2.2 General approach

For the purpose of modelling the wastewater treatment systems in order to carry out life-cycle assessments and life-cycle-cost calculations, and in order to be able to design suitable treatment systems for other wastewaters than those studied experimentally, the unit processes of the treatment systems are divided into two categories:

- Core processes, defined as processes which are part of the treatment system as such. Their performance may be affected by changes of the properties of the treated wastewater.
- Peripheral processes, i.e. processes which deliver commodities such as energy or chemicals to the wastewater treatment system (upstream processes), or processes which treat wastes or other outputs from the treatment

system (downstream processes). The performances of the peripheral processes are not affected by changes of the properties of the treated wastewater. They only react to changes by supplying more or less of their commodity or treating larger or smaller quantities of outputs.

The distinction between core and peripheral processes is illustrated in Figure 4.1 with an example from Chapter 1, namely treatment of landfill leachate in a sequencing batch biofilter granular reactor (SBBGR).

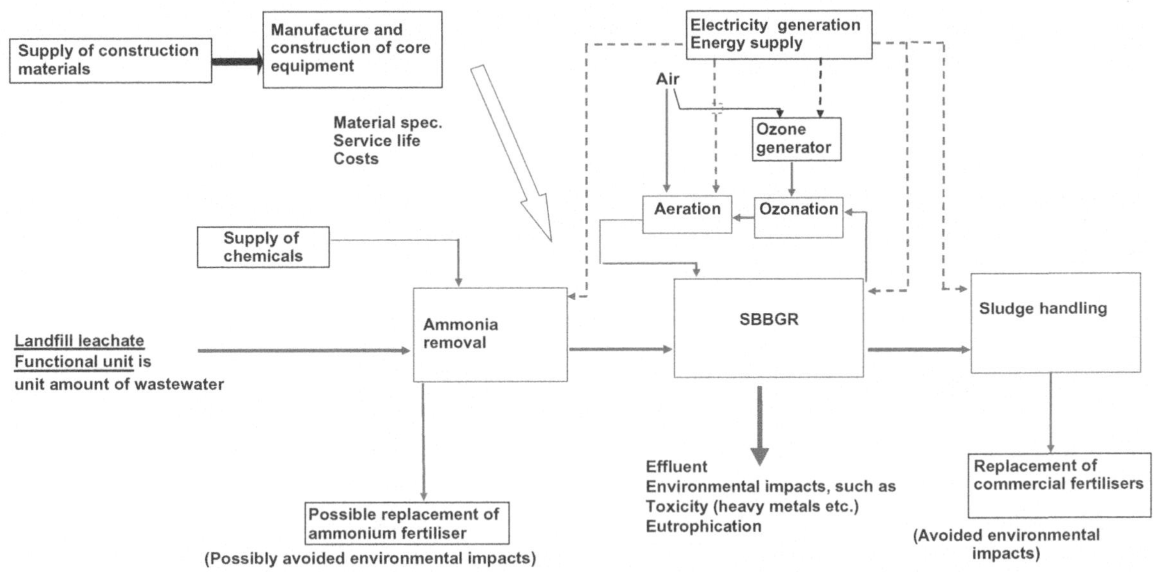

Figure 4.1 Modelling example from Chapter 1. Treatment of landfill leachate in a SBBGR. Core and peripheral processes. Red text and lines: Core processes modelled mechanistically or statistically. Black text and lines: Peripheral processes described by linear life-cycle inventory data

4.2.3 Data and software requirements

4.2.3.1 Core processes

The aim of the modelling of the core processes is to be able to predict their performance as a function of the conditions and the composition of the wastewater. This requires software with mathematical functions more advanced than simple first-degree linear equations. A graphical interface is also required. Unless a mechanistic model of the process can be used to derive a mathematical model, the latter must be derived from experimental data collected during statistically planned experiments. In either case, the data collected during the experimental investigations should be adequate for the purpose of confirming or deriving a mathematical model and must be adequate for the purposes of life-cycle assessment and life-cycle costing.

For each treatment technology studied the assessment work starts by preparing a simple flowchart of the process and a list of suggested parameters to be measured during experimental investigations. In those cases where modelling is possible the model is built with the software MatLab with Simulink. Each part of the treatment process is developed as a separate process module which can be removed or added anytime and anywhere in the treatment process. The Simulink model calculates the core process data, which are then exported to the selected LCA software for inventory calculations.

In those cases where modelling is not possible the experimental data are used to calculate preliminary designs of real treatment plants. These calculations are quite simply performed in Excel and the result exported to the selected LCA software.

For the life-cycle assessments it is convenient to use available specialised LCA software. They have convenient tools for compilation of inventories and for characterisation and impact-assessment calculations.

The software GaBi (GaBi 4.2) was selected as the tool for calculating life-cycle assessments. This software can import data from MatLab via an Excel table. An entire core-process train can be aggregated to one module and exported as such to GaBi. GaBi can also be connected to relation databases and is delivered with a reasonably extensive LCI (life-cycle inventory) database. Additional databases may be purchased. Finally it offers some non-linear mathematics.

For the life-cycle assessment of the entire treatment process the module or the modules describing the core processes are combined in GaBi with modules describing the peripheral modules to an assessment model of the entire process.

4.2.3.2 Peripheral processes

The aim of the modelling of the peripheral processes is to be able to assess the environmental impacts of supplying commodities to and treating outputs from the wastewater treatment system. For this purpose standard LCA inventory methodology is adequate, i.e. the processes are assumed to be linear and adequately described by first-degree equations and by data calculated as averages over a certain period of time. For daily well-known commodities, like electricity, fuels and basic chemicals, generic LCA data collected from databases or from the literature are sufficient. If generic data are not available or not applicable, inventories of the production chains are carried out following standard LCA methodology, i.e. average data for the latest available year is collected from the producer.

The task of the peripheral process models is to transform flows of commodities, calculated by the core process models, into sets of environmental impact parameters. The peripheral processes are assumed to be independent of changes of the core processes.

The modules describing peripheral processes are most conveniently compiled directly in GaBi. The mathematical versatility of MatLab is not needed. In many cases the required data for the peripheral processes are available as ready-made modules in the databases of GaBi.

4.2.4 Environmental assessment methodology

The assessment methodology used is the life-cycle assessment (LCA) methodology as standardised by ISO (2006) and described for instance by Guinée (ed.) (2002). The practical procedures have been agreed and co-ordinated with the parallel EU funded project NEPTUNE (2006).

4.2.4.1 Goal and scope of the assessments

The goal of the LCA: s in both INNOWATECH and NEPTUNE is holistic environmental performance ranking (optimisation) of different wastewater treatment technologies. In INNOWATECH the objective is to assess and to find sustainable treatment technologies for a variety of widely different industrial effluents. In NEPTUNE the main objective is to assess post-treatment of municipal wastewaters to remove focus micro pollutants (and pathogens) as defined in the Water Framework Directive.

Both projects have one common scope of their LCA: s, namely to assess the new treatment technologies in comparison to existing ones. In INNOWATECH this is done by assessing the new technologies against models of existing treatment systems which treat the investigated wastewaters according to current best-available technologies. In NEPTUNE the base-line case is a municipal wastewater treatment plant with best-available technology today but without any post-treatment.

Both projects furthermore compare induced impacts to avoided impacts in order to establish the net environmental benefits of the treatments.

In INNOWATECH the LCA models of the newly developed treatment systems are also used to detect, as early as possible, environmentally weak points ("hot spots") of the new treatment designs.

4.2.4.2 Functional unit

The functional unit, i.e. the basis of calculation, is 1 m^3 of "standard" wastewater, i.e. effluent water as it is discharged today from the plant generating the wastewater.

4.2.4.3 System boundaries

The upstream system boundary is the influent to the wastewater treatment plant, and the extraction from nature of the resources necessary to produce the commodities needed for the treatment. The downstream system boundary is the receiving water, to which the effluent from the treatment is assessed as an emission, and the final disposal of residual materials, such as sludge, generated by the treatment.

4.2.4.4 Temporal boundaries

The data for the new technologies obviously represent contemporary technology on the verge of becoming operational. The data for reference technologies represent currently used technology. Data for the side stream processes are generic data and thus considerably older. As a general rule the datasets represent technology of the year 2000 $\pm$ 10 years. The age of each peripheral dataset is specified in the inventory of the peripheral processes in Appendix 2.

4.2.4.5 Geographical boundaries

The wastewater treatment plants are assumed to be located in Europe and to be supplied by European suppliers including electricity which is assumed to be average European (EU-25). In the cases of the standard wastewater used for a comparative study the geographical setting is Southern Europe, and the electricity is assumed to be supplied by the Spanish grid. Effluent water streams from the wastewater treatment processes are assumed to be discharged to watercourses with fresh water.

4.2.4.6 Impact assessment

The following existing impact categories are used:

- Global warming
- Acidification
- Nutrient enrichment/eutrophication
- Photochemical ozone formation
- Ecotoxicity (aquatic and terrestrial; via soil, water and air)
- Human toxicity (via soil, water and air)
- Resource consumption
- Waste generation

All these impact categories are included according to the EDIP97 methodology (Wenzel *et al.*, 1997).

When assessing advanced wastewater treatment processes avoided ecotoxicity from micro pollutants is an important impact category. The EDIP methodology was used to assess this impact category as well. The recently developed and released Usetox methodology (Rosenbaum *et al.*, 2008) could not be used.

In the EDIP methodology (as in the USES-LCA methodology) the principle of "most sensitive species" (PNEC = predicted no-effect concentration) is applied. Fate and distribution of emitted substances are accounted for in a rather simple and approximate way in EDIP, but the necessary data are available. For this reason the EDIP methodology was selected also for the impact category toxicity. It makes it possible to at least demonstrate complete assessments with numerical examples.

The aquatic characterisation factor for ecotoxicity is in EDIP calculated from the $PNEC_{water}$ value with the following equation.

$EF_w = f * BIO * 1/PNEC_{water}$

EF_w = characterisation factor for aquatic ecotoxicity (m^3 of water/kg of substance)

f = distribution factor water/air/soil

BIO = biodegradation factor

The impact category "Human toxicity" is characterised with factors based on the NEC approach of EDIP97.

4.2.4.7 Assessment of environmental cost/benefit

The basis of the assessment of each technology per se is to compare the impacts caused by the wastewater treatment to the impacts avoided by the treatment. The avoided impacts are calculated by assuming that the wastewater is discharged untreated to a receiving watercourse. The impacts that would be caused by such a discharge are calculated from the composition of the wastewater using the EDIP97 methodology. The avoided impacts are then calculated as the difference between the impacts from the discharge of the untreated water and from the discharge of the treated water. Figure 4.2 illustrates the calculation.

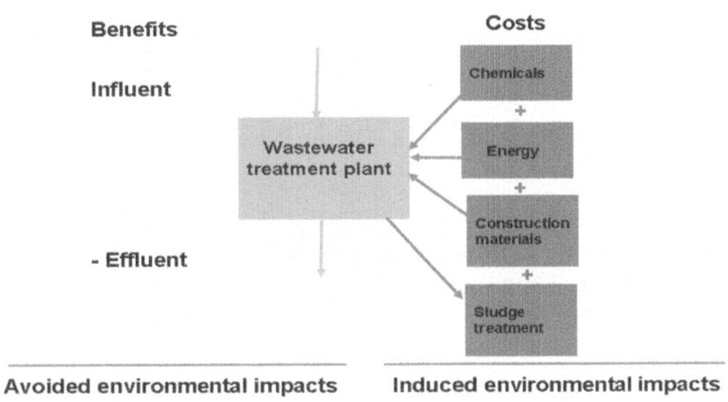

Figure 4.2 Environmental efficiency – Environmental cost/benefit assessment

Avoided impacts are primarily nutrient enrichment, oxygen consumption and toxicity from organic matter and toxicity from metals in the wastewaters. In those cases where organic compounds in the wastewater have been synthesised from fossil raw materials (crude oil and natural gas) their degradation to carbon dioxide contributes to global warming as well.

Induced impacts are impacts of all the types listed in the section Impact assessment. The cost/benefit assessment thus requires recalculation of all impacts to a common scale.

4.2.4.8 Normalization

The purpose of normalization is to recalculate the different environmental impact potentials of the categories listed in section Impact assessment to a common scale to facilitate comparisons between them. In the EDIP97 methodology this is done by dividing each impact potential from the life cycle of a product or a process, e.g. the global warming potential as kg CO_2 equivalents caused by released greenhouse gases, by the entire potential impact of that category from a specified region during a specified reference year (Wenzel *et al.*, 1997). The potential impacts from entire regions are given as kg impact equivalents per person and year. The normalized values of all impact potentials from the life cycle of the product or process are thus expressed as person equivalents (pe) · years. The person equivalent can be regarded as the annual impact of an average person in the specified region.

Table 4.1 lists the potential impacts at present used by the EDIP97 system as normalization references. The table also includes weighting factors. They are a measure of the relative importance of each impact category. How they are derived in the EDIP97 system is briefly explained below.

Table 4.1 Normalization references and weighting factors for environmental impacts according to EDIP97

Impact category	Unit	Normalization reference	Reference year	Weighting factor	Reference year	Reference region
Environmental impacts						
Global						
Global warming	kg CO_2-eq/pers/year	8.70E+03	1994	1.1	2004	World
Ozone depletion	kg CFC-11-eq/pers/ar	0.103	1994	63	2004	World
Regional and local						
Photochem. oz. form.	kg C_2H_4-eq/pers/year	25	1994	1.3	2004	EU-15
Acidification	kg SO_2-eq/pers/year	74	1994	1.3	2004	EU-15
Nutrient enrichment	kg NO_3^--eq/pers/year	119	1994	1.2	2004	EU-15
-N-equivalents	kg N-eq/pers/year	24	1994	1.4	2004	EU-15
-P-equivalents	kg P-eq/pers/year	0.4	1994	1	2004	EU-15
Ecotoxicity						
-water acute	m^3 water/pers/year	2.91E+04	1994	1.1	2004	EU-15
-water chronic	m^3 water/pers/year	3.52E+05	1994	1.2	2004	EU-15
-soil chronic	m^3 soil/pers/year	9.64E+05	1994	1	2004	EU-15
Human toxicity						
-via air	m^3 air/pers/year	3.06E+09	1994	1.1	2004	EU-15
-via water	m^3 water/pers/year	5.22E+04	1994	1.3	2004	EU-15
-via soil	m^3 soil/pers/year	1.27E+02	1994	1.2	2004	EU-15
Waste						
-bulk Waste	kg/pers/year	1350	1991	1.1	2000	Denmark
-hazardous waste	kg/pers/year	20,7	1991	1.1	2000	Denmark
-slag and ashes	kg/pers/year	350	1991	1.1	2000	Denmark
-nuclear waste	kg/pers/year	0.035	1989	1.1	2000	Sweden

The consumption of non-renewable resources are normalized to pe · years by dividing the consumption of each resource by the world annual consumption per person of that resource. Normalization values for some resources are given in Appendix 1.

For renewable resources the normalization is done by dividing by the total consumption per person of each resource in the relevant region. The normalization is done on a regional level, because the rate of regeneration of renewable resources varies from region to region.

Weighting, i.e. assessing the relative importance of the impact categories, is in the EDIP methodology done by gauging the necessary reductions in environmental impacts to reach emission targets set for a reference year ten years later than the reference year for normalization. The weighting factor is calculated with equation 2.

$$WF(i) = ER(i)_y / ER(i)_{y+10} \quad (2)$$

WF(i) = the weighting factor of impact category i.

$ER(i)_y$ = the normalization reference of impact category i for the reference year y.
$ER(i)_{y+10}$ = the targeted normalization reference of impact category i for the reference year y + 10.

In EDIP97 the reference year for normalization is still 1994 and the reference year for weighting 2004. Since we shall use EDIP97 the values in Table 4.1 will be used.

For resource consumptions quite different weighting factors are suggested by EDIP. A supply horizon is calculated by dividing the known reserves of a resource by the annual global consumption of that resource. The weighting factor is then taken as the reciprocal value of the supply horizon. This weighting factor consequently has the quantity $year^{-1}$. Mathematically this definition means, that the weighted normalized consumption of a certain resource to produce a product or provide a service will be the quotient between that resource consumption in kg (or MJ) and the entire known global reserves per person of that resource, a very uncertain number.

4.2.5 Economic assessment

For the purpose of cost assessments a life-cycle cost (LCC) approach is used. The core processes are assessed with capital cost calculations and with calculations of maintenance, labour and other operating costs. The costs are periodized as lifetime annuities, i.e. as a regularly paid constant annual amount over a limited period of time. It is the same as the net present value multiplied by an annuity factor. The net present value is the sum of all costs and receipts at a given point of time, discounted with a specific discount rate.

First, the net present value is calculated:

$$LCCtot = \sum_{t=0}^{t=n} \frac{C_t}{(1+r)^t} \tag{3}$$

where n is the considered economic life, r is the interest rate and C_t are the calculated total costs at time t.

These costs are then multiplied by an annuity factor:

$$LCC/year = LCCtot * \frac{r}{1-(1+r)^{-fn}} \tag{4}$$

where fn is the economic life of the core process facility.

The annuity can then be related to the functional unit, which in all assessments is 1 m^3 of wastewater to be treated, i.e. effluent water directly from industries.

For the peripheral processes no cost calculations are carried out. Their costs are part of the operating costs and enter the assessments as prices of commodities or fixed treatment costs for wastes, like sludge.

4.2.6 Assessment strategy

4.2.6.1 Progress compared to existing technology

The progress compared to available technology is assessed by comparing the environmental and economic performance of the new technologies to reference cases. In these reference case studies treatment plants are designed with available technology to treat the wastewaters used in the R&D work described in Chapters 1–3. These (fictitious) plants are then assessed with LCA and LCC methodology in the same way as the new technologies. The references cases are thus modelled cases, not data from actually existing treatment plants. Table 4.2 summarises the reference cases.

Table 4.2 The reference case studies and the corresponding new technologies

New technology	Existing (reference) technology	Wastewater studied
Sequencing Batch Bioreactor with granular biomass (Nereda) (Ch.1)	Activated sludge design by DHV and IVL based on actual plant data, costing by AnoxKaldnes	Food industry wastewater
Struvite precipitation + SBBGR + Ozone (Ch. 1) SBBGR = Sequential Batch Biofilter Granular Reactor	Activated sludge with nitrification/denitrification, design by AnoxKaldnes	Landfill leachate from Italy
MBR + Ozone (Ch. 2) MBR = Membrane Bioreactor Photo-Fenton/IBR (Ch. 2) Membrane contactor + IBR (Ch. 2 + 3) IBR = Immobilised Biomass React.	Moving-Bed Bioreactor (MBBR) + Powdered activated carbon removal, design by AnoxKaldnes	Pharmaceutical industry wastewater

4.2.6.2 Finding the most efficient technology in a real case – The standard wastewater case

A wastewater from a pharmaceutical industry is used as a standard treatment case for comparative assessments of treatment technologies. The wastewater has a matrix of synthetic but biodegradable organic compounds, but also contains a non-biodegradable substance known as nalidixic acid (1-ethyl-1,4-dihydro-7-methyl-4-oxo-1,8-naphthyridine-3-carboxylic acid). Nalidixic acid is an antibiotic used to treat urinary infections in both animals and humans. It has an estimated low PNEC value (predicted no-effect concentration) in the environment, which means that it is assigned a high ecotoxicity potential in the EDIP97 methodology. The wastewater is furthermore saline and acidic. Table 4.3 describes its properties.

Table 4.3 The standard wastewater treatment case

Parameter	Value	Unit
Wastewater flow	5	m^3/day
Nalidixic acid	0.045	kg/m^3
N-tot (N in the nalidixic acid)	0.005	kg/m^3
DOC	0.78	kg/m^3
COD	2.7	kg/m^3
TSS	0.50	kg/m^3
VSS	0.26	kg/m^3
PO_4^{3-}	0.01	kg/m^3
pH	4	
Cl^-	2.8	kg/m^3
SO_4^{2-}	0.16	kg/m^3
Na^+	2	kg/m^3
Ca^{2+}	0.02	kg/m^3
Conductivity	7	mS/cm

The treatment technologies, which could be tested on the standard case, are listed in Table 4.4. Some of them are designs based on experimental results from the R&D work described in Chapters 1–3. Other designs are put together from literature data or experience or by combining data modules from different work packages. In all cases treatment data are calculated for full-scale plants which treat exactly the wastewater of Table 4.3.

All treatments were assumed to take place in the same geographical setting as follows:

- Supply of chemicals: Europe in general
- Supply of electricity: Southern Europe (Spain)
- Receiving water for the wastewater: Freshwater course.

Since the plant designs are preliminary, and there is some uncertainty concerning the material standard of ordinary equipment like holding tanks, aeration tanks, pumps etc. the best approximation for the purpose of this comparative assessment is to exclude the construction material for these items and just include the construction material for special equipment like solar collectors, ozone generators + destructors and membrane contactors. The error introduced in the comparisons by this approximation is in our judgment within the uncertainty of the data at this point.

4.3 FOOD INDUSTRY WASTEWATER

4.3.1 The assessed treatment case

4.3.1.1 The wastewater

The food industry wastewater, for which treatment with a reference technology could be assessed, originates in a brewery. It is subjected to an anaerobic treatment before being passed on to an activated-sludge treatment which is the existing treatment today, i.e. the reference treatment. Table 4.5 gives the analytical parameters as reported by the partner DHV.

4.3.1.2 The treatment systems

The new technology tested for the brewery wastewater is a biological sequencing batch reactor (SBR), Nereda^{TM}. It operates in a fill-and-draw mode, combining the aeration and the settling stages of an activated sludge system into one process stage. A process cycle starts with the sludge in a settled stage in a partly empty reactor. A batch of the

Table 4.4 The technologies tested on the standard case

Fictitious case MBBR + PAC
Reference case
Design with existing technology

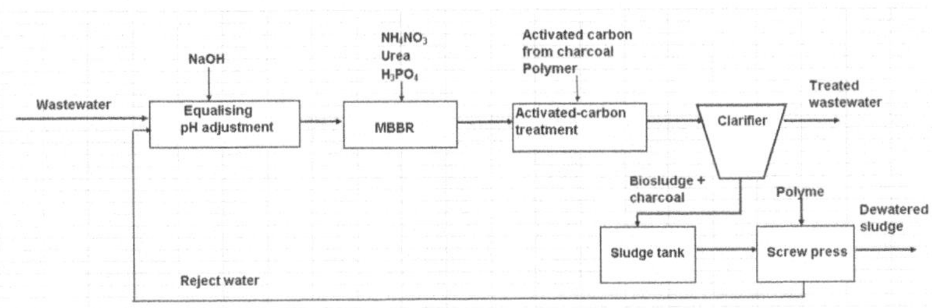

Studied case from Ch. 2
Photo-Fenton + IBR

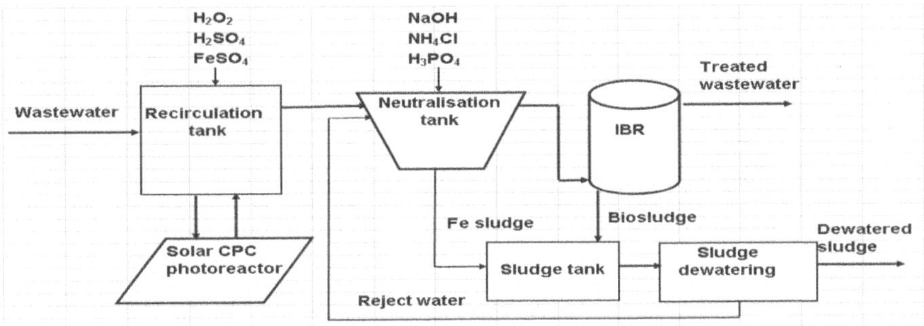

Studied case from Ch. 2
IBR + Photo-Fenton

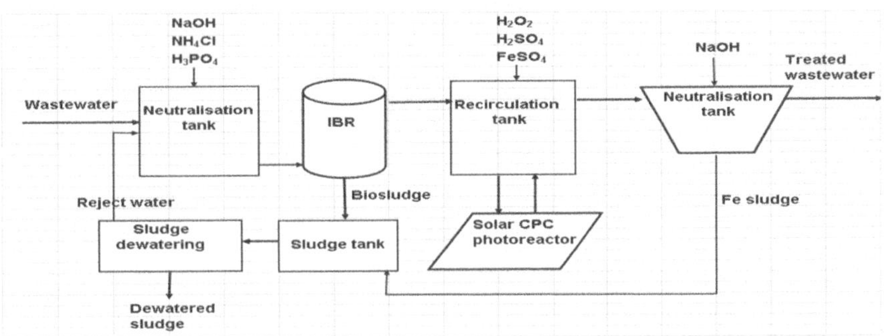

Studied case from Ch. 2
MBR + Ozone

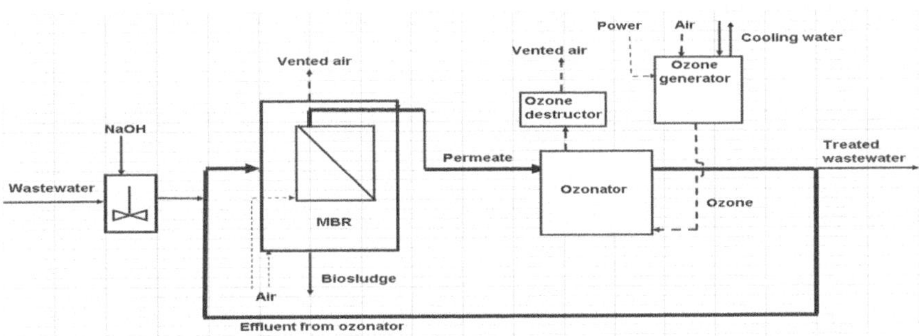

Fictitious case
Membrane contactor + IBR
MC model from Ch. 3
IBR data from Ch. 2

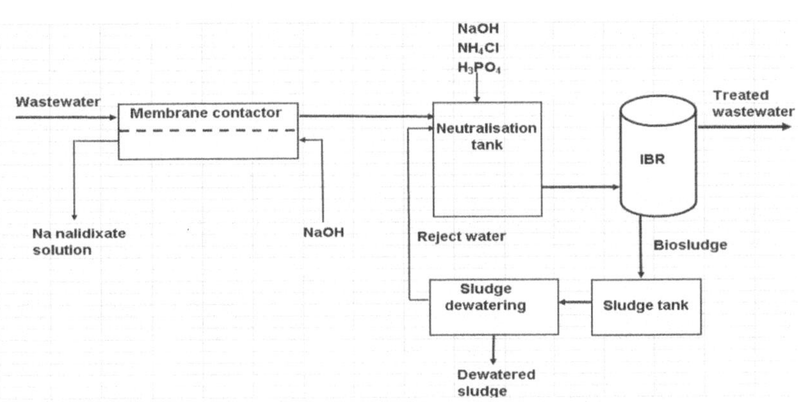

wastewater to be treated is added, the sludge is dispersed, and the mixed liquor is aerated. After a suitable period the aeration is terminated and the sludge is allowed to settle. The supernatant is withdrawn as treated water. Excess sludge is wasted as settled sludge.

Table 4.5 Reported analytical parameters of the anaerobically treated wastewater used for the pilot-plant test with a Nereda™ SBR (annual average data reported by DHV 2008)

Parameter	Value	Unit
Flow	4776	m^3/day
COD tot	650	mg/l
COD dissolved (GF/A)	440	mg/l
P-tot	12.4	mg/l
PO_4-P	7.7	mg/l
N-tot	35	mg/l
NH_4-N	5.4	mg/l
TSS (GF/A)	240	mg/l
pH	6.3	
Temperature	25	°C

The Nereda™ SBR developed by DHV has two special features:

- The sludge consists of granular aerobic biomass.
- The wastewater is added through the settled bed of biomass without dispersing the biomass until the reactor has been filled up and aeration starts.

A SBR with granular sludge potentially offers several advantages compared to a standard activated sludge reactor:

- Efficient integrated removal of nitrogen and phosphorus may be achieved inside granules.
- Granular sludge settles fast. External settling tanks are not necessary.
- High COD conversion capacities in combination with low sludge production can be achieved in granular sludge.

Figure 4.3a shows a flowchart of a Nereda™ treatment system. A flowchart of a standard activated sludge system is depicted in Figure 4.3b. In the studied application both systems use simultaneous precipitation with ferric chloride. A cationic polyelectrolyte is used to improve the dewatering properties of the sludge.

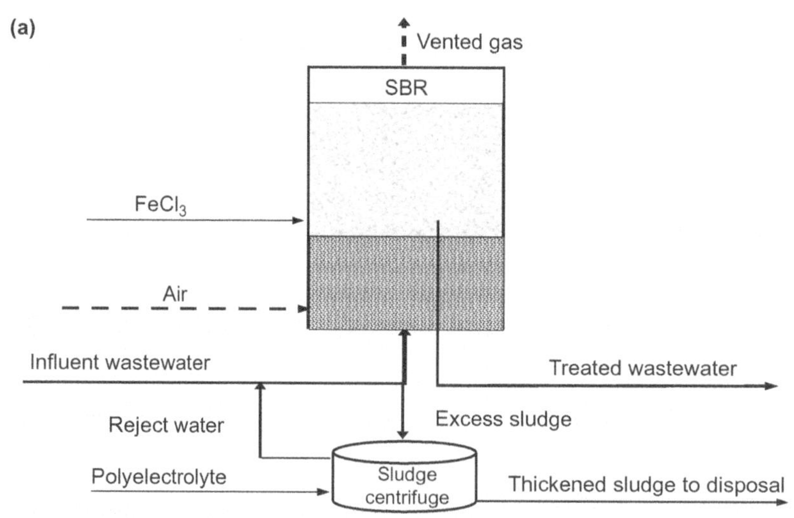

Figure 4.3a Principle of the Nereda™ treatment system

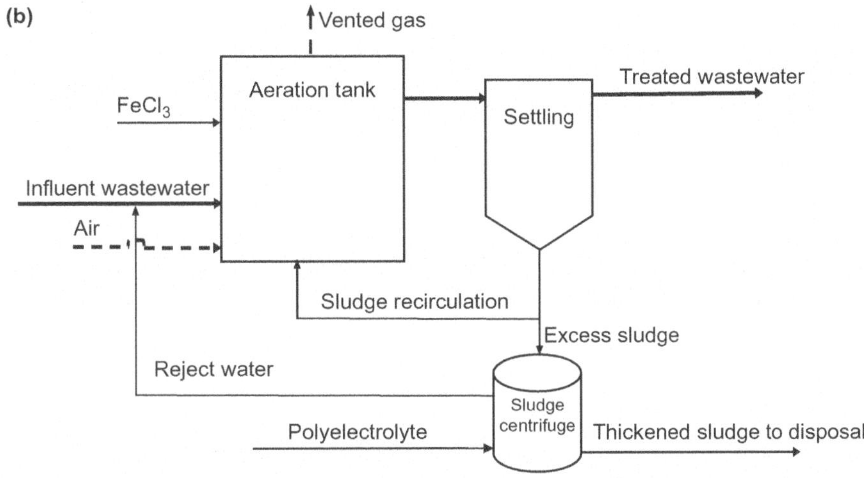

Figure 4.3b Activated sludge system, the reference technology for treating the brewery wastewater

The reported treatment efficiencies of the two systems when treating the brewery wastewater are compiled in Table 4.6. All values are annual averages. Table 4.7 summarises the design data for a full-scale Nereda™ plant for the treatment of the brewery wastewater and the data for the existing activated sludge plant.

Table 4.6 Projected treatment efficiency of the Nereda™ plant for the treatment of the brewery wastewater and the actual treatment efficiency of the existing activated sludge plant. Annual averages (Giesen, 2008) unless otherwise noted

Parameter	Nereda™ Effluent values	Activated sludge Effluent values	Unit
COD total	46	46	mg/l
COD dissolved	41	41	mg/l
P-tot	2	2.0	mg/l
PO_4-P	1	1,0	mg/l
N-tot	3.7	3.7	mg/l
NH_4-N	0.33	0.3	mg/l
TSS	12	12	mg/l
pH	7.7	7.7	
Temperature	25	25	°C
Sludge generation	3.1	3.7	kg thickened/m^3 influent
Sludge total solids	20	20	% of thickened sludge
Ash content in sludge	50	50	% of total solids
P-tot in sludge	15[2]	15[3]	g/kg total solids
N-tot in sludge	5.6	*42*[1]	g/kg total solids

[1] *Calculated by IVL from a mass balance for nitrogen, assuming that no nitrogen is lost by nitrification/denitrification.*
[2] *A mass balance of phosphorus indicates a P-content of 17 g/kg total solids.*
[3] *A mass balance of phosphorus indicates a P-content of 14 g/kg total solids.*

Table 4.7 Projected plant data for a Nereda™ plant. Actual plant data for the activated sludge plant (Giesen, 2008). (Some minor items have been left out)

Parameter	Nereda™	Activated sludge	Unit
Wastewater flow	4776	4776	m^3/day
Activated sludge tank	–	10800	m^3
Settling basin	–	Ø 30	m

(Continued)

Table 4.7 (*Continued*)

Parameter	Nereda™	Activated sludge	Unit
Reactor vessels	2 × 2500	–	m^3
Wastewater feed pump	8	10	m^3/h
Sludge dewatering pump	8	10	m^3/h
Sludge recirculation pump	–	300	m^3/h
Sludge centrifuge capacity	8	10	m^3/h
Aerators	1 × 70	3 × 180	kW
Power use	1239	2539	kWh/day
$FeCl_3$ (40% solution), consumption	272	540	kg/day
Polyelectrolyte (50% solution), high-molecular, cationic, consumption	20	25	kg/day
Excavation for buildings and structures	1571	9049	m^3

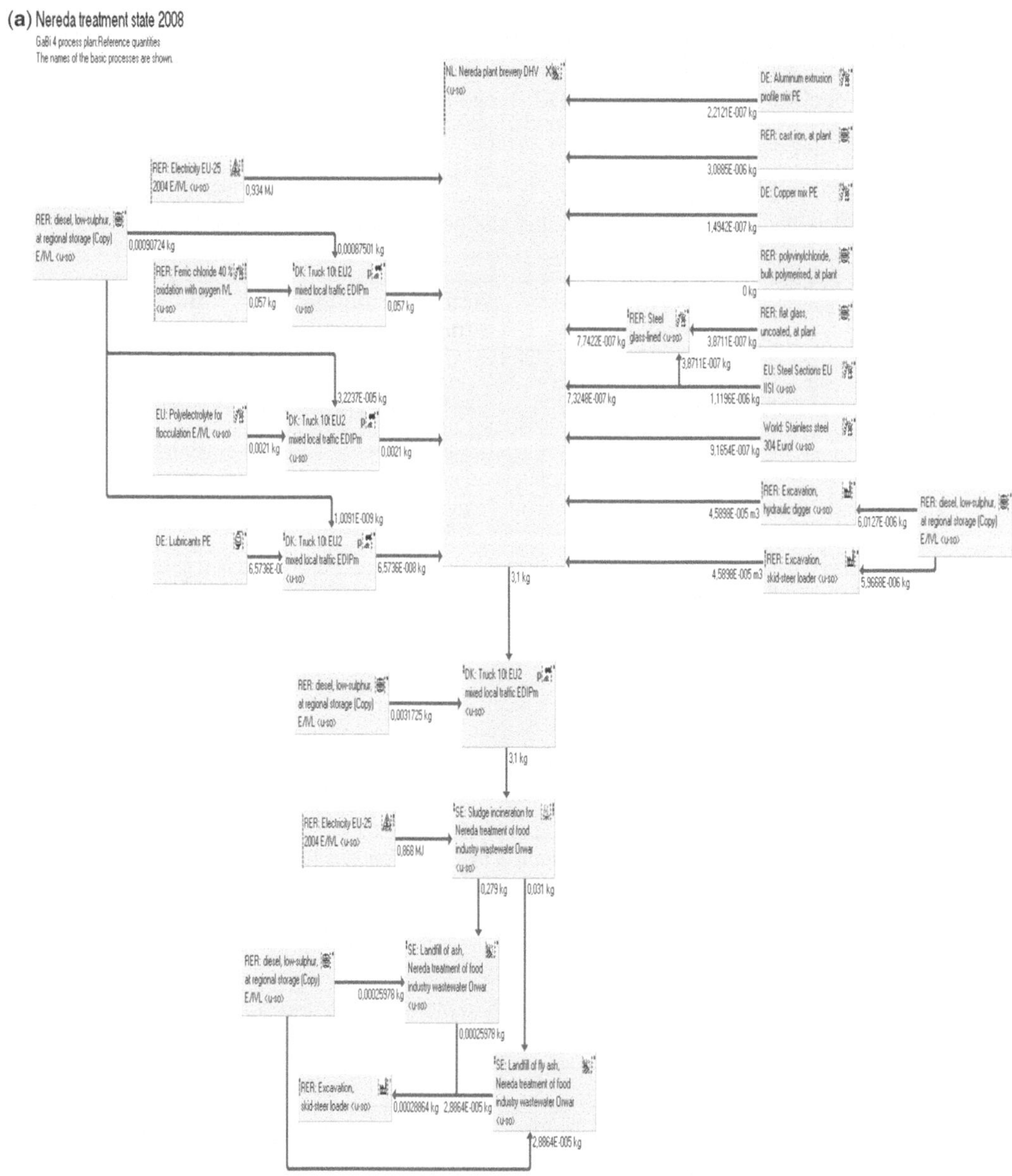

Figure 4.4a GaBi flowchart of the inventory model of the Nereda™ treatment system. RER, UCTE and CH modules: Data from the data base Ecoinvent 1.2 (Ecoinvent). DE modules: Data from the professional data base of GaBi (GaBi). DK modules: Data from the data base EDIP

4.3.1.3 The inventory

The plant and wastewater data reported in the system description are in this case study transformed directly into data sets suitable to enter into the LCA software GaBi. All data are recalculated to the functional unit, 1 m^3 of wastewater to be treated. The weights and the material compositions of the pumps used in the treatment systems have been deduced by collecting available data for suitable pumps from a supplier. For the sludge centrifuges and the aerators no inventory data are reported. Since sludge centrifuges of about equal capacities are used in both the Nereda^{TM} and the activated sludge systems, the data gap concerning the centrifuges should not significantly influence the comparative assessment of the two technologies. Flow charts from the GaBi models are shown in Figures 4.4a and 4.4b.

Inventories of the sludge disposal, which is incineration and landfilling of the ashes, and the peripheral processes are reported in Appendix 2.

In Figure 4.4a the module "NL: Nereda plant brewery DHV" calculates the inflows and the outflows of the treatment plant itself. Below this core module are the modules describing the sludge treatment and the landfilling of ashes. The inventories of the peripheral processes electricity, ferric chloride and polyelectrolyte have been put into the flowchart aggregated to modules and connected to the core module via transport modules. They are located to the left of the core module. To the right of the core module are modules containing the in- and outflows from the manufacture of the construction materials.

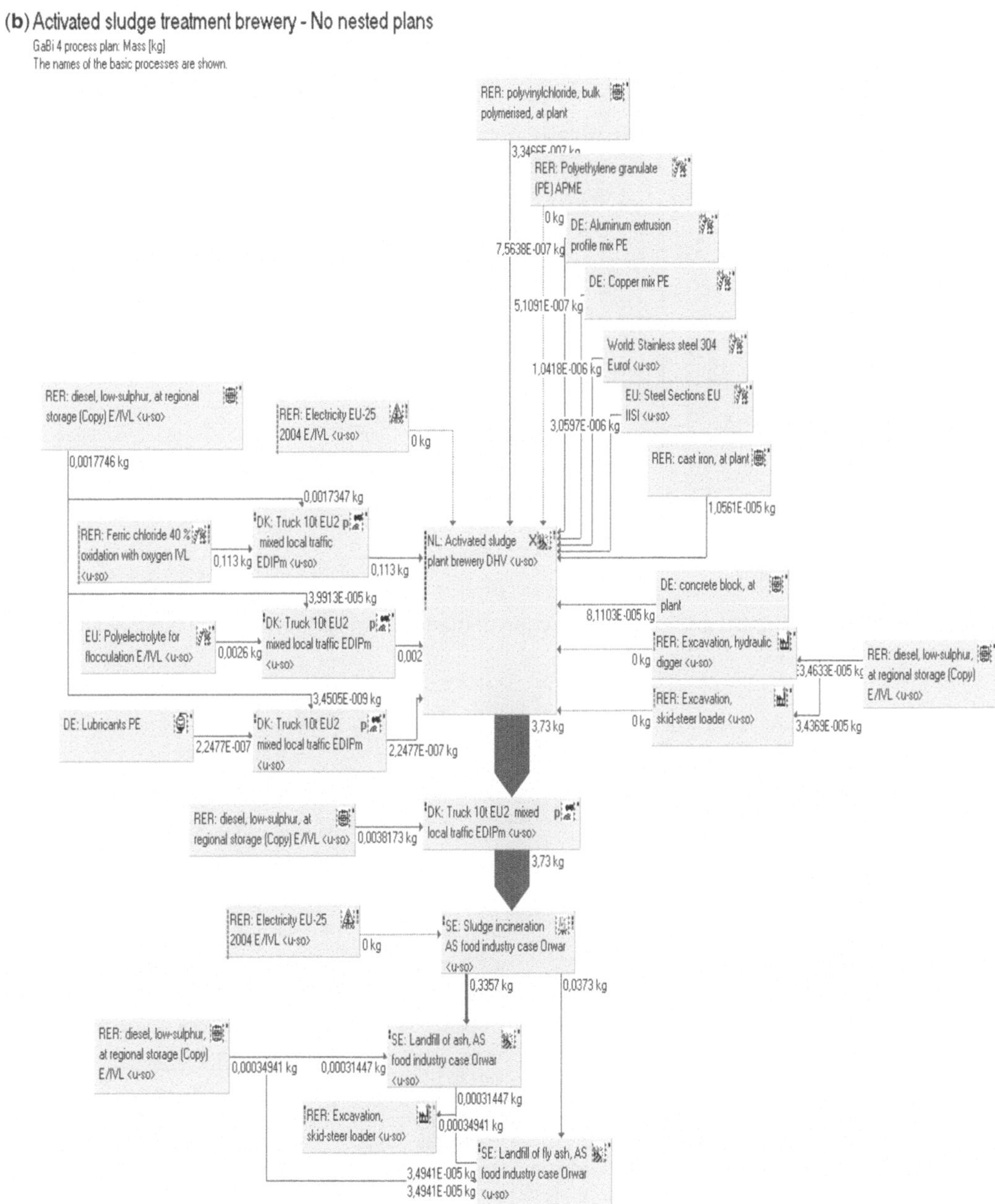

Figure 4.4b GaBi flowchart of the inventory model of the activated sludge treatment of a food brewery wastewater

When a quantity of wastewater is specified in the core module, all flow parameters specified somewhere in the inventory are calculated.

The inventory of the activated-sludge treatment of the brewery wastewater in Figure 4.4b is constructed in a similar way.

4.3.2 Results of the assessment

4.3.2.1 Comparison of the Nereda^{TM} and the activated-sludge technology

The environmental profiles of the two treatment systems are compared in a set of bar charts (Figures 4.5 – 4.9). In each chart one or a few of the induced impacts (impacts caused by the construction and operation of the treatment system) are compared for each of the following process stages of the activated sludge and the Nereda^{TM} processes:

- *Power*. Supply of electric power for the wastewater and the sludge treatment.
- *Chemicals*. Supply of the required chemicals, including manufacture and transport.
- *Construction*. Supply of the main structural materials for the wastewater treatment plants and excavation works on site, excluding transports to the site.
- *Sludge*. Treatment by incineration and landfilling of ashes, including transport of the dewatered sludge from the wastewater treatment plant, but excluding supply of electric power.

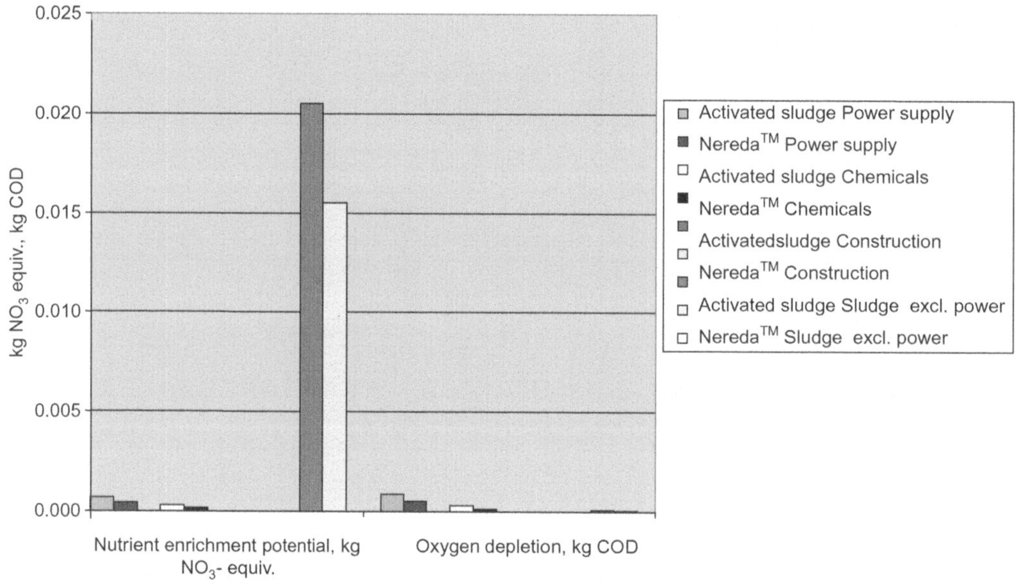

Figure 4.5 Nutrient enrichment potential and chemical oxygen demand per m^3 wastewater. The impact from discharge of the treated wastewater is not included

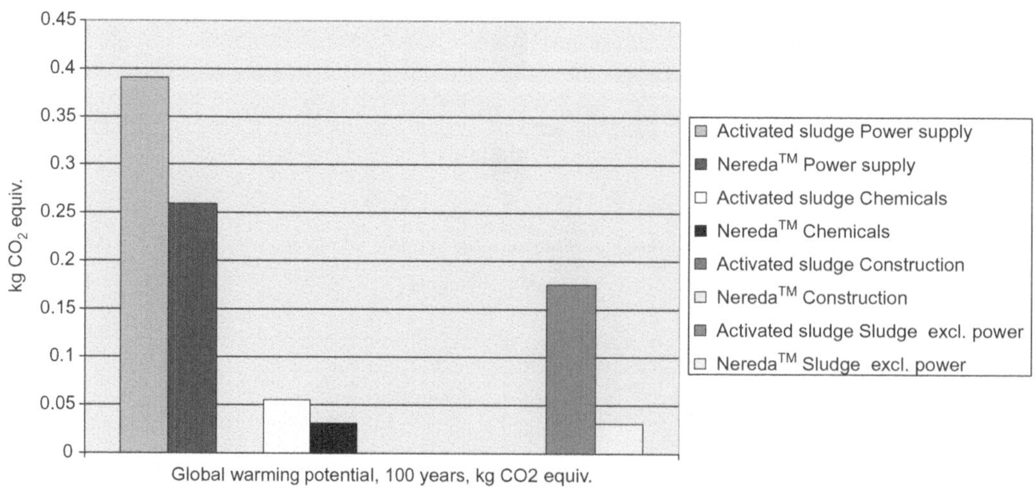

Figure 4.6 Global warming potential per m^3 wastewater. No emissions to the air from the wastewater treatment plants, e.g. nitrous oxide (laughing gas), are included, because of lack of data

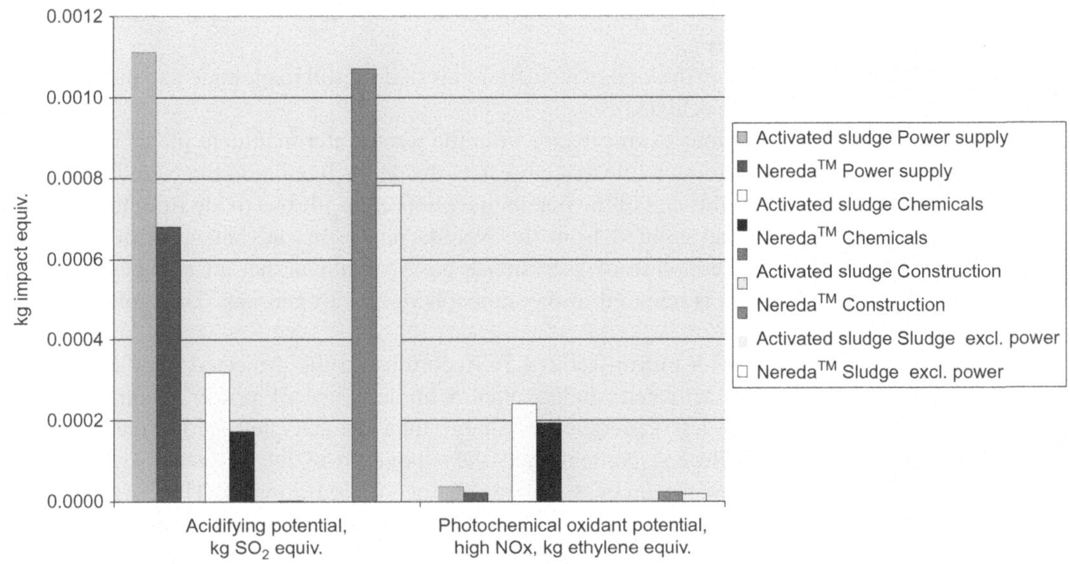

Figure 4.7 Acidifying and photochemical oxidant potentials per m^3 wastewater

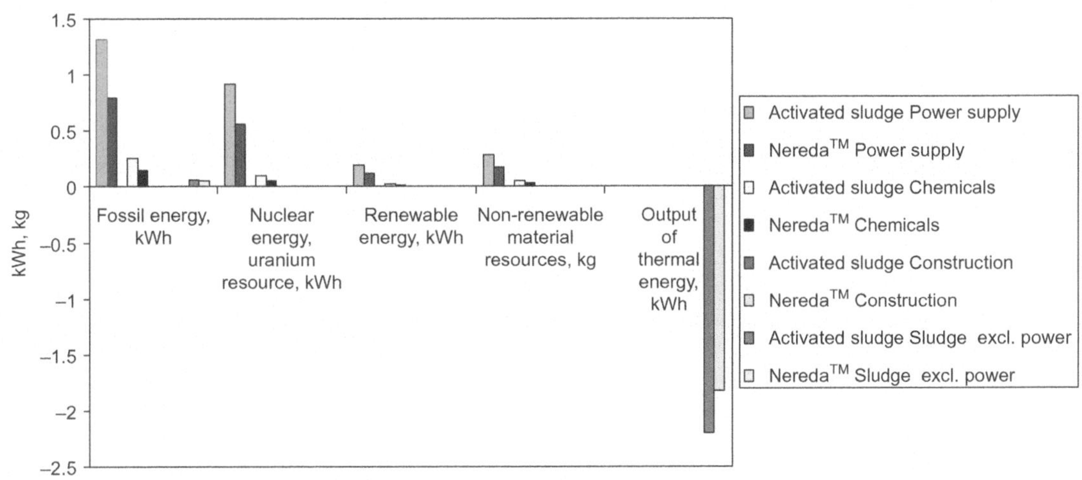

Figure 4.8 Use of fossil, nuclear and renewable energy resources and of non-renewable material resources. Potential generation of thermal heat from sludge incineration. Per m^3 wastewater

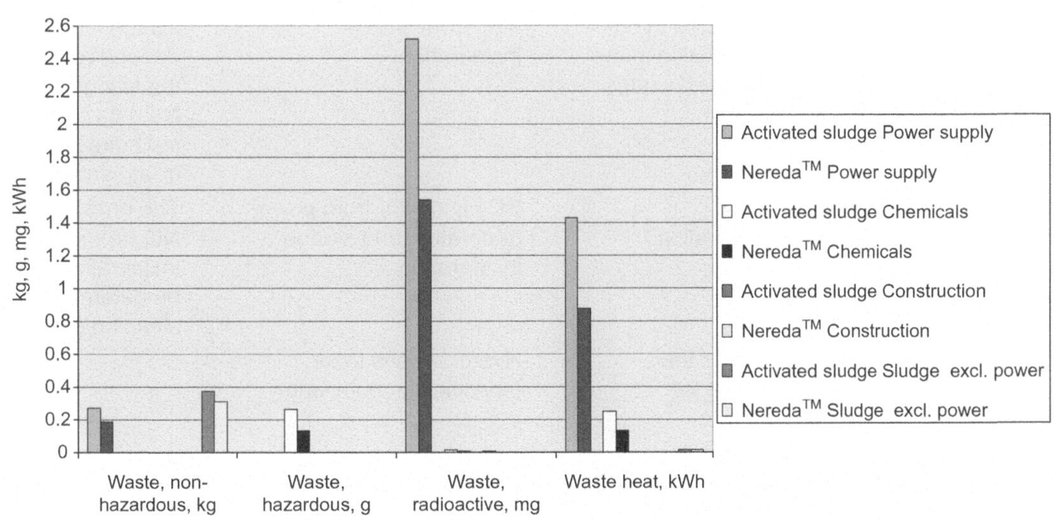

Figure 4.9 Generation of non-hazardous, hazardous and radioactive waste and generation of waste heat. Per m^3 wastewater

Since both treatment systems achieve (or are designed to achieve) the same result the impact from the discharge of the treated water is not included in the comparison.

The potential output of thermal energy from the sludge incineration is of doubtful magnitude and use and no credit in the form of replaced other energy resources is calculated.

The lack of any information about emissions to air directly from the wastewater treatment plants is a serious data gap, especially regarding the Nereda plant. According to the available data the denitrification in the Nereda plant decreases the nitrogen content of the sludge and thus the emission of the potent greenhouse gas nitrous oxide from the incineration of the sludge. However, if nitrous oxide is instead emitted from the Nereda plant, this advantage is of course diminished. Figure 4.6. Global warming potential, i.e. emission of greenhouse gases, assumes that all nitrogen removed from the wastewater and not contained in the sludge is released to the atmosphere as nitrogen gas. There are however so far no data to support this assumption.

The results reported in the Figures 4.5–4.9 mirror Table 4.7. According to the projected data the Nereda plant can achieve the same treatment result as the activated sludge plant with less use of power, chemicals and structural materials, with a smaller footprint and with less generation of sludge than the activated sludge plant. The sludge from the Nereda plant also has much lower nitrogen content than the sludge from the activated sludge plant. All this translates into lower impacts at all stages of the process and for all the studied impacts. The magnitude of the saved impacts and the relative importance of the different steps of the process chain will depend on the sources of power and chemicals and on the technology of sludge disposal in each specific case. When the power is drawn from the average European electricity grid and sludge is disposed of by incineration, power supply and sludge disposal will be the main environmental aspects to consider.

The contribution to the environmental impacts of the manufacture of structural materials for buildings and equipment is negligibly small calculated per m^3 of treated wastewater during the service life of the equipment.

Table 4.8 lists the main causes of each assessed impact.

Table 4.8 Main contributing emissions to and causes of each impact from the treatment of the brewery wastewater

Impact	Main cause	Main contributing emissions	Comment
Nutrient enrichment potential	Flue gas condensate from the sludge incineration	Phosphorus compounds to water	The result in Figure 4.5 is based on analysis data reporting less sludge from Nereda but the same P content in the sludge DS from both treatments. This does not seem plausible, since the same amount of P is reportedly removed by both treatments.
Global Warming potential	Power supply for treatment and sludge incineration Sludge incineration in the case of activated sludge	CO_2 from power generation N_2O from sludge incineration	The emission of N_2O from sludge incineration depends on the N content of the sludge. No data on emissions of N_2O directly from the treatment plants.
Acidifying potential	Power supply Sludge incineration	SO_2 and NO_x from power generation and sludge incineration	The emissions of SO_2 and NO_x from sludge incineration are assumed to be determined by flue gas cleaning standards.
Photochemical oxidant potential (near-ground ozone creation potential, "chemical smog")	Manufacture of the polyelectrolyte for flocculation	Hydrocarbons to air (specifically propylene)	
Waste, non-hazardous	Power supply Sludge incineration	Overburdens (from mining) Ashes	
Waste, hazardous	Manufacture of ferric chloride	Inorganic sludge	

4.3.2.2 Environmental efficiency of the Nereda^{TM} treatment as such

The sustainability of a treatment process is assessed by comparing the environmental impacts that are caused by the treatment process, to the impacts, that are avoided by the treatment and to those that remain after the treatment. Avoided and remaining impacts are calculated as described in Section 4.2.4.7 Assessment of environmental cost/benefit. The comparison for the Nereda system, when it is applied to the brewery wastewater, is shown in Table 4.9.

Table 4.9 Avoided, remaining and induced impacts, when the brewery wastewater is treated in a Nereda plant. Per m^3 wastewater

	Avoided impacts	Remaining impacts	Induced impacts
Nutrient enrichment potential, kg NO_3^- equiv.	0.48	0.081	0.016
Oxygen depletion, kg COD	0.60	0.046	0.00074
Global warming potential, 100 years, kg CO_2 equiv.			0.32
Acidifying potential, kg SO_2 equiv.			0.0016
Photochemical oxidant potential, high NO_x, kg ethylene equiv.			0.00024
Fossil energy, kWh			0.99
Nuclear energy, uranium resource, kWh			0.61
Renewable energy, kWh			0.13
Non-renewable material resources, kg			0.20
Waste, non-hazardous, kg			0.50
Waste, hazardous, kg			0.00013
Waste, radioactive, kg			1.55E-06
Waste heat, kWh			1.0
Potential output of thermal energy, kWh			1.8

The assessed avoided and remaining impacts from the wastewater are nutrient potential, and potential oxygen depletion, the latter impact measured as COD. The nutrient enrichment potential is caused by the nitrogen and phosphorus compounds in the wastewater.

The organic compounds in the wastewater are assumed to be entirely of organic origin. Thus the carbon dioxide, which will be formed if the organic compounds in the wastewater are allowed to be degraded in a receiving water, is not considered to contribute to global warming. Consequently there is no avoided or remaining global warming impact from the wastewater.

Other possible impacts from the wastewater, like possible ecotoxic impacts from heavy metals or trace amounts of biocides or detergents, cannot be assessed for lack of data.

Both the avoided and the remaining nutrient and oxygen depletion potentials exceed the same impacts caused by the treatment. In order to balance the benefits of the avoided nutrient and oxygen depletion potentials against the disadvantages of the other impacts from the treatment process, normalization as described in Section 4.2.4.8 Normalization is used.

In Table 4.10 the impacts from emissions (except COD) and wastes (except waste heat) in Table 4.9 are divided by the corresponding normalization references collected from Table 4.1. The quantities of primary fossil and nuclear energy in Table 4.10 are split up into the energy wares used to generate the energy. The quantities of energy wares are then divided by normalization references for crude oil, hard coal etc. collected from Appendix 1.

Figures 4.10a–c show the normalized induced emission impacts, the normalized values of waste generation and the normalized use of non-renewable resources in the form of bar charts. Since the available normalization references for wastes do not have the same reference year and reference region as the normalization references for emissions, the normalized values of these two impacts on the physical environment should not be directly compared.

If the results in Table 4.11 and Figures 4.10a–c are accepted at face value it seems that the treatment of the wastewater is beneficial from a sustainability point of view, as long as the scope of the assessment is limited to emissions. The total normalized impact of the induced emissions is one twentieth of the total avoided impact. The impact category where the treatment process infringes most on the available "sustainability space" is however the very impact the treatment is designed to remove, namely nutrient enrichment potential.

If the scope is broadened to resource use and waste generation the picture becomes more difficult to interpret. One observation is that if lack of landfilling space and the need to economise on scarce resources are important concerns, the need to landfill ashes and the use of non-renewable resources for power generation may be more important aspects than emissions.

Table 4.10 Avoided, remaining and induced impacts, when the brewery wastewater is treated in a Nereda plant. Normalized values per m³ wastewater

	Normalized values, pe · years		
	Avoided impacts	**Remaining impacts**	**Induced impacts**
Nutrient enrichment potential, kg NO_3^- equiv.	4.03E-03	6.81E-04	1.35E-04
Oxygen depletion, kg COD			
Global warming potential, 100 years, kg CO_2 equiv.			3.69E-05
Acidifying potential, kg SO_2 equiv.			2.21E-05
Photochemical oxidant potential, high NO_x, kg ethylene equiv.			939E-06
Total emissions	4.03E-03	6.81E-04	2.03E-04
Waste, bulk			1.40E-04
Waste, ashes			8.86E-04
Waste, hazardous			6.34E-06
Waste, radioactive			4.43E-05
Total wastes			1.08E-03
Hard coal			7.99E-05
Lignite			1.88E-04
Crude oil			2.42E-05
Natural gas			6.62E-05
Uranium			6.97E-04
Total non-renewable resources			1.06E-03

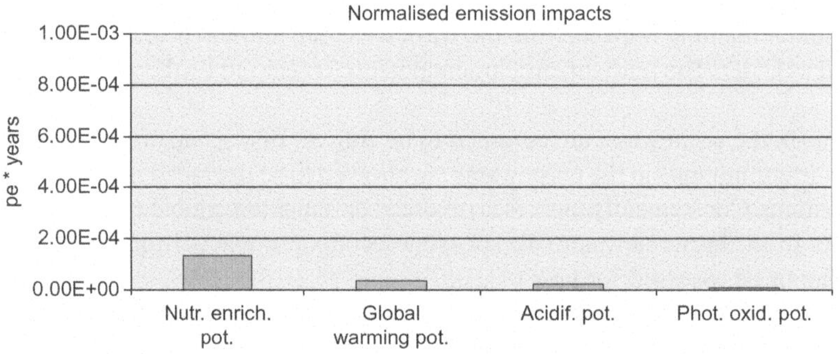

Figure 4.10a Induced emission impacts, when the brewery wastewater is treated in a Nereda plant. Normalized values per m³ wastewater

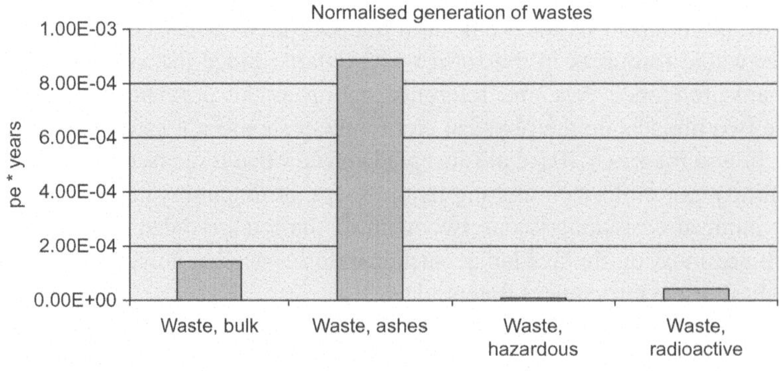

Figure 4.10b Waste generation, when the brewery wastewater is treated in a Nereda plant. Normalized values per m³ wastewater

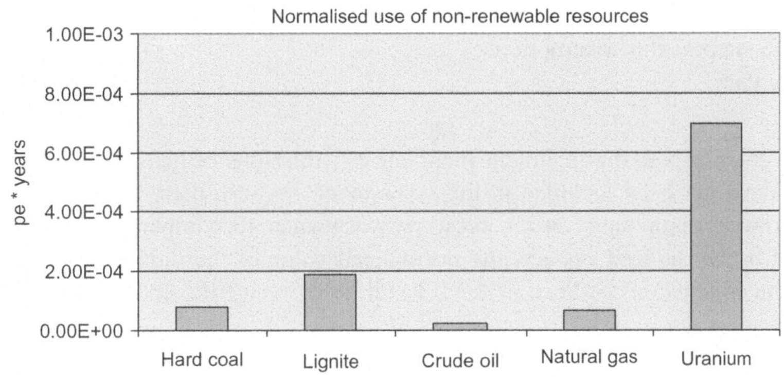

Figure 4.10c Use of non-renewable resources, when the brewery wastewater is treated in a Nereda plant. Normalized values per m^3 wastewater

The observation from the study of the environmental profiles of the treatment methods, namely that the main environmental issues are power supply and sludge disposal, still holds after normalization. However, the focus is shifted. The main sustainability concerns may be use of limited resources and disposal of long-lasting wastes rather than emissions.

4.3.3 Conclusions of the assessment of the food wastewater treatment

4.3.3.1 Environmental profiles of the Nereda^{TM} and the activated sludge treatment systems

1. The main environmental aspects are power supply and sludge disposal.
2. If the Nereda^{TM} and the activated sludge treatment use the same type of power and the same chemicals and have similar emissions to air from the treatment, and if the Nereda^{TM} treatment requires less energy and chemicals and generates smaller quantities of sludge with a lower nitrogen content than the activated sludge treatment to achieve the same result, the Nereda^{TM} treatment will obviously be more sustainable than the activated sludge treatment.

4.3.3.2 Sustainability of the Nereda^{TM} treatment system

1. If only the four major emission categories, namely nutrient enrichment, global warming, acidification and photochemical oxidant creation, are considered, the Nereda^{TM} treatment achieves a sustainable treatment result.
2. If the use of non-renewable resources and generation of wastes are included in the assessment, the question of sustainability becomes an issue of weighting of emission impacts and resource and waste impacts relative each other. Within their respective categories, use of non-renewable resources for power consumption and landfilling of ashes are important sustainability aspects.

4.3.4 Uncertainties

4.3.4.1 Uncertainties and data gaps in the process data

The assessment of the Nereda^{TM} process is based on one single data set. The data are reported to be averages of data collected from a pilot-plant over an extended period of operation, but no standard deviations are given. Nor are any data available, which relate the treatment performance to changes of the composition of the wastewater or to other changes of the treatment conditions.

The induced impacts of the Nereda treatment are generally 50–75% of the corresponding impacts of the equivalent activated sludge treatment. Since these savings are due to savings of energy and chemicals and to decreased quantities of sludge, parameters which should all be accurately measured, one would expect the calculated percentage savings to be outside the standard deviation of the data.

There are two gaps in the reported data, however:

1. Lack of analysis of the vented gas from the treatment plants. Altogether the calculated emissions of greenhouse gases from the Nereda plant are about 50% lower than the greenhouse gas emissions from the activated sludge treatment. About half of this saving is due to a calculated decrease of emissions of nitrous oxide from the sludge

incineration. The calculation assumes that the denitrification process in the Nereda plant emits no nitrous oxide. There are no data to support this assumption.
2. Lack of ecotoxicity data.

The supply of energy and chemicals to the treatment plant causes emissions of heavy metals at mines and power plants. These induced emissions have not been included in the assessment, because there are no data on avoided emissions of heavy metals or other ecotoxic compounds in the brewery wastewater to compare with. If the induced emissions of heavy metals are included in the induced impacts the normalized value of the induced total emissions in Table 4.11 is $3.6 \cdot 10^{-4}$ pe · years, i.e. the conclusions of Section 4.3.2.1 still hold, even if no ecotoxic impact at all is avoided by the treatment of the brewery wastewater.

4.3.4.2 Uncertainties in the normalization references

No standard deviations are given for the normalization references in Table 4.1. In order to check the robustness of conclusion (1) in Section 4.3.2.2 a standard deviation of $\pm 50\%$ will be assumed. In Table 4.11 a case has been, calculated, where the normalization reference for nutrient enrichment has been multiplied by 1.5 and the other references divided by 2. This is the most "unfavourable" assumption for the sustainability of the treatment system.

Table 4.11 Avoided, remaining and induced emission impacts, when the brewery wastewater is treated in a Nereda plant. Normalized values per m^3 wastewater, assuming that the normalization reference for nutrient is 50% higher and the other emission references 50% lower than the values in Table 4.1

	Normalized values, pe · years		
	Avoided impacts	**Remaining impacts**	**Induced impacts**
Nutrient enrichment potential, kg NO_3^- equiv.	2.69E-03	4.54E-04	9.00E-05
Oxygen depletion, kg COD			
Global warming potential, 100 years, kg CO_2 equiv.			7.37E-05
Acidifying potential, kg SO_2 equiv.			4.42E-05
Photochemical oxidant potential, high NO_x, kg ethylene equiv.			1.88E-05
Total emissions	2.69E-03	4.54E-04	2.27E-04

Conclusion (1) of Section 4.3.2.2 still holds, although the normalized total induced impact is increased from one twentieth to one twelfth of the avoided impact.

Introduction of weighting factors for the emission impacts would not significantly change the result.

4.3.4.3 Supply and sludge-disposal scenarios

The technologies used in the supply chain and for sludge disposal have a decisive influence on the sustainability of the treatment process. If the sludge, instead of being incinerated to ashes, could be used as a fertiliser and replace commercial fertilisers, this would reduce the induced emission impact and the waste impact as well as the net impact of the use of non-renewable resources.

It is also obvious that a change of the source of power will have a large effect. A switch from the European grid to the French grid (nuclear power) for instance, would mean a reduced emission impact but a strongly increased impact from the use of non-renewable resources.

4.4 LANDFILL LEACHATE

4.4.1 The assessed treatment case

4.4.1.1 The leachate

The leachate, for which treatment with both a new technology and a reference technology could be assessed, is the effluent from a landfill in Italy. The composition reported in Table 4.12 is used for the assessment of the two technologies described below.

Table 4.12 Reported characteristics of the leachate from the Italian landfill (data from IRSA)

Parameter	Value	Unit
Wastewater flow	90	m^3/day
COD total	9700	mg/l
COD dissolved (GF/A)	9100	mg/l
DOC (0, 45 μm)	2830	mg/l
BOD_7	1500	mg/l
LAS	11	mg/l
VOC	1.5	mg/l
AOX	4	mg/l
TSS (GF/A)	390	mg/l
VSS (GF/A)	220	mg/l
N-tot	3200	mg/l
NH_4-N	2900	mg/l
$NO_{2,3}$-N	0	mg/l
P-tot	27	mg/l
pH	8,0	
Conductivity	28	mS/cm
Cl^-	4800	mg/l
Alkalinity	16 100	mg $CaCO_3$/l
Ca	240	mg/l
Cd	100	μg/l
Cr	1300	μg/l
Cu	1000	μg/l
Fe	12	mg/l
Mg	170	mg/l
Mn	200	μg/l
Ni	600	μg/l
Pb	700	μg/l
Zn	800	μg/l

4.4.1.2 The treatment systems

The landfill leachate to be treated is rich in oxygen consuming substances and nitrogen compounds. The first step of the new-technology treatment is a struvite precipitation, where nitrogen compounds are precipitated as magnesium ammonium phosphate, struvite. It is assumed that the struvite will be sold as a fertiliser and replace nitrogen and phosphorus fertilisers. The leachate is then treated in a sequential batch biofilter granular reactor (SBBGR). The reactor is supplied with oxygen via an aerator on a recirculation line and via direct introduction of oxygen into the reactor. Ozone is applied intermittently to the aerator. Figure 4.11 gives an outline of the system.

Based on laboratory results obtained from experiments with the leachate of Table 4.13 (IRSA) a treatment system for a 90-m^3/day flow of leachate is designed. The entire system is operated in a batch mode. The struvite precipitation is operated once a day. The SBBGR treats three batches per day. The plant is operated in the following way:

1. Starting position

The aerator is empty and shut off from the SBBGR and the ozonator.

2. Filling

30 m^3 of landfill leachate from the struvite precipitation is pumped into the aerator during half an hour. The inlet is closed.

3. Aeration with biological degradation

The circulation between the aerator and the SBBGR is started and the aeration is switched on. Aeration with circulation is continued for 6 hours while pure oxygen is dosed in the bottom of SBBGR. During this time the circulation between the aerator and the ozonator remains closed.

4. Ozonation

The circulation between the aerator and the ozonator and the ozone dosing are started and operated for one hour. During this time O_2 addition in the bottom of SBBGR, the aeration and circulation between the aerator and the SBBGR are continued.

5. Drawing off

Aeration, ozonation and both circulations are shut off. The aerator is sealed off from the SBBGR and the ozonator. 30 m^3 of water are emptied from the aerator as treated water.

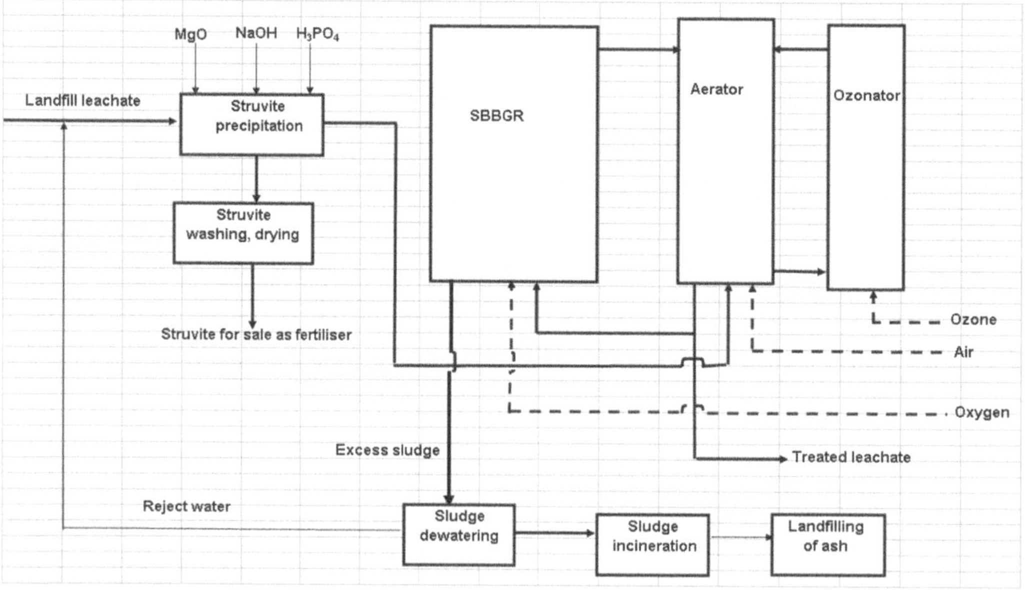

Figure 4.11 Struvite + SBBGR + ozone treatment

Table 4.13 Basic plant data for the struvite – SBBGR system

Parameter	Value	Unit
Wastewater	90	m^3/day
COD removal	98	%
N removal	99	%
P removal	70	%
Struvite precipitation, constructional steel	90	m^3
SBBGR reactor, reinforced concrete	270	m^3
Aerator, reinforced concrete	120	m^3
Ozonator, reinforced concrete	10	m^3
Ozone generator	60	kg O_3/h
Oxygen, liquid, including ozone generation	28	kg/m^3 treated water
Ozone	2.0	kg/m^3 treated water
Magnesium oxide	15	kg/m^3 treated water
Sodium hydroxide, 100%	5.3	kg/m^3 treated water
Phosphoric acid	19	kg/m^3 treated water
Sulphuric acid, 100%	3.7	kg/m^3 treated water
Polyelectrolyte	0.00032	kg/m^3 treated water
Electric power, including ozone generation	21	kWh/m^3 treated water
Produced struvite, pure	51	kg/m^3 treated water
Sludge generation	0.32	kg DS/m^3 treated water
Labour	5	person hours/week

Table 4.13 reports some operational and constructional data, all referred to 1 m^3 of treated water. Table 4.14 gives the treatment performance.

Table 4.14 Treatment performance of the struvite – SBBGR - ozone system, based on laboratory results from IRSA

Parameter	Influent leachate	Treated leachate	Sludge	Unit
COD total	9700	160		mg/l
COD dissolved (GF/A)	9100	158		mg/l
DOC (0, 45 μm)	2830	145		mg/l
BOD_7	1500	4		mg/l
LAS	11	0		mg/l
VOC	1.5	0		mg/l
AOX	4	0		mg/l
TSS (GF/A)	390	20		mg/l
VSS (GF/A)	220	12		mg/l
N-tot	3200	30		mg/l
NH_4-N	2900	0		mg/l
$NO_{2,3}$-N	0	3		mg/l
P-tot	27	8		mg/l
PO_4-P	–	5		mg/l
pH	8.0	7.8		
Conductivity	28	34		mS/cm
Cl^-	4800	4800		mg/l
Alkalinity	16 100	6500		mg $CaCO_3$/l
Ca	240	48		mg/l
Cd	100	100		μg/l
Cr	1300	100		μg/l
Cu	1000	1000		μg/l
Fe	12	1.3		mg/l
K	–	1150		mg/l
Mg	170	15		mg/l
Mn	200	100		μg/l
Ni	600	800		μg/l
Pb	700	100		μg/l
Zn	800	100		μg/l
Ash in sludge			60	% of DS
N-tot in sludge			7	% of DS
P-tot in sludge			6	% of DS

The activated sludge plant, to which the performance of the new technology is compared, is a fictitious plant with an estimated and calculated performance. The treatment consists of a pre-denitrification step, a nitrification step and a post-denitrification step with recirculating wastewater from the nitrification step to the pre-denitrification step. Methanol is added as a carbon source in the post-denitrification step. Air is added in the nitrification step. Phosphorus is removed by precipitation as a post-treatment and a polyelectrolyte is added in the settling volume for improved sludge characteristics. A schematic flowchart of the treatment system is given in Figure 4.12. Some operational and constructional data of the plant are given in Table 4.15. Table 4.16 reports its projected performance.

4.4.1.3 The inventory

The plant and wastewater data reported in the system description are in this case study transformed directly into data sets suitable to enter into the LCA software GaBi. All data are recalculated to the functional unit, 1 m^3 of wastewater to be treated. The use of construction materials per m^3 treated water is calculated from the estimated service life of various parts of the plant (see Tables 4.21 and 4.22 below in the Section Cost calculations). The weights and the material compositions of the pumps used in the treatment systems have been deduced by collecting available data for suitable pumps from a supplier.

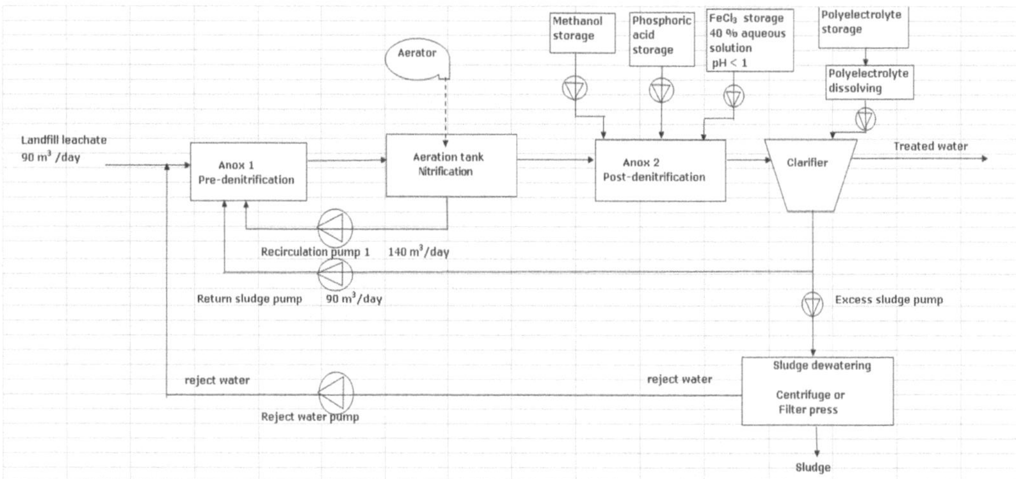

Figure 4.12 Projected activated-sludge plant for treatment of the Italian landfill leachate described in Table 4.12

Table 4.15 Basic plant data for the activated sludge system

Parameter	Value	Unit
Wastewater	90	m^3/day
COD removal	31	%
N removal	99+	%
P removal	93	%
Anox 1	225	m^3
Aeration tank	1200	m^3
Anox 2	400	m^3
Clarifier	50	m^3
Methanol	6.6	kg/m^3 treated water
Ferric chloride, 40% solution	0.033	kg $FeCl_3$/m^3 treated water
Polyelectrolyte	0.005	kg/m^3 treated water
Electric power	4.8	kWh/m^3 treated water
Sludge generation	6.8	kg DS/m^3 treated water
Labour	10	person hours/week

The SBBGR + ozonation system requires an ozonator with a capacity of 60 kg of ozone/hour. The model PDO 4500 from Wedeco Ozone Systems has the necessary capacity. Operational and constructional details were provided by Kampmann (2009).

It is presumed that the precipitated struvite is used as fertiliser and saves other mineral fertilisers. The saved resources and avoided emissions are calculated from production data from monoammonium phosphate, available from Ecoinvent (Ecoinvent, 2007), assuming that nitrogen and phosphorus in struvite can replace these nutrients in monoammonium phosphate on a 1:1 basis. The ammonium phosphate is produced from ammonia and phosphoric acid. The data are given as two data sets per kg N and per kg P_2O_5, resp. The multi-output process monoammonium phosphate delivers the co-products monoammonium phosphate as N and monoammonium phosphate as P_2O_5. Allocation factors are based on the energy requirements of the respective nutrients for the production process: 45% for monoammonium phosphate as N and 55% for monoammonium phosphate as P_2O_5. The allocated inventories must always be used simultaneously. This is done in the LCA model of the struvite-SBBGR-ozone system.

Since the impacts of ordinary construction materials are small or negligible compared to impacts from consumables the constructional materials are disregarded in the environmental assessment of the activated sludge system as a first approximation.

Table 4.16 Projected treatment performance of the activated-sludge plant when treating the Italian landfill leachate

Parameter	Influent leachate	Treated leachate	Sludge	Unit
COD total	9700	6740		mg/l
COD dissolved (GF/A)	9100	6720		mg/l
DOC (0, 45 μm)	2830	–		mg/l
BOD_7	1500	10		mg/l
LAS	11	0		mg/l
VOC	1.5	1.5		mg/l
AOX	4	1.6		mg/l
TSS (GF/A)	390	20		mg/l
VSS (GF/A)	220	16		mg/l
N-tot	3200	10		mg/l
NH_4-N	2900	0		mg/l
NO_3-N	0	5		mg/l
P-tot	27	2		mg/l
PO_4-P	–	–		mg/l
pH	8,0	9		
Conductivity	28	–		mS/cm
Cl^-	4800	4800		mg/l
Alkalinity	16 100	–		mg $CaCO_3$/l
As	0,0005	0,0005		mg/l
Ca	240	–		mg/l
Cd	0,1	–		mg/l
Cr	1.3	0.78		mg/l
Cu	1.0	0.6		mg/l
Fe	12	2		mg/l
K	–	–		mg/l
Mg	170	–		mg/l
Mn	0.2	0.1		mg/l
Ni	0.60	0.48		mg/l
Pb	0.70	0.56		mg/l
Zn	0.80	0.56		mg/l
Ash in sludge			25	% of DS
N-tot in sludge			4,0	% of DS
P-tot in sludge			0,37	% of DS

Inventories of the sludge disposal, which is incineration and landfilling of the ashes, and the peripheral processes are reported in Appendix 2.

4.4.2 Results of the environmental assessment

4.4.2.1 Summary of avoided and induced impacts

Tables 4.17 and 4.18 summarise the performance of the treatment methods as environmental cost/benefit analyses. The tables also show the contributions of supply of chemicals and electricity, of sludge treatment, of the supply of construction materials and of the impacts of saved mineral fertilisers to the environmental costs, i.e. to the induced impacts.

Inspection of Tables 4.17 and 4.18 indicates that both treatment technologies are about equally efficient at removing nutrients. In addition the SBBGR technology saves further emissions of nutrients from saved manufacture of mineral fertilisers. The SBBGR technology furthermore efficiently removes oxygen depleting substances. However, this increased COD-removal efficiency obviously comes with a cost. The SBBGR technology causes higher emissions of greenhouse gases, acidifying substances and photo-oxidants, and uses more resources, than the activated-sludge technology.

The increased COD-removal efficiency of the SBBGR technology does not translate into a proven efficient removal of toxic substances. Neither technology seems to be very efficient at removing toxic substances. According to the available

Table 4.17 Avoided and remaining impacts compared to induced impacts when landfill leachate is treated in a 90 – m^3/day struvite + SBGGR + ozonation plant. All data per m^3 treated wastewater. Negative values for the struvite by-product mean that impacts are saved when mineral fertilisers are replaced by struvite

	Avoided impacts	Rem. impact	Induced impacts						
			Total	Chemicals	Ozone	Electricity	Materials	Sludge treatment	Struvite
Nutrient enrichment potential, kg NO_3^- equiv.	15	0.39	−5.5	2.6	0.05	0.0026	1.2E-4	0.031	−8.2
Oxygen consumption, recalculated to kg NO_3^- equiv.	9.5	0.16	0.51	0.52	0.032	0.0035	6.5E-5	1.3E-4	−0.049
Ecotoxicity potential, acute, water, m^3 water	110	0.00	1694	7800	390	7.88	0.406	0.38	−6500
Ecotoxicity potential, chronic, water, m^3 water	220	0.00	17 000	79 000	1 900	77	4,0	4.4	−64 200
Ecotoxicity potential, chronic, soil, m^3 soil	0.00	0.00	−15	26	6.7	0.051	0.013	0.016	−47
Human toxicity potential, air, m^3 air	0.00	0.00	3.3E+7	2.7E+7	6.7E+6	1.1E+5	1.7E+4	1.6E+4	−1.1E+6
Human toxicity potential, water, m^3 water	0.028	0.00	−3131	5100	83.4	4.59	1.13	0.203	−8300
Human toxicity potential, soil, m^3 soil	0.00	0.00	−0.81	8.4	1.3	0.036	0.0044	0.0025	−11
Global warming potential, 100 years, kg CO_2 equiv.	0.00	0.00	40.8	50.2	18.3	1.55	0.037	0.17	−30
Acidifying potential, kg SO_2 equiv.	0.00	0.00	0.19	0.680	0.076	0.0044	1.1E-4	4.1E-4	−0.57
Photochem. oxidant potential, high NO_x, kg ethylene equiv.	0.00	0.00	0.0089	0.013	0.0027	1.4E-4	1.3E-05	1.1E-05	−0.0073
Bulk waste, kg	0.00	0.00	7.1		6.0	1.1	0.0074	0.032	
Hazardous waste, kg	0.00	0.00	2.8E-9				2.8E-09		
Slags and ashes, kg	0.00	0.00	1.9E-1					0.19	
Nuclear waste, kg	0.00	0.00	6.6E-5		5.6E-5	1.0E-5	1.1E-7	3.0E-7	
Fossil energy, including feedstock, MJ			330	490	240	18	0.32	0.60	−410
Nuclear energy, uranium resource, MJ			260	150	150	13	0.065	0.40	−57
Renewable energy, MJ			50	27	28	2.5	0.018	0.078	−8.7

analytical data the SBBGR technology actually appears to cause more ecotoxicity than it removes. The reason for this apparent lack of efficiency may quite simply be that, apart from LAS, the analytical protocol does not comprise specific organic toxicants, only metals. Neither technology is designed to remove metals.

Table 4.18 Avoided and remaining impacts compared to induced impacts when landfill leachate is treated in a 90 – m^3/day activated-sludge plant with nitrification/denitrification. All data per m^3 treated wastewater. The contribution to global warming from the treatment process is CO_2 from the part of the added methanol, which is biologically oxidised

	Avoided impacts	Remaining impacts	Induced impacts				
			Total	Chemicals	Electricity	Sludge treatment	Treatment process
Nutrient enrichment potential, kg NO_3^- equiv.	15	0.11	0.085	0.015	0.0039	0.066	
Oxygen consumption, recalculated to kg NO_3^- equiv.	3.0	6.7	0.014	0.0057	0.0051	0.0026	
Ecotoxicity potential, acute, water, m^3 water	110	0.00	76.4	59	12	5.8	
Ecotoxicity potential, chronic, water, m^3 water	220	0.00	433	260	120	56.6	
Ecotoxicity potential, chronic, soil, m^3 soil	0.00	0.00	2.88	1.79	0.76	0.34	
Human toxicity potential, air, m^3 air	0.00	0.00	2.5E+6	2.0E+6	1.6E+5	3.2E+5	
Human toxicity potential, water, m^3 water	0.028	0.00	24	14	6.8	3.1	
Human toxicity potential, soil, m^3 soil	0.00	0.00	0.60	0.50	0.054	0.046	
Global warming potential, 100 years, kg CO_2 equiv.	0.00	0.00	19	5.3	2.3	6.7	4.5
Acidifying potential, kg SO_2 equiv.	0.00	0.00	0.028	0.012	0.0065	0.010	
Photochem. oxidant potential, high NO_x, kg ethylene equiv.	0.00	0.00	0.0042	0.0030	2.1E-4	3.1E-4	7.5E-04
Bulk waste, kg	0.00	0.00	2.3	4.1E-4	1.6	0.70	
Hazardous waste, kg	0.00	0.00	7.7E-5	7.7E-5			
Slags and ashes, kg	0.00	0.00	1.7			1.7	
Nuclear waste, kg	0.00	0.00	2.1E-5	4.9E-9	1.5E-5	6.5E-6	
Fossil energy, including feedstock, MJ			320	280	27	13	
Nuclear energy, uranium resource, MJ			32	4.1	19	8.5	
Renewable energy, MJ			6.6	1.2	3.8	1.6	

4.4.2.2 Comparison of the treatment technologies – Assessment by normalization

Tables 4.19 and 4.20 report the avoided and the induced impacts of the SBBGR and the activated-sludge technologies normalized according to the methodology described in Section Normalization. Since the removal of metals is random and not very efficient the toxic impacts of metals have been excluded from the characterisation of the untreated and treated leachate, but not from the assessment of the background processes. For the leachate only the impact of removed LAS (linear alkyl benzene sulfonates) is considered.

The tables indicate only a marginal advantage of the SBBGR technology compared to the activated-sludge technology regarding avoided and induced emissions and induced waste generation. The reason is that the achievement of the ozonation is just evaluated as removed oxygen depletion characterised as removed nutrient enrichment. This gives the increased COD removal a rather low assessment value. The recalculation ratio COD to nutrient enrichment is

0.22 kg NO_3^- equivalents/kg COD. There are no data to assess the impact of the removed COD in terms of removed toxicity.

Table 4.19 Normalized values of avoided and induced impacts when landfill leachate is treated in a 90 – m^3/day struvite + SBGGR + ozonation plant. All data per m^3 treated wastewater. Negative values mean that saved emissions from struvite replacing NP-fertilisers are higher than the total emissions from other sources of induced impacts

		Normalized values, mpe · years	
		Avoided impacts	**Induced impacts**
Nutrient enrichment potential, kg NO_3^- equiv.		120	−46
Oxygen consumption, recalculated to kg NO_3^- equiv.		18	0.94
Ecotoxicity potential, acute, water, m^3 water		3.8	58
Ecotoxicity potential, chronic, water, m^3 water		0.6	49
Ecotoxicity potential, chronic, soil, m^3 soil		0.00	−0.02
Human toxicity potential, air, m^3 air		0.00	11
Human toxicity potential, water, m^3 water		0.00	−60
Human toxicity potential, soil, m^3 soil		0.00	−6.4
Global warming potential, 100 years, kg CO_2 equiv.		0.00	4.7
Acidifying potential, kg SO_2 equiv.		0.000	2.5
Photochemical oxidant potential, high NO_x, kg ethylene equiv.		0.00	0.36
	Total emissions	150	14
Bulk waste, kg			5.3
Hazardous waste, kg			1.3E-07
Slags and ashes, kg			0.55
Nuclear waste, kg			1.9
	Total wastes		7.7
Crude oil, kg			4.8
Hard coal, kg			7.3
Lignite, kg			28
Natural gas, kg			0.33
Uranium, kg			82
	Total non-renewable energy resources		120

The advantage of the SBBGR technology is paid for by an increase of the use of energy resources, notably uranium ore and coal to generate electricity for the production of ozone and chemicals. The main use of energy wares for the activated-sludge process is the consumption of natural gas to produce the methanol, which is added as a carbon source.

The data of the Tables 4.17 and 4.18 are illustrated by the Figures 4.13–4.16. In Figures 4.13 and 4.14 we compare the avoided impacts excluding toxicities to the induced impacts of the same categories for the SBBGR technology and the activated-sludge technology respectively. The induced impacts are split between the commodities required for running the processes. In the Figures 4.15 and 4.16 we have added toxicities but excluded toxicities caused by metals in the leachate, since data on removal of metals are uncertain and the technologies are not designed to remove metals anyway. Figure 4.17 finally illustrates a scenario, where no ozonation is applied and the COD removal of the SBBGR technology is assumed to be the same as that of the activated-sludge technology.

Several points can be made from the results presented above.

- The EDIP characterisation and normalization data put heavy emphasis on toxicity and nutrient enrichment. While their data are open to discussion, this emphasis seems reasonable when assessing advanced and costly treatment technologies. In order to be sustainable such technologies should be applied to eliminate pollutants with a high and serious impact. To use resource-demanding technologies to remove biologically recalcitrant compounds with little impact on the environment will hardly be sustainable.

- The environmental performance of the struvite-SBBGR technology is determined by the use and source of chemicals in the struvite precipitation and by the possibility to use the struvite as a fertiliser which replaces NP-fertilisers. Without this latter possibility the total normalized values of the induced impacts from emissions from the SBBGR technology would be about five times higher than the avoided impacts, as far as we can assess them. The sustainability of the treatment technology would be very doubtful, with or without ozone treatment.
- With the present probably incomplete evaluation of the impact of the COD additionally removed by the ozonation one cannot claim that the ozonation is environmentally efficient. A comparison of the Figures 4.15 and 4.17 suggests that the marginal decrease of the normalized avoided impacts from the use of ozone is less then the marginal increase of the normalized induced impacts caused by running the ozonation.
- Likewise, given the present probably incomplete evaluation of the impact of removed COD it seems doubtful if the struvite + SBBGR technology is more efficient than the activated-sludge technology. True enough, the struvite + SBBGR technology performs marginally better than the activated-sludge technology in terms of avoiding nutrient enrichment without causing higher net induced impacts. However in the case of the struvite + SBBGR technology the net induced impact is a difference between high actual emissions and high theoretically saved emissions. The gross induced impact from supplying the chemicals to the struvite precipitation is very high indeed. Furthermore we have not yet taken the use of resources into account. The struvite + SBBGR technology uses about 80% more primary energy resources than the activated-sludge technology after deduction of the energy saved when NP-fertilisers are replaced by struvite.

Table 4.20 Normalized values of avoided and induced impacts when landfill leachate is treated in a 90 – m^3/day activated-sludge plant with nitrification/denitrification. All data per m^3 treated wastewater

		Normalized values, mpe · years	
		Avoided impacts	**Induced impacts**
Nutrient enrichment potential, kg NO_3^- equiv.		130	0.72
Oxygen consumption, recalculated to kg NO_3^- equiv.		5.5	0.02
Ecotoxicity potential, acute, water, m^3 water		3.8	2.6
Ecotoxicity potential, chronic, water, m^3 water		0.63	1.23
Ecotoxicity potential, chronic, soil, m^3 soil		0.00	0.00
Human toxicity potential, air, m^3 air		0.00	0.82
Human toxicity potential, water, m^3 water		0.00	0.46
Human toxicity potential, soil, m^3 soil		0.00	4.7
Global warming potential, 100 years, kg CO_2 equiv.		0.00	2.16
Acidifying potential, kg SO_2 equiv.		0.00	0.38
Photochemical oxidant potential, high NO_x, kg ethylene equiv.		0.00	0.17
	Total emissions	140	13
Bulk waste, kg			1.69
Hazardous waste, kg			0.00
Slags and ashes, kg			4.83
Nuclear waste, kg			0.61
	Total wastes		7.1
Crude oil, kg			0.43
Hard coal, kg			1.2
Lignite, kg			2.8
Natural gas, kg			18
Uranium, kg			10
	Total non-renewable energy resources		32

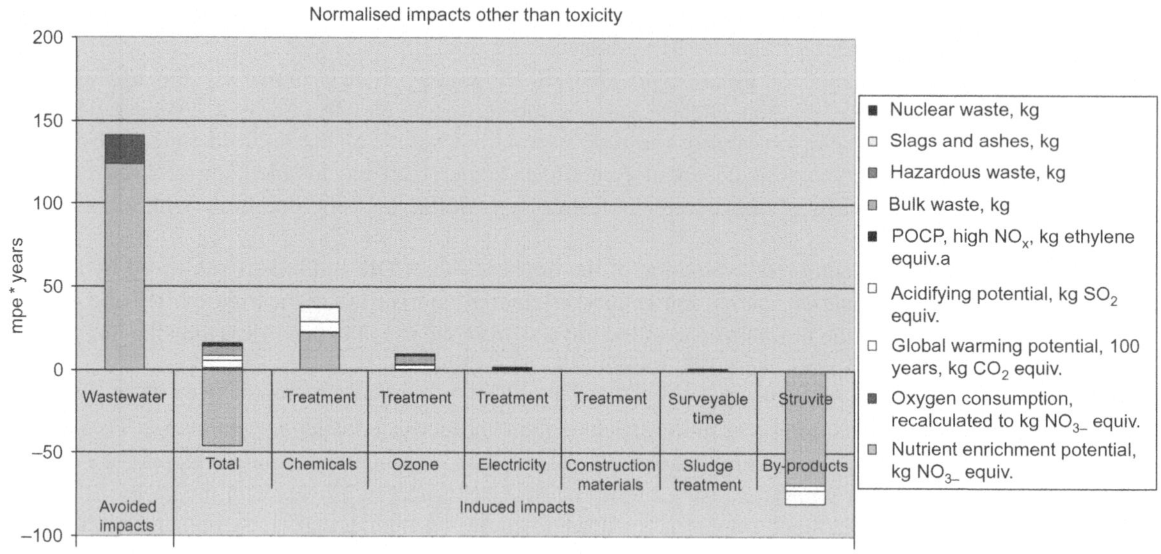

Figure 4.13 Normalized values of avoided and induced impacts from emissions other than toxic emissions and from waste when landfill leachate is treated in a 90 – m^3/day struvite + SBGGR + ozonation plant. All data per m^3 treated wastewater. Negative values mean that saved emissions from struvite replacing NP-fertilisers are higher than the total emissions from other sources of induced impacts

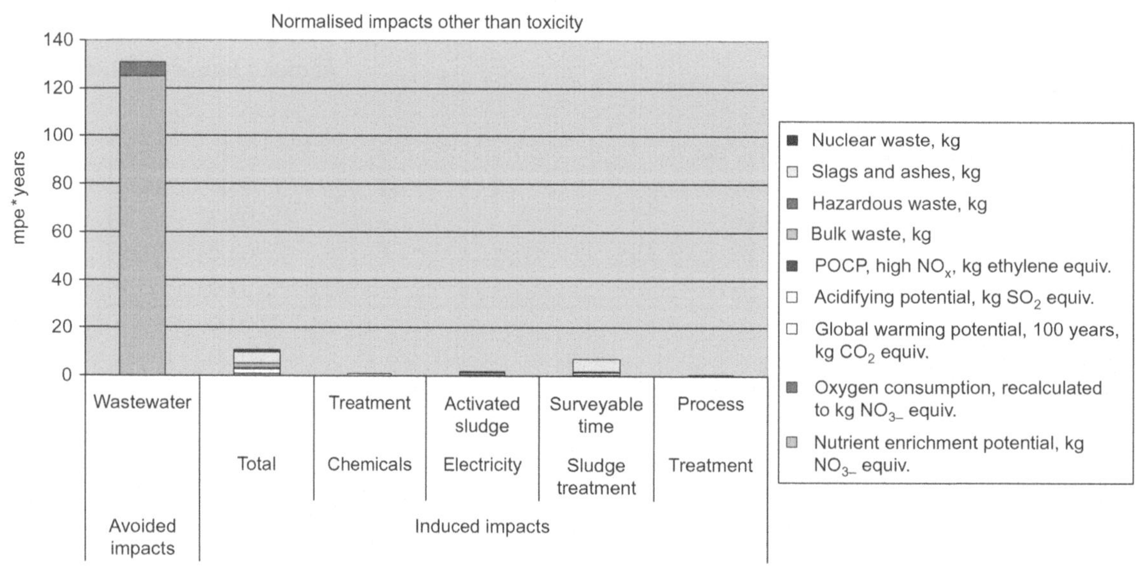

Figure 4.14 Normalized values of avoided and induced impacts from emissions other than toxic emissions and from waste when landfill leachate is treated in a 90-m^3/day activated-sludge plant with nitrification/denitrification. All data per m^3 treated wastewater

4.4.3 Economic assessment

4.4.3.1 Costing data

Costing data for equipment, constructions and commodities have been obtained in Euros, Swedish Crowns (SEK) and US Dollars. They have been recalculated to Euro with the following rates of exchange: 1 Euro = 9.2 SEK = 1.34 US $.

For the struvite + SBBGR + ozonation plant equipment costs have been given by IRSA. The cost of purchase of the ozone generator has been obtained from Wedeco (Kampmann, 2009). Costs for engineering, construction and start-up

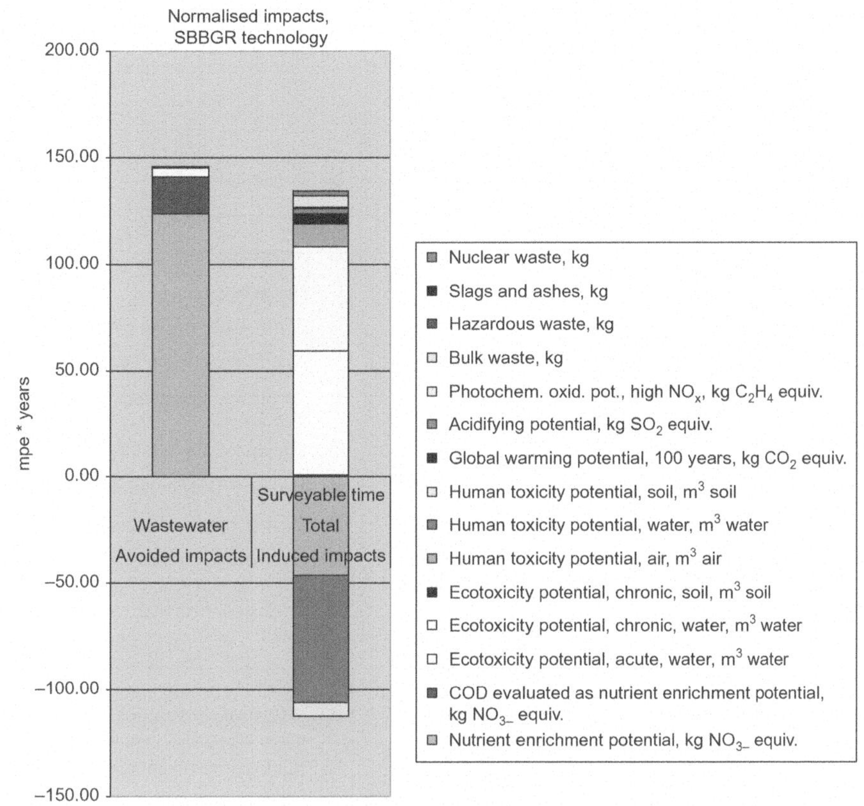

Figure 4.15 Normalized values of avoided and induced impacts from emissions and waste when landfill leachate is treated in a 90 – m^3/day struvite + SBGGR + ozonation plant. All data per m^3 treated wastewater. Negative values mean that saved emissions from struvite replacing NP-fertilisers are higher than the total emissions from other sources of induced impacts

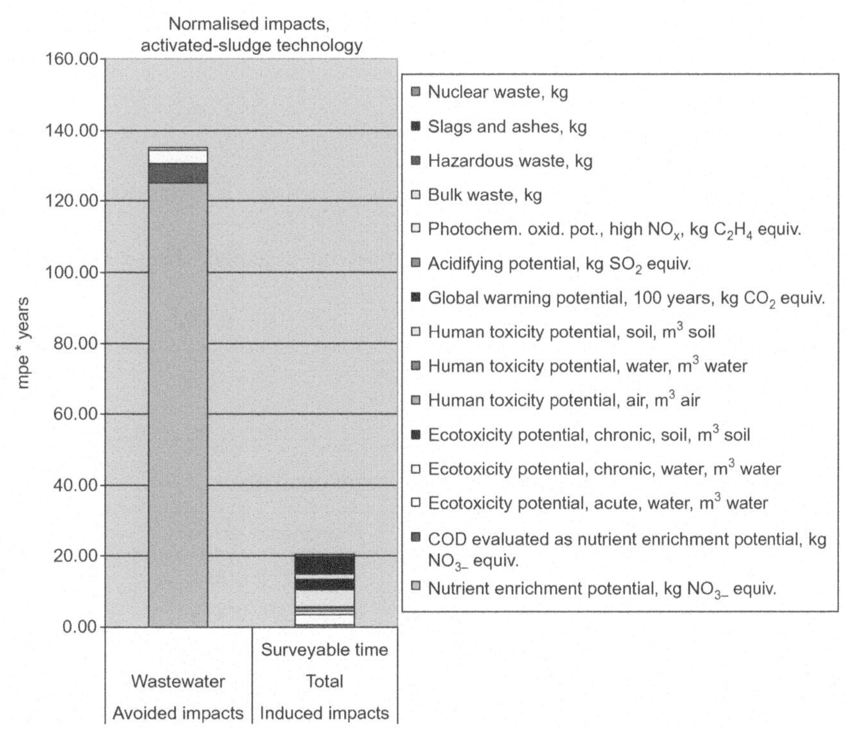

Figure 4.16 Normalized values of avoided and induced impacts from emissions and waste when landfill leachate is treated in a 90 – m^3/day activated-sludge plant with nitrification/denitrification. All data per m^3 treated wastewater

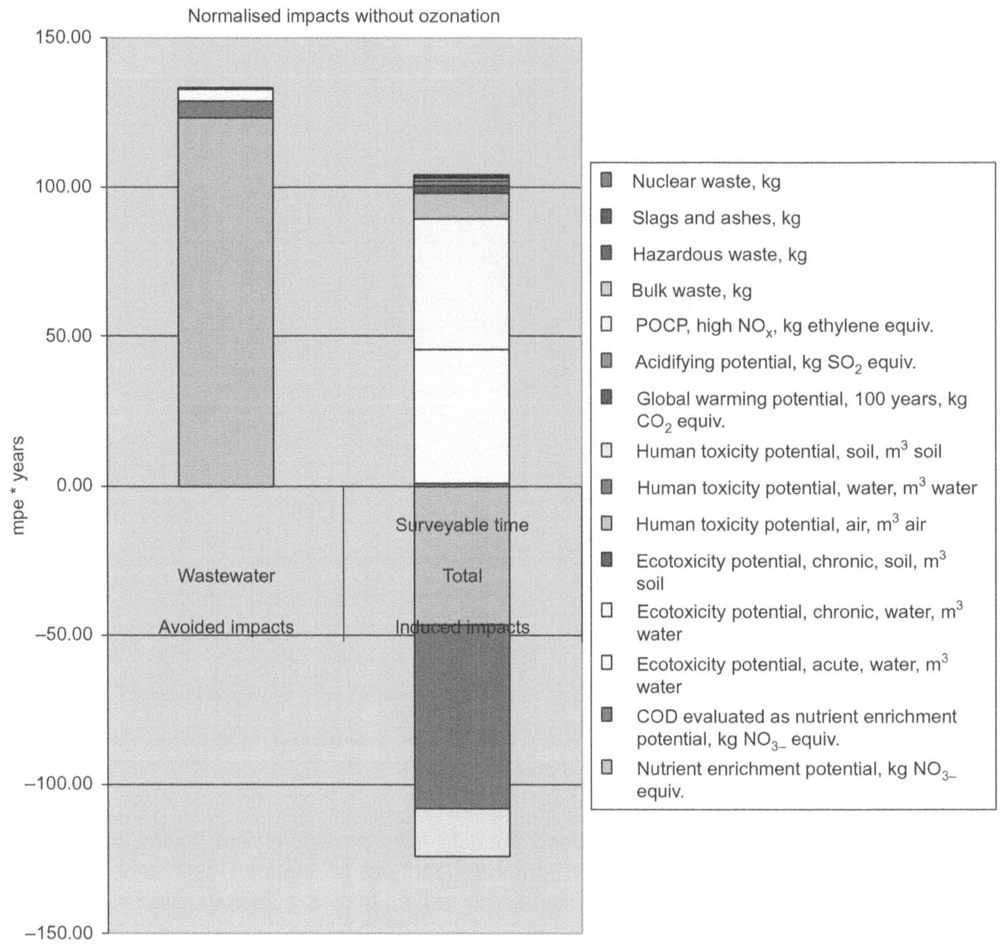

Figure 4.17 Normalized values of avoided and induced impacts from emissions and waste when landfill leachate is treated in a 90 – m^3/day struvite + SBGGR plant without ozonation. The COD removal efficiency is assumed to be the same as for the activated sludge plant. All data per m^3 treated wastewater. Negative values mean that saved emissions from struvite replacing NP-fertilisers are higher than the total emissions from other sources of induced impacts

have been estimated by IVL by analogy to the corresponding costs for the activated sludge system, which were calculated by AnoxKaldnes. The investment is reported in Table 4.21.

For the activated sludge plant a complete cost calculation was carried out by AnoxKaldnes based on the design by AnoxKaldnes. The result is reported in Table 4.22.

When calculating the results, some basic economic data must be assumed. These assumptions are presented in Table 4.23.

Prices of commodities are given in Appendix 3. Costs of maintenance, labour and sludge incineration and the price of struvite are reported in Table 4.24.

Table 4.21 Investment costs for the struvite + SBBGR + ozonation plant

Cost item	Purchase price, €	Service life, years
Engineering, construction, start-up	272 000	20
Struvite reactor	345 000	10
SBBRG + ozonator system, including power supply, control equipment, piping, fittings, dosing and measuring equipment	460 000	10
Ozone generator (Wedeco PDO 4500)	680 000	10
Total initial investment	1 757 000	

Table 4.22 Investment costs for the activated sludge plant

Cost item	Purchase price, €	Service life, years
Engineering, construction, start-up,	282 000	20
Concrete basins	394 000	20
Machinery	171 000	10
Power supply, control equipment	109 000	8
Piping, fittings, dosing and measuring equipment	59 000	8
Total initial investment	1 015 000	

Table 4.23 Assumptions of basic economic prerequisites

Parameter	Assumption
Interest for costing purposes	4%
Economic life of the facility	20 years
Annual operating time	Continuous operation 24 hours/day, 365 days/year

Table 4.24 Costs of maintenance, labour and sludge incineration and the market value of struvite

Cost item	Price	Source, comment
Struvite	250 €/tonne	Market price 2010 (Di Iaconi, 2010)
Maintenance		
Consumables, spare parts, SBBGR	10% of investment/year	Di Iaconi (2010)
Consumables, spare parts, ozone generator	1.5% of investment/year	Wedeco (Kampmann, 2009)
Consumables, spare parts	2% of investment/year	Other installations (IVL)
Labour		
Technical staff	28.4 €/h	25.79 €/h in EU15 in 2006. Eurostat, epp.eurostat.ec.europa.eu 10% added (our estimate)
Sludge incineration		
Sludge	300 €/tonne DS	Minimum for monocombustion and maximum for co-incineration of dewatered sludge, including landfill costs (Ries, 2003), as cited by Kügler *et al.* (2004).

4.4.3.2 Results of the economic assessment

Table 4.25 is a compilation of the results of the cost calculations in absolute values per m^3 wastewater to be treated. The costs are split into cost items.

The major cost item of the struvite + SBBGR + ozonation treatment is the cost of power, consumables and maintenance. Figure 4.18 shows a break-down of this cost item.

Obviously the costs of the struvite + SBBGR + ozonation are dependent on the prices of the chemicals oxygen, magnesium oxide and phosphoric acid, and on the market value of struvite. For the economy of the process it will be very important to optimise the use of chemicals. In the present design magnesium oxide is added in almost 100% excess of the stoichiometric demand, and the cost of magnesium oxide constitutes 60% of the net costs of consumables. To that can be added that we may well have underestimated the cost of magnesium oxide as a commodity in Europe.

Table 4.25 Costs of treatment of a landfill leachate with a struvite + SBBGR + ozonation system and alternatively by an activated sludge plant. All costs are in €/m^3 wastewater to be treated

	Struvite + SBBGR + ozonation	Activated sludge
Initial investment	3.9	2.3
Replacement investments during the plant's service life	2.2	0.73
Power, consumables and maintenance	10[1]	4.4
Sludge disposal (incineration)	0.096	2.0
Labour	0.23	0.45
Total costs	16	9.9

[1] *The market value of produced struvite has been deducted.*

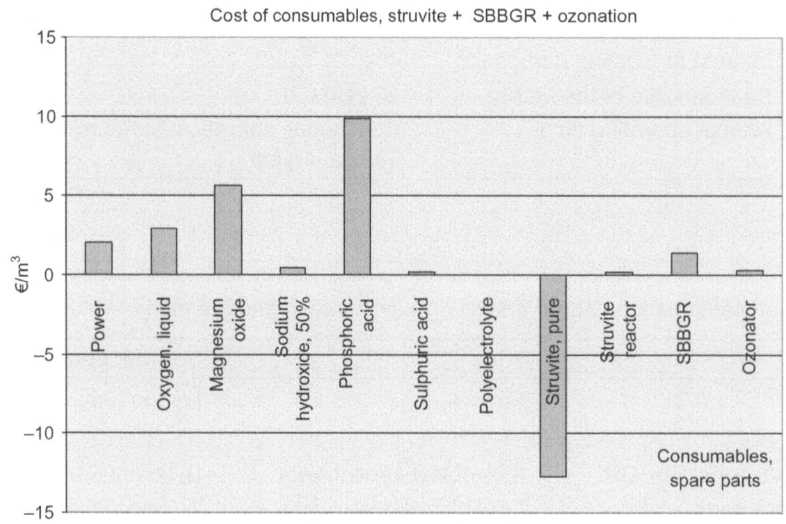

Figure 4.18 The costs of power, consumables and maintenance for the struvite + SBBGR + ozonation treatment

High investment costs also automatically mean high costs for maintenance according to the method of calculation used here.

4.4.4 Uncertainties

The major uncertainties in our assessment are:

1. The real impact on the environment of the COD removed by the ozone treatment. Is it just nutrient enrichment (eutrophication) or is toxicity also removed, and if so, to what extent?
2. The technical uncertainty. The inventories for the LCA:s have not been collected from real plants but from projected plants based on data from laboratory and pilot-plant experiments (struvite + SBBGR + ozone technology) and from experience of similar plants (activated-sludge technology).
3. There are always uncertainties in the normalization references and the toxicity data.

Despite the uncertainties the main findings, that the struvite + SBBGR + ozone technology causes too high induced impacts if the avoided impact is just nutrient enrichment, and that the environmental efficiency of the process depends on whether or not struvite can replace NP-fertilisers, seem plausible and robust.

4.4.5 Conclusions

- The main proved advantage of the struvite + SBBGR + ozonation system over the activated sludge system is the much better COD removal. Regarding nutrients removal both systems are comparable with a slight edge on the activated sludge system which uses simultaneous chemical coagulation to remove phosphorus.

- The preliminary assessment of the process shows, that removal of organic matter from wastewaters must be evaluated in terms of removal of toxic compounds and toxicity potentials, if the use of advanced treatment processes shall be proved environmentally efficient.
- The net cost of the chemicals used and produced in the struvite precipitation ($MgO + H_3PO_4 + NaOH$ – struvite) is 3.28 Euros/m^3 treated water. With the prices of chemicals used in this assessment this is equal to the cost of the methanol used for denitrification in the activated sludge process. The investment cost for the struvite reactor may be somewhat higher than the investment cost for the two anoxic reactors, and struvite precipitation uses some electric power for drying, but given the present prices of chemicals struvite precipitation and nitrification/denitrification should be cost-neutral on comparison $\pm$1 Euro/m^3 treated water. However, prices of chemicals may be very volatile. The market price of struvite as a fertiliser should fluctuate as the price of phosphoric acid but not necessarily as the price of magnesium oxide or methanol.
- The cost difference between the two systems thus mainly reflects the higher energy costs for electric power and oxygen supply and the higher maintenance costs of the struvite + SBBGR + ozonation system, which is the price paid for the increased COD removal.

4.5 PHARMACEUTICAL INDUSTRY WASTEWATER – THE STANDARD TREATMENT CASE

The effluent from manufacture of nalidixic acid, an antibiotic, is used as a standardised treatment case to compare the performance of treatment technologies. The characteristsics of the wastewater are reported in Table 4.3. The treatment technologies compared are summarised in Table 4.4. Their performance when treating the standardised wastewater is calculated from the experimental results reported in Chapters 2 and 3. In two cases mathematical models are deduced empirically and theoretically and used to design the treatment systems.

4.5.1 Properties of nalidixic acid

Some properties of nalidixic acid are collected in Table 4.26.

Table 4.26 Properties of nalidixic acid (1-ethyl-1,4-dihydro-7-methyl-4-oxo-1,8-naphthyridine-3-carboxylic acid)

Parameter	Value	Unit
Melting Point	229.5	° C
pKa Dissociation Constant	6[1]	–
log P (octanol-water)	1.59	–
Water Solubility at 23°C	100	mg/l in pure water
Vapour Pressure at 25°C	$3.56 \cdot 10^{-7}$	mm Hg
Henry's Law Constant at 25°C	$5.12 \cdot 10^{-16}$	atm-m^3/mole
Atmospheric OH Rate Constant	$2.15 \cdot 10^{-11}$	cm^3/molecule-sec

[1]*(Timmers and Sternglanz 1978).*

The assessment of the environmental efficiency of the treatment technologies is highly dependent on the toxicity potential of the nalidixic acid. No toxicity data have been found for nalidixic acid that would enable us to directly determine a PNEC-value for that substance. However, data are available for a group of analogous substances (fluoroquinolones), namely flumequine, oxolinic acid, ofloxacine, levofloxacine and moxifloxacine. Based on the toxicity (growth inhibition) of these substances to cyanobacterium, green algae and blue-green algae, a $PNEC_{water}$ value for nalidixic acid in freshwater in the range of 5–10 µg/l, can be estimated (Woldegiorgis, 2009).

For acute toxicity in water f in equation 2 is set to 1. Given the persistence and low Henry's law constant of nalidixic acid (much lower than 10^{-3} atm $\cdot$ m^3/mol according to Table 4.26) we make the conservative estimate that f = 1 for chronic aquatic toxicity as well, although the substance may adsorb onto sediment.

Since the compound is not biodegradable, we will set BIO = 1 for acute and chronic toxicity in an aquatic environment.

This PNEC value is used to calculate characterisation factors for chronic ecotoxicity of nalidixic acid in water in the range of $EF_{wc} = 1 * 1 * 1/(10 * 10^{-6}) = 10^5$ m^3/kg to $1 * 1 * 1/(5 * 10^{-6}) = 2 * 10^5$ m^3/kg.

4.5.2 Mathematically modelled treatments – Photo-Fenton + IBR and membrane contactor + IBR

4.5.2.1 Modelling the treatment of the nalidixic acid wastewater in the photocatalytic reactor, case photo-Fenton + Immobilised Biomass Reactor (IBR)

Due to the complexity of the photo-Fenton process, it has not been possible to derive stoichiometric relations for the reactions taking place within the reactor. The equations derived are instead based on experiments carried out at PSAor on assumptions based on experience within wastewater treatment.

In connection to the compound parabolic collector the photocatalytic system also has a conditioning container where chemicals can be added.

Degradation of nalidixic acid: In the case where the photo-Fenton reactor precedes the IBR the degradation of nalidixic acid follows a zero order kinetic reaction, according to the results of the experiments carried out by Malato *et al.* (2009).

$$C = C_0 * (-0.0049t + 0.8342), \quad R^2 = 0{,}9704 \tag{5}$$

C = Concentration of nalidixic acid at time t
C_0 = Initial concentration of nalidixic acid (at t = 0)
t = illumination time (min)

In the models, a 98% removal of nalidixic acid is assumed, based on recommendations from PSA With a 98% removal, the illumination time is calculated to 166 minutes.

Consumption of H_2O_2: Based on the experiments carried out, a polynomial correlation is found between the consumed concentration of H_2O_2 and the irradiation time at the standard intensity of 30 W/m^2, see Figure 4.19. The equation below should only be considered as a mathematical description for the specific laboratory data received. It should not be considered as a mechanistic description of the photo-Fenton process, since the photo-Fenton process is much more complex and cannot be described by these kinds of equations.

$$H_2O_2 \text{ consumed } (\text{mg/l}) = -6 * 10^{-6} * t_{30W}^3 + 0.018 * t_{30W}^2 + 2.5419 * t_{30W} + 475.88 \tag{6}$$

The time (t_{30W}) is given in minutes.

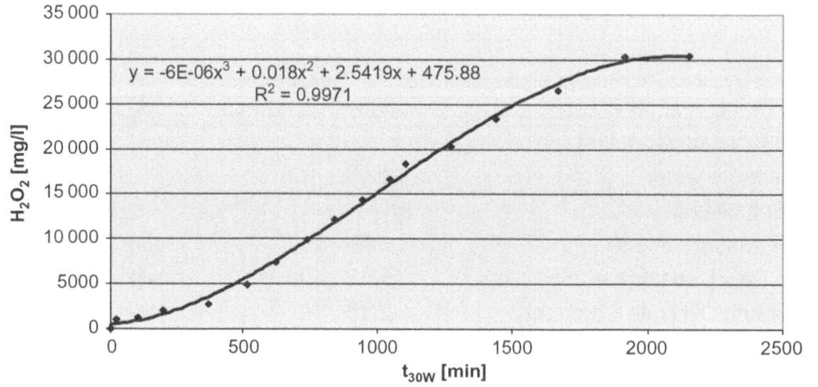

Figure 4.19 Correlation between consumed H_2O_2 (mg/l) and illumination time, t_{30W} (min)

Reduction of TOC: From the experimental data a polynomial correlation can been found between the removed concentration of TOC and the added dosage of hydrogen peroxide, see Figure 4.20. The equation below should only be considered as a mathematical description for the specific laboratory data received. It should not be considered as a mechanistic description of the photo-Fenton process, since the photo-Fenton process is much more complex and cannot be described by these kinds of equations.

$$\text{TOC consumed} = -4 * 10^{-6} * (H_2O_2 \text{ consumed})^2 + 0.0156 * (H_2O_2 \text{ consumed}) \tag{5}$$

Reduction of COD: The reduction of COD is found to be fairly well correlated to the amount of hydrogen peroxide consumed over the entire investigated range for untreated wastewater. Equation 6 was found by linear regression, see Figure 4.21. The equation below should only be considered as a mathematical description for the specific laboratory

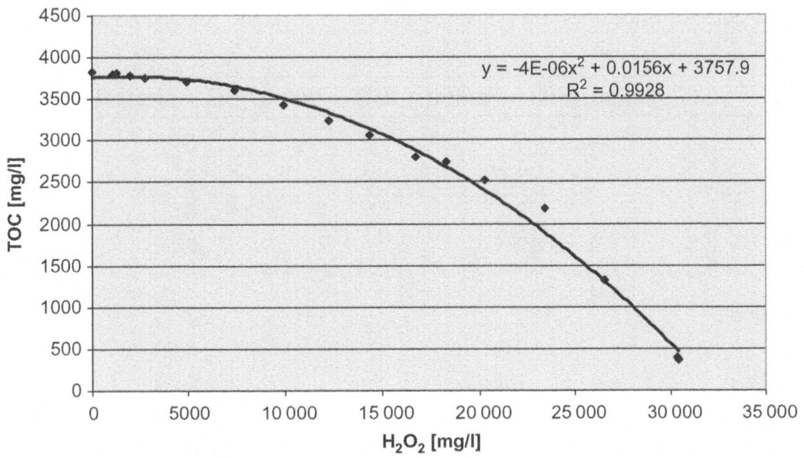

Figure 4.20 Correlation between TOC reduction (mg/l) and consumed H_2O_2 (mg/l)

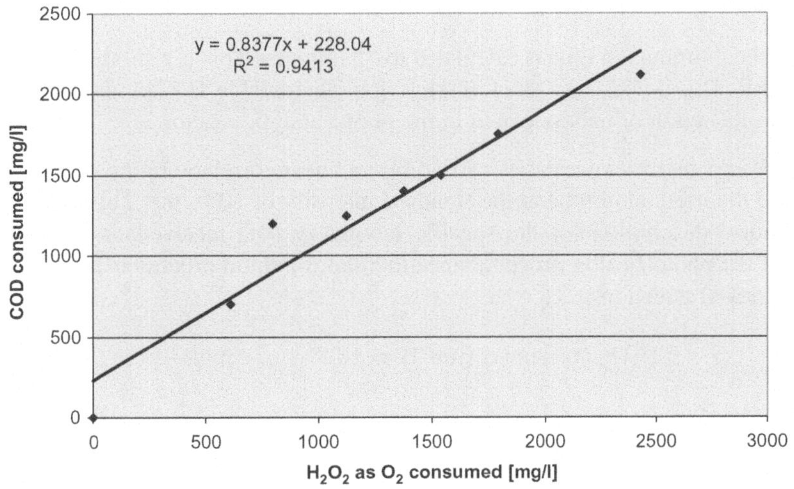

Figure 4.21 Correlation between consumed COD (mg/l) and consumed H_2O_2 as O_2 (mg/l)

data received. It should not be considered as a mechanistic description of the photo-Fenton process, since the photo-Fenton process is much more complex and cannot be described by these kinds of equations.

$$\text{COD consumed} = 0.8377 * (H_2O_2 \text{ as } O_2 \text{ consumed}) + 228 \quad (6)$$

Table 4.27 gives some design data for the photo-Fenton + IBR and the IBR + photo-Fenton treatment systems.

4.5.2.2 Modelling the treatment of the nalidixic acid wastewater in the photocatalytic reactor, case Immobilised Biomass Reactor (IBR) + photo-Fenton

Degradation of nalidixic acid: In this case the equation used to describe the degradation of nalidixic acid is not the same as in the case where photo-Fenton precedes the IBR. When the photo-Fenton reactor is used to treat the effluent from the IBR, first order kinetics applies instead, according to the results of the experiments carried out at PSA Treatment to 98% removal of the nalidixic acid is assumed in this case, too.

$$\ln(C/C_0) = -0.1164t + 0.2178R^2 = 0.9867 \quad (7)$$

C = Concentration of nalidixic acid at time t
C_0 = Initial concentration of nalidixic acid (at t = 0)
t = illumination time (min)

Table 4.27 Some projected data for photo-Fenton plants designed to treat the nalidixic acid wastewater to 98% removal of nalidixic acid

Parameter	Value, photo-Fenton + IBR	Value, IBR + photo-Fenton	Unit
Wastewater flow	5	5	m^3/day
COD/DOC removal	98	99	%
N removal	8	8	%
P removal	99	96	%
IBR		10	m^3
Area, compound parab. collector	130	28	m^2
Electric power	5.0	4.5	kWh/m^3 treated water
Hydrogen peroxide, as 100%	1.4	0.53	kg/m^3 treated water
Sodium hydroxide, as 100%	0.060	0.28	kg/m^3 treated water
Sulphuric acid, 100%	0.073	0.25	kg/m^3 treated water
Phosphoric acid, as 85%	0.0074	0.026	kg/m^3 treated water
Ammonium chloride	0.10	0.20	kg/m^3 treated water
Ferrous sulphate, as pentahydrate	0.1	0.1	kg/m^3 treated water
Sludge generation, incl. TSS_{in}	0.81	1.1	kg/m^3 treated water

With a 98% reduction, the illumination time is calculated to 35 minutes. This is a shorter illumination time compared to the case photo-Fenton + IBR. This is because a lot of interfering biodegradable TOC has already been degraded in the IBR. This gives a more efficient reduction of nalidixic acid in the photocatalytic reactor.

Consumption of H_2O_2: Based on the experiments carried out, a linear correlation may be found between the consumed concentration of H_2O_2 and the irradiation time at the standard intensity of 30W/m^2. The equation below should only be considered as a mathematical description for the specific laboratory data received. It should not be considered as a mechanistic description of the photo-Fenton process, since the photo-Fenton process is much more complex and cannot be described with these kinds of equations.

$$H_2O_2 \text{ consumed } (\text{mg/l}) = 10.7 \cdot t_{30W} \text{ (min)} + 151 \tag{8}$$

The equation was found to be applicable in the interval of 0–600 minutes of UV irradiation at 30W/m^2.

Reduction of TOC: From the experimental data a linear correlation can been found between the removed concentration of TOC and the added dosage of hydrogen peroxide.

$$\text{TOC consumed } (\text{mg/l}) = 0.034 \cdot H_2O_2 \text{ consumed } (\text{mg/l}) + 4.2 \tag{9}$$

This equation was found to be valid for H_2O_2 dosages between 0 and 2400 mg/l of consumed H_2O_2.

Reduction of COD: No data on the reduction of COD by the photo-Fenton process has been reported for the treatment IBR + photo-Fenton. We assume that the oxidation processes are mineralization of organic compounds of the same average oxidation state and put the percentage removal of COD equal to that of TOC.

Table 4.27 gives some design data for the photo-Fenton + IBR and the IBR + photo-Fenton treatment systems.

4.5.2.3 Modelling the treatment of the nalidixic acid wastewater with a membrane contactor

The design of the membrane contactor + IBR treatment of the nalidixic acid water is based on a theoretical model of the membrane contactor. The IBR has been designed in analogy with the photo-Fenton + IBR treatment technology. This application was not tested experimentally during the INNOWATECH project.

A membrane contactor may essentially be a hollow-fibre or a flat-sheet membrane module. The membranes consist of an active layer on one side of a supporting carrier material. The active layer shall ideally be penetrable for non-polar organic compounds but not for water or ionic compounds. Thus if an aqueous solution of organic acids, like phenols or carboxylic acids, is passed over the active layer at a pH value, where the acids are not protolysed, they will penetrate the active layer. If an alkaline solution of a pH value well above the pK_a value of the acid is now passed through the module at the other side of the membrane, the acid will immediately be protolysed to its salt as it passes through the membrane and thus prevented from passing back through the active layer again. The organic acids will be extracted as their salts from the treated water stream to the alkaline water stream. Figure 4.22 shows a schematic flowchart of the membrane contactor as applied to the nalidixic acid case.

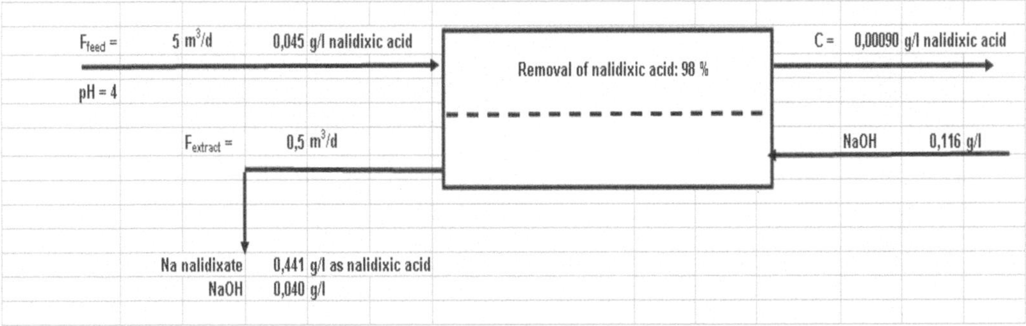

Figure 4.22 Stoichiometry of the extraction of nalidixic acid from the pharmaceutical effluent

In order to design the membrane cell the necessary length of the extraction path is calculated on the basis of the removal requirement and the extraction equilibrium. The following equation is used for a hollow-fibre cell (Bayer 2008):

$$C(L) = C_0 \cdot e^{(dv - \sqrt{(d2v2+16kdDeff}) \cdot L/2dDeff} = C_0 \cdot e^{xL} \quad (10)$$

C(L) = residual concentration of the extracted compound at length L of the membrane cell.
C_0 = original concentration of the compound to be extracted.
d = internal fibre diameter.
v = linear velocity of the feed solution.
k = transfer coefficient
D_{eff} = diffusion coefficient

Some calculated design data for the membrane contactor + IBR treatment system are compiled in Table 4.28. An environmental optimisation model for treatment of a wastewater containing phenol and bisphenol-A is presented by Bayer *et al.* (2010).

Table 4.28 Some projected data for the membrane contactor + IBR system designed to treat the nalidixic acid wastewater to 98% removal of nalidixic acid

Parameter	Value	Unit
Wastewater flow	5	m^3/day
COD/DOC removal	96	%
N removal	8	%
P removal	96	%
Total membrane area	226	m^2
Active layer material	Polydimethylsiloxane (PDMS)	
Electric power, membr. contactor	0.044	kWh/m^3 treated water
Electric power, total	2.6	kWh/m^3 treated water
Sodium hydroxide, as 100%	0.092	kg/m^3 treated water
Phosphoric acid, as 85%	0.026	kg/m^3 treated water
Ammonium chloride	0.20	kg/m^3 treated water

4.5.3 Treatment by the reference technology and by the membrane bioreactor + ozonation (MBR + AOP)

The reference system with existing technology, to which the performance of the new technologies shall be compared, is a fictitious plant with an estimated and calculated performance. It has been designed by AnoxKaldnes (and consists of a pH-adjustment and equalisation step, followed by an aerobic moving-bed bioreactor (MBBR). The effluent from the MBBR is treated with powdered activated carbon (PAC) from charcoal. The carbon is separated and collected together with the biosludge. The combined sludge and carbon is dewatered in a screw press. A 2% excess of nutrients is used. Since the treated flow is very small, the plant is operated in a batch-wise mode. Aeration is continuously maintained to safe-guard the biosludge. A schematic flowchart of the treatment system is shown in Table 4.4. Basic data are given in Table 4.29.

Table 4.29 Basic plant data for the MBBR + PAC system

Parameter	Value	Unit
Treatment capacity	5	m^3/day
COD/DOC removal	88	%
Nalidixic-acid removal	90	%
N removal	−120	%
P removal	70	%
MBBR	45	m^3
Moving-bed carrier volume	17	m^3
Screw press	1–2	m^3/h
Sodium hydroxide, as 100%	0.08	kg/m^3 treated water
Phosphoric acid, as 85%	0.0052	kg/m^3 treated water
Nutriol-N (urea + ammonium nitrate)	0.074	kg/m^3 treated water
Powdered activated carbon	2	kg/m^3 treated water
Polyelectrolyte	0.002	kg/m^3 treated water
Electric power	23	kWh/m^3 treated water
Sludge generation	2.8	kg DS/m^3 treated water
Labour	5	person hours/week

The membrane bioreactor has a membrane module submerged in a reactor tank. The permeate is treated with ozone and recirculated from the ozonation tank to the bioreactor. The recirculation flow is four times the flow of wastewater through the plant. The ozone generator is operated on air. The vented gas from the ozonator is passed through an ozone destructor. The wasted biosludge is dewatered in a simple static sludge thickener. A schematic flowchart of the treatment system is given in Table 4.4 and some basic data in Table 4.30. The plant is designed from experimental data reported by IRSA.

Table 4.30 Basic plant data for the MBR + ozone system

Parameter	Value	Unit
Wastewater	5	m^3/day
COD/DOC removal	98	%
Nalidixic-acid removal	96	%
N removal	0	%
P removal	ca. 30	%
Bioreactor tank	22	m^3
Membrane area	64	m^2
Ozonation tank	0.21	m^3
Ozone generator, capacity	25	g/h
Used ozone generation capacity	15	g/h
Sodium hydroxide, as 100%	0.082	kg/m^3 treated water
Ammonium chloride	0.041	kg/m^3 treated water
Sodium hypochlorite, as 100%	0.0018	kg/m^3 treated water
Citric acid	0.0027	kg/m^3 treated water
Electric power, including ozone generation	10.7	kWh/m^3 treated water
Sludge generation	0.77	kg DS/m^3 treated water
Labour	5	person hours/week

4.5.4 Result of the environmental assessment

The treatment results are reported in a traditional way in Table 4.31. The important function of the treatments is to remove organic compounds in general and nalidixic acid especially. Removal of nitrogen and phosphorus is not an important issue, and in this respect the results are varying. The increase of nitrogen in the MBBR + AC case is caused by an intentional excess dosage of nitrogen nutrient for the biology.

Table 4.31 Treatment results for the nalidixic acid wastewater – Traditional reporting

	Nalidixic acid removal, %	COD/TOC removal, %	Nitrogen removal, %	Phosphorus removal, %
MBBR + PAC	90	88	−120	70
Ph-F + IBR	98	98	8	98
IBR + Ph-F	98	99	8	96
MBR + ozone	96	98	0	ca. 30%
MC + IBR	98	96	8	96

In Figure 4.23 the treatment results are reported in a holistic way, i.e. both avoided impacts and impacts induced by operating the treatment plants and supplying them with commodities are taken into account. Other impacts than emissions of nutrients and COD with the treated wastewater, such as the greenhouse gas effect (global warming) and acidification are also considered. All impacts have been normalized to a common scale, milliperson equivalents · years following the procedures and using the normalization references of EDIP97 as described in the Section Normalization. The avoided impacts have been calculated as illustrated in Figure 4.23.

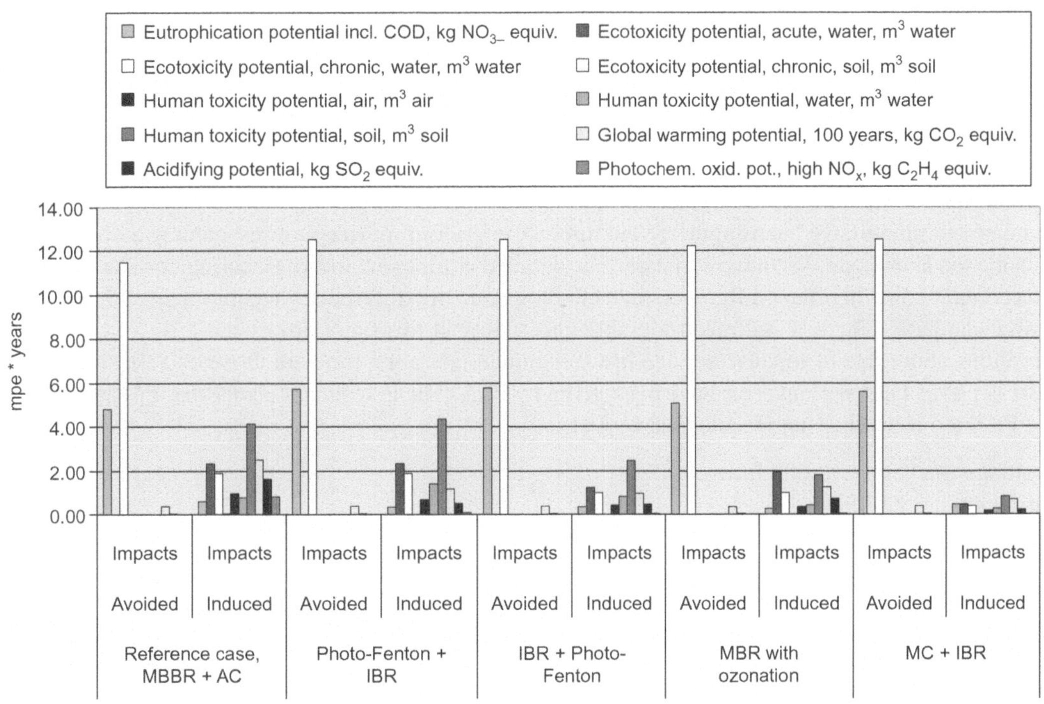

Figure 4.23 Treatment results for the nalidixic acid wastewater – Holistic reporting

Only impacts from emissions have been taken into account in Figure 4.23. Data on quantities and classification of generated waste from the commodity supply are uncertain, as are the normalization references for wastes. All treatment technologies generate inert wastes to landfills, mostly overburdens from mining, and radioactive wastes. Use of the data from the inventories and the EDIP normalization references for wastes seems to put an unreasonably high emphasis on wastes, however. We have thus limited this assessment to a comparison of avoided and induced emissions.

All the treatments avoid potential eutrophication from removed COD and potential ecotoxicity from removed nalidixic acid. A small potential greenhouse gas effect is also avoided. The organic compounds in the wastewater are synthetic and presumably of fossil origin. Their degradation in nature would thus contribute to the greenhouse gas effect. This avoided greenhouse gas effect is, however, induced again, when the wastewater is treated biologically or oxidatively, and when the biosludge is incinerated.

If the estimate of the ecotoxicity potential of nalidixic acid is accepted, the highest-valued benefit of the treatments is the removal of an ecotoxic emission to water. This does not seem unreasonable since that is the reason for special treatment of this wastewater in the first place. Figure 4.23 also suggests that if nalidixic acid would have been a completely harmless compound use of the advanced treatment technologies would hardly have been environmentally warranted. Much or most of the benefit of high COD removal would have been neutralised by the impacts of potentially toxic emissions from the commodity supply. Potentially ecotoxic emissions to water and impacts of potential human toxicity in soil seem to be the kind of induced impacts, which infringe most on the sustainability, if we regard all environmental impacts as equally important. The normalized sums of induced acute and chronic ecotoxic emissions to water from the supply of electricity and chemicals amount to between 10% and 40% of the avoided ecotoxic impact, if we consider all kinds of ecotoxic emissions at any location equally important. The reported ecotoxic emissions from generation of electricity and manufacture of chemicals are mostly metals. The impact on soil from substances, which are potentially toxic to human beings, is rather highly emphasized by the EDIP normalization. The origin of this induced impact is emission of various hydrocarbons, notably benzene, to air. The inventories of emissions of organic compounds from the commodity supply chain are probably not complete. The induced toxic emissions may thus well be underestimated.

There is at this point a technical uncertainty in the results. The treatment plant designs are based on laboratory or short-period pilot-plant trials. The MBBR + PAC (reference case) and the MC + IBR cases are calculated from projected data from other applications. There is obviously a potential for improvements and optimisation. The MBBR reactor for instance is designed primarily with safe operation of the biology in mind, but could probably be better adapted to such small flows as 5 m^3/day, which would save a lot of electric power.

4.5.5 Economic assessment

Three technologies are assessed with the LCC methodology, namely the reference technology MBBR + PAC, the IBR + photo-Fenton technology and the MBR + ozone treatment.

4.5.5.1 Costing data

Costing data for equipment, constructions and commodities have originally been calculated or obtained in Euros, Swedish Crowns (SEK), US Dollars and British Pounds (GBP). They have been recalculated to Euro with the following rates of exchange:

1 Euro = 9.76 SEK = 1.34 US $ = 0.89 GBP.

The three treatment plants have been engineered in three different European countries, namely Sweden, Spain and Italy. Sweden is outside the Euro zone. We have compared the detailed equipment and engineering costs and tried to harmonize them, so that a comparison will reflect differences in technology and not differences in equipment and entrepreneurial costs and engineering standards. There is still a considerable uncertainty at this point however.

Our assumptions about rate of interest, service life and annual operating time are those of Table 4.23.

For the IBR + photo-Fenton plant costs have been given by PSA. The investment is reported in Table 4.32. To the costs calculated by PSA the following items are added by IVL:

- Static sludge thickener: 10 000 Euro
- Air-flow meter: 3500 Euro
- Carrier material, 10 m^3: 4300 Euro

Table 4.32 Investment costs for the IBR + photo-Fenton plant

Cost item	Purchase price, €	Service life, years
Engineering, construction, start-up	5000	20
Tanks, IBR, CPC collector, mountings	50 700	20
Machinery, pH-control equipment	12 000	10
Piping, fittings, air-flow meter	31 500	8
Power supply, control equipment	20 000	8
Carrier material, pH electrodes	5300	3
Total initial investment	124 500	

For the MBR + ozone treatment plant the costing was originally done by IRSA In order to adapt the plant to the standard of the reference plant the following items are added by IVL:

Stirrer for equalising tank: 3000 Euro
Mountings for dosing pumps: 2600 Euro
Air-flow meter: 3500 Euro

The investment is reported in Table 4.33. The cost of the ozonator was given by Wedeco

Table 4.33 Investment costs for the MBR + ozone plant

Cost item	Purchase price, €	Service life, years
Engineering, construction, start-up, unforeseen, documentation	101 900	20
Tanks, sludge thickener, building	53 000	20
Ozonator, GSA 20 (Wedeco)	8500	10
Machinery, pH-control equipment	36 700	10
Power supply, control equipment	45 500	8
Piping	16 600	8
Membrane module	12 500	5
Probes	9400	3
Total initial investment	284 100	

For the MBBR + PAC plant a complete cost calculation was carried out by AnoxKaldnes based on the design by AnoxKaldnes. The original engineering, entrepreneurial and construction costs were adjusted to the level of those calculated by IRSA for the MBR + ozone plant. There is no reason why these costs should be different in these two cases. The cost of the blowers for air supply to the biology was also originally very different for no apparent reason. The cost given by IRSA was selected for both cases. The result is reported in Table 4.34.

Table 4.34 Investment costs for the MBBR + PAC plant

Cost item	Purchase price, €	Service life, years
Engineering, construction, start-up, unforeseen, documentation	102 400	20
Tanks, building	93 100	20
Machinery, sludge dewatering equipment, pH-control equipment	93 500	10
Power supply, control equipment	43 600	8
Piping, fittings	23 000	8
Carrier material, pH-electrodes	7 400	3
Total initial investment	363 000	

Prices of commodities are given in Appendix 3. Costs of maintenance, labour and sludge incineration are reported in Table 4.35.

Table 4.35 Costs of maintenance, labour and sludge incineration

Cost item	Price	Source, comment
Maintenance		
Consumables, spare parts, ozone generator	1.5% of investment/year	Wedeco
Consumables, spare parts Labour	2% of investment/year	Other installations (IVL)
Technical staff	28.4 €/h	25.79 €/h in EU15 in 2006. Eurostat, epp.eurostat.ec.europa.eu 10% added (our estimate)
Sludge incineration		
Sludge	300 €/tonne DS	Minimum for monocombustion and maximum for co-incineration of dewatered sludge, including landfill costs (Ries, 2003), as cited by Kügler *et al.*, 2004).

4.5.5.2 Result of the economic assessment

Table 4.36 is a compilation of the results of the cost calculations in absolute values per m^3 wastewater to be treated. The costs are split into cost items.

Table 4.36 Costs of treatment of the nalidixic-acid wastewater with three technologies. All costs in €/m^3 treated wastewater

	IBR + photo-Fenton	MBR + ozone	MBBR + PAC, ref. technology
Initial investment	5.0	11	15
Replacement investments during the plant's service life	3.8	7.0	7.1
Power, consumables and maintenance	2.6	4.0	13
Sludge disposal (incineration)	0.32	0.23	0.83
Labour	4.0	4.0	4.0
Total costs	16	26	40

The major cost items are the initial investments and the replacement investments. This is not unexpected considering the low capacity of the plants. For the MBR + ozone and the MBBR + PAC plants engineering, entrepreneurs, start-up and documentation are major cost items. For the IBR + photo-Fenton plant these costs are conspicuously low. We assume that the reason for this is that the components of the plant are ready off- the-shelf and can be put together and put into operation without any engineering and documentation, and that start-up will be quick and easy. If so, the IBR + photo-Fenton plant clearly have a competitive edge. The need for replacements, like exchange of carrier materials and membrane modules are also major cost items to consider. Here to, the IBR + photo-Fenton system seems to be competitive, according to the information available to us. The IBR is smaller and has less carrier materials than the MBBR, and the major components seem to last the service-life of the plant.

The MBBR + PAC technology in its existing form does not seem to be well suited for this kind of special small-scale application. If it is designed with the present knowledge and for safe operation, the costs are higher and the projected performance somewhat inferior to the other two technologies. The batch-wise operation and the assumed need to aerate the bio bed between batches cause high costs of power per m^3 of treated water. There is, however, an obvious potential for improvement and optimisation of this technology, which could not be explored during the INNOWATECH project.

4.5.6 Conclusion

Despite the uncertainties discussed above, the results in Figure 4.23 and Table 4.35 look promising. It seems that the new technologies can improve treatment results for toxic and recalcitrant water impurities without higher environmental costs in the form of higher induced impacts from emissions and without incurring higher financial costs. The results also demonstrate how data from different experiments can be used for modelling "tailor-made" solutions. The "tailor-made" membrane contactor + IBR case (never studied experimentally in INNOWATECH) utilises the fact, that the problematic pollutant is an organic acid, to design and assess a treatment with data from one work package and a model from another work package.

REFERENCES

Bayer C. (2008). Rheinisch-Westfälische Technische Hochschule Aachen, Chemical Engineering Department, personal communication.

Bayer C., Follmann M., Melin T., Wintgens T., Larsson K. and Almemark M. (2010). The ecological impact of membrane-based extraction of phenolic compounds – a life cycle assessment study. *Water Science & Technology*, **62.4**, 915–919.

Di Iaconi C. (2010). Istituto di Ricerca sulle Acque, Sede di Bari, personal communication.

Ecoinvent (2007). Life Cycle Inventories of Agrcultural Production Systems, final report Ecoinvent v. 2.0.

Emmerson R. H., Morse G. K., Lester J. N. and Edge D. R. (1995). The life-cycle analysis of small-scale sewage treatment processes. *Journal of the Chartered Institute of Water and Environmental Management*, **9**(3), 317–325.

Fahner S., Bührer H., Grabski C. and Heim T. (1998). Die ökologische Bewertung einer Kläranlage – Vor- und Nachteile von zwei Ökobilanzmethoden. *Korrespondenz Abwasser*, **45**(7), 1309–1313.

GaBi (4.2) PE International GmbH, Leinfelden-Echterdingen, Germany www.gabi-software.com.

García-Montaño J., Ruiz N., Muñoz I., Domènech X., García-Hortal J. A., Torrades F. and Peral J. (2006). Environmental assessment of different photo-Fenton approaches for commercial reactive dye removal. *Journal of Hazardous Materials*, **A138**, 218–225.

Giesen A. (2008). DHV B.V., Amersfoort, The Netherlands, personal communications.

Guinée J. B. (Ed.) (2002). Handbook on Life Cycle Assessment. Operational Guide to the ISO Standards. Kluwer Academic Publishers, Dordrecht/Boston/London.

Hospido A., Moreira T. M., Férnandez-Couto M. and Feijoo M. G. (2004). Environmental performance of a municipal wastewater treatment plant. *Int. J. LCA*, **9**(4), 261–271.

ISO (2006). EN ISO 14044:2006 Environmental Management – Life Cycle Assessment – Requirements and Guidelines, European Committee for Standardization, Brussels.

Jørgensen K. R., Villanueva A. and Wenzel H. (2004). Use of life cycle assessment as decision-support tool for water reuse and handling of residues at a Danish industrial laundry. *Waste Management & Research*, **22**, 334–345.

Kampmann M. (2009). ITT Water & Wastewater Herford GmbH, technnical descriptions and personal communications.

Kügler B., Öhlinger A. and Walter B. (2004). Dezentrale Klärschlammverbrennung, *Berichte BE-260*, Umweltbundesamt, Wien.

Larsen H. F. and Hauschild M. (2007). Evaluation of Ecotoxicity Effect Indicators in LCIA., *Int. J. Life Cycle Assess.*, **12**(1), 27–33.

Lundin M., Bengtsson M. and Molander S. (2000). Life cycle assessment of wastewater systems: influence of system boundaries and scale on calculated environmental loads. *Environ. Sci. Technol.*, **34**, 180–186.

MatLab/Simulink distributed by The MathWorks, www.mathworks.com.

NEPTUNE (2006). New sustainable concepts and processes for optimization and upgrading municipal wastewater and sludge treatment. EU Commission, contract No. 036845.

Ødegaard H. (1995). An evaluation of cost efficiency and sustainability of different wastewater treatment processes. *Vatten*, **51**, 291–299.

Ortiz I. M. (2006). Life Cycle Assessment as a Tool for Green Chemistry: Application to Different Advanced Oxidation Processes for Wastewater Treatment. Ph.D. Thesis, Universitat Autònoma de Barcelona.

Ortiz M., Raluy R. G., Serra L. and Uche J. (2007). Life cycle assessment of water treatment technologies: wastewater and water-reuse in a small town. *Desalination*, **204**, 121–131.

Ries T. (2003). Kommunaler Klärschlamm - Wohin damit? Aktuelle Verfahren zur Klärschlammbehandlung und-entsorgung, Vortrag, Forschungszentrum Karlsruhe, Fortbildungszentrum für Technik und Umwelt, www.aew-betrieb.de.

Roeleveld P. J., Klapwijk A., Eggels P. G., Rulkens W. H. and van Starkenburg W. (1997). Sustainability of municipal wastewater treatment. *Water Science & Technology*, **35**(10), 221–228.

Rosenbaum R. K., Bachmann T. M., Gold L. S., Huijbregts M. A. J., Jolliet O., Juraske R., Koehler A., Larsen H. F., McLeod M., Margni M., McKone T. E., Payet J., Schumacher M., van de Ment D. and Hauschild M. Z. (2008). *Int. J. Life Cycle Assess.*, **13**, 532–546.

Strömberg L. and Paulsen J. (2002). LCA application to Russian conditions. *Int. J. Life Cycle Assess*, **7**(6), 349–357.

Tangsubkul N., Beavis P., Moore S. J., Lundie S. and Waite T. D. (2005). Life cycle assessment of water recycling technology. *Water Resources Management*, **19**, 521–537.

Tarantini M. and Ferri F. (2001). LCA of Drinking and Wastewater Treatment System of Bologna City: Final Results. *4th IRCEW Conference*, Fortaleza, Brazil.

Timmers K. and Sternglanz R. (1978). Ionization and divalent cation dissociation constants of nalidixic and oxolinic acids. *Bioinorganic Chemistry*, **9**, 145–155.

Vidal N., Poch M., Martí E. and Rodríguez-Roda I. (2002). Evaluation of the environmental implications to include structural changes in a wastewater treatment plant. *J. Chem. Technol. Biotechnol.*, **77**, 1206–1211.

Wenzel H., Hauschild M. Z. and Alting L. (1997). Environmental assessment of products. Vol. 1 - Methodology, tools, techniques and case studies, Chapman & Hall, United Kingdom, ISBN 0 412 80800 5, Kluwer Academic Publishers, Hingham, MA, USA, 544 pp.

Woldegiorgis A. (2009). Miljöriskbedömning (nalidixic acid 389-08-2) baserat på READ-ACROSS ("Environmental Risk assessment (nalidixic acid 389-08.2) based on READ-ACROSS"), personal communication.

CHAPTER 4 APPENDIX 1

Normalization references for some resources (EDIP 97, Technical University of Denmark)

Non-renewable elements and energy wares	Normalization reference	Unit	Standard deviation	Reference year	Reference region	Reference
EDIP 1997, Aluminum	4.52E + 00	kg/pe/year	0	2004	World	1
EDIP 1997, Antimony	1.75E-02	kg/pe/year	0	2004	World	1
EDIP 1997, Beryllium	2.55E-05	kg/pe/year	0	2004	World	1
EDIP 1997, Cadmium	2.69E-03	kg/pe/year	0	2004	World	1
EDIP 1997, Cerium		kg/pe/year	0			
EDIP 1997, Cobalt	7.33E-03	kg/pe/year	0	2004	World	1
EDIP 1997, Copper	2.27E + 00	kg/pe/year	0	2004	World	1
EDIP 1997, Crude oil	6.04E + 02	kg/pe/year	0	2004	World	
EDIP 1997, Gold	3.B6E-04	kg/pe/year	0	2004	World	1
EDIP 1997, Hard coal	6.02E + 02	kg/pe/year	0	2004	World	2, 4, 5
EDIP 1997, Iron	9.77E + 01	kg/pe/year	0	2004	World	1
EDIP 1997, Lanthanum		kg/pe/year	0			
EDIP 1997, Lead	4.92E-01	kg/pe/year	0	2004	World	1
EDIP 1997, Brawn coal	2.64E + 02	kg/pe/year	0	2004	World	2, 4, 5
EDIP 1997, Manganese	1.72E + 00	kg/pe/year	0	2004	World	1
EDIP 1997, Mercury	2.73E-04	kg/pe/year	0	2004	World	1
EDIP 1997, Molybdenum	2.17E-02	kg/pe/year	0	2004	World	1
EDIP 1997, Natural gas	3.53E + 02	kg/pe/year	0	2004	World	
EDIP 1997, Nickel	2.19E-01	kg/pe/year	0	2004	World	1
EDIP 1997, Palladium	2.97E-05	kg/pe/year	0	2004	World	1, 7
EDIP 1997, Platinum	3.41E-05	kg/pe/year	0	2004	World	1, 7
EDIP 1997, Silver	3.05E-03	kg/pe/year	0	2004	World	1
EDIP 1997, Tantalum	1.98E-04	kg/pe/year	0	2004	World	1
EDIP 1997, Tin	3.91E-02	kg/pe/year	0	2004	World	1
EDIP 1997, Zinc	1.42E + 00	kg/pe/year	0	2004	World	1
EDIP 1997, Uranium	5.63E-03	kg/pe/year	0	2002	World	
EDIP 1997, Chromium	8.28E-01	kg/pe/year	0	2004	World	1
EDIP 1997, Selenium	2.34E-04	kg/pe/year	0	2004	World	1
EDIP 1997, Strontium	7.97E-02	kg/pe/year	0	2004	World	1
EDIP 1997, Tellurium	1.48E-05	kg/pe/year	0	2004	World	1
EDIP 1997, Thallium	2.34E-06	kg/pe/year	0	2004	World	1
EDIP 1997, Titanium	4.88E-01	kg/pe/year	0	2004	World	1
EDIP 1997, Tungsten	9.78E-03	kg/pe/year	0	2004	World	1
EDIP 1997, Vanadium	6.88E-03	kg/pe/year	0	2004	World	1
EDIP 1997, Ytrium	2.97E-04	kg/pe/year	0	2004	World	1
EDIP 1997, Zirconium	1.34E-01	kg/pe/year	0	2004	World	1

References:

/1/ *Mineral Commodity Summaries 2005. U.S. Department of the Interior. U.S. Geological Survey (USGS): http://minerals.usgs.gov/ minerals/ pubs/mcs/ 2005/mcs2005.pdf*

/2/ *BP Statistical Review of World Energy June 2005. "Putting energy in the spotlight": http://www.bp.com/liveassets/bp_internet/globalbp/ globalbp_uk_english/publications/energy_reviews_2005/STAGING/local_assets/downloads/pdf/statistical_review_of_world_ energy_full_ report_2005.pdf*

/3/ *HNG: http://www.hng.dk/naturgas/noegletal/tekniskenoegletal.asp (June 28 2005)*

/4/ Wenzel H, Hauschild M, Alting L (1997). Environmental Assessment of Products, Vol. 1. First ed. Chapman & Hall, London, Great Britain.

/5/ Annual Energy Review 2003 (EIA): http://www.eia.doe.gov/ aer/pdf/aer.pdf

/6/ UN (2005): http://www.un.org/ News/Press/docs/2005/pop918.doc.htm

/7/ *Personal communication with Michael George. Precious Metals Specialist. U.S. Geological Survey (USGS): mgeorge@usgs.gov (30/06-2005).*

/8/ *AIEA/IAEA/OECD (2004). Uranium 2003. Resources, Production and Demand. "Red Book": http://www1.oecd.org/ publications/e-book/ 6604081E.PDF*

/9/ *Mobbs P. (2005). Uranium Supply and the Nuclear Option. Published in Oxford Energy Forum, the quarterly journal of the Oxford Institute for Energy Studies, Issue 61, May 2005: http://www.fraw.org.uk/mobbsey/papers/oies_article.html*

APPENDIX 2

INVENTORY OF PERIPHERAL PROCESSES

Ammonia

For ammonia LCI data from Ecoinvent (Ecoinvent (2004), report No. 8) have been used. The data represent European average production during the period 1995–2000. Ammonia is produced by steam reforming of natural gas (85% of the production) and by partial oxidation of heavy fuel oil (15% of the production).

Ammonium chloride

Ammonium chloride is recovered as a by-product from the Solvay soda process. The environmental impacts of that process are allocated to the sodium carbonate product. The ammonium chloride is assigned the environmental impacts of the production of ammonia and of sodium chloride according to the following stoichiometry:

$$\underset{17{,}031}{NH_3} + \underset{58{,}444}{NaCl} + H_2O \longrightarrow \underset{53{,}492}{NH_4Cl} + NaOH$$

Bisphenol A

A data set is available in Ecoinvent. The functional unit represents 1 kg of solid bisphenol A at the factory gate. The uncertainty is large due to weak data on the production process and missing data on process emissions.

Building an construction of equipment

For equipment only the impacts from the supply of construction materials are considered. The impacts of the processes of building and assembling the equipment are neglected. They are difficult to assess with any certainty at this stage of the process development, and their influence on the overall assessment is probably not crucial. Transports of construction materials to the site are also disregarded.

Compound parabolic collector

The material composition of the compound parabolic collector used at PSA has been obtained from Ortiz (*Ortiz I.M., Life Cycle Assessment as a Tool for Green Chemistry: Application for Different Advanced Oxidation Processes for Wastewater Treatment, 2006, Dissertation, University of Barcelona*). The composition is reported in Table 1.

The LCI model of the supply of construction materials for the CPC is shown in Figure 1 as a copy of the flowchart form the GaBi software. The upstream data are from the Ecoinvent database and the professional database of GaBi (PE International).

The supply of anodised aluminium profiles is modelled by three modules, representing manufacture of aluminium metal, extrusion of aluminium profiles and anodising of aluminium sheet (5.57 m^2/m^2 collector area).

The cumulative transport work of materials to the site, 46 tkm of truck and 14 tkm of rail transport, is modelled as 1000 kg of cargo transported 46 km by truck and 14 km by rail.

The electricity for the rail transport is assumed to be supplied by the European average grid.

The construction and assembly of the plant is not considered.

Table 1 Material composition of the compound parabolic collector at PSA. All data per m^2 of collector area

Material	Quantity, per m^2 collector area
Aluminium extrusion profile (anodised)	9.68 kg
Stainless steel	7.81 kg
Steel reinforcing	31 kg
Steel, tinplated coil	0.17 kg
Concrete, normal	0.34 m^3
Glass, borosilicate	6.72 kg
Total transport demand for materials by road	46 tkm
Total transport demand for materials by rail	14 tkm
Service life	15 years continuous operation

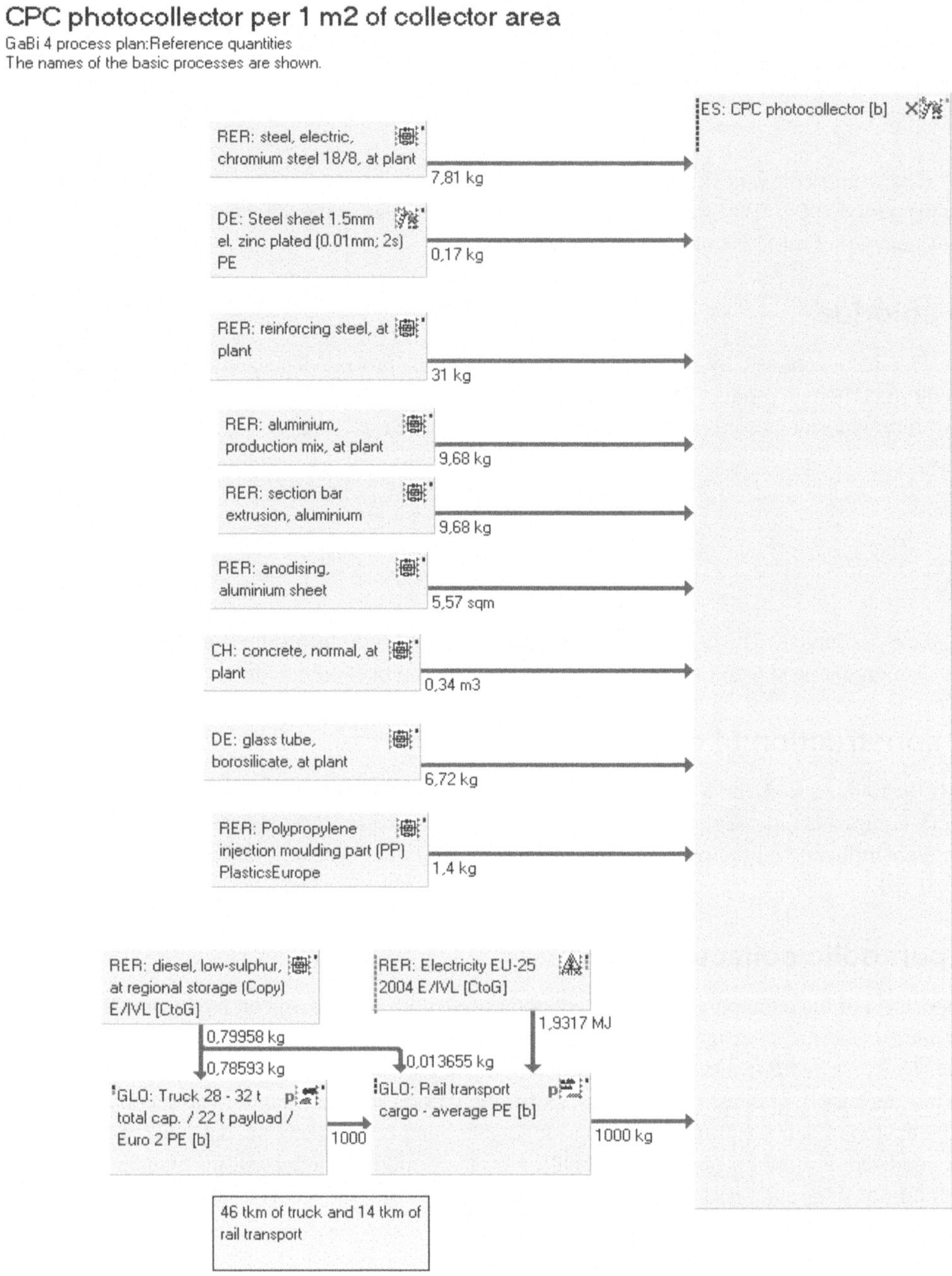

Figure 1 GaBi model of the supply of the materials for the CPC

Cooling water

Ozone generators require cooling water of a certain quality. We assume that it is fed from the public water supply. As a first approximation we have used available data for public waterworks in Switzerland, available from Ecoinvent. The inventory assumes that the necessary energy is generated in Germany.

Electricity

When the treatment plants and the suppliers are assumed to be somewhere in Europe without further specification we assume that they are supplied with average European electricity from the European grid, that is to say we do not assume that the existing or the new treatment technologies will constitute marginal loads more than other power users. Table 2 shows the composition of European electricity we have used for all assessments of plants not specifically located in a special part of Europe.

The life-cycle inventory of electricity generation is illustrated by Figure 2.

Modules, which describe the environmental impacts of generating electricity from specific energy sources, like coal, uranium etc., have been selected from two databases available in GaBi. The modules have been used to construct an inventory of European electricity, based on the data in Table 1. In Table 3 we report some of the environmental impacts calculated for generation of European electricity.

Table 2 Electricity production mix EU-25 (OECD statistics 2004)

Type of generated power	% of production mix
Coal (of which 20% is assumed to be lignite, based on older data from EURPROG)	30.55
Oil	4.11
Gas	18.95
Nuclear	30.91
Hydropower	10.55
Biomass	1.58
Waste incineration	1.23
Geothermal power	0.17
Solar photovoltaic	0.022
Solar thermal	0.016
Other sources (assumed to be wind power by us)	1.91
Distribution losses (assumed by us for distribution to industrial customers)	3[1]

[1] *The OECD statistics give the total distribution losses as 6.7% of the generated power.*

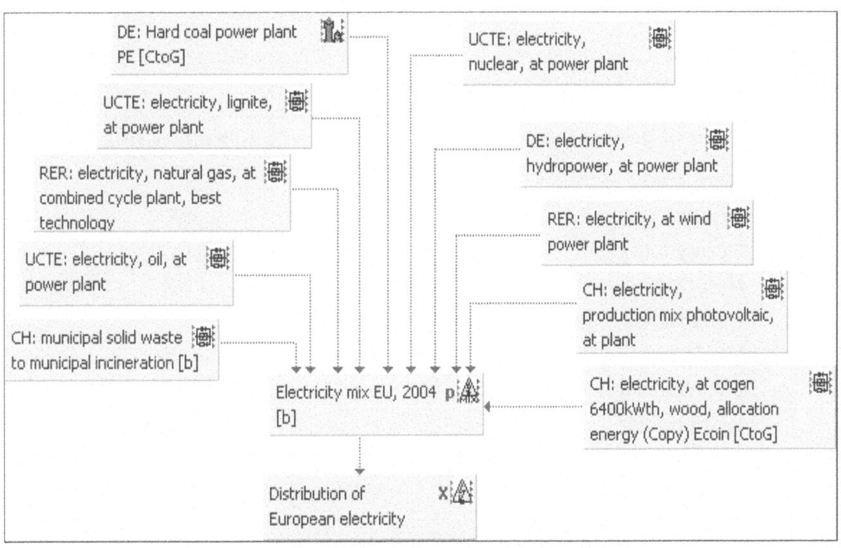

Figure 2 GaBi flowchart of the electricity inventory RER, UCTE and CH modules: Data from the data base Ecoinvent 1.2 (Ecoinvent). DE modules: Data from the professional data base of GaBi (GaBi)

In the case of the nalidixic-acid treatment the plants are assumed to be located in southern Europe. The electricity is supposed to be supplied by the Spanish average grid. Inventory data for the Spanish supply mix in 2002 are available from the professional database of the supplier of the GaBi software (PE International). Table 4 gives the composition of the mix. The data set considers the whole supply chain of the fuels from exploration over extraction and preparation to transport of fuels to the power plants as well as the country-specific technology standards and the grid losses. The modelling principle is illustrated in Figure 3.

Ferric chloride

Specific data for the manufacture of a commercial ferric chloride product have been obtained from a major supplier of chemical coagulants. The product is specified in Table 5. The ferric chloride is produced by dissolution of magnetite in hydrochloric acid and oxidation of the solution with oxygen. The product is shipped and used as a 40% aqueous solution. We assume that this product is used wherever ferric chloride is needed.

Table 3 Some emissions and primary energy resource consumption for electricity generation, per kWh generated and distributed electricity EU-25 mix

	Emissions during surveyable time (100 years)	Long term emissions
Nutrient enrichment potential, kg NO_3^- equiv	0.000801	$8.89 * 10^{-5}$
Oxygen consuming emissions, kg COD	0.000265	0.000829
Ecotoxicity potential, acute, water, m^3 water	3.14	125
Ecotoxicity potential, chronic, water, m^3 water	30.9	920
Global warming potential, 100 years, kg CO_2 equiv	0.477	–
Acidifying potential, kg SO_2 equiv.	0.00201	0.000567
Photochemical oxidant potential, high NO_x, kg ethylene equiv.	$4.40 * 10^{-5}$	–
Fossil energy resources, kWh	1.48	–
Uranium, primary energy resource, kWh	1.12	–
Renewable energy resources, kWh	0.852	–

Table 4 Composition of the Spanish electricity supply mix in 2002 (data from PE International)

Type of generated power	% of supply mix
Hard coal	28.2
Lignite	4.7
Heavy fuel oil	11.6
Natural gas	13.2
Blast furnace gas	0.5
Nuclear power	25.6
Hydropower	10.7
Wind power	3.5
Biomass, solid and gaseous	1.6
Waste incineration	0.2
Other sources	0.2
Total	100

Data for the raw materials, chemicals and energy supply necessary to produce the ferric chloride have been collected from Ecoinvent and from the professional data base of GaBi.

Ferrous sulphate

Ferrous sulphate is obtained as a by-product from the manufacture of titanium dioxide. No environmental impacts from the production process to the ferrous sulphate have been assigned. The product from the TiO_2 production contains heavy metals and sulphuric acids. This is neglected in this inventory.

Glass

Data for manufacture of glass tubes (20% boron silicate, 80% soda glass) are available (Ecoinvent).

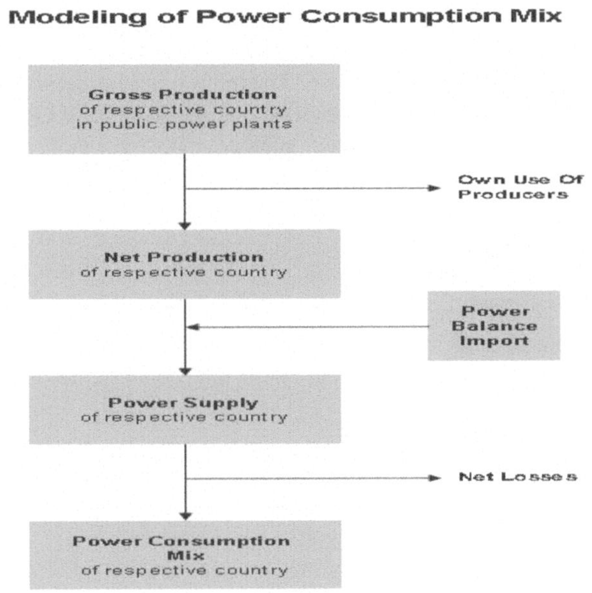

Figure 3 Modelling principle of the electricity mix (Figure from PE-International)

Table 5 Commercial ferric chloride coagulant (PIX 111)

Density of the aqueous ready-to-use solution	1.42 kg/l
Fe(III)	13.7 ± 0.2%
Active substance	2.45 mol(Me^{3+})/kg
pH	<1
Free acid	5–15 g/kg
Chloride	260–280 g/kg

Hydrogen peroxide

Life cycle inventory data for hydrogen peroxide from extraction of raw materials to the factory gate have been taken from the Ecoinvent database (Ecoinvent (2004), report No. 8). The data are average data from eight European manufacturers and represent the technology around 1995.

Hydrogen peroxide is produced with the help of an alkylanthraquinone, which is reduced by hydrogen to the corresponding hydroanthraquinone in the presence of a catalyst. The catalyst is removed and the hydroanthraquinone oxidised by oxygen back to anthraquinone and hydrogen peroxide. The reaction is carried out in an organic solvent mixture. The produced hydrogen peroxide is extracted from the solution and supplied as a 50% aqueous solution.

Magnesium oxide

Data have been collected from Ecoinvent (Ecoinvent, 2007a). They pertain to manufacture of MgO from magnesite ($MgCO_3$) in Europe. The data on the production process are very uncertain. Data from iron ore mining and beneficiation as well as from lime crushing and milling have been used as approximations. The reported energy use is very low, and the reported emission of CO_2 is approximately equal to the stoichiometric amount from calcining of magnesite.

Methanol

Methanol is produced from synthesis gas, produced by steam reforming of natural gas. Data from Ecoinvent 2000 are used. They have been collected from various plants of different locations and are thus not specific for Europe.

Mineral fertiliser replaced by struvite

It is presumed that struvite precipitated from the treatment of landfill leachate is used as fertiliser and saves other mineral fertilisers. The saved resources and avoided emissions are calculated from production data from monoammonium phosphate, available from Ecoinvent (Ecoinvent, 2007) assuming that nitrogen and phosphorus in struvite can replace these nutrients in monoammonium phosphate on a 1:1 basis. The ammonium phosphate is produced from ammonia and phosphoric acid. The data are given as two data sets per kg N and per kg P_2O_5, resp. The multi-output process monoammonium phosphate delivers the co-products monoammonium phosphate as N and monoammonium phosphate as P_2O_5. Allocation factors are based on the energy requirements of the respective nutrients for the production process: 45% for monoammonium phosphate as N and 55% for monoammonium phosphate as P_2O_5. The allocated inventories must always be used simultaneously. This is done in the LCA model of the struvite-SBBGR-ozone system.

Oxygen

Liquid oxygen is produced by cryogenic separation of air. Co-products are liquid nitrogen and liquid argon. Allocation between the products is done based on heats of vaporisation (Ecoinvent, 2000). The commodities used are electricity and cooling water.

Phosphoric acid

It is assumed that industrial-grade phosphoric acid with 85% H_3PO_4 is used. Data for production of phosphoric acid via the dihydrate process are available from Ecoinvent (Ecoinvent (2004), report No. 8).

In the dihydrate process phosphate rock is treated with sulphuric acid to produce a 70% fertiliser-grade phosphoric acid. Gypsum ($CaSO_4 \cdot 2H_2O$) is formed as a by-product. Most of the gypsum is discarded as a waste. A second by-product is fluosilicic acid, which is formed in minor amounts and has a value.

The fertiliser-grade phosphoric acid is purified and concentrated to the industrial grade product by extraction with an organic solvent and sulphuric acid, followed by evaporation.

The inventory by Ecoinvent assumes that fertiliser-grade phosphoric acid is produced globally and transported to Europe, where it is refined to industrial-grade phosphoric acid. The global production is modelled as a mixture of 63% produced in the USA and 37% produced in Morocco. For the purification incomplete data from one site in England could be obtained. The environmental impacts of the dihydrate process are allocated 98.2% to the fertiliser-grade phosphoric acid and 1.8% to the fluosilicic acid.

The data represent the technology in the 1990′s.

Polyamides

Data for nylon 66 (a co-polymer of hexamethylene diamine and adipic acid) are available from Ecoinvent. This dataset is used to assess the environmental impacts of producing polyamides. The data are originally from the Eco-profiles of the Association of Plastic Manufacturers of Europe (APME). Not included are the values reported for: recyclable wastes, amount of air $/N_2/O_2$ consumed, unspecified metal emissions to air and to water, mercaptan emission to air, and unspecified CFC/HCFC emissions to air. The amount of "sulphur (bonded)" is assumed to be included in the amount of raw oil.

Polydimethylsiloxane (PDMS)

For silanes and siloxanes average data are available in the data base Ecoinvent on resource consumption and emissions caused by the manufacture of these materials. The inventory has been calculated as a weighted mean of two production facilities which both produce a range of about 2000 different silicone products. It is noted that the input data contain no recorded input of any silicon source.

Polyelectrolyte as dewatering aid

Polyelectrolytes for aiding flocculation may be copolymers of acryl amide and sodium acrylate (anionic polymers) or copolymers of acryl amide and a tertiary or quaternary amine, such as N,N-dimethyl-2-aminoethyl acrylate (cationic polymer). Acryl amide is synthesised from acrylonitrile by reaction with water at 100–150°C in the presence of a copper catalyst.

No LCA data for polyelectrolytes have been found. We use a mixture of equal parts w/w of acrylic acid and acrylonitrile as a model for polyelectrolytes. The modules describing acrylic acid and acrylonitrile have been collected from Ecoinvent.

Polymers

The environmental impact of supplying polymers as construction materials is described with average European data from the inventory of the European plastics industry, as available from Ecoinvent. For polyester data for polyethylene terephtalate have been used. Polyether sulphone has environmentally been modelled as polyether polyol.

PVC

The environmental impact of supplying PVC as a construction material is described with average European data from the inventory of the European plastics industry, as available from Ecoinvent (bulk PVC).

Sludge treatment

The sludge treatment scenario is incineration with landfilling of the bottom ash and the fly ash. The mass and energy balances around the sludge incinerator are for now in each case separately calculated from the composition of the sludge leaving the wastewater treatment plant.

The equations used for the calculations were originally worked out for waste incineration by Björklund (Björklund, 1998). The emission factors have since then been extensively modified during the course of the ORWARE project (ORWARE, 2000).

The sludge from the treatment of brewery wastewater and landfill leachate is considered to be biogenic. The carbon dioxide emitted from the incinerator in this case is thus not regarded as a contributor to the global warming potential.

The emission of NO_x is calculated from the EU daily average limit in flue gas from new waste incinerators, 200 mg/Nm^3 at 11% residual oxygen (*Directive 2000/76/EG, 12/04/2000*). The volume of flue gas is calculated from the carbon content of the sludge with the following stoichiometry:

$$\begin{array}{lcl} C + 1.12O_2 + 4.48N_2 & \longrightarrow & CO_2 + 0.12O_2 + 4.48N_2 \\ 12\ \text{kg} & & 125.4\ \text{Nm}^3 \end{array}$$

The emissions of SO_2 and HCl to air are calculated from the EU daily average limits in flue gas from waste incinerators, 50 mg/Nm^3 and 10 mg/Nm^3 respectively.

The use of the chemicals $Ca(OH)_2$, $CaCO_3$, polyflock, TMT and urea is not considered in the inventory.

For bottom ash and fly ash from sludge incineration incoming pollutants in the ashes are divided into emissions to air and water during the surveyable time (100 years) and as remaining in the landfill after the surveyable time. Emission factors are from Björklund (*Björklund, A (1998), "Environmental systems analysis of waste management with emphasis on substance flows and environmental impact", licentiate thesis, Royal Institute of Technology, Stockholm*). Emission factors for ammonia, COD, dioxins and nitrate come from ORWARE (ORWARE, 2000).

Sodium chloride

Sodium chloride is assumed to be supplied as brine from solution mining. The data were collected in 1994 from one source and are thus somewhat uncertain. They have, however, been checked against more recent data from a European producer in 2000 ((Ecoinvent (2004), report No. 8).

Sodium hydroxide

Sodium hydroxide is produced from sodium chloride (salt from solution mining, rock salt or solar salt) in the chlor-alkali electrolysis process, which also delivers chlorine and hydrogen. Data for an average chlor-alkali technology in Europe in the year 2000 are used (Ecoinvent (2004), report No. 8). In 2000 55% of the European chlor-alkali capacity consisted of mercury cells, 22% were diaphragm cells and 20% membrane cells (which leaves 1% unaccounted for).

The environmental impacts of the chlor-alkali process are allocated by mass between chlorine, sodium hydroxide and hydrogen according to the stoichiometry of the overall cell reaction.

$$NaCl + H_2O \longrightarrow NaOH + \tfrac{1}{2}Cl_2 + \tfrac{1}{2}H_2$$

This allocation principle puts 52.3% of the environmental burden of the electrolysis on NaOH. The sodium hydroxide is delivered as a 50% aqueous solution.

Sodium hypochlorite

For sodium hypochlorite we use data for a process, where chlorine emissions are captured in sodium hydroxide solution (Ecoinvent (2004)). There is thus no environmental burden from the supply of the chlorine.

Steel

For the environmental impact of supplying steel data from Ecoinvent are used. For low-alloyed steel a data set representing an average of European and world steel production has been selected. This is assumed to reflect the average use of steel in Europe.

The stainless steel is assumed to be produced in Europe from scrap with addition of alloying metals

Sulphuric acid

Sulphuric acid is in Western Europe produced by the contact process, i.e. by combustion of sulphur or sulphur-containing materials to sulphur dioxide. The sulphur dioxide is further oxidised by air over a catalyst to sulphur trioxide, which is absorbed in concentrated sulphuric acid, where it reacts with water to sulphuric acid.

The most important sulphur sources in Europe were in 2002 elemental sulphur from crude oil, natural gas or coal flue gas (50% of the production), smelter gas (35%), pyrites (8,5%) and spent acids (5.5%) .

In Ecoinvent data are available for a European average production considering the contributions of the sulphur sources mentioned above (Ecoinvent (2004), report No. 8). Data from the period 1990–2000 have been used. Sulphuric acid produced from smelter gas can be considered a by-product from metal-ore processing. This part of the sulphuric acid production is thus considered to be "free of environmental burdens".

The oxidation of sulphur to sulphur dioxide is exothermic, and the contact process is a net energy generator, except in case the sulphur source is spent acids. The excess energy is recorded as waste heat by Ecoinvent.

Transports

For transport of chemicals to the wastewater treatment plants, a distance of 300 km has been assumed as a default. The trucks are assumed to return empty, which means that the total utilisation of the load capacity is 50%. For transports of dewatered sludge to incineration a distance of 10 km has been assumed.

As a default transports of chemicals to and sludge from wastewater treatment plants are presumed to be carried out by 10-tonne Euro2-trucks with a maximum capacity of 5200 kg. Fuel consumption and emission data, originating from TEMA2000, have been collected from the EDIP data base (*EDIP (UMIP) database, FORCE Technology, Lyngby, Denmark, www.lca-centre.dk*). The fuel is low-sulphur diesel oil (0.005% sulphur) and the driving pattern is mixed driving, 24% motor road at 80 km/h, 44% main road at 70 km/h and 32% at 25 km/h. The load factor equation, relating the fuel consumption to the utilisation ratio, is taken from the GaBi professional data base. The emission factors per kg diesel are calculated from the original data of EDIP.

APPENDIX 3

Costs of energy and commodities

Cost item	Price	Source, comment
Energy		
Electric power	0.0941 €/kWh	EU 27, medium-sized industrial customers, annual consumption 500–2000 MWh, 2009, excluding taxes. (Eurostat, http://epp.eurostat.ec.europa.eu)
Chemicals		
Activated carbon	35 SEK/kg	Price quotation 30 – 35 SEK/kg from a Swedish supplier 2008. No later quotation given.
Ammonium chloride	300 GBP/tonne	Default price Chemicals Cost Guide 2002
Citric acid	900 GBP/tonne	Default price Chemicals Cost Guide 2002
Ferric chloride, 40% solution	110 €/tonne	Solution ready to use (major European supplier, March 2008)
Ferrous sulphate, $FeSO_4 \cdot 7H_2O$	0.3 €/kg	Rough estimate of the average market price by a Swedish supplier.
Hydrogen peroxide, 50%	510 GBP/tonne	Bulk chemical, default price Chemicals Cost Guide 2002.
Methanol	500 €/tonne	Estimate for delivery as a commodity. The contract price was 165 €/tonne fob Rotterdam in September 2009 (ICISpricing).
Oxygen, liquid	105 €/tonne	Default price from Chemicals Cost Guide 2002.No price quote from major suppliers was offered.
Magnesium oxide	505 US$/tonne	US$480–500/tonne for caustic calcined magnesia powder, 95% MgO, fob China. Industrial Minerals, 9 March 2010
Phosphoric acid	700 US$/tonne	Spot price quoted as US$740/tonne Sept. 2009 Purchasing.com
Polyelectrolyte	3 €/kg	Solid powder, cationic electrolyte (major European supplier, March 2008)

(*Continued*)

Costs of energy and commodities (*Continued*)

Cost item	Price	Source, comment
Sodium hydroxide	120 US$/dry tonne	Liquid, fob North-West Europe, spot price (ICISpricing, September 2009)
Sodium hypochlorite	300 GBP/dry tonne	Default price Chemicals Cost Guide 2002
Sulphuric acid	60 €/tonne	€45–65/tonne cfr NW Europe, World Sulphuric Acid Weekly, May 2009
Urea	205 €/tonne	Spot price 200–205 €/tonne, fca North-West Europe, ICISpricing Sept. 2009